Chemistry of Fragrant Substances

Chemistry of Fragrant Substances

by
Paul José Teisseire

translated by
Peter A. Cadby

Paul José Teisseire, formerly
Director of Research and Administrator
Rouré-Bertrand-Dupont Society

Translator: Peter A. Cadby
Firmenich S.A.
1, route des jeunes
CH-1211 Genève 8
Switzerland

Library of Congress Cataloging-in-Publication Data

Teisseire, Paul José.
[Chimie des substances odorantes. English]
Chemistry of fragrant substances / Paul José Teisseige: translated by Peter A. Cadby.
p. cm.
Includes bibliographical references and index.
ISBN 1-56081-610-4
1. Perfumes. I. Title
TP983.T44513 1994
668'.54—dc20 93-4441
CIP

This book is printed on acid-free paper. ♾

Originally published under the title: "Chimie des substances odorantes" by Lavoisier, TEC & DOC, 11 rue Lavoisier, F.75384 Paris, France
Printed in the United States of America
ISBN 1–56081–610–4 VCH Publishers

Printing History:
10 9 8 7 6 5 4 3

Published jointly by

VCH Publishers, Inc.
220 East 23rd Street
New York, New York 10010

VCH Verlagsgesellschaft mbH
P.O. Box 10 11 61
69451 Weinheim
Federal Republic of Germany

VCH Publishers (UK) Ltd.
8 Wellington Court
Cambridge CB1 1HZ
United Kingdom

Contents

1. Introduction and Historical Background 1

2. Terpenoid Chemistry 19
Introduction—Nomenclature 19
Acyclic Monoterpenes 24
Monocyclic Monoterpenes 36
Bicyclic Monoterpenes 63

3. Industrial Syntheses Starting from α- and β-Pinene 85
Acid-Catalized Reactions 85
Pyrolysis Reactions of the Pinenes 92
Hydroperoxidation Reactions 99

4. Total Synthesis of Aliphatic Terpenoids of Industrial Importance 109
Introduction 109
Ethynylation reactions 111
Methods for the Synthesis of Methylheptenone 120
Industrial Chemistry of Dehydrolinalool 125
Synthesis of Terpenoids 136

5. Biogenesis of Natural Substances 145
Introduction 145
Some Aspects of Enzymology and Biochemistry 150
Biosynthetic Mechanisms Leading to Terpenoids 173

6. The Chemistry of Sesquiterpenes and of Some Macrocyclic Diterpenes 193
Introduction 193
Cyclizations and Interconversions in the Sesquiterpene Series 193
Detailed Study of Some Syntheses of Sesquiterpenes of Major Interest 242
The Chemistry of Diterpenes with 14–Membered Rings 281

7. Substances with a Musk Odor 289

Introduction 289

Methods of Synthesis of Macrocyclic Ketones 292

Methods of Synthesis of Macrocyclic Lactones 325

Aromatic musks 333

Appendix 1. The Concept of Prochirality 359

Appendix 2. The Wittig Reaction 365

Appendix 3. The Chemistry of Enol Ethers 385

Appendix 4. Functionalization of Unactivated Carbon Atoms 401

Appendix 5. Methods for Identifying Natural and Synthetic Organic Compounds 417

Appendix 6. Summary of Some Physico-Chemical Methods Used in the Analysis of Natural Odorant Products 425

Bibliography 433

Index 451

CHAPTER

1

Introduction and Historical Background

Perfumes have been intimately associated with man's history. There is evidence that they have been used since furthest antiquity in religous ceremonies and in the treatment of sickness. They not only have pleasing odors but numerous aroma chemicals possess useful bacteriostatic and antiseptic properties. Moreover, the rarity and high cost of aromatic resins, spices and perfumes have made them useful as currencies and as gifts to be exchanged between rulers. Yet, we have no information regarding the invention of the first process for elaborating perfumes; this is common to all of the major basic technical advances that are at the origin of our civilization.

The stages in the creation of natural perfumes would presumably have been passed through slowly with the first manipulation of perfumes being made probably by primitive therapists utilizing the available natural resources and who, as their experience developed, founded the still profane craft of *aromaticists*. Thus, they created the first perfume formulations and contributed to the knowledge which later led to the techniques of perfumery as we know them today. Over the centuries, all progress in the knowledge of perfumes was the result of the work of anonymous artisans. Only recently has this progress led to the concentration in Grasse, in the South of France, of an initially artisanal but later industrial dominance in the manufacture of natural perfumery materials.

The history of perfumery is generally identified first with the history of natural perfumes. The introduction of synthetic products in this domain is extremely recent and did not occur until the end of the nineteenth century. For the ancient civilizations, perfumes were synonymous with crude aromatic raw materials. These were important objects of barter, being both rare and precious in small quantities. Perfumes also added to the mystique of religious ceremonies. In China, the first reports of the use of perfumes in this way go beyond the first thousand years before our era. The same can be said for India. As with the later Buddist liturgy, statues of the gods were required to be annointed with perfumed water.

Similar rites were found in the religious customs of ancient Egypt. It is here that we see the art of perfumery finally progressing from the use of crude natural objects to become a relatively sophisticated element in the art of healing. The civilization of the Nile has passed on to us, through the stone of its monuments, a complete iconography of the processes used for preparing oils, balsams, and fermented liqueurs. One has to wait however, until the period between the seventh and fourth centuries B.C. before we find proof of the existance of the technique of expressing flowers as represented in the relief: "Harvest and Pressing of Lilies" which can be seen in the Louvre Museum in Paris.

From Hippocrates (the greatest physician of antiquity) and Herodotus, who both described all possible aspects of the state of knowledge of their times, we learn what the ancient Greeks knew of perfumes. A century later, around 300 B.C., Theophrastus, the "divine speaker" according to Aristotle, published 9 volumes entitled: "Enquiry into Plants" followed by another 16 volumes on plants. At the dawn of our era, Pliny the Elder, the Roman historian, published his "Natural History" in 37 volumes. Borrowing the same terminology as Theophrastus for describing perfumes, he clarified and added precision and detail to the work of his predecessor by describing for the first time, among other things, the procedure for elaborating perfumes using extraction with fats.

It is evident that the Romans became familiar with perfumes through their conquests. Under the empire of the first Caesars, knowledge about perfumes was already quite substantial and often touched the field of medicine as well. Gradually, the luxury of perfumes spread throughout the Roman world. Horace, among others, recounted that the rich conserved perfumed oils in earthenware vases and rhinoceros horns. The Romans, according to Pliny, considered the use of perfumes as "one of the most honest pleasures of man." Later, towards 330–350 A.D. at the birth of the Eastern Roman Empire, we hear of the appearance of numerous gums: myrrh, olibanum, opopanax, and many others, which arrived principally from Yemen via the Red Sea. This commerce was maintained by this route up until 1453 when the fall of Constantinople marked the end of the Byzantine Empire.

From the year 1000 onwards, two phenomena effectuated the appearance of a large number of aromatic materials in Europe, and particularly in France. The *first phenomenon* results from the eight Crusades which took place from the eleventh to the thirteenth centuries. During this time, the School of Salerno was founded at one of the principal ports of call of the Crusaders. It was from this school that the science of the Arabs, who were at that time masters of half of the Mediterranean littoral, penetrated the Occident. There was also another center of knowledge in Toledo, Spain which had only just reestablished Christianity during the eleventh century. From these two intellectual centers, techniques such as distillation became known in the South of France, particularly in the region of Montpellier.

The *second phenomenon* was the important role that Venice was beginning to play in the commerce of spices. Venice was already a center of exchange and trade when the changes brought about by the Crusades provided an opportunity for it to assume a dominant role in exchanges with the Levantine. It was through Venice that spices arrived in Europe: cinnamon, pepper, nutmeg, cloves as well as numerous aromatic raw materials such as musk, civet, benzoin, sandalwood, olibanum, camphor, the balsams and the gums. In brief, one could find in Venice all the elements of luxury for which the Crusaders, who had often departed crude and unlearned but who had returned cultivated and art-loving, had an acquired taste.

The domination by Venice of this commerce lasted almost until 1797 when this city-state was ceded to Austria. From the start of the Renaissance, competition began to come from elsewhere. The Portuguese of Vasco da Gama's time were bringing galleons filled with aromatic raw materials and spices directly from Asia to Lisbon. In 1523, such products were sold for a total of 700,000 cruzados in one day alone in Lisbon. Nothing on a similar scale had been seen elsewhere before. The route discovered by the Portuguese had indeed changed the pattern of the world's commerce and the "spice wars" had started. They would last for another two centuries.

"The Sun rises in the East!" Symbolically, this old expression glorifies the fact that the East is the Cradle of Civilization. It was also in the East that the history of distillation began. The first basic principles for the preparation of essential oils had, in effect, been developed in the Orient; in India, Persia, and in Egypt. Information is extemely vague on the methods, the objectives, and the results of distillation techniques used in ancient times. Nevertheless, it seems that turpentine is the only essential oil for which the method of preparation has been clearly described. The Greek historian, Herodotus, and the Roman natural historian, Pliny the Elder, as well as his contemporary, Dioscorides, all mention turpentine oil and give some limited information on the methods used in its preparation; in contrast, camphor oil, which is known to have been made since much further antiquity, lacks such details.

Until the Middle Ages, however, the art of distillation was used mainly for the preparation of distilled waters. When an oil appeared on the surface of these waters, it was regarded as an undesirable by-product. The first description of the distillation of a true essential oil is generally attributed to a Catalonian physician, Arnold de Villanova (1235–1311). Yet, de Villanova was apologetic about the medicinal qualities of his distillates, which were nonetheless used in all of the pharmacopoeias at the end of the Middle Ages and frequently later than that.

Towards 1500, at the beginning of the Renaissance, some Venetian manuscripts mention distilled oils (and not waters) of sage and rosemary. Approaching the end of this period when distillation became known as a means of separating the "essential" from the "nonessential," the great Swiss physician, Paracelsus (1493–1541), expounded his theory of the "Quinta essentia" (quintessence). For Paracelsus, this quintessence was the truly effective element in each medicinal preparation and its isolation was the goal of all pharmaceutical science.

There is little doubt that this theory was the foundation for considerable subsequent research into the preparation of essential oils. The present use of this term "essential oil" still evokes the concept of "Quinta essentia" of Paracelsus. It is, however, quite certain that during the same period, provincial artisans were also developing their techniques for obtaining these essential oils. In 1500 and 1507, Brunschwig, a physician from Strasburg, published his two famous volumes on distillation entitled "Liber de arte distillandi" which mentions four essential oils: turpentine, juniperwood, rosemary and spike lavender. Brunschwig described how spike lavender was produced in the Provence region of France. This is also confirmed by another physician from Strasburg, Walter Ryff, in a book published in Frankfurt in 1556 which mentions the French essential oil industry: "The oils of spike lavender and lavender are supplied to us from Provence in small bottles at a high price." Other distilled essential oils, namely clove, mace, nutmeg, anise and cinnamon are described by Ryff as "exquisite."

During this period, under the influence of Catherine de Medici, Regent to the Kingdom of France, a taste for soaps, perfumed gloves, and fragrant oils and pommades developed within the royal court. At this epoque, however, all of these products were still produced outside of France. Soap had to be imported from Genoa, Bologne and Alicante until 1565, when the first factory was established at Chaillot. The French Statutes of 1582 contributed to the development of the artisanal trade of Glovers-Perfumers which eventually became distinct from the corporation of Apothicaries-Spicers.

In addition, at Grasse in the South of France, there were numerous corporations for the different arts and crafts among which were those of the *tanneurs et curatiers* which had gained major importance. This guild, under the patronage of St. Claude, had been codified by a first statute dated November 20, 1260 and written in the "provençal" dialect as were the later amendments dated November 28, 1304 and January 14, 1322. The tanning industry, however, declined dramatically in Grasse from the start of the eighteenth century due to competition from Nice and the adjacent Italian coast. A certain number of artisans from this corporation joined another, that of the Glovers-Perfumers, which further transformed the leather made by the tanners. The statutes of this corporation date from 1724 and were homologated by a letter patent of Februaury 12, 1729. The present syndicate of the fragrance manufacturing industry is descended directly from this institution. It was this body, composed of artisans distributed among a small number of families, which created the floral perfume industry based on the culture of odoriferous plants and the production of essential oils.

As previously noted, some of the most odoriferous flowers provided no essential oil on steam distillation. In his "Cours de Chymie," published in Paris in 1701, Nicholas Lemery made this observation for the flowers of jasmin and violet. As already explained, the observation and employment of the affinity of fats for odors particularly at elevated temperatures, dates from antiquity. From this invention, the technique of "enfleurage" was born in the eighteenth century. This is low temperature extraction of flowers by direct contact with, rather than full immersion in, fats. Previously, this technique had been used at an experimental level, being the one procedure which most faithfully reproduced the true odor of the flowers.

At the start of the nineteenth century, the technique of enfleurage had reached its current degree of perfection; however, it was so labor-intensive that it was, to all intents and purposes, abandoned after World War II.

In 1809, Bertrand indicated that the malaxation of pommades with spirits of wine (esprit-de-vin 3/6, which today we would call 85°/86° alcohol) provided products he called "Extraits de Pommades." It was also at this time, thanks to the work of Chaptal under the impetus of Napoleon I, that the first distillery was created in the South of France which was capable of obtaining the azeoptrope of alcohol (95°/96°). The availability of increasingly neutral and odorless qualities of alcohol opened the way for entirely new applications for the industry in Grasse both for the washing of pommades and the production of infusions.

Around 1860 it became common to define different concentrations of pommades and their washings by a number; however, these dilutions were difficult to manipulate and transport. Consequently, around 1870, Louis-Haximin Roure, an accomplished technician, considered concentrating alcoholic washings obtained with 96° alcohol to form for the first time, "Essences concretes de Pommades" which were entirely soluble in alcohol.

In September 1875, the "Moniteur Scientifique" of Dr. Quesneville wrote very favorably of these products: "The essence concretes of the company Roure-Bertrand Fils, on the whole, broaden the domain of essences. They represent for the industry the introduction of new materials which are entirely soluble in alcohol, taken to their maximum degree of concentration, and which can be recommended for their natural qualities and their ease of use."

Being most attentive of all the work being done at that time on natural products, Louis-Maximin Roure was impressed by the results obtained by Robiquet, Buchner, and Favrot in France and by Hillon and Ferrand in Algeria on the extraction of several types of fragrant flowers (daffodil, acacia, rose, jasmin, tuberose, narcissus, etc.) using diethyl ether and carbon disulfide. However, these solvents seemed far too dangerous to justify their industrial use. From 1870 onwards, Roure still made some studies on the use of other volatile solvents; unfortunately, the technology was not sufficiently advanced at that time to enable him to succeed in his endeavors.

It was not until the important work of Naudin and Schneider in 1879 and of Massignon in 1880, that petroleum ether, obtained from the infant crude oil-refining industry, could be used for the extraction of floral products. Thanks to this research, Louis-Maximin Roure, just before his death in 1898 and at almost the same time as the Societé Chiris, developed an installation for hydrocarbon extraction; a technique which has not ceased to grow in importance even to this day. Louis-Maximin Roure also had the idea of using the same technique on these concretes that he had developed in 1870 for extracting pommades with alcohol. He called these products "solids" to distinguish them from the concretes: He thus obtained, once again for the first time, the essences which today are called "absolutes."

These studies brought the production of essential oils and extracts obtained with fats and volatile solvents to a state of near perfection. They did not, however, contribute in any way to our knowledge of their inherent composition. In any case, analytical techniques for this work were still very rudimentary and adulteration, especially of essential oils, was relatively easy.

In fact, although they knew nothing about the composition of the essential oils they were dealing with, certain old hands in Grasse had already recognized that similarities existed between, on the one hand, essential oils having principally terpenic constitutions and turpentine and, on the other hand, essential oils rich in sesquiterpenic constituents and the essence of Gurjun balsam. When used as adulterents, these two essences were designated, respectively, by two very evocative names in the provencial dialect: "la musica" and "lou faber."

Not until the studies of Wallach, carried out between 1880 and 1914, was there any real understanding of the composition of essential oils. It is interesting to recall what actually happened in 1880. Otto Wallach, an assistant of Kekule, had discovered in the latter's laboratory a number of abandoned flasks containing essential oils. This happened just at the time when his interest in such matters had been raised by a series of lectures he had been giving on pharmaceutical chemistry. He had noted the near total ignorance of this subject by fellow chemists and it is interesting to try to envision the main questions with which he was confronted when he embarked on his studies.

It was already known that there were hydrocarbons with the molecular formula $C_{10}H_{16}$ which had been named "terpenes" by Kekule because of their presence in turpentine. Substances with the molecular formulas $C_{10}H_{16}O$ and $C_{10}H_{18}O$, which were obviously related to terpenes, were also known under the generic name "camphor"; the prototype of this group was camphor itself, which was known since antiquity in the Orient and at least since the eleventh century in Europe.

These terpenes were regarded as rather enigmatic substances which were difficult to study and which were subject to alteration as they are, of course, still today. Kekule had from the outset regarded any attempt to study them as a difficult, and not particularly tempting, venture. At that time, obscurity and confusion reigned over this domain and many terpenes carried different names based on their various botanical sources even though they were often chemically identical.

Wallach's plan, as he described it himself, was as follows:

- to disentangle the basic properties which permit the differentiation of the individual chemicals based on the principles of "immediate analysis" defined 50 years earlier by Eugene Chreveul;
- to study any pure chemicals he could isolate;
- to understand their precise chemical constitutions.

From his first publication in 1884, Wallach revealed that several terpenes described under different names were, in fact, the same substances. In 1891, he established a list of nine different terpenes, each of which was easily characterized. These were pinene, camphene, fenchene, limonene, dipentene, phellandrene, terpinolene and sylvestrene. The last of these was not actually a natural product but an "artifact" derived from the transformation of carene under the influence of acids. This was only confirmed much later by Simonsen and Rao.

Subsequently, he recognized that what he had termed pinene, terpinene and phellandrene were mixtures of isomers. Hence, this original list had to be lengthened. The analytical method employed by Wallach for separating these substances was systematically a single fractional distillation of the essential oils. In order to characterize them, he let the different fractions react with simple inorganic reagents such as hydrochloric acid, bromine, nitrosyl chloride (used for the first time in this context by Tilden), the oxides of nitrogen and the like. This frequently resulted in crystallizable substances for which, incidentally, he was also able to shed light on their mechanism of formation. Thus, in the course of treating 3-carene with hydrochloric acid, he obtained sylvestrene.

From 1884 to 1914, Wallach wrote 180 treatises which were assembled into a single book entitled "Terpene und Camphor." These reports were obviously not only devoted to terpene hydrocarbons, but also to the oxygenated compounds in essential oils, particularly with regard to the relationships existing between the principal constituents of essential oils. Even at that time, he was able to draw more or less exact conclusions regarding their biogenic interrelationships. As early as 1887, he realized that these compounds "must be constructed from isoprene units." Thirty years later, Robinson (Nobel prize in chemistry in 1947) was able to perfect Wallach's isoprene rule by indicating that the joining of these units must take place in a head to tail fashion. (This will be dealt with later in more detail.)

Wallach eventually extended his research to two sesquiterpenes he had isolated (cadinene and caryophyllene), but he rapidly abandoned this new direction, almost certainly because he felt that it was not sufficiently "ripe" for that era. He explained that "the study of essential oils by one man alone can only embrace an extremely modest territory in proportion to the inexhaustible work that the world of plants places at his disposal." Nonetheless, it is undeniable that he was still able to accomplish successfully the task which his master Kekule had considered to be a virtual impossibility.

The work of Wallach is of such importance that he was honored in 1910 with the Nobel prize in chemistry "in recognition of the impetus he gave to the development of Organic Chemistry and the Chemical Industry by his pioneering work in the domain of alicyclic substances."

The contributions of other scientists in this field should not be forgotten. The foremost among them is certainly Adolf von Baeyer who was Wallach's elder by 12 years and who, like Wallach, had been trained in the "crucible" of Kekule's laboratory. In the presence of his master, von Baeyer had experienced the birth of structural formulas as well as the proposition, in 1865, of the celebrated "Kekulian" formula for benzene. He was one of the first chemists to become convinced of the merits of structural chemistry and was adept in its development. He applied it systematically to all of his work, which was extremely vast and covered the most varied fields in organic chemistry. Those which have particular relevance to the development of aromatic chemistry can be cited as follows:

- his study of natural colors, notably indigo, for which he devised several total-synthetic procedures, one of which was later industrialized by BASF;
- his study of synthetic colors, in particular the triphenylmethanes and the rhodamines. He also discovered phenolphthalein, fluorescein and eosine.

His research in this area led him to attribute, in 1902, the concept of "carbonium valency," which expressed the reality of the ionic state of the triphenylmethane compounds. His descriptions of organometallic complexes were also quite close to present concepts.

For about 30 years, all of this work was founded on his understanding of the fundamental aspects of aromaticity. His central idea of the basic symmetry of the formula of benzene is still close to that which is maintained by modern theoretical chemistry. At the same time, he did not lose sight of the remarkable work of his co-disciple, Wallach, which increasingly attracted his attention. In this way, he established a link between the two types of chemistry: the chemistry of aromatic compounds and that which would develop into the chemistry of alicyclic compounds. He did this by passing from artificial hydrobenzenes to their natural analogues in the form of the cyclic terpenes. He devoted over 10 years of work to this field, creating a vast body of knowledge with co-workers such as Henrich, Ipatiev and especially Villiger.

Along with Wallach, he devoted considerable work towards elucidating the structure and properties of pinene, which they were sure contained a cyclobutyl unit. His study of its properties demonstrated the instability of small rings for which the double bond of ethylene could be taken as the limiting case of carbocyclic smallness. He further concluded that in polymethylenic carbocycles of 3, 4, 5, 6 and 7 atoms (which were not all known at that time), the regular arrangement of carbon atoms would have, as a consequence, the deformation of the carbon-carbon bonds—the mechanical energy of this deformation being commensurate with the fragility of these systems. This was called the "theory of strain" (1885) and has been expanded and modified by conformational analysis as developed by Hassal and Barton.

These considerations incited von Baeyer to study thuyone and its isomerization in sulfuric acid. The structures of the products obtained would be partially defined by Haller in 1905 (carvotanacetone). The action of alkali on the bromohydrate of dihydrocarvone permitted von Baeyer to obtain a non-natural bicyclic ketone: carone, for which the structure was unknown except for his recognition of the presence of a cyclopropyl moiety. In fact, the structures he assigned to carone and dihydrocarvone were incorrect. Tiemann and Semmler rectified them some months later in the same year (1894) and Wagner definitively established the correct structure of carone. This was probably only the second example of the synthesis of a cyclopropyl derivative. In 1887, Perkin obtained methyl cyclopropyl ketone using an identical chemical mechanism.

The study of all of these products induced von Baeyer to play an important role in the development of the stereochemical doctrines which originated in 1874 and which still encountered much skepticism 10 years later. He had in fact been imbued with these doctrines from very early times. It should not be forgotten that it was Kekule who, nine years after Couper had demonstrated in Paris in 1858 the tetravalence of carbon, suggested for the first time that this atom could be represented by a tetrahedron. From 1888 to 1890, von Baeyer developed theories on the isomerism (called "Baeyerian isomerism") of closed chains by proposing the notions of *cis* and *trans* isomers with regard to the plane of these cycles.

Returning to pinene, although it had been known for a long time that it contained a cyclobutyl ring, the presently accepted stuctural fomula of α-pinene was proposed for the first time in 1894 by Wagner. The confirmation of this structure was due to the rigid experimentation of von Baeyer between 1896 and 1907 as well as the contemporary work of Tiemann and Semmler and the later work of Perkin and Thorpe (1913).

The oxidation of camphor to its corresponding lactone under the action of persulfuric acid, originally studied in 1899 by von Baeyer and Villiger, has now been generalized as the "Baeyer-Villiger Reaction." All of this work was honored by the attribution to von Baeyer of the 1905 Nobel prize in chemistry "in recognition of the merit he has shown in the development of Organic Chemistry and Industrial Chemistry, by his work on coloring matters and hydroaromatic bonding."

Alongside these two illustrious scholars and co-disciples to whom we incontestably owe our thanks for the creation and development of the chemistry of terpenes, we have already had opportunities to name others such as Tiemann, Semmler, Wagner and Perkin. We should also add the names of Gildemeister, Bertram, Tilden, Walbaum, Hoffmann and others, some of whom were students or rivals of Wallach. These names constitute what in essence was the German school of the end of the nineteenth century. As for the French school, it is possible to cite the names of Haller, Barbier, Bouveault, Grignard, Blanc, Veze and Dupont. Later, from 1920 onwards, Ruzicka in Switzerland became the real perpetuator of Wallach's work.

There seems to have been one area in which Wallach lost interest and on which von Baeyer concentrated very little work, which is the chemistry of camphor. With the exception of camphor itself, one of the oldest known terpenoid compounds was the so-called "camphor of Borneo" which we know today as borneol. Crystalline at room temperature, it was therefore easy to purify and study. Thus, it was possible for Pelouze, in as early as 1840, to discover that its oxidation gave an identical product to the camphor which could be obtained from "oil of camphor."

In 1858, Berthelot accomplished for the first time the synthesis of camphor by direct oxidation of camphene. In 1859, he also established the alcoholic functionality of borneol. In 1870, Baubigny reduced camphor with sodium in moist toluene and obtained two alcohols, one of which was identified as borneol. The exact relation between these two alcohols and camphor was finally established in 1892 by Haller, but it was not until 1893, however, that Bredt suggested the currently accepted structural formula of camphor as being the only one which was consistent with all of its chemical properties known at that time. The total synthesis of camphoric acid (a product of the degradation of camphor) from diethyl β,β-dimethylglutarate was carried out in 1903 by Komppa. Prior to this in 1896, Haller had obtained camphor by pyrolysis of the calcium salt of natural homocamphoric acid. Thus, this series of reactions by different workers constitutes the total synthesis of camphor.

A last point, but not the least, is the examination of the consequences of the science of molecular chemistry. In 1899, Wagner was the first to recognize the rearrangement to which various bicyclic terpenes were subject. In fact, the history of the rearrangement had already started about 40 years earlier when in 1860 in Gottingen, Fittig noticed that when pinacol was briefly heated under reflux in

30% sulfuric acid, a product of dehydration, which he called pinacolone, was obtained. Fittig had discovered the pinacol-rearrangement! The structure of this ketone was not established until 1873 when Butlerov in St. Petersburg recognized an analogy with the conversion of ethylene glycol into acetaldehyde. Wagner, a student of Butlerov and Markownikoff in St. Petersburg, was therefore particularly well placed to recognize that this same rearrangement occurred in the bicyclic terpenes that he had chosen to study.

Meerwein generalized the views of Wagner and established that these rearrangements which are now known as "Wagner-Meerwein rearrangements" were part of a much more general phenomenon of organic chemistry, appearing in numerous substitution, addition and elimination reactions. He was also the first to suggest the participation of carbenium ion intermediates in these rearrangements. Along with the results of Whitemore, which were obtained only in 1930, he can thus be credited with preparing the way for the work of Robinson, Ingold, and Hughes on the modern theories of mechanistic organic chemistry.

Despite these considerable efforts, only a relatively limited number of monoterpenic substances were really well known at the start of the twentieth century. It was not until 1910 that Semmler determined the correct structure of the first sesquiterpene compound: β-santalene. It was necessary to wait another three years for Kerschbaum to establish in 1913 the structure of trans–2, trans–6-farnesol, which was only the second sesquiterpene structure to be described with any precision. Since then, the number of structures established for natural terpenic compounds has grown prodigiously and if one counts the mono-, di-, tri-, sesqui- and polyterpenoids, this number is currently near to 5000.

In as far as the monoterpenes are concerned, about 200 are currently known and these can be divided into about 15 families on the basis of their carbon skeletons. This number excludes compounds such as the "irroids" which contain a cyclopentyl chain and which were initially isolated as defense substances in some species of ant. Several hundreds of other monoterpenes are known and have been isolated, generally in the form of β-D-glucosides, from the leaves, fruit, seeds, roots, and bark of dicotyledons. The water-solubility of these substances permits their distribution in all tissues of the plants in contrast with the free terpenoids, which accumulate in specialized glands. These free monoterpenes are nearly always odoriferous, whereas their gylcosides are generally odorless. Yet, some glycosides can participate with intensity in the flavor of some plants or parts of plants. Saffron owes its taste, particularly in the stamens of its flowers, to picrococine, the β-D-glucoside of a cyclic monoterpene aldehyde, safranal, which has a characteristic odor in the free form.

Only around 30 different sesquiterpenes were known 20 years ago and these had nearly 15 different carbon skeletons. Ten years later, about 300 were known and these could be divided into about 40 skeletal families. There are now almost 1,000 known sesquiterpenoid compounds belonging to about 200 different skeletal families. This almost exponential growth in our knowledge in this field is essentially due to considerable advances over the last 20 years in methods of direct analysis. Advances in chromatographic methodology in all its forms have been remarkable. This is particularly true for gas chromatography, which is especially well suited to the analysis of volatile odorants.

It can be said that the terpenoids only provide what one would call the "bare bones" of the odor of the principal essential oils. This can be best illustrated in the following example. The concentration of the mixture of linalool and its acetate in the different essential oils—lavender, lavandin, clary sage, bergamot, petitgrain and neroli—attains and often passes 70%. Despite this basic similarity, the odors of these essential oils often show extremely large differences that are due to the presence of minor but powerful components which differ from one essential oil to the next and which must therefore have a critical impact on their odors. The analytical methods just referred to have also permitted the identification of several of these minor components which demonstrate just about all of the functional groups known to organic chemistry: hydrocarbons and saturated aliphatic and mono- and polyethylenic esters, alcohols, aldehydes, sulfur compounds, phenols, amines and aromatic nitriles, not to mention numerous derivatives of oxygen and nitrogen heterocycles. There are even halogenated derivatives in the products obtained by extraction of lichens with volatile solvents as well as nitroaromatic substances in the essential oil of ylang-ylang just to mention a few examples.

One frequently finds norterpenoids such as methylheptenone, cryptone and norpatchoulenol in natural products, but it is almost certain that these compounds are "artifacts" formed during hydrodistillation or even by enzymatic degradation in the plant itself. The industrial production of essential oils by steam distillation is carried out in metallic stills and, under these conditions, organic acids can in fact be liberated from the plant material. In the hydrodistillation of geranium, for example, the pH of the residual waters in the still is between 2.8 and 3.0! It is evident that, under these conditions, major chemical transformations such as hydrolysis, cyclization, decarboxylation and rearrangements can occur. For this reason, the chemical composition of the essential oil is often different from the fragrance which characterizes the natural source material. All the same, these essences are still the basis of commercial transactions in the fragrance industry even though they may contain numerous transformation products that can be responsible for notable odor differences with the natural plant.

Extraction with volatile solvents respects the odor of the original plant material far more—especially if it is carried out at low temperatures. Nonetheless, when making biochemical studies on plant materials, it is necessary to cool them to the temperature of liquid nitrogen, to grind them down and then to extract them with appropriate solvents. This is often the only way to avoid secondary reactions. In particular, the action of phosphatases, which are among the most active enzymes, is completely inhibited under these conditions, whereas at –30°C they are still capable of hydrolyzing phosphate esters, thereby liberating free alcohols which can often give a characteristic odor to essential oils.

The discussion has so far centered on a class of natural products which are based on the terpene skeleton. Their formation in plants is the result of an extremely uniform and ubiquitous biosynthetic mechanism passing through mevalonic acid and isopentenyl pyrophosphate (discussed in more detail later).

Despite their importance in many essential oils, the terpenoids only constitute minor components of a great number of other essential oils. This is the case, for example, in the essential oils of anise, star anise, basil, bay, cinnamon leaves and bark, clove buds, leaves and stems, pimento, pepper, sassafras, thyme, oregano, ylang-ylang, and many others. The essential oils obtained from these plant sources contain substances which are called the "phenylpropanoids" as their principal aromatic constituents.

The formation of these constituents by the plants is the result of a biosynthetic mechanism which is totally different from that for the terpenoids. These constituents are derived from the phosphate of the enol form of pyruvic acid and shikimic acid. This "cyclogenetic" mode of biosynthesis plays an important role in nature, particularly in the formation of the lignin of trees: lignin from which, incidentally, the near totality of anethole and vanillin is made—two "synthetic" substances of great importance to the flavor and fragrance industries.

Another type of aromatic constituent of essential oils, the aromatic components of lichens (which will be discussed in more detail later), are obtained from the cyclization of polyketone chains which in turn result from the condensation of acetic and malonic units. This different mode of biosynthesis involves, nonetheless, the same fundamental element as that of the isoprenoids: acetyl coenzyme A.

In the case of plants which contain phenylpropanoids, cinnamic acid is not only frequently a constituent of their essential oils, but along with p-hydroxycinnamic acid, it is one of the most important metabolic intermediates in the process leading to the volatile phenylpropanoids in their essential oils. These compounds, which are frequently phenolic, exist in the plant in the form of nonvolatile glucosides. Following the enzymatic cleavage which occurs during the "postmortem" catabolic processes, the aglycone is liberated and becomes a component of the essential oil.

Finally, it should be mentioned that a certain number of plant and animal products contain more or less large quantities of macrocyclic substances having a musk odor. These will be dealt with later, but it can be noted now that the chemistry of these substances was studied principally by Ruzicka and his school. In 1939, the Academy of Sciences in Stockholm honored this work by awarding him the Nobel prize in chemistry "for his research on polymethylene compounds." Thus, we see that in less than 30 years, three Nobel prizes in chemistry were awarded to chemists for their research in the field of odorants.

If progress in methods of direct analysis have permitted the isolation and purification of hundreds of new odoriferous compounds, it is thanks to the parallel progress in methods of structural analysis that the constitution of these substances could be established. In particular, the development of spectroscopic methods such as ultraviolet, infrared, nuclear magnetic resonance, mass spectrometry, X-ray diffraction, optical rotatory dispersion and circular dichroism have provided the chemist with irreplacable tools.

An equally fundamental problem is still far from being totally resolved. It concerns the synthesis of all of these substances, some of which have, of course, extremely complex structures. This problem raises a real challenge to organic chemists because in numerous cases it is only by stereospecific synthesis that a structure can be definitively established.

The era of synthesizing odorant substances started long ago. In fact, it was as far back as 1856 that Chiazza made synthetic cinnamic aldehyde by condensing benzaldehyde and acetaldehyde in the presence of hydrochloric acid. The current industrial synthesis of this aldehyde has not changed except that hydrochloric acid has been replaced by an alkaline catalyst. Crude cinnamic aldehyde was obtained for the first time in 1833 by Blanchet when he steam distilled Ceylonese cinnamon bark. In 1834, Dumas and Peligot obtained this product in its pure form and were then able to study its properties and establish its molecular formula. In 1853, Bertagnini discovered that cinnamic aldehyde gave a bisulfite derivative which permitted its purification. This extremely simple method has since then been largely generalized to the extraction and purification of numerous carbonyl compounds.

Cinnamic aldehyde is currently produced in large quantities (500–700 metric tons/year). It is used directly in perfumes and flavors, but it also finds use as an intermediate in the synthesis of numerous substances including perfumery raw materials such as cinnamic alcohol, phenylpropyl alcohol and their esters.

Cinnamic acid has perfumery and flavoring use as a synthetic raw material and is present in its free state or in the form of esters such as cinnamyl cinnamate and benzyl cinnamate, in the balsams of Peru, Tolu, and Styrax. The essences obtained from extraction of these balsams with volatile solvents, have extremely agreeable odors which are warm, sugary, balsamic, and particularly tenacious.

Cinnamic acid was discovered before 1800, but it was only definitively identified in 1835 by Dumas and Peligot and its first synthesis was made in 1853 by Bertagnini. Its exact structure, however, was not determined until 1866, by Erlenmeyer. The synthesis consisted of heating benzaldehyde with acetic anhydride in the presence of sodium acetate. This method, which has general use in the synthesis of α, β-unsaturated acids, is now known as the "Perkin method." It is interesting to note that while cinnamic aldehyde and cinnamic acid were rapidly adopted by perfumers, cinnamic alcohol, which has an agreeable hyacinth odor (and which is a natural component of that essence), was only used much later after Barbier and Leser had synthesized it in 1905.

Another important product is the substance known as heliotropine or piperonal. This substance has an odor resembling that of heliotrope, but has never been found in this plant although it has been found in small quantities in the false acacia. It is one of the rare chemical products whose discovery was the result of research in an area totally unconnected with perfumery. During studies on the alkaloid of pepper (piperine) in 1852, Anderson found that nitric acid oxidation produced an odoriferous product. This observation was confirmed five years later by Babo and Keller and, in 1869, Fittig and Nielch described with precision the production of "piperonal" by permanganate oxidation of piperic acid. They also established the molecular formula and the physical properties of this substance shortly before its exact structure was confirmed by Barth in 1871.

Heliotrope was extremely costly in 1880 because it could only be obtained by oxidation of pipenic acid prepared by hydrolysis of pipenine from pepper. A new synthesis by Eykman brought about a veritable commercial revolution. In 1885, out of interest in the chemical constitution of a Japanese by-product from the manufacture of camphor called "Shikimol," he performed a permanganate oxidation on it and obtained a product with the familiar odor of heliotropine. The following year, Polek obtained heliotropine by permanganate oxidation of isosafrole which had been prepared by the alkaline isomerization of safrole—a substance readily obtained from oil of sassafras.

Finally, Eykman was able to show that shikimol and safrole were, in fact, one and the same product. This study brought about a considerable reduction in the price of heliotropine, thus making this substance widely accessible to the perfumery industry.

In 1820, Vogel reported that coumarin was a major constituent of the tonka bean. Since then, its presence has been established in several natural products. Its synthesis was realized for the first time by Perkin in 1877 from salicylic acid, the preparation of which had already been described by Piria in 1848. The industrial synthesis of salicylic acid uses the method of Kolbe who succeeded in 1860 in obtaining it by heating sodium phenolate in the presence of anhydrous carbon dioxide at 180°–200° C. Perkin's synthesis of coumarin was followed by another by von Pechmann which consisted of condensing a phenol with a β-ketoester. One can therefore obtain coumarins which are substituted on the lactone ring or even on the aromatic ring.

Vanilla is incontestably one of the most popular flavorings. Its consumption in the form of vanilla beans, their extracts, or simply as vanillin is almost universal. It is indispensible for the production of foods such as cookies, sweets, chocolate and ice cream, but it is also used in the perfumery industry where its odor is greatly appreciated. It has been used in Europe since the sixteenth century, when it was introduced by the Spanish Conquistadors. The phenomenon by which vanillin crystallizes out of and covers the beans had permitted its isolation a long time ago. However, it was only in 1858 that Gobley obtained it in a sufficiently pure form to permit Carles, in 1872, to make an exact analysis. In 1874, Tiemann and Haarmann determined its structure and, in 1876, Reimer achieved its synthesis from guaiacol, thereby furnishing definitive proof of its structure.

The patents of Tiemann and Haarmann provided an enormous impetus to chemical research in directions which are still being pursued today. This synthesis of vanillin, preceded by that of salicylaldehyde, is known today as "The Reimer-Tiemann reaction" and is still used as a general method for making phenolic aldehydes. It consists of chloroform reacting with a phenol in the presence of sodium. On close inspection, one can consider that this is the first use of carbenes in synthesis ($CHCl_3 + NaOH \rightarrow CCl_2$). The dichlorocarbene thus generated reacts with the mesomer of the phenolate ion to give the precursor of vanillin.

The discovery of nitrated substances with musky odors is certainly one of the oldest cases of a completely fortuitous discovery made by research workers having no links with the fragrance industry. It was in the "Acts of the Berlin Academy" in 1759 that a publication reported that treatment of the "oil of amber" with fuming nitric acid gave a product which possessed a musky odor. This observation was confirmed a century later (in 1857), by Vohl, who showed that treating hydrocarbons from brown coal with fuming nitric acid also yielded products with musky odors.

It was only in 1880 that Kelbe finally obtained a chemically defined product: trinitro-para-cycmene, which, in fact, only possesses a very weak odor of musk. It was the work of Baur, starting around 1885, which really created the family of nitro-musk chemicals such as those still used today. From the start, the exact structures of the numerous products obtained by Baur were not known with any certainty even though these products were perfectly well defined. It was not until around 1930 that the results of the research of Tchichibabine cast more light on this field of fundamentally interesting substances. (This will be discussed later in more detail.)

The few compounds just described constituted the near totality of the arsenal of synthetic fragrance substances available to the perfumer at the end of the last century. Apart from these, however, there are a few others that should not be forgotten. The ionones, for example, were synthesized by Tiemann and Kruger in 1893 and they rapidly conquered perfumery with their fine odor of violets. Amyl salicyclate, possessing a clover-like odor with a chocolate intonation, had already been known for a long time, but it was Darzens who recommended it as a fragrance ingredient. The quinoleines, discovered by Skraup in 1880, also brought interesting "touches" to perfumes thanks to their varied odors evocative of leather, oak, moss, and wood.

The start of the twentieth century saw a sequence of important discoveries. In 1903, Moureu and Delange found methyl heptine carbonate and methyl octine carbonate which possess intense odors reminiscent of violet leaves. Darzens, in 1904, discovered the "glycidic method" of synthesizing aldehydes and used it to prepare methylnonyl acetaldehyde from methylnonyl ketone which had been extracted from rue oil. The success of this product was immediate. Another discovery of this period, still rather obscure, was hydroxycitronellal which was prepared from citronellal isolated from citronella oil. When one considers the fine "lily of the valley" odor of this substance, it is easy to imagine the success it must have had then, particularly as its popularity has not diminished since. As one can imagine, this discovery stimulated intense developments in the synthesis of hydroxyaldehydes.

Since this era, the number of synthetic fragrance materials has increased to the extent that today's perfumer has several thousand synthetic products on their "palette." Moreover, every day the search goes on for original perfumery notes which will enrich even more this "palette" among the many new substances generated by chemists in the research laboratories of French, Swiss, German, Dutch, English, American and Japanese fragrance manufacturers.

The syntheses discussed so far are all extremely simple ones with only one or two steps and starting generally from cheap and abundent raw materials. For this reason they obviously have great industrial interest, but this does not detract in any way from the challenge of the search which exists at two levels.

The first level concerns syntheses which can be qualified as academic in that they employ the most sophisticated synthetic methods and often involve a very large number of steps. In short, no economic considerations can be said to intervene in this approach for which only the final objective counts. If such syntheses were to be industrialized, they would lead to prices exceeding even those of natural products.

An example that illustrates this relates to the essence of sandalwood (to be discussed in more detail later). The odor of this essence is due principally to the predominance of two sesquiterpene alcohols: α- and β-santalol. Stereospecific syntheses of the academic type have been realized for both alcohols. In the syntheses of α-santalol by E. J. Corey and M. Julia, both used the same starting material, (+) -camphor, via π-bromotricyclene. Thus, if we consider the previous total synthesis of camphor discussed earlier, then the syntheses of Corey and Julia can be considered total syntheses of α-santalol. Counting the preparation of intermediates, the syntheses of Corey and Julia both involved 13 steps. The main difference resides in the fact that Corey's was linear, whereas Julia's was convergent and hence economically advantageous.

One can readily imagine the difficulty of extrapolating such syntheses to the industrial scale. In fact, if we were only to take into account the notion of final yield of the whole operation and exclude all consideration of the labor costs, reagents, energy, etc., and were to accept an average industrial yield of 80% for each step, the total yield would be approximately $0.8^{13} \times 100 = 5.5\%$ for the synthesis of Corey and 11% for the synthesis of Julia. These results speak for themselves. Under these conditions, the cost price of α-santalol would clearly be superior to that of sandalwood oil.

The second level of challenge concerns the *industrial* synthesis of terpenoids. We will not consider here those syntheses which use natural substances as relays. Such manipulations, which are actually only "half-syntheses," will be discussed later. Some of these have been used industrially since the start of the century, but they are competely different from total industrial syntheses. It was only in 1955 that the first total industrial synthesis of linalool at a cost price below that of natural linalool was made in Grasse. Syntheses such as this, which are often called "acetylenic syntheses," presently give access, in addition to linalool, to the near totality of aliphatic mono-, di- and sesquiterpenoids and they permit the total synthesis of vitamins A, E and K as well as numerous carotenoids.

Despite these incontestable successes, it is not possible to foresee in the near future the total industrial syntheses of the complex substances which are found as components of essential oils such as those of sandalwood, cedarwood, costus, patchouly and vetyver, to name just a few. For this reason, all of the natural fragrance materials will continue to hold their place of choice in the composition of perfumes. They will remain indispensible to the perfumery industry for a long time to come providing their price does not rise to the extent that they pass a threshold beyond which the synthesis of their components becomes profitable. In other words, the synthesis of aroma chemicals can be considered as an element in the economic regulation of the international markets for these natural products. Under these conditions, a harmonious development between natural substances and synthetic products will continue to be observed. We have seen that the latter occupy almost all of the areas of organic chemistry. They have utility in perfumery as long as they possess a sufficiently high vapor pressure so that enough molecules reach the receptors in the olfactive mucosae. This condition is not sufficient in itself. It is still necessary for the compound to have odor and for the perfumer to consider that this odor is sufficiently interesting for inclusion in their compositions!

CHAPTER

2

Terpenoid Chemistry

Introduction—Nomenclature

A. Introduction

The terpenes constitute a widely represented and chemically interesting group of substances. Although they show wide structural diversity, they do, however, share a common characteristic: they can be disconnected back to isoprene units. For this reason, a system of classification is possible based on the number of isoprene units they are seen to contain.

Hemiterpenes: contain only one C_5 isoprene unit (2-methyl–1,3-butadiene) like prenol (isopentenol or 3-methyl–3-butenol) [2.01].

[2.01]

$(CH_3)_2C{=}CH{-}CH_2OH$ prenol

$(CH_3)_2C{=}CH{-}CH{=}O$ prenal

Monoterpenes: contain 10 carbon atoms and can be cleaved into *two* isoprene units. This is the case for three isomeric allylic alcohols of great interest [2.02].

[2.02]

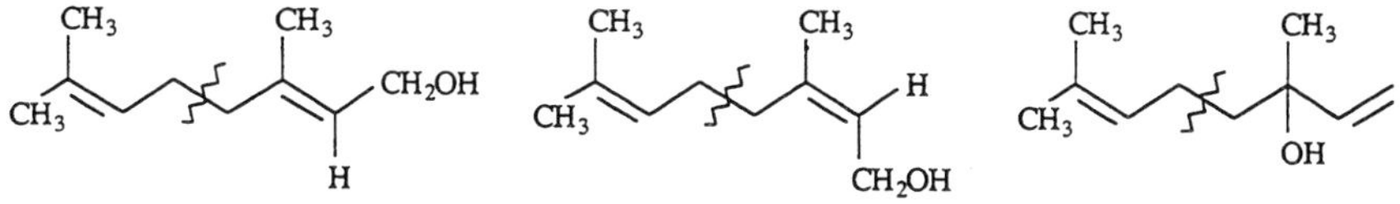

CH₃ CH₃ CH₃ CH₂OH H — geraniol

CH₃ CH₃ CH₃ H CH₂OH — nerol

CH₃ CH₃ CH₃ OH — linalool

Sesquiterpenes: 3 isoprene units (C_{15})
Diterpenes: 4 isoprene units (C_{20})
Sesterterpenes: 5 isoprene units (C_{25})
Triterpenes: 6 isoprene units (C_{30})
Carotenoids: 8 isoprene units (C_{40})
Rubber: n isoprene units [(C_5) n]

The great majority of known terpene compounds result from a so-called "head to tail" coupling of isoprene units [2.03].

[2.03]

head tall

Such a structure is called "regular," as in the case of geraniol [2.04]

[2.04]

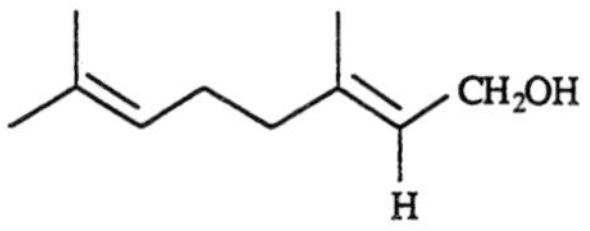

CH_2OH geraniol

H

and predominates in the sesquiterpenes, diterpenes and sesterterpenes. The triterpenes and carotenoids, however, are formed respectively from *two groups* of *three* or *four* isoprene units in which the structures are regular but which are coupled at the mid-point of the molecule in the "*head to head*" manner. Squalene constitutes the most basic example of this in the triterpenes [2.05].

[2.05]

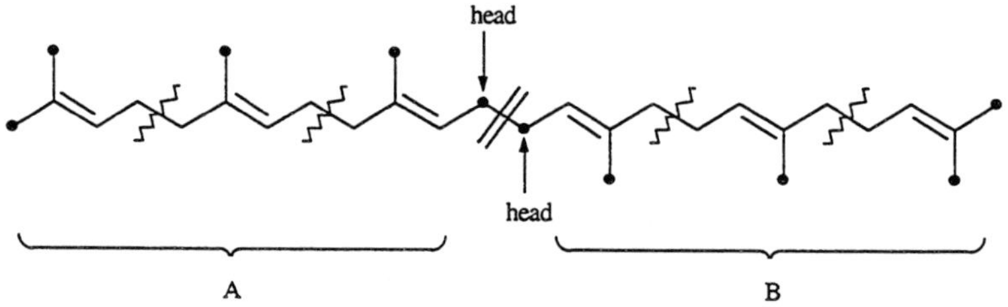

head

head

A B

In the carotenoid family, the isoprenic homologue of squalene is *lycopene* in which each group A and B contains *4* isoprene units instead of *3*. A rather more complex example is provided by β-carotene in which the chain is entirely conjugated. This is not the case for squalene and lycopene [2.06].

[2.06]

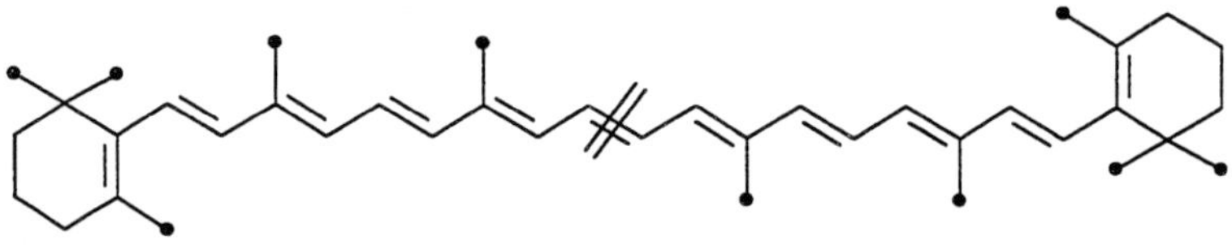

There exist a number of exceptions to the regular isoprenoid coupling. These can be explained by rearrangement of their regular isoprenic precursors either during the course of their biosynthesis or as the consequence of the manner in which they were physically isolated (particularly when isolated by steam distillation).

The isoprene units are also found in numerous other natural products. This is the case for certain indolic alkaloids such as vindoline (containing a monoterpene fragment), the lysergic acid family (an isoprene unit) and certain natural quinones such as vitamins K and F_1.

Aside from their importance, the chemistry of the terpenoids has posed numerous problems that have quite literally stimulated the development of fundamental organic chemistry. The Wagner-Meerwein rearrangement, research into non-classic carbonium ions and the theory of conformational analysis owe much to the chemistry of terpenoids. Moreover, the problems raised by the total synthesis of numerous polycyclic terpenoids have contributed largely to the launching of synthetic organic chemistry.

Finally, the development of experimental biosynthesis certainly owes much to the hypotheses of biogenesis suggested in 1950 by Ruczicka and his school. Their "isoprene biogenesis rule," which is a development of a hypothesis of Wallach, constitutes a fundamental contribution to our knowledge of the biogenesis of terpenes. In addition, they have given a stereochemical interpretation to this based on Barton's work on conformational analysis. This rule stipulates that the *terpenoids* are derived from aliphatic precursors such as *geraniol* for the monoterpenes, *farnesol* for the sesquiterpenes, *geranylgeraniol* for the diterpenes and *squalene* for the triterpenes. These precursors then lead to a whole series of mono- and polycyclic terpenoid compounds as the result of "reasonable transformations" which are based on reactions which can be carried out *in vitro*.

Using this rule of isoprene construction as inspiration, Karrer and collaborators were able to rapidly shed light on the domain of the carotenoids. The structure of β-carotene, isolated from carrots by Wackenrodder in 1837, was thus elucidated by Karrer in 1930.

The study of carotenoids has also contributed to the development of chromatographic techniques, which were virtually revolutionized by analyzing natural substances; notably by the development of gas-phase chromatography by James and Martin. At the same time, spectroscopic techniques have also shown a spectacular development since 1945. The conjunction of these two "tools" has made possible a veritable explosion in our understanding of the chemistry of natural products and the study of their structures. Consequently, we have seen an enormous material and intellectual investment in the elucidation of problems of syntheses.

Research started around 1950 and continues today in numerous laboratories. In this field, one cannot ignore the work of Barton, Woodward, Corey, Ireland, Stork, Nakanishi—to name only a few.

B. Nomenclature

First, it is necessary to describe the principal skeletons.

Monoterpenes [2.07]
Sesquiterpenes [2.08]
Diterpene skeleton labdane [2.09]

A symbolic representation of the structures of terpenoids was proposed in 1954 by Klyne. This allows one to represent the mode of cyclization of cyclic compounds by identifying the carbon atoms of the isoprene unit in the following way [2.10]:

As this matrix is symmetrical with regard to the principal diagonal, it is only neces-

[2.07]

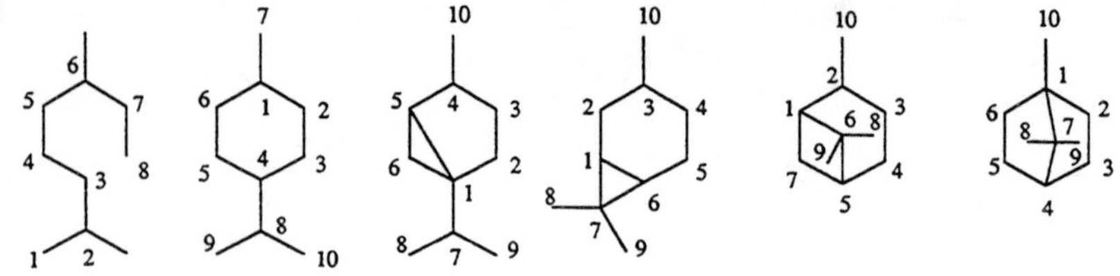

2,6 dimethyl-octane | para-menthane | thujane | carane | pinane | bornane (or octane

[2.08]

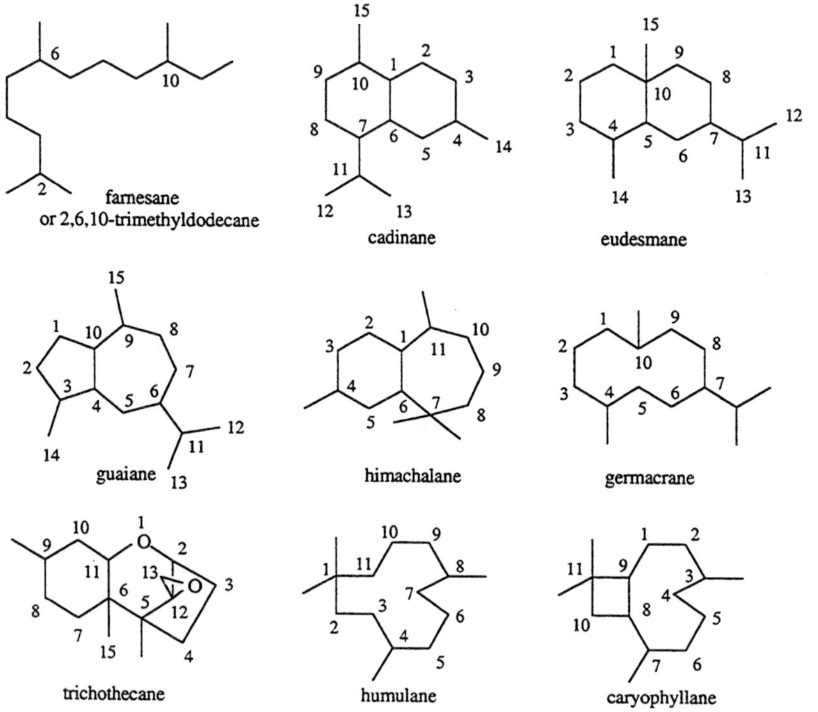

farnesane or 2,6,10-trimethyldodecane | cadinane | eudesmane

guaiane | himachalane | germacrane

trichothecane | humulane | caryophyllane

[2.09]

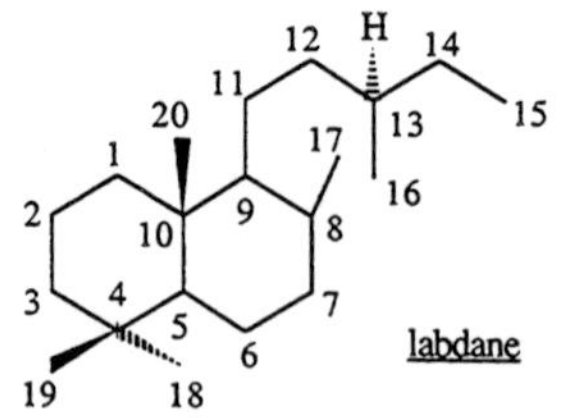

labdane

sary to consider the possibilities shown either above or below this line. For the construction of aliphatic monoterpenes, there are 10 possible modes of linking [2.11].

Two regularly linked units will thus be symbolized by (Z-W) as exemplified by geraniol, which is constructed with a 2,6-dimethyloctane skeleton [2.12].

If three isoprene units are associated in this way, they will be designated $(Z\text{-}W)_2$

[2.10]

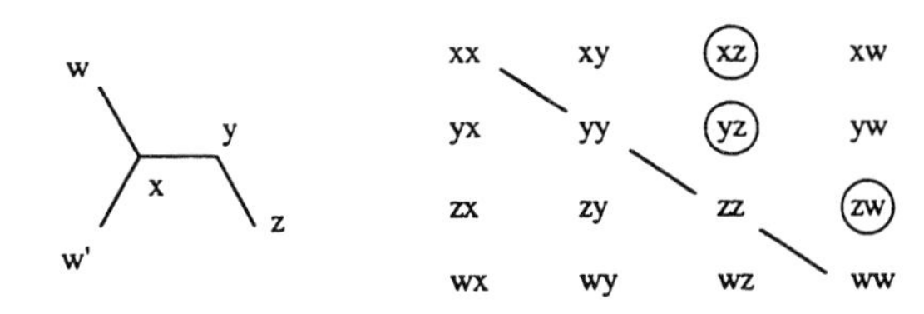

[2.11]

ZW = regular structure
YZ = lavandulyl structure
XZ = artemisyl structure

lavandulol

artemisia ketone

[2.12]

as shown here for farnesane [2.13].

[2.13]

Cyclic structures can be described in the following manner [2.14].

[2.14]

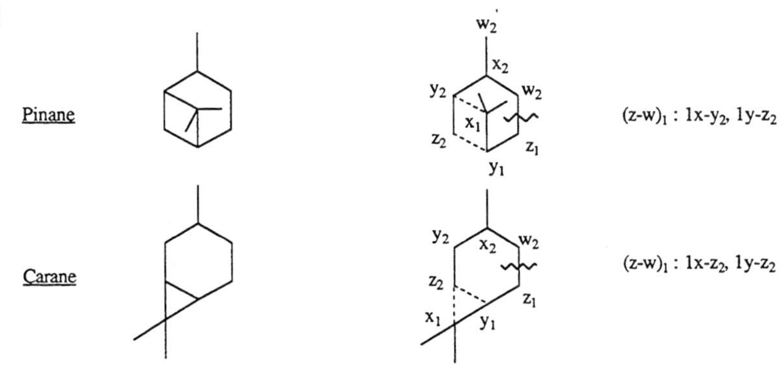

Thujane

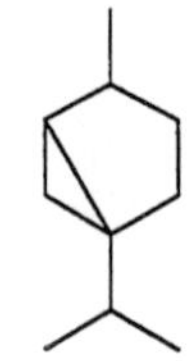

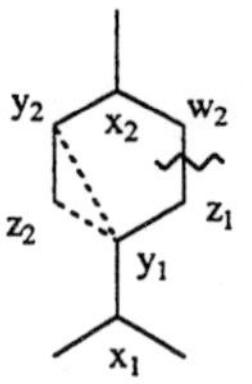

$(z\text{-}w)_1 : 1y\text{-}y_2, 1y\text{-}z_2$

But one can also write

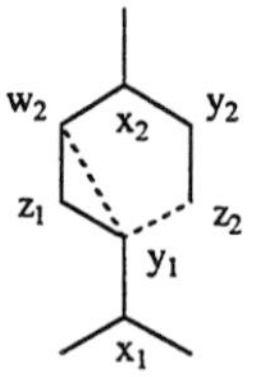

$(z\text{-}w)_1 : 1y\text{-}z_2, 1y\text{-}w_2$

The same problem is encountered for certain sesquiterpene structures such as eudesmane [2.15].

[2.15]

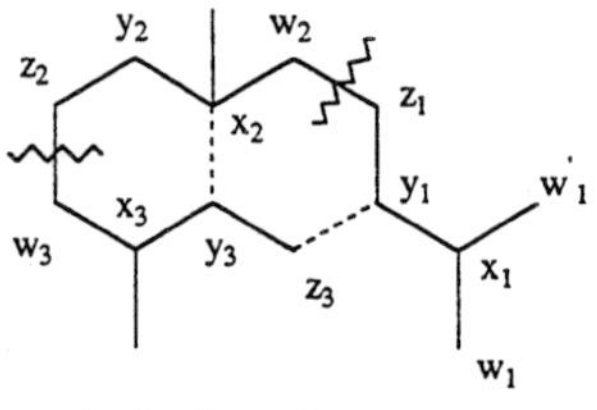

$(z\text{-}w)_2 : 1y\text{-}z_3, 2x\text{-}y_3$

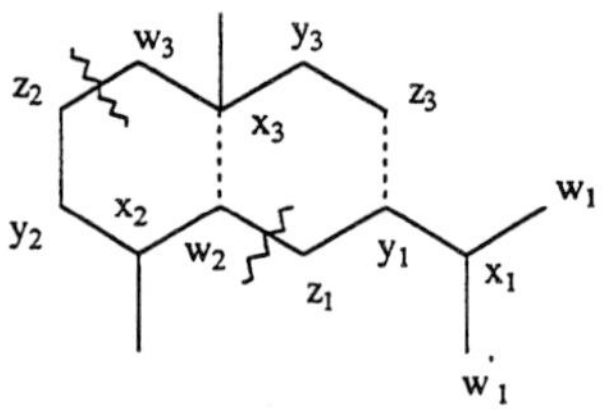

$(z\text{-}w)_2 : 1y\text{-}z_3, 3x\text{-}w_2$

This equivocal situation raises the question of the way in which nature carries out the cyclization of the isoprene chains. Biosyntheses undertaken with labeled compounds permit us to see the correct way.

Monoterpenes: the monoterpenes (C_{10}) can be classed as acyclic, monocyclic, and bicyclic. There is even a tricyclic monoterpene with the obvious name of "tricyclene." In each series there are oxygenated derivatives such as alcohols, aldehydes, ketones, epoxides, esters, etc., which have been known for about a century (especially following the work of Wallach) and which compose the principal constituents of a very large number of essential oils drawn from flowers, leaves or fruits.

Acyclic Monoterpenes

It is not possible to provide here an extensive discussion of all of those represented in this series. By limiting ourselves to the more characteristic members, we can constitute a vocabulary which will be useful in the subsequent study of industrial syntheses.

The acyclic monoterpenes are almost all derived from 2,6-dimethyloctane. There are three principal hydrocarbons: β-myrcene 1, cis β-ocimene 2 and trans β-ocimene *3* [2.16].

[2.16]

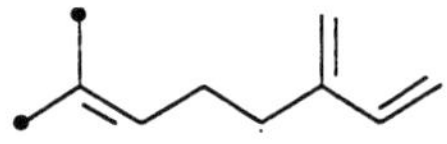

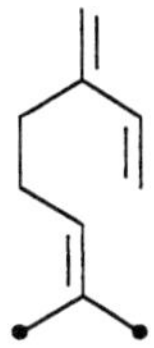

1 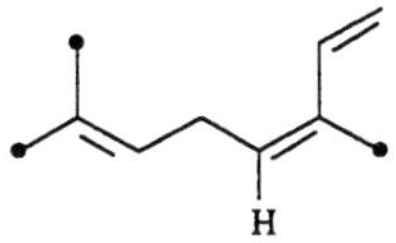 or

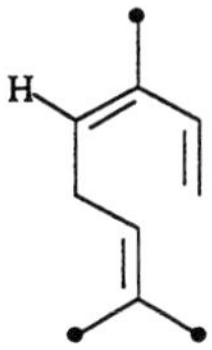

2 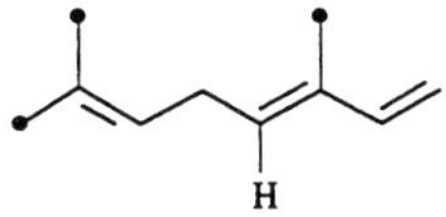 or

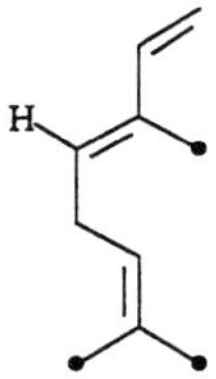

H H

3 or

H H

The prefix "β" indicates the isopropylidene form as opposed to the isopropene form, which is qualified as "α." [2.17]

[2.17]

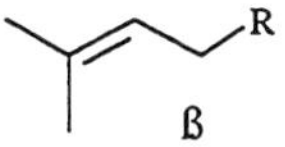

R β R α

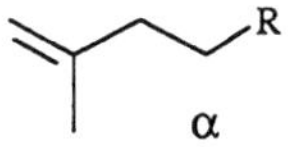

Indeed, some reactions give rise to α-myrcene 1' as well as the cis- and trans-ocimenes (2' and 3'). From their infrared spectra, it is evident that the majority of natural products exist in the "β"-form (isopropylidene) [2.18].

[2.18]

1' 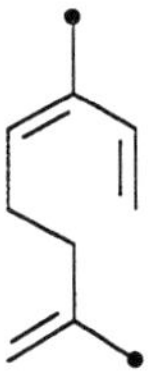 2' 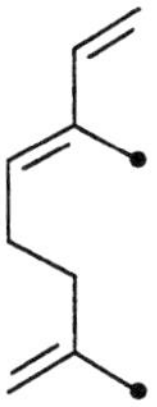 3'

These highly unsaturated hydrocarbons are very sensitive to oxygen and elevated temperatures. By virtue of their conjugated diene structures, when stereochemistry permits (if the cisoid conformation is easily accessible), they will undergo Diels-Alder reactions. In this way, myrcene reacts with dienophiles such as acrolene and acrylonitrile to give adducts 4A and 4B, respectively [2.19].

It is also possible to observe dimerization reactions. One molecule of myrcene can play the role of diene and a second molecule can act as the dienophile, there-

[2.19]

A(B)

A(B)

Δ

4 A = - CH=O
4 B = - CN

1

by obtaining α-camphorene 5 [2.20].

[2.20]

Δ

5

Heating cis β-ocimene to 185°–190° C gives rise to a completely different reaction belonging to the group of sigmatropic rearrangements; in this case a thermally permitted (cf. Woodward-Hoffmann rules) 1,5 H-migration [2.21].

[2.21]

≡

6

The product obtained is 4-(E)-, 6-(Z)-allo-ocimene 6 in which the three double bonds are conjugated. The stereochemistry of the product can be explained by the structure of the two possible transition states I and II [2.22].

[2.22]

I

II

The transition state I, in which the isobutenyl group is equatorial, is less "strained" than II, in which the same group is axial. The rules which control pericyclic reactions (Woodward-Hoffmann, Dewar-Zimmermann) also provide an explanation of the stereospecificity of the reaction. Allo-ocimene 6 possesses an ultraviolet spectrum characteristic of conjugated trienes.

λ_{max} (cyclohexane): 268, 277, 289 nm
ε: 27600, 36500, 28000

The β-ocimenes possess an activated $-CH_2-$ group since they are intercalated between two double bonds (allylic H). Under the action of bases, it forms a delocalized anion which when reprotonated gives a mixture of the four allo-ocimene isomers in which the (E)-4-(Z)-6-isomer 7 predominates (the isomers 8 and 9 are practically absent) [2.23].

[2.23]

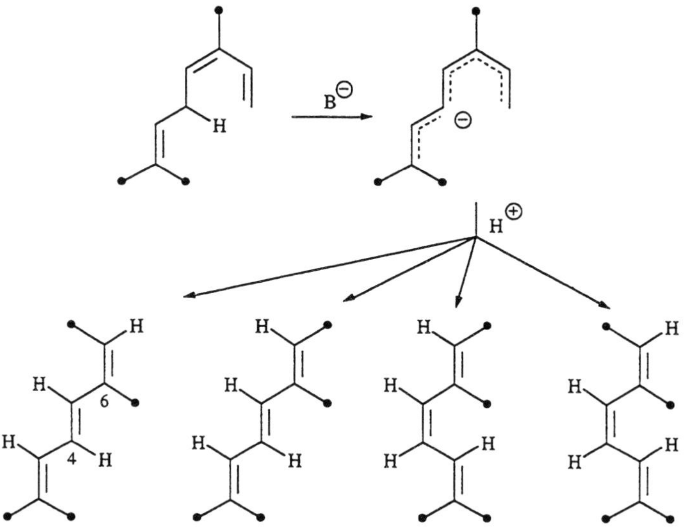

These hydrocarbons can also undergo photochemical reactions. For example, myrcene 1 gives (via a photochemically permitted $\pi_s^2 + \pi_s^2$ process) the three products 10, 11 and 12. In the presence of photosensitizers, compound 12 1-vinyl–6,6-dimethylbicyclo[2.2.1]hexane constitutes the principal product [2.24].

β-Myrcene especially is of considerable importance to the perfumery industry as an intermediate for the preparation of numerous oxygenated terpenes—the principal ones being linalool 13, geraniol 14, nerol 15, citronellol 16 and the corresponding aldehydes (citral 17 and citronellal 18) [2.25].

The three allylic alcohols 13, 14 and 15 rearrange in acidic media to yield an equilibrium mixture (allylic migration). Kinetic studies on this reaction show that geraniol 14 and nerol 15 are consumed at roughly the same rate; however, geraniol gives linalool while nerol gives α-terpineol 19 at a rate which is 18 times

[2.24]

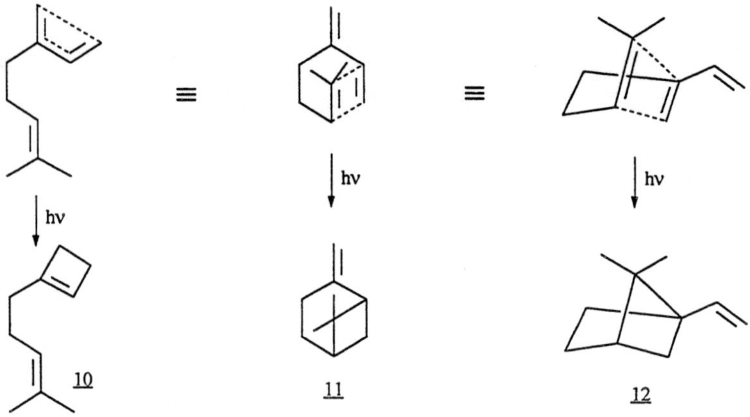

hν hν hν

10 11 12

[2.25]

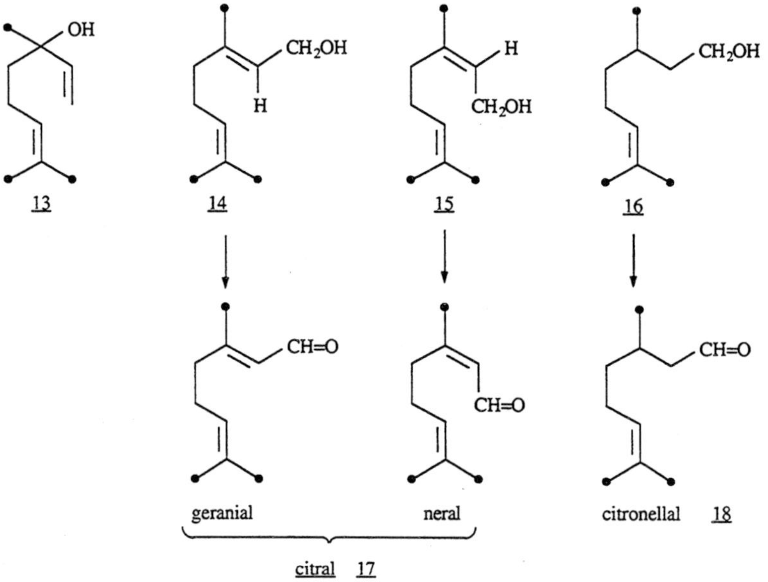

OH CH$_2$OH H CH$_2$OH CH$_2$OH

H

13 14 15 16

CH=O CH=O CH=O

geranial neral citronellal 18

citral 17

faster than for geraniol (more favorable stereochemistry of the double bond) [2.26].

[2.26]

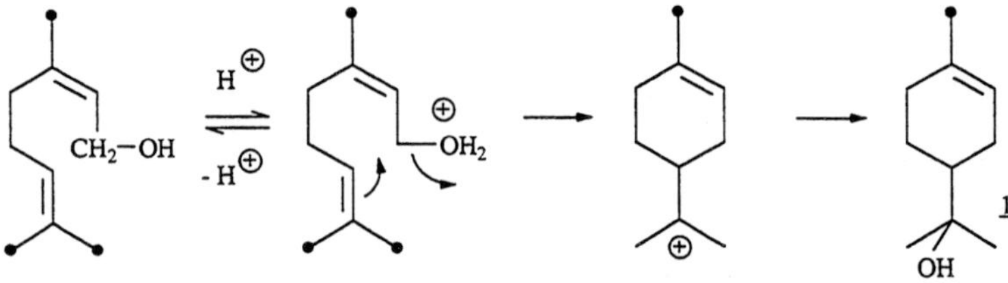

H$^{\oplus}$ $\oplus$

CH$_2$–OH OH$_2$

- H$^{\oplus}$

$\oplus$ 19

OH

When the alcohol functional group of geraniol is blocked (for example, by acetylation), it is also possible to carry out a cyclization in acidic media to obtain a mixture of α- and β-cyclogeranyl acetates 20 and 21 [2.27].

[2.27]

CH_2OAc $H^{\oplus}$ CH_2OAc $-H^{\oplus}$ CH_2OAc

20

CH_2OAc

21

This type of acid-catalyzed cyclization is very important because of its ability to provide a number of derivatives such as the ionones and the methylionones, which are of great importance to the perfumery industry. In this way, citral 17 (geranial + neral) is first condensed with acetone or methylethylketone in basic media. Aldolization followed by crotonization gives pseudoionones 22 or pseudomethylionones 23 depending on the ketone used. When methylethylketone (an unsymmetrical ketone capable of giving two different enolates) is used, two isomers called "normal" and "iso," and having structures 23 and 23', respectively, are obtained [2.28].

[2.28]

CH=O + $CH_3—C(=O)—CH_3$ $B^{\ominus}$ / BH CH=O $CH_2^{\ominus}—C(=O)—CH_3$

BH

22 (4, 5, 3, 6) - H_2O β-aldol (OH, O, CH_3)

Although the 3,4 or 4,5 double bond will be (E), the 5,6 or 6,7 double bond will be a mixture of (E) and (Z) geometry, reflecting the initial proportions of geranial and neral which have different geometry in this regard [2.29].

The cyclization of the "pseudo" derivatives in acidic media gives: (a) α- and β-ionones 24 and 25 from the pseudoionones (b) four derivatives (normal and iso

[2.29]

5 6 4 7 23 23'

variants of both α-methylionone and β-methylionone) 26, 28, 27 and 29 depending on whether 23 or 23' are cyclized [2.30].

[2.30]

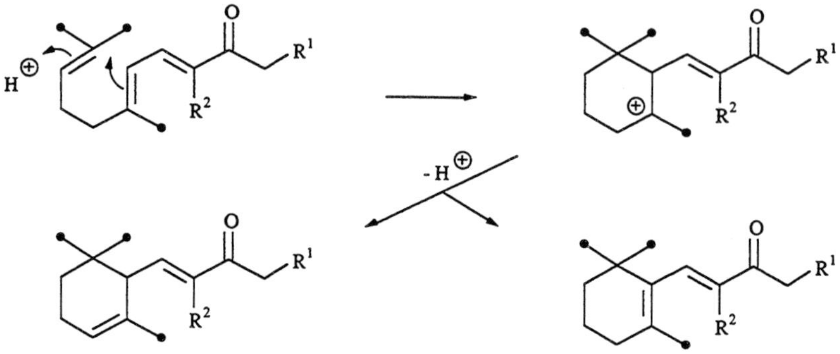

H R¹ R² -H Forms α Forms β

$R^1 = R^2 = H$: 24
$R^1 = CH_3, R^2 = H$: 26
$R^1 = H, R_2 = CH_3$: 28

$R^1 = R2 = H$: 25
$R^1 = CH_3, R^2 = H$: 27
$R^1 = H, R_2 = CH_3$: 29

In acidic media, linalool and its acetate lead to complex mixtures in which compositions are strongly influenced by the experimental conditions. On the other hand, if linalool is heated to high temperatures in rigorously neutral media, it forms a mixture of alcohols with a cyclopentanyl skeleton. These are called "plinols" and they result from an intramolecular ene-reaction [2.31].

[2.31]

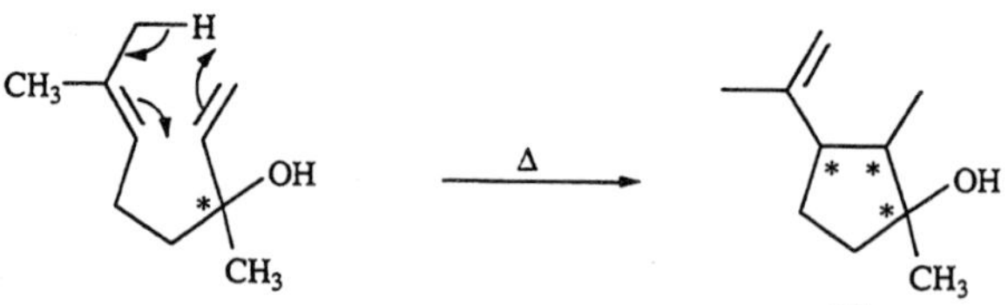

CH_3 H OH CH_3 Δ OH CH_3 30

In this way, from a compound having only one asymmetrical carbon, a product containing three is obtained. This permits the possible existance of $2^3 = 8$ optical isomers of which four will be dextro-rotatory and four will be levo-rotatory. However, as is frequently the case in concerted reactions, asymmetrical induction occurs. In this way, if optically pure linalool 13 is used, the main product is the epimer 30a which is favored over its stereoisomers 30b, 30c and 30d [2.32].

[2.32]

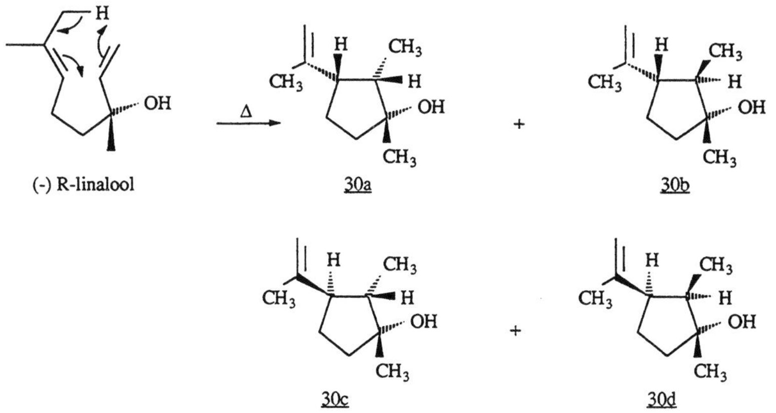

(-) R-linalool $\xrightarrow{\Delta}$ 30a + 30b + 30c + 30d

Linalyl acetate has been used as a starting point for the synthesis of the recently discovered alcohols of lilac 36 [2.33].

[2.33]

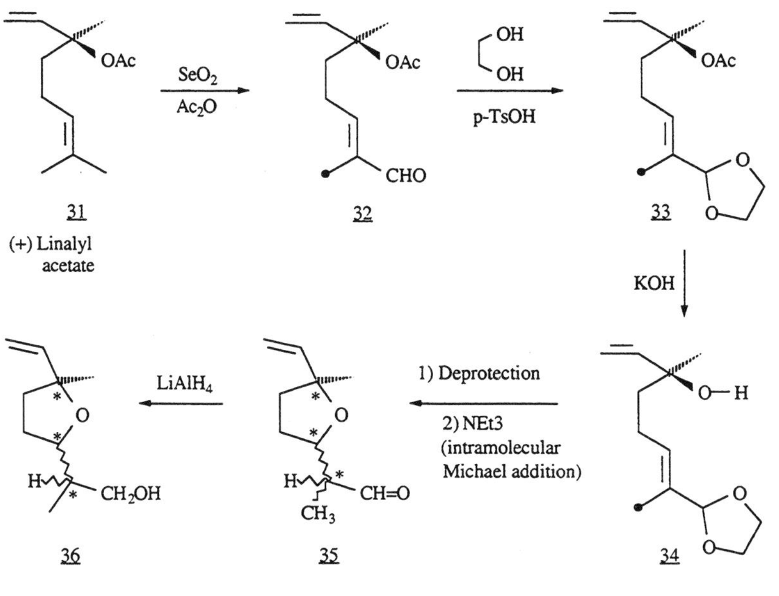

31 (+) Linalyl acetate $\xrightarrow[Ac_2O]{SeO_2}$ 32 $\xrightarrow[p\text{-TsOH}]{HOCH_2CH_2OH}$ 33 $\xrightarrow{KOH}$ 34 $\xrightarrow{\text{1) Deprotection; 2) NEt3 (intramolecular Michael addition)}}$ 35 $\xrightarrow{LiAlH_4}$ 36

In this sequence, (+) linalyl acetate first undergoes allylic oxidation by selenium oxide. The aldehyde function thus obtained is then protected in the form of its ethylene acetal 33 by the action of the glycol in the presence of p-toluenesulfonic acid, and the hydroxyl function is liberated by hydroxide-induced saponification.

Hydrolysis of the ethylene acetal in acidic media followed by treatment with base (Et_3N) gives rise to compound 35 by an intramolecular 1,4-addition (Michael addition). Reduction of this aldehyde 35 with $LiAlH_4$ gives the alcohol 36 as a mixture of four diastereoisomers. When (–) linalyl acetate is used, the four mirror images of these diastereoisomers are obtained.

The monoterpenic compounds just discussed all have a regular isoprenic linkage. This is not always the case, however. A number of natural products possess irregular linkages: for example, lavandulol 37, one of the constituents of lavender; artemesia ketone 38; and artemisia alcohol 39, which are both present in the essential oil of "*Santolina chamaecyparissus*" [2.34].

[2.34]

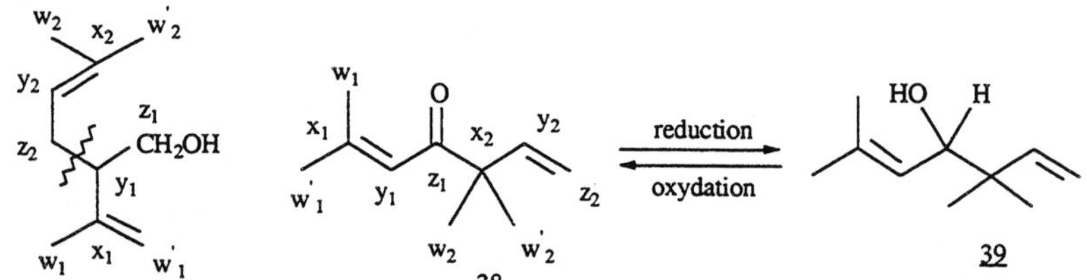

37

lavandulol (type yz)

38

artemisia ketone

(type xz)

reduction

oxydation

39

artemisia alcohol

This essential oil also contains another C_{10} substance, santolinatriene 40, which has a structure that does not respect the rule of isoprenic construction. It is also noteworthy that Morroccan chamomile oil contains as a major constituent an alcohol with the same carbon skeleton—2,5-dimethyl–3-vinyl–4-hexene–2-ol 41. A third example of this type of "erroneous" structure is lyratol 42 found in the essential oil of *Cyathoclina lyrata* [2.35].

[2.35]

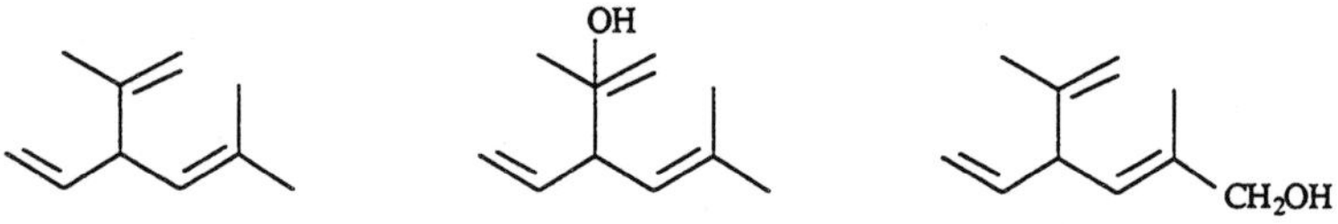

40 41 42

"santolinyl structures"

Yomogi alcohol 43 (essential oil of *Artemisia feddei*) is another example of the artemisyl structure. The allylic rearrangement of this alcohol 43 gives artemisia alcohol 39 [2.36].

[2.36]

43 → 39

A classical synthesis of yomogi alcohol 43 uses prenyl chloride 44 as starting material. This is coupled in the presence of magnesium to give 2,5,5-trimethyl–2,6-heptadiene 45 which by photosensitized oxidation followed by reduction with sodium sulfite yields yomogi alcohol along with an isomeric allylic secondary alcohol 46 [2.37].

[2.37]

CH_2Cl —Mg, Et_2O→ ; 1) $O_2h\nu$ photosens. 2) Na_2SO_3

44 45 43 OH 46 OH

Under the action of peracids, yomogi alcohol is converted into a mono-epoxide 47 which is transformed by treatment in acidic solution into derivatives of the santolinyl group. The mechanism of this isomerization can be understood by invoking the intervention of a non-classic "bicyclobutonium" cation. This transitory species with delocalized charge provides an explanation for the nature of the products of this reaction as arising from a "homoallylic" migration which is well known to organic chemists [2.38].

Protonation of the methylenic carbon A can lead by intramolecular cyclization to two oxygen heterocycles 50 and 51 [2.39].

This type of intramolecular cyclization is well known in the chemistry of terpenes where the reactive sites are conveniently disposed (preferably γ or δ to each other). For example, the epoxidation of linalool does not at ordinary temperatures yield an epoxy-alcohol. Instead, an equilibrium mixture of diastereoisomeric alcohols, the "pseudo-epoxylinalools" which have tetrahydrofuran 52 and tetrahydropyran 53 structures, are obtained [2.40].

In the case of artemisia alcohol 39, a recent synthesis makes use of diprenyl ether 56, obtained by condensing prenyl chloride 54 with the alcoholate of prenol 55 [2.41]. Treatment of diprenyl ether with phenyl lithium at low temperatures gives a mixture of four products in which artemisia alcohol 39 is largely predominant [2.42]. The formation of artemisia alcohol under these conditions could result from a [2.3] sigmatropic migration occurring in the anion 60 (formed by attack of the base on the ether 56), thereby giving the alcoholate of artemisia alcohol [2.43].

A mechanism involving intramolecular nucleophilic substitution by the anion would explain the formation of 57 [2.44].

It would no doubt be preferable, however, to invoke a radical process involving homolytic rupture of the C=O bond which unites the non-anionic carbon to the oxygen in the carbanion 60. It is indeed known that radicals are formed in the Wittig and related reactions (e.g., Stevens reaction). This has been shown in stud-

[2.38]

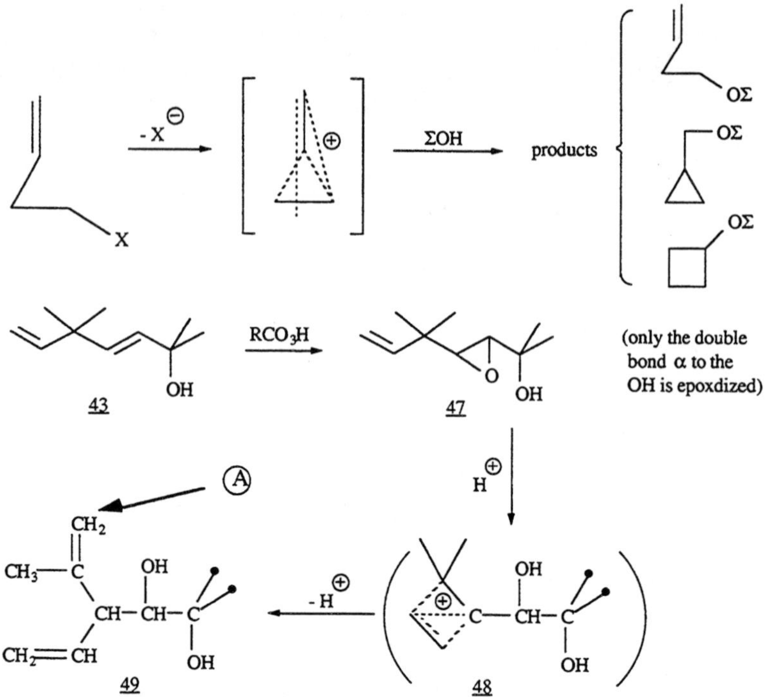

(only the double bond α to the OH is epoxdized)

[2.39]

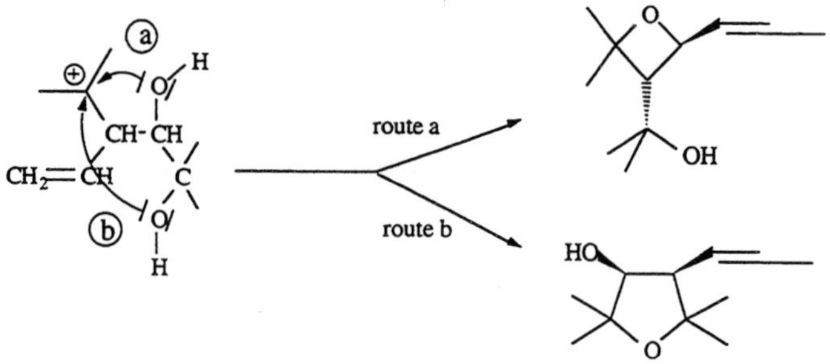

ies using chemically induced nuclear polarization. It is therefore possible to imagine the following scenario in order to explain the formation of products 57, 58 and 59: the "cetyl" radical anion resulting from the rupture

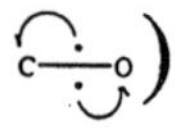

can be described as a resonance hybrid of three contributing structures A, B and C. The other fragment, the allylic radical, is also delocalized (resonance hybrid of two contributing structures D and E) [2.45].

The recombination, in the solvent cage, of the radical fragments through forms B and E would give artemisia alcohol 39 and through forms B and D, its isomer 57 [2.46].

[2.40]

OH H OH RCO$_3$H a b a OH - H$^{\oplus}$ O

13 52

b HO O - H$^{\oplus}$ HO O–H

53

[2.41]

CH_2Cl + $Na^{\oplus}$ $O^{\ominus}$ O

54 55 56

[2.42]

O=CH O=CH

HO H OH

39 (x-z) 57 (z-z) 58 (x-x) 59 (x-z)

67% 14% 10% 8%

[2.43]

C_6H_5Li

(- H)

2 3 O 2 $Li^{\oplus}$ 1 1 $O^{\ominus}$ $Li^{\oplus}$

56 60

[2.44]

$O^{\ominus}$ $Li^{\oplus}$ protonation 57

$Li^{\oplus}$

[2.45]

A B C

60

D E

[2.46]

E

B

39

57

For the formation of aldehydes 58 and 59, it is necessary to recombine form C with forms D or E [2.47].

[2.47]

hydrolysis

enol

CH=O

59

D C

E

hydrolysis

enol

CH=O

58

Monocyclic Monoterpenes

A. Skeletons based on six carbon atoms

The large majority of monocyclic terpenes possess a six-membered ring derived from para-menthane 60 [2.48].

[2.48]

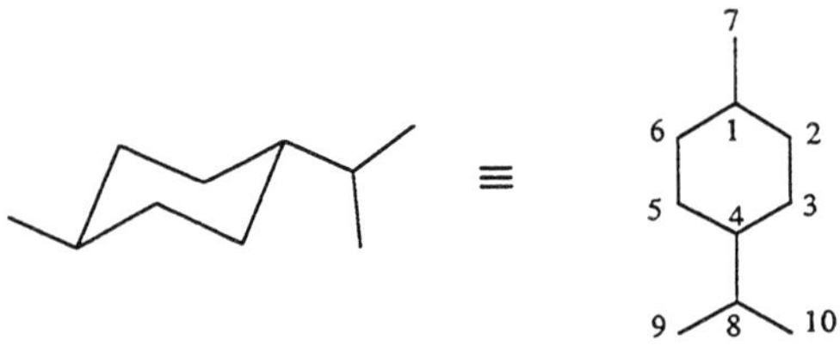

7 6 1 2 5 4 3 60 9 8 10

Six dienic hydrocarbons have been found in nature: limonene 61, terpinolene 62, α-terpinene 63, γ-terpinene 64, α-phellandrene 65 and β-phellandrene 66 [2.49].

[2.49]

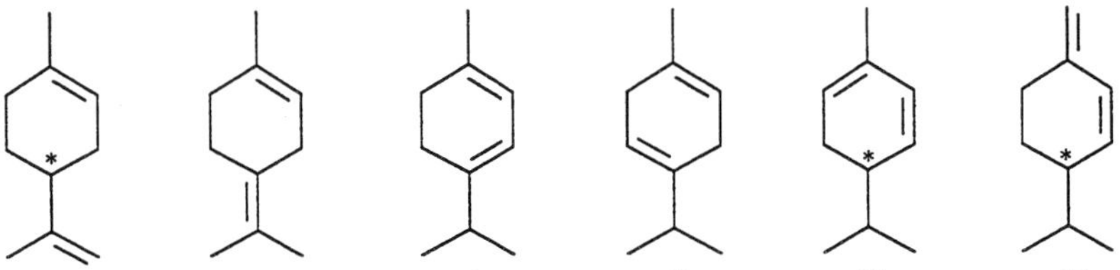

61 62 63 64 65 66

Limonene 61 is the principal constituent of the essential oils of orange, lemon and bergamot. It is present in lesser quantities in many of other essential oils and is the most important and one of the most widely found monoterpenes. It possesses an asymmetrical carbon atom (C*) at position 4 and can therefore exist in the dextro-rotatory (+), levo-rotatory (–) and racemic (±) forms. The last of these is known under the name of "dipentene."

The same situation exists with α- and β-phellandrenes which also have a C* in position 4. In fact, a total of 14 dienic hydrocarbons with a para-menthanyl skeleton can theoretically exist. All have been synthesized, but only the six mentioned here have been found in nature. Pyrolysis of limonene 61 gives two molecules of isoprene as the result of a retro Diels-Alder reaction. This process is reversible under conditions of constant volume and, when carried out in an autoclave, will lead to the racemization of limonene to give dipentene. If, however, the pyrolysis is carried out at ordinary pressures, it is possible to recuperate the isoprene as it is formed [2.50].

As a conjugated diene similar to cyclohexadiene, α-terpinene 63 can undergo both thermal and photochemical electrocyclic opening ($\pi^2 + \pi^2 + \sigma^2$ thermally disrotatory and photochemically conrotatory according to the Dewar-Zimmermann rules) [2.51].

[2.50]

Δ
retro Diels-Alder
(±)

61

[2.51]

Δ
or h ν
(reversible reaction)
z
y

63 67

It can be noted that the resulting conjugated triene has an isoprenic function of the (y-z) type which is characteristic of the "lavandulyl" system.

Acid-catalyzed hydration of limonene occurs principally at the vinylic double bond ($\Delta^{8,9}$) and furnishes α-terpineol 68. A small quantity of β-terpineol 69 is also formed by hydration of the endocyclic ($\Delta^{1,2}$) double bond.

Prolonged acid treatment of α-terpineol 63 leads to isoterpinolene 70. The addition of anhydrous HCl to limonene brings the same carbocations into action, but in this case the reaction is rapid and the two double bonds react to give the dichloride 71. The scheme shown below illustrates these results [2.52].

The phellandrenes 65 and 66 behave differently under the effect of dry HCl due to the conjugation of their unsaturation. The same monochloride 72 as a mixture of two epimers, is obtained from both. Hydrolyzation with calcium hydroxide obtains four allylic alcohols, 73, 74, 75 and 76 (two pairs of epimers)—the latter pair being designated by the trivial name "piperitols." Sulfo-chromic oxidation of this mixture is initially accompanied by an allylic rearrangement of the first pair of epimers 73 and 74 to give a single ketone, "piperitone" 77. The scheme shown below summarizes this situation [2.53].

In 1860, Tilden discovered the formation of crystalline nitrosochlorides from the addition of nitrosyl chloride to double bonds. Tetrasubstituted alkenes give stable nitrosochlorides 78 having an intense blue color. By contrast, the nitrosochlorides formed from trisubstituted alkanes 79 rapidly rearrange to chloro-oximes 80 [2.54].

Tilden's discovery played a very important role in the chemistry of monoterpenes. It can be compared to the role played in the chemistry of sugars by Fischer's discovery of phenylhydrazines.

Tilden himself used limonene in an early but important application of this reaction. The nitrosochloride 81 isomerizes to the chloro-oxime 82 which is converted by alkaline hydrolysis into a ketone of great interest, carvone 84. This procedure provided numerous industrial developments with many patents being

[2.52]

accorded. Carvone is used abundantly in the flavoring of consumable items such as chewing gums [2.55].

Both enantiomers of carvone exist in nature. The levo-rotatory form is the principal constituent of spearmint while the dextro-rotatory form is the principal constituent of caraway. Carvone 84 has been the subject of intensive studies, some examples of which can be cited here. Hydrobromic acid added rapidly to the terminal exocyclic double bond gives a bromo derivative 85 which is solvolized in alkaline ethanol to obtain, through an intermediate bicyclic ketone 86 (carenone), a seven-membered cyclic ketone, eucarvone 87 [2.56].

In alcoholic solution, carvone is transformed into carvonecamphor 88 under the action of sunlight. This reaction (reported by Ciamician and Silber in 1908)

[2.53]

HCl

65 72 66

$Ca(OH)_2$ H_2O

73 + 74 + 75 + 76

CrO_3 H_2SO_4

(-) piperitone

77

[2.54]

NOCl

78

NOCl

79 80

marks the start of photochemistry. The structure of the "photoproduct" has only recently been established [2.57].

On passage through alumina, carvonecamphor is transformed into isocarvone 89. If, on the other hand, the irradiation is effected by high frequency ultraviolet light (scission), a ketene 90 is obtained which reacts with alcohol to give the ester 91.

[2.55]

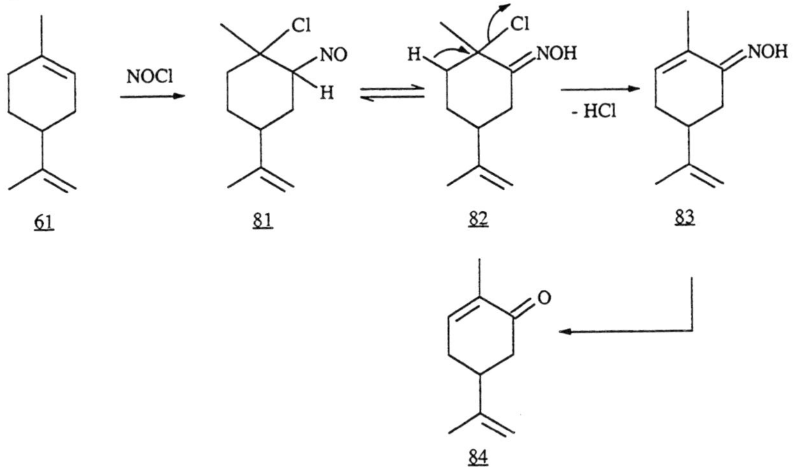

61 —NOCl→ 81 ⇌ 82 —(- HCl)→ 83 → 84

[2.56]

84 —HBr→ 85 —OH⁻→ [86] → —$H^{\oplus}$→ 87

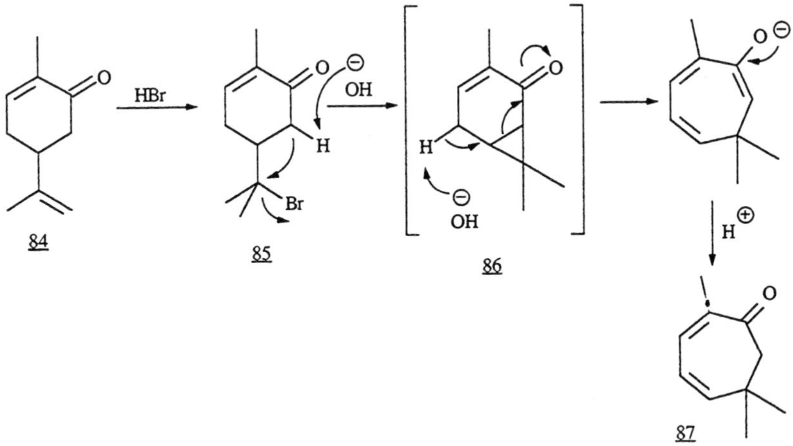

Photosensitized oxidation of limonene gives a mixture of hydroperoxides which on reduction yields six alcohols 92, 93, 94, 95, 96 and 97. This result is in compliance with the generally admitted mechanism of photosensitized oxygenation of olefins (purely formal analogy with an ene-reaction) [2.58].

In the case of limonene 61, the allylic hydrogens attached to carbons 3, 6 and 7 can participate in the reaction [2.59].

When photo-oxygenation is carried out in the absence of a photosensitizer, the two hydroperoxides which are precursors to alcohols 96 and 97 are not formed. Furthermore, the two carveols, trans 94 and cis 95, are obtained as racemates, which suggests the mediation of allylic radicals, whereas the two alcohols 93 [cis–2,8 (9)-p-menthadien–1-ol] and 92 (trans) are still optically active. This can be explained in the following scheme [2.60].

[2.57]

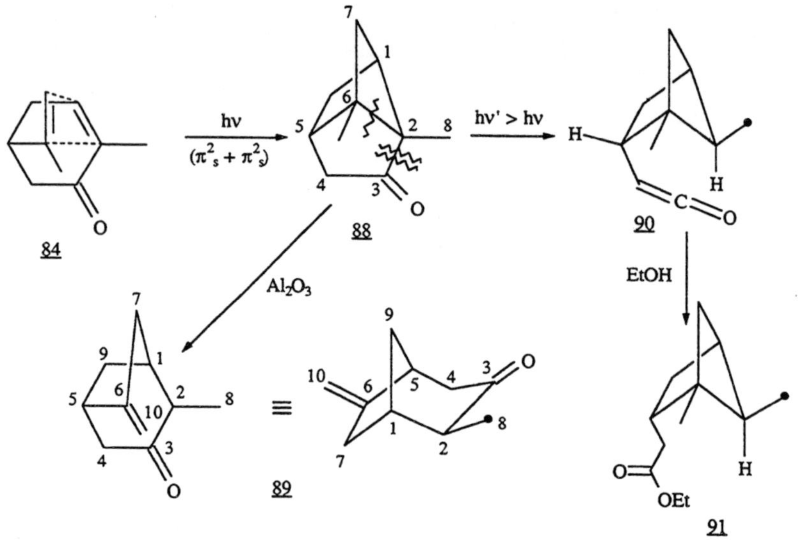

[2.58]

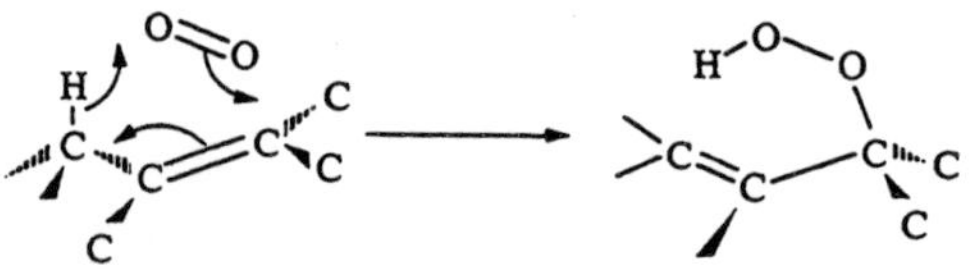

While the chemical oxidation of cis and trans carveols (95 and 94) gives carvone 84, the same reaction of the p-menthadienols 92 and 93 produces isopiperitenone 98 by an allylic migration [2.61].

The most important and widely used member of the family of cyclic monoterpenes is menthol 99. Annual world production of this chemical exceeds 1,000 metric tons with the largest part coming from the crystallization of "*Mentha arvensis*" (extreme Orient and Brazil). Natural menthol is levo-rotatory.

Of the four possible isomers, the (–) menthol is the most thermodynamically stable with the three substituents of the six-membered ring [-OH,-CH_3, -$CH(CH_3)_2$] all being equatorially disposed [2.62].

Several syntheses of menthol are known. One of the most important involves (+) citronellal 18 which can be either treated thermally or heated in the presence of acids (HCO_2H, H_3PO_4, Ac_2O . . .). The principal product obtained from either treatment is (–) isopulegol 100 along with three isomers 101–103 [2.63].

The preference for (–) isopulegol 100 is in accord with the structure of the transition state. This can be represented as shown here in similarity with the mechanism of the ene-reaction [2.64]. It is noteworthy that the presence of Lewis acids such as $ZnCl_2$ can dramatically increase the yield of (–) isopulegol, up to 89% [2.65].

[2.59]

(-) limonene

7

6

3

O_2, hν, photosensitizer

(3) (6) (7)

O—OH

OH HO OH OH OH OH

H H H H

92 93 94 95 96 97

[2.60]

(+) limonene

$[O_2]$

hν

enantiomers

OH + OH

(±) 94 (±) 95

no loss of asymmetry

OH + OH

92 93

[2.61]

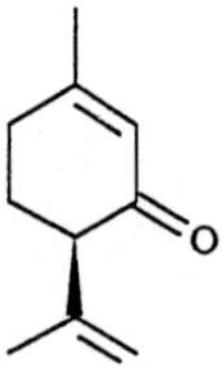

98

[2.62]

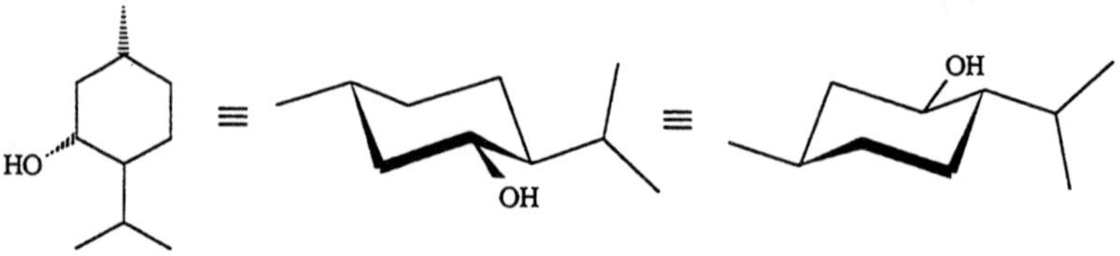

HO ≡ OH ≡ OH

99 (-) menthol

[2.63]

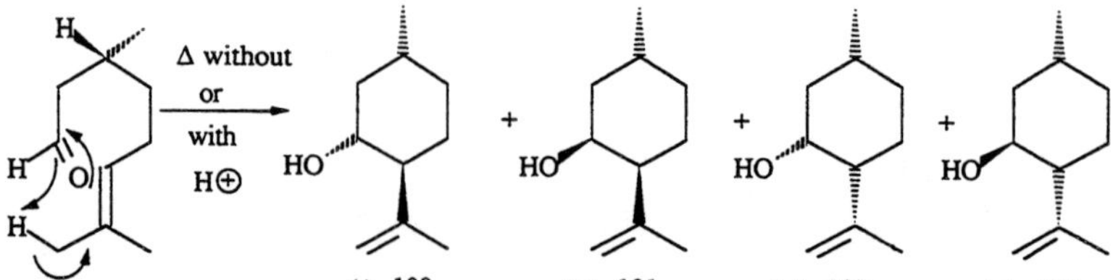

H H O H Δ without or with H⊕

(+) 18

HO (-) 100 isopulegol + HO (+) 101 neoisopulegol + HO (+) 102 neoisoiso + HO (+) 103 isoiso

[2.64]

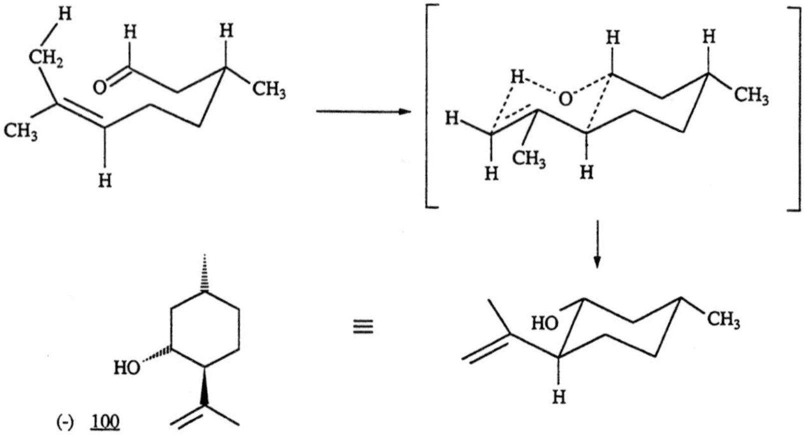

H CH_2 CH_3 H O H CH_3 H → [H H O H CH_3 H CH_3 H H CH_3]

HO (-) 100 ≡ HO CH_3 H

[2.65]

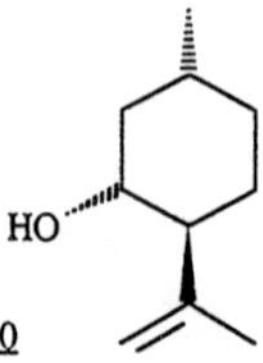

HO (-) 100

Catalytic hydrogenation of (–) isopulegol gives (–) menthol, whereas the other isomers are converted as follows [2.66].

[2.66]

(+) 101 ⟶ (+) neomenthol 104

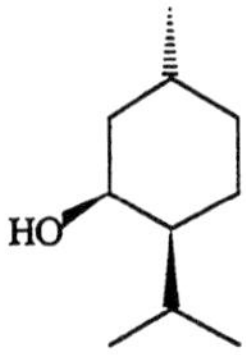

(+) 102 ⟶ (+) neoisomenthol 105

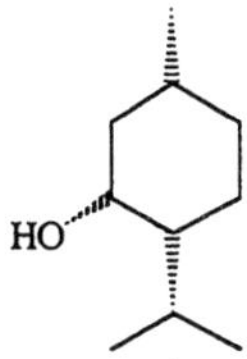

(+) 103 ⟶ (+) isomenthol 106

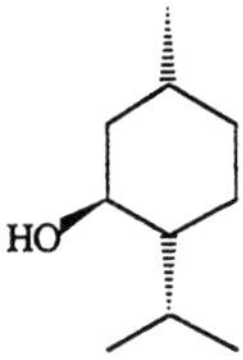

According to the Cahn, Ingold, Prelog nomenclature, these monocyclic alcohols have the following configurations:

(–) menthol 101: lR, 3R, 4S
(+) neomenthol 104: lR, 3S, 4S
(+) neoisomenthol 105: lR, 3R, 4R
(+) isomenthol 106: lR, 3S, 4R
(–) Menthone 107 corresponds to (–) menthol and to (+) neomenthol [2.67].

[2.67]

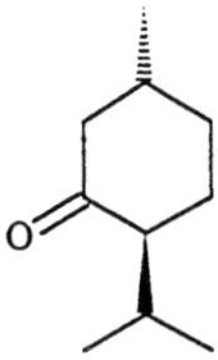

107

(+) Isomenthone 108 corresponds to (+) isomenthol and to (+) neoisomenthol [2.68].

[2.68]

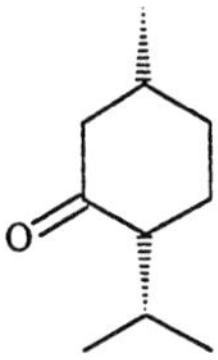

108

In the same way, the pair (–) isopulegol and (+) neoisopulegol correspond to (–) isopulegone 109, and the pair (+) isoisopulegol and (+) neoisoisopulegol correspond to (+) isoisopulegone 110. These two ketones can isomerize to (+) pulegone 111 [2.69].

[2.69]

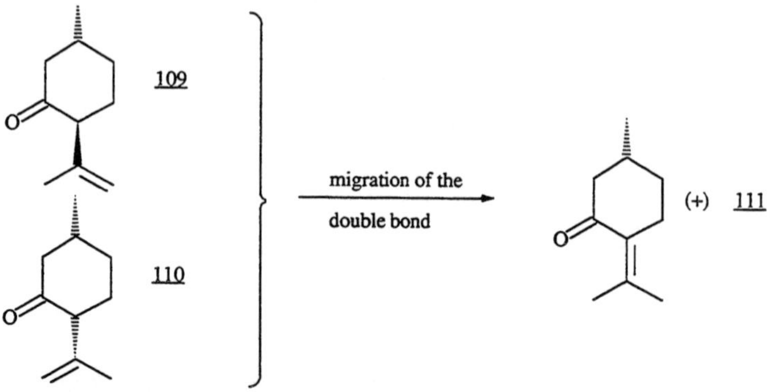

109

110

migration of the double bond

(+) 111

Pulegone 111 is the principal constituent of pennyroyal oil.

Along with the production of crystalline (–) menthol (MP: 45°C), large quantities of (±) menthol as well as mixtures of the different stereoisomeric menthols are produced industrially. One of the first methods used consisted of the hydrogenation of thymol 112 extracted from thyme oils or obtained from metacresol [2.70].

[2.70]

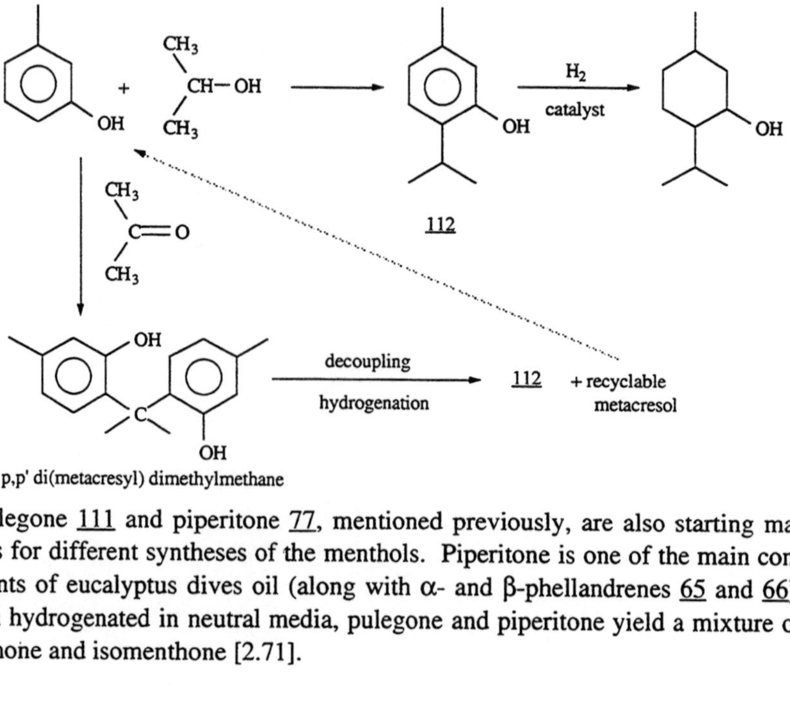

OH + CH_3 CH—OH CH_3 → OH H_2 catalyst → OH

112

CH_3 C=O CH_3

OH

OH

p,p' di(metacresyl) dimethylmethane

decoupling hydrogenation → 112 + recyclable metacresol

Pulegone 111 and piperitone 77, mentioned previously, are also starting materials for different syntheses of the menthols. Piperitone is one of the main constituents of eucalyptus dives oil (along with α- and β-phellandrenes 65 and 66). When hydrogenated in neutral media, pulegone and piperitone yield a mixture of menthone and isomenthone [2.71].

[2.71]

111 → 107 + 108 ← 77

It is evident that when the oil of eucalyptus dives is used for obtaining menthols, it would be advantageous to also use the phellandrenes also contained in this oil. In fact, as shown, both of these hydrocarbons give a single hydrogen chloride addition product which can be converted into piperitone. Reduction of menthone to menthols is usually achieved by chemical means (Na + C_2H_5OH, for example).

All of the industrial methods for producing the menthols are aimed at obtaining the natural form 99. These extremely elaborate procedures are based on two principles: a) Epimerization to convert the isomers other than 99 into this natural form, which is thermodynamically the most stable due to the equatorial positions of all of the groups; b) Purification by formation of crystalline derivatives such as benzoates or p-nitrobenzoates, which has faster rates of formation for the natural form. The relative rates of formation of esters on reaction with p-nitrobenzoyl chloride illustrate this.

neomenthol: 1; menthol: 22.5; neoisomenthol: 4.2; isomenthol: 16.5

As a general rule, it is evident that the alcohols with equatorial -OH groups react more rapidly—menthol being the most reactive.

The mechanism of the epimerization of the different isomers is not yet fully known, although this process would seem to imply the presence of small quantities of the ketone in the reaction medium. The Verley-Meerwein-Ponndorf reaction (treatment of a ketone with isopropanol in the presence of aluminum isopropylate), when applied to menthone, leads selectively to the axial alcohol (i.e., neomenthol), while under the same conditions, isomenthone gives neoisomenthol.

In the Verley-Meerwein-Ponndorf reaction, it is known that the aluminum isopropylate functions as a hydride-donor. Complex hydrides such as $LiAlH_4$ will also reduce the menthones, although in the case of $LiAlH_4$, the reaction temperature is of critical influence: at low temperatures neomenthol is selectively obtained while around 30°C, there is preference for menthol.

There are numerous mono- or bicyclic secondary terpenic alcohols which can, like menthol, exist in four diastereoisomeric forms. Determination of the configurations of members of groups of this type has been the subject of several studies which have determined a certain number of rules. With regard to elimination reactions, it may be possible to say:

- that the dehydrohalogenation in basic media tends to give an olefin with the most substituted double bond (thermodynamically most stable olefin or "Saytzeff olefin");

- that the elimination in basic media of sulfonium salts or of quarternary ammonium hydroxides (and related salts) gives an olefin with the least substituted double bond ("Hoffmann olefin") [2.72].

[2.72]

$$R-CH_2-CH(\overset{\oplus}{N}(CH_3)_3)-CH_3 \xrightarrow{\ominus OH} R-CH_2-CH=CH_2 \quad \text{(Hoffmann)}$$

$$R-CH_2-CH(X)-CH_3 \xrightarrow{\ominus OH} R-CH=CH-CH_3 \quad \text{(Saytzeff)}$$

This is, of course, a rule that is not always respected. Literature on advanced organic chemistry gives explanations for these exceptions.

It has been established that dehydrohalogenation and Hoffmann-elimination reactions occur under kinetic control. Electronic effects seem to present the most important cause for the different orientations when the leaving groups are changed.

For E_1 or E_2 elimination of halides or sulfonates, the double bond is almost completely formed in the transition state (transition state resembles product), and under these conditions it is clear that the most rapid reaction (lowest energy transition state) will be that which leads to the most thermodynamically stable olefin ("Saytzeff olefin") [2.73].

[2.73]

$$CH_3-C(H)(H_3C)-CH(Br)-CH_3 \xrightarrow{B^{\ominus}} \left[CH_3-C(H\cdots B)(H_3C)\overset{---}{=}\overset{\ominus}{C}H(\cdots Br)-CH_3 \right]^{\ddagger} \longrightarrow (CH_3)_2C=CH-CH_3$$

In the E_2 elimination reactions of "onium" salts (quarternary ammonium and sulfonium salts), the relative acidities of the β-CH groups will control the outcome and orientation of the reaction. An accumulation of alkyl groups (electron donors compared to hydrogen) on the β-carbon will diminish the acidity of the attached H as well as rendering more difficult any approach by the base. It will therefore be the H attached to the lesser substituted β-carbon which will be preferentially attacked and this will lead to the least substituted olefin [2.74].

[2.74]

$$CH_3-C(CH_3)(H)-CH(\overset{\oplus}{N}(CH_3)_3)-CH_2-H \xrightarrow{\ominus OH} (CH_3)_2CH-CH=CH_2$$

the least "acidic" (H on C); the most "acidic" (H on CH_2)

"Hoffmann" product

It is clearly impossible to discuss here all that is known regarding the influence of solvents and of the steric bulk of the base, although these factors play a role in these reactions.

It is now possible for us to apply these notions to the derivatives of menthol and its isomers.

The following table summarizes the results obtained from various derivatives of menthol and neomenthol. It can be noted that neomenthyl chloride is far more reactive than menthyl chloride; the latter being stable in the presence of aniline or quinoleine and giving 2-para-menthene under the action of sodium ethylate. On the other hand, neomenthyl chloride undergoes a facile elimination with the formation of a mixture of 2- and 3-para-menthene. The ease of this elimination is linked to the trans-diaxial (antiperiplanar) situation of the H atom which is attacked by the base and the leaving groups in the neomenthol derivatives [2.75].

[2.75]

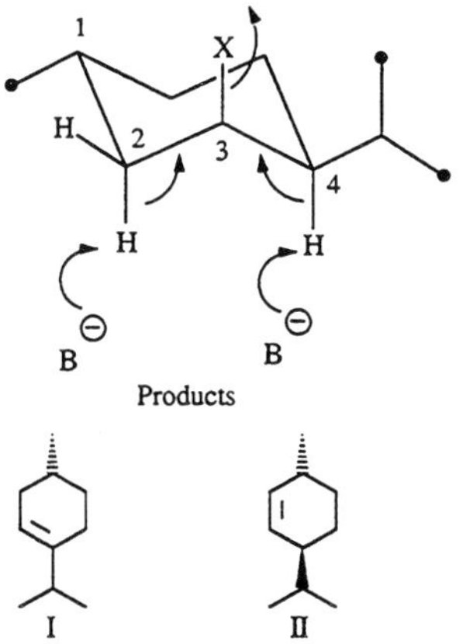

Compounds	Conditions	Products I	Products II	Type of elimination
1. (–) menthyl chloride	heating in EtOH	70	30	E_1
2. (–) menthyl toxylate	heating in EtOH	70	30	E_1
3. (–) menthyl chloride	EtONa	0	100	E_2
4. (–) menthyl toxylate	EtONa	0	100	E_2
5. (–) menthyl trimethyl ammonium	heating without H_2O	0	100	E_2
6. (+) neomenthyl chloride	heating without EtOH	±99	1	E_1
7. (+) neomenthyl chloride	EtONa	75	25	E_2
8. (+) neomenthyl trimethyl ammonium	heating without H_2O	80	20	E_2

The principal product is (+) -3-para-menthene (the most substituted olefin), in agreement with the predictions made following the rule of Saytzeff. The production of a significant percentage of the "anti-Saytzeff" olefin can be linked to the presence of a sterically hindering group (isopropyl) in position 4 which hinders the approach to the H atom on carbon 4.

It is more difficult to explain the practically exclusive formation of 2-p-menthene in the E_2 elimination reactions of the menthol derivatives. Eliel examined the energy difference which exists for any given substituent between its axial and its equatorial conformations. Here are some values for these energy differences ("A" values in Kcal. mole^{-1}) for some substituents [2.76].

[2.76]

- CH_3 = 1,70
- OH = 0,52 - 0,87
- Cl = 0,43
- $CH(CH_3)_2$ = 2,15
- $N(CH_3)_2$ = 2,10
- OCH_3 = 0,60
- $C(CH_3)_3$ > 4,4

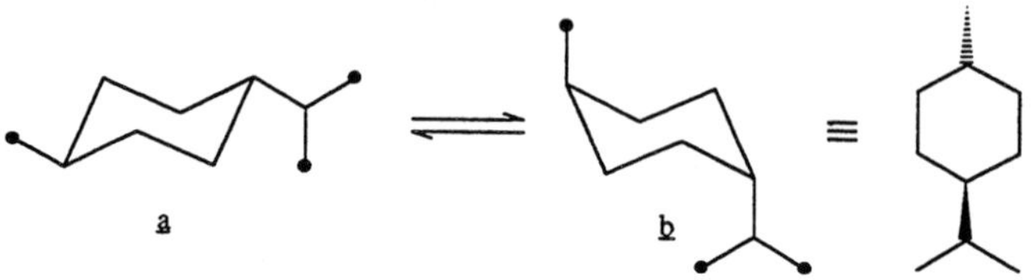

a ⇌ b ≡

$\Delta G = 1{,}70 + 2{,}15 = 3{,}85$ Kcal mole^{-1}

trans p-menthane

Example

The ratio of the populations at equilibrium of the two conformers is expressed as:

$$K = \frac{n_a}{n_b} = e^{-\Delta G/RT}.$$

In the case of trans-para-menthane, the population of the all-equatorial conformer represents over 99.9% of the total. For the 1,2 and 1,3 derivatives, there are supplementary difficulties due to interactions between the substituents.

For the 1,2 derivatives, there are "gauche" interactions a,e = e,e. For the 1,3 derivatives, the interactions are essentially limited to the diaxial conformation which destabilizes the a,a form. Some energy values for 1,3-diaxial interactions are indicated in the following table:

X	–OH	$-OA_c$	–OH	$-CH_3$
Y	–OH	$-OA_c$	$-CH_3$	$-CH_3$
DG in Kcalmole^{-1}	1.9	2.0	2.1–2.4	3.7

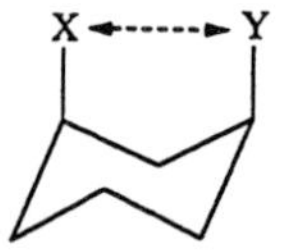

X ◄------► Y

In the case of menthol derivatives, by considering all of the diverse interactions, it is possible to estimate that the difference in standard free enthalpy between the two conformers e and a is above 6 Kcal·mole^{-1} [2.77].

[2.77]

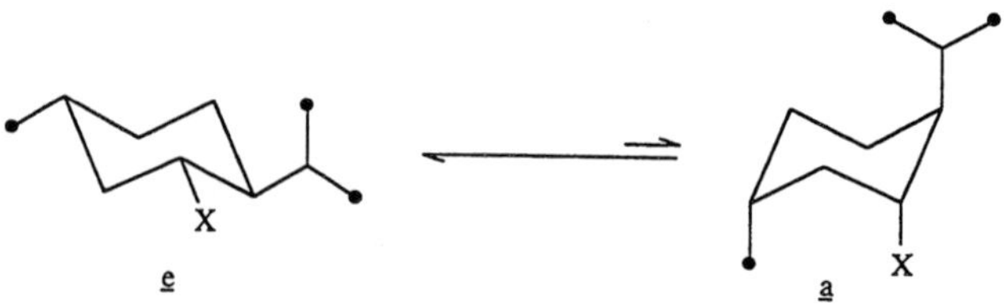

X e ⇌ X a

With the population of conformer a being approximately zero, it is difficult to invoke a mechanism explaining the formation of (+) 2-p-menthene by trans-diaxial E_2 elimination [2.78].

[2.78]

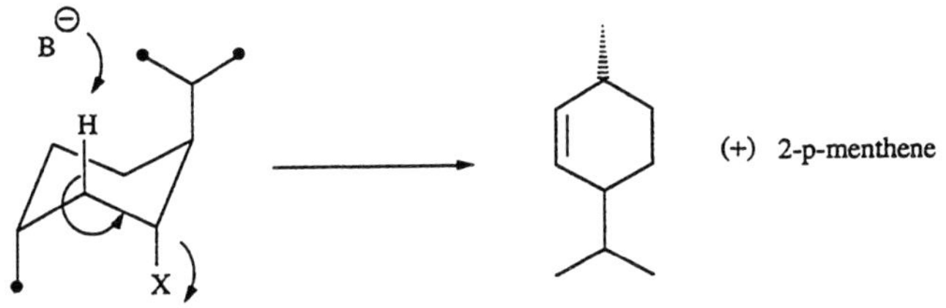

(+) 2-p-menthene

It is an accepted fact that the chair-chair interconversion in cyclohexane involves a sequence that passes through intermediate conformations [2.79].

[2.79]

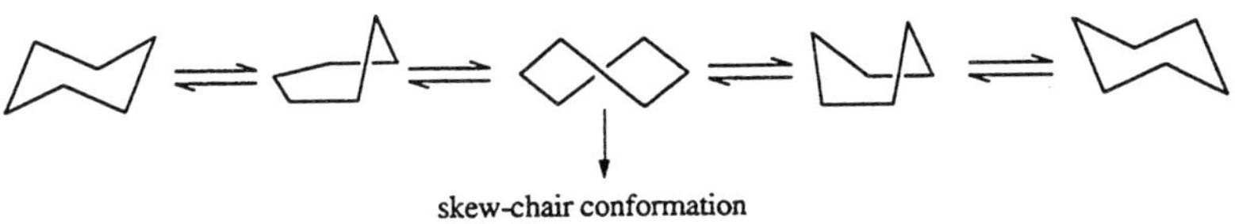

skew-chair conformation

Rather than pass through the triaxial conformation a of the menthol derivatives, it is preferable to envisage E_2 elimination occurring in the skew-chair conformation which is 2 to 4 Kcal·mole^{-1} less strained than the triaxial conformer. This only permits trans diaxial elimination by attack on the H atom at the 2 position, thereby giving exclusively 2-p-menthene [2.80].

[2.80]

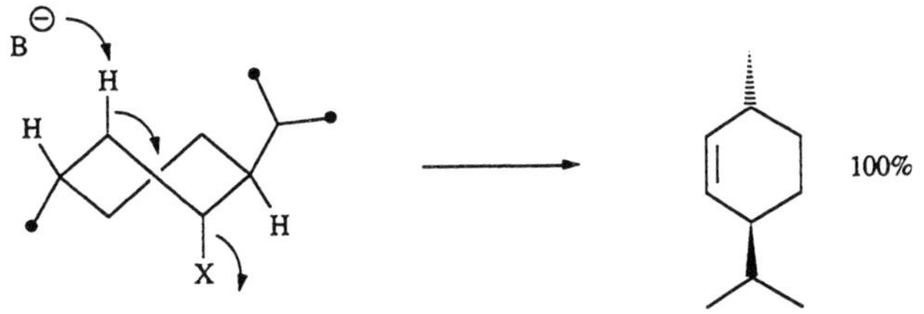

100%

It is important to revert back to the reactions 1–6 in the previous table, which have been the subject of kinetic studies. Solvolyses in H_2O-EtOH 20:80 have been carried out at a temperature of 124.9°C and the following values for the rate constants have been found:

(–) menthyl chloride $K_1 = 0.402 \cdot 10^{-5}$ sec^{-1}
(+) neomenthyl chloride $K_1 = 16.3 \cdot 10^{-5}$ sec^{-1}.

In E_1 mechanisms, neomenthyl derivatives are much more reactive than menthyl derivatives (about 40 times in the cited case).

- Menthyl chloride yields 69% of elimination products composed of: (+) 3-p-menthene 68% (Saytzeff dominant) and (+) 2-p-menthene 32%.

- With neomenthyl chloride, 96% of the elimination products are obtained and these are composed of: (±) 3-p-menthene 99% (super Saytzeff) and (+) 2-p-menthene, about 1%.

Under these conditions, 3-p-menthene is racemized as the consequence of the subsequent acidification of the reaction medium. This is clearly demonstrated when the reaction is carried out in the presence of sodium acetate (which does not significantly affect either the kinetics or the distribution of products). Under these conditions, (+) 3-p-menthene with an optical purity over 85% is recuperated.

Given that unimolecular (E_1) elimination proceeds through a carbocation intermediate, this should be the same for the two epimeric substrates and therefore should lead to the same elimination products [2.81].

[2.81]

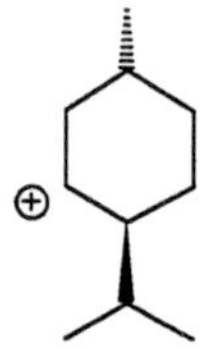

⊕

As this is not the case, it is necessary to admit that the initial conformations are different for each epimeric precursor. The initial conformation of the cation derived from menthyl chloride *must permit* the elimination of the proton at position 2. The racemization of 3-p-menthene can be explained by the transfer of a hydride ion to give a tertiary carbocation having a plane of symmetry and giving rise to the racemate [2.82].

[2.82]

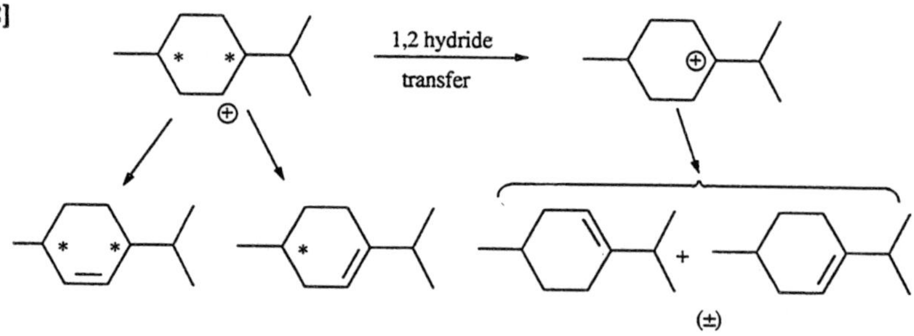

1,2 hydride transfer

(±)

The reactions just described are so-called ionic reactions. However, other reactions also play a role in helping to determine the relative configurations of terpenic derivatives. These are the thermal eliminations, characterized by their cyclic transition states and by the concerted reorganization of bonds—some of which are ruptured while others are simultaneously formed. These reactions develop via six-, five-, or four-membered transition states and have negative activation entropies (ΔS-reduction in the system's degrees of freedom). They are not particularly sensitive to the effects of solvents, of acids or bases or, for that matter, of electronic effects. They are, however, strongly influenced by steric effects. All are "syn-eliminations" with the two groups eliminated being on the same side of the nascent double bond.

1. Six-centered thermal eliminations

The pyrolsis of xanthogenates or the Chugaev reaction belongs to this group [2.83].

[2.83]

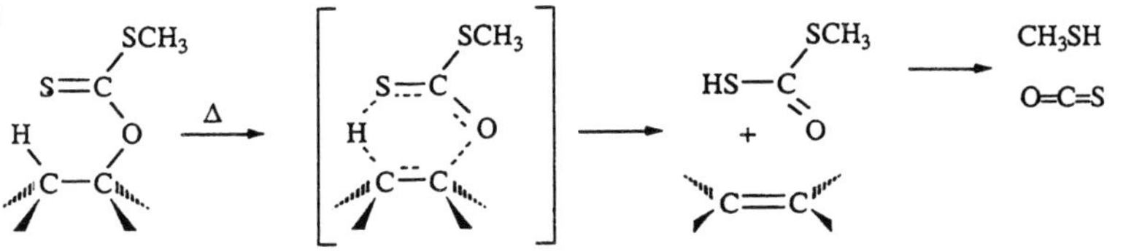

In cases where a "syn-elimination" is possible in several directions, the favored transition state is the one with minimal steric compressions. When applied to (–) menthol and to (+) neomenthol, this reaction affords, respectively:

–70% of (+) 3-p-menthene and 30% of (+) 2-p-menthene;
–20% of (+) 3-p-menthene and 80% of (+) 2-p-menthene.

For (–) menthol it is possible to invoke a mechanism that passes through the skew-chair conformation [2.84].

[2.84]

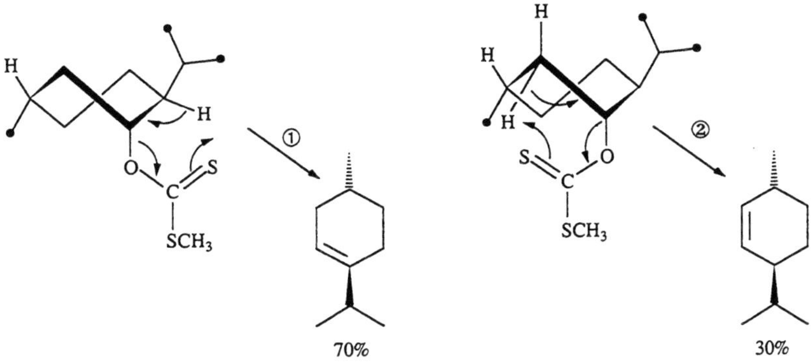

In six-centered reactions, the conformation of the transition state assumed by the participating atoms can resemble the gauche form of cyclohexane. Therefore, in the processes (1) or (2), the steric strains are virtually the same. For this reason, the Saytzeff elimination giving 3-p-menthene will predominate. With derivatives of (+) neomenthol, two conformations can lead to "syn-elimination":

a) A relatively unstrained deformed boat conformation which leads to the formation of (+) 2-p-menthene as the principal product (Hoffmann-type elimination) [2.85].

[2.85]

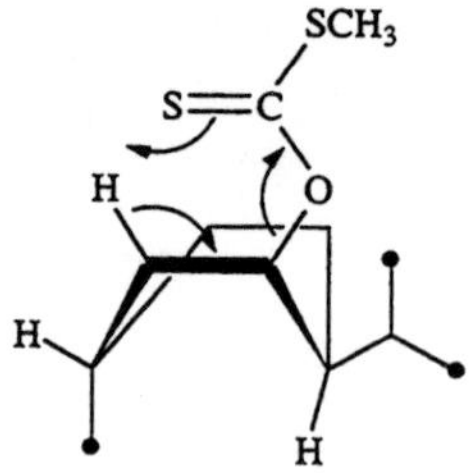

SCH₃ S=C O H H H

b) A far more strained and therefore less populated conformation which eliminates towards (+) 3-p-menthene [2.86].

[2.86]

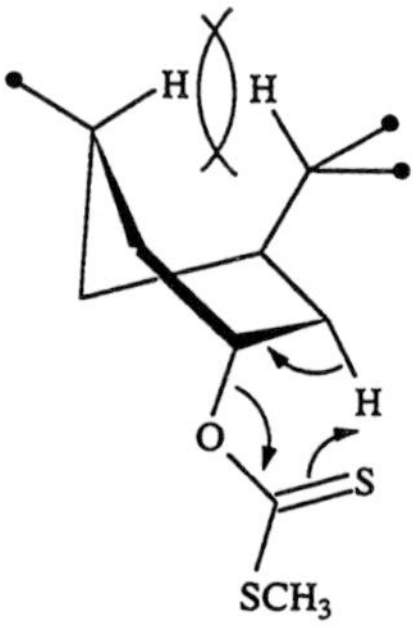

H H H O S SCH₃

2. *Five-centered thermal syn-eliminations*

A typical example of this is the Cope reaction (pyrolysis of amine oxides) [2.87].

[2.87]

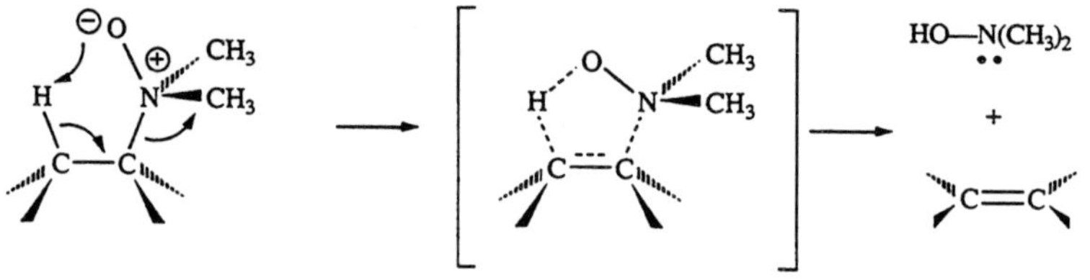

O CH₃ H N CH₃ C C → H O CH₃ N CH₃ C C → HO—N(CH₃)₂ + C=C

The thermal elimination of sulfoxides which occurs at lower temperatures can also be placed in this class [2.88].

[2.88]

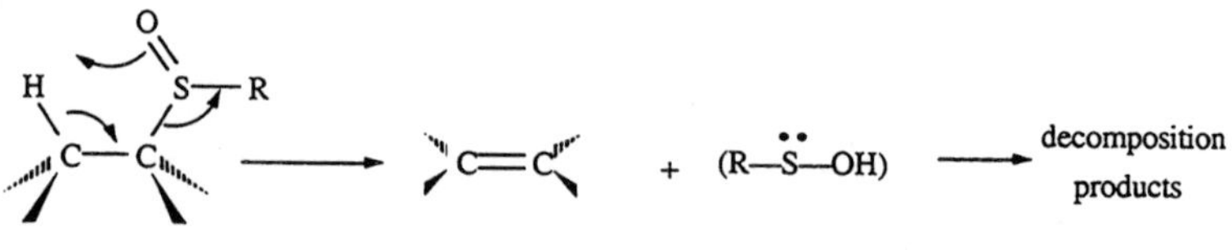

O H S R C C → C=C + (R—S—OH) → decomposition products

In the case of menthol derivatives, two C-H bonds are conveniently situated and the Hoffmann-type elimination predominates [2.89]. With the corresponding derivative of neomenthol, only a conformation of type A (deformed boat) permits the establishment of a planar five-membered transition state, thereby

[2.89]

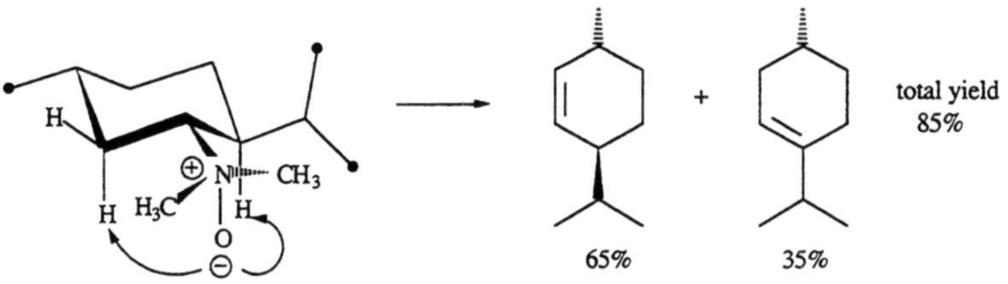

total yield 85%

65% 35%

giving rise to the exclusive formation of the anti-Saytzeff (Hoffmann) product [2.90].

[2.90]

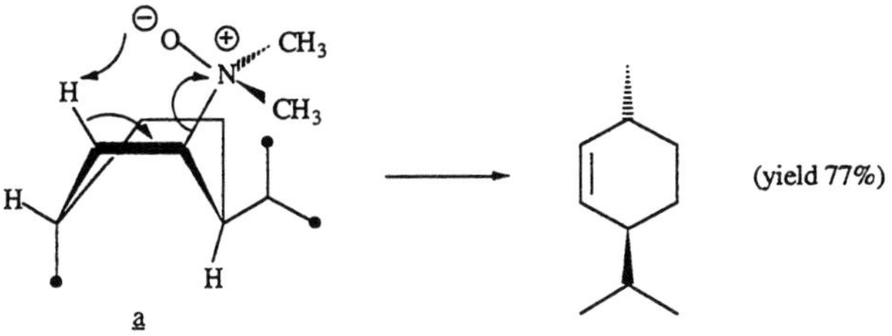

(yield 77%)

3. *Four-centered thermal syn-eliminations*

Thermal dehydrohalogenation is one example [2.91].

[2.91]

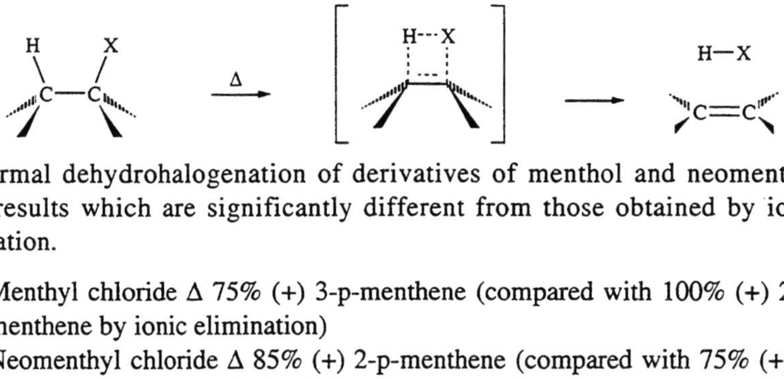

Thermal dehydrohalogenation of derivatives of menthol and neomenthol gives results which are significantly different from those obtained by ionic elimination.

- Menthyl chloride Δ 75% (+) 3-p-menthene (compared with 100% (+) 2-p-menthene by ionic elimination)
- Neomenthyl chloride Δ 85% (+) 2-p-menthene (compared with 75% (+) 3-p-menthene by ionic elimination).

The p-menthane derivatives are not the only cyclohexane derivatives in the monoterpene series. Derivatives of 1,1,2,3-tetramethylcyclohexane derivatives are also found in nature. Safranal 113 is the principal odorant of saffron and the heteroside corresponding to picrocrocine 114 is its bitter principle [2.92].

As recently as 1969, a 1,1,2,5-tetramethylcyclohexane derivative was isolated from the roots of *Ferula hispanica.* This substance, ferulol 115, was in fact extracted in the form of its ester 116 with angelic acid [2.93]. It can be noted that ferulol shows an irregular isoprenic linking.

Derivatives of metamenthane and orthomenthane are also found in nature

[2.92]

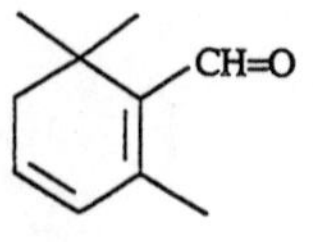

CH=O

glucose—O

<u>113</u> 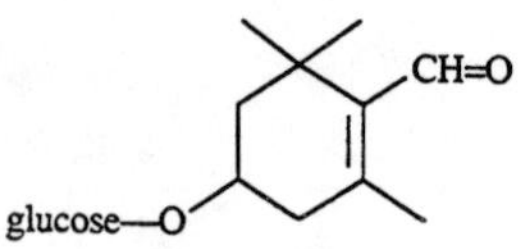 <u>114</u>

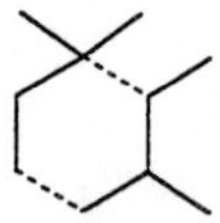

(z-w)$_1$, 1x-y$_2$

[2.93]

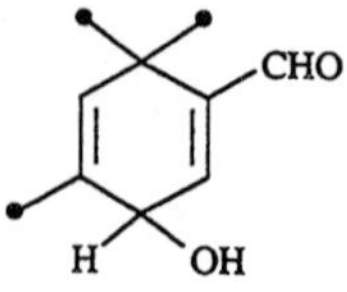

CHO

H OH

<u>115</u>

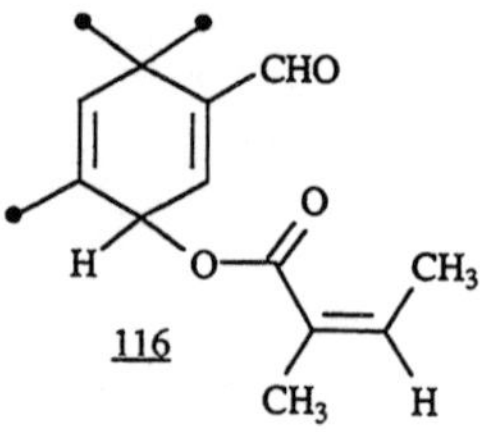

CHO

H O O CH$_3$

<u>116</u>

CH$_3$ H

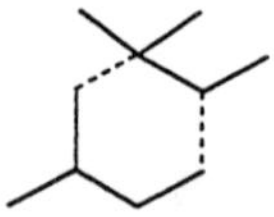

1x-w$_2$, 1y-z$_2$

although the former are generally considered to be artifacts derived from Δ–3-carene. Carquejol <u>117</u>, which has been isolated from several *Baccharis* species, also has the orthomenthane skeleton [2.94].

[2.94]

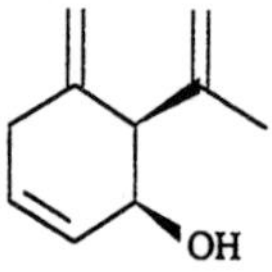

OH

<u>117</u>

B. Iridoids (cyclopentanyl monoterpene derivatives)

Members of this series are photocyclocitral <u>118</u> (obtained from the photochemical rearrangement of citral) and the "plinols" <u>30</u> [2.95].

[2.95]

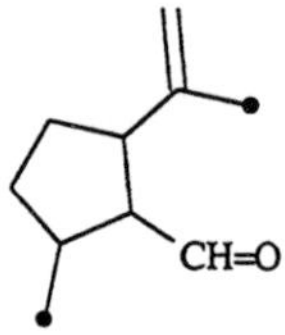

CH=O

<u>118</u>

CH$_3$

OH

CH$_3$

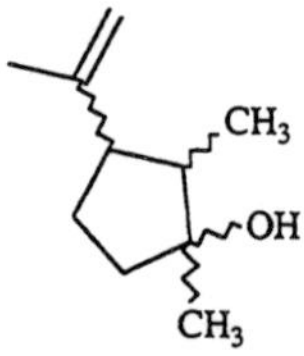

<u>30</u>

Dehydrolinalool, the key intermediate in the total synthesis of some terpenoids, also results in a cyclopentanyl derivative by an intramolecular ene-reaction [2.96].

A group of important natural heterosides has been known for a long time. One of these, verbenaline <u>119</u>, was isolated in 1835 in crystalline form. Similarly, asperuloside <u>120</u> was isolated in 1851. However, it was only in 1958 that Halpern and Schmid demonstrated for the first time the presence of a cyclopentanyl terpenic aglycone in plumieride <u>121</u> [2.97].

Since then, these substances have been called "iridoid glucosides" by Briggs

[2.96]

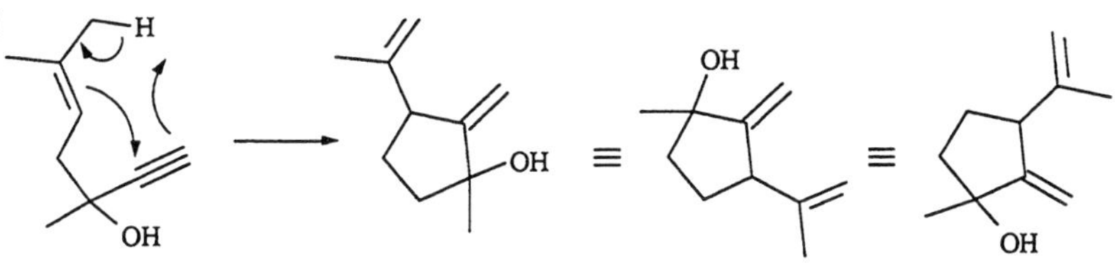

[2.97]

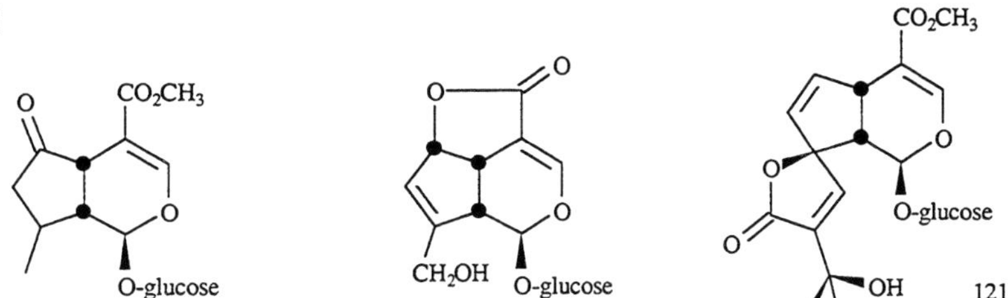

because of their relations with iridomyrmecine and iridodial. These iridoid glucosides are chemically characterized by the instability of their aglycônes in strong acids which form a variety of colors. These color-producing reactions have drawn the attention of chemists and botanists to plants containing these substances. A consequence of this instability, however, has been the problems encountered in the determination of the structures of the aglycones.

These iridoids are characterized by the presence of a regular terpenic linking in the nucleus A [2.98].

[2.98]

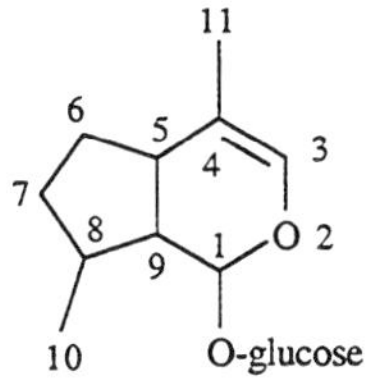

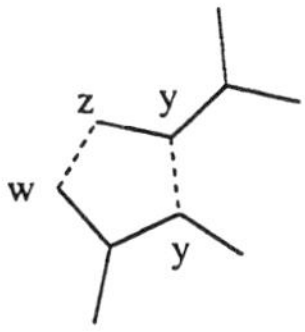

Nearly 50 compounds of this type are currently known, but not all contain the glucosidic residue. This is the case, for example, with genipine 122, which has been isolated from the juice of *Genipa americana* L. along with genipic acid 123 and genipinic acid 124 [2.99].

Contact of these substances with an amino acid or even the skin produces a blue coloration. This property has been used by Indians in Mexico and elsewhere in Latin America to color their skin by direct application of the juice of *Genipa*.

Another example, loganine 125, was extracted in 1884 from the fruit of *Strychnos nux vomica* L. (Loganiaceæ family) in which two important alkaloids, strychnine and brucine, are found [2.100].

Another important group of iridoids is composed of the defense substances of arthropods. These compounds were reported for the first time in 1950 by Pavan,

[2.99]

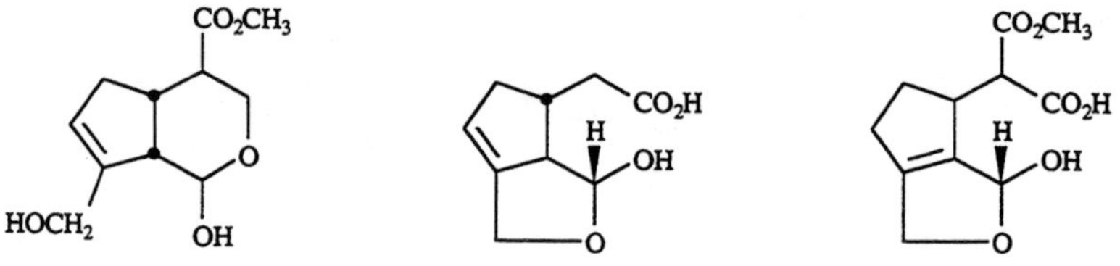

CO_2CH_3 HOCH$_2$ OH CO_2H H OH O

122 123 124

[2.100]

CO_2CH_3 HO O-glucose

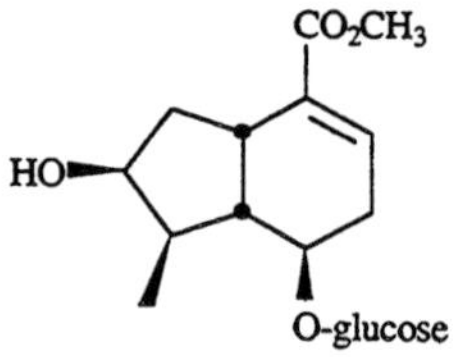

125

who isolated from the anal glands of the Argentinian ant (*Iridomyrmex humilis*) a monoterpenic lactone having antibiotic and insecticidal properties. He named it "iridomyrmecine." The definitive structure of this substance 126 was demonstrated in 1955 by Fusco et al., who were able to link it to the structure of nepetalactone 127 which had just been determined by Meinwald and McElvain. Nepetalactone is the major element of the labiate *"Nepeta cataria"* or "catnip," the essential oil of which is called "essence of cataire."

The work of Pavan led to the discovery of other compounds similar to iridomyrmecine: iridodial 128, dolichodial 129, etc. The existance of these substances in arthropods is not limited only to ants [2.101].

[2.101]

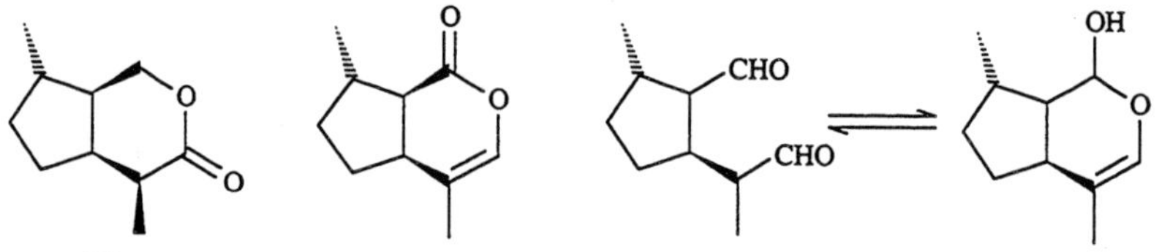

O O CHO CHO OH

126 127 128

CHO CHO

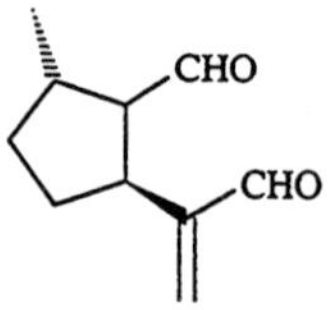

129

The total synthesis of nepetalactone 127 was undertaken starting with (+) pulegone 111 obtained from pennyroyal oil. In fact, this gives the mirror image of natural nepetalactone, which permits its correlation with (–) pulegone and hence to (+) limonene and (–) citronellal.

The scheme of the synthesis of nepetalactone follows [2.102]:

[2.102]

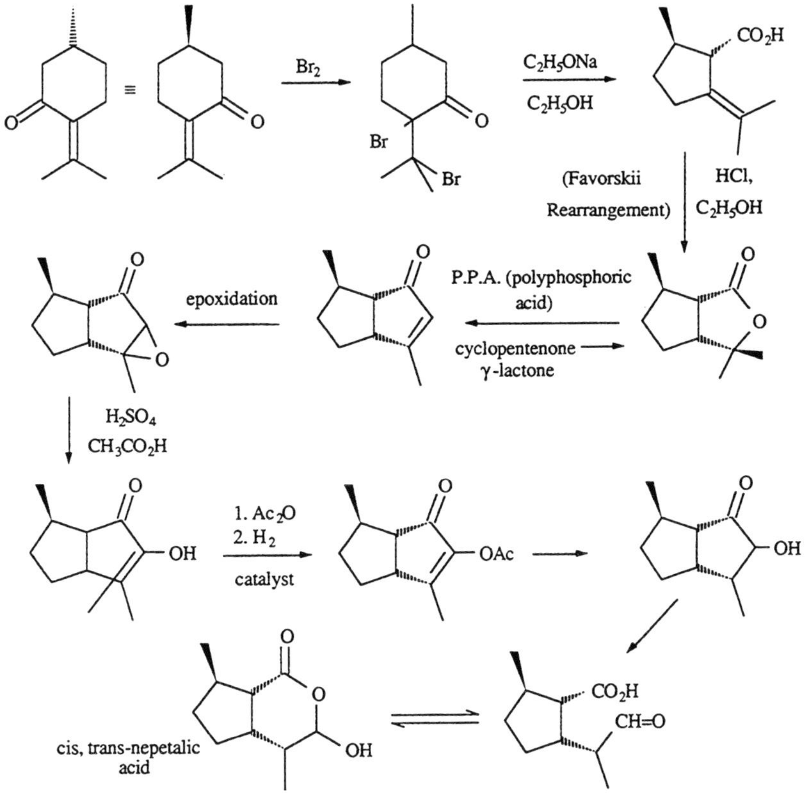

Br_2 C_2H_5ONa C_2H_5OH (Favorskii Rearrangement) HCl, C_2H_5OH P.P.A. (polyphosphoric acid) cyclopentenone γ-lactone epoxidation H_2SO_4 CH_3CO_2H 1. Ac_2O 2. H_2 catalyst cis, trans-nepetalic acid

Pyrolysis of cis,trans-nepetalic acid gives a partially racemized mirror image of natural nepetalactone along with a small quantity of the trans,cis-isomer 127a [2.103].

[2.103]

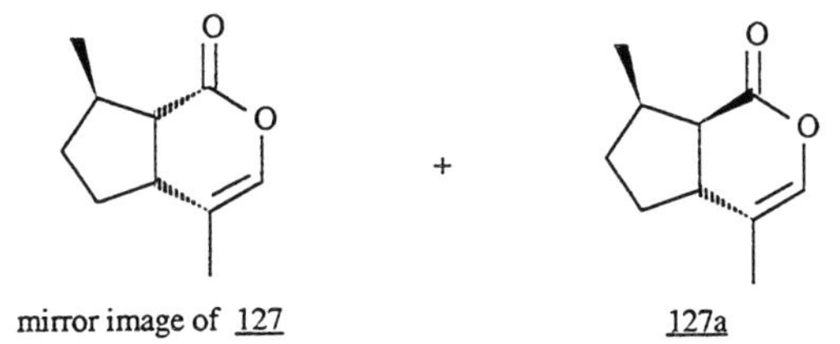

mirror image of 127 + 127a

C. Example of cyclobutyl monoterpenes

No single representative of monocyclic terpenes containing a cyclobutyl ring was known prior to 1969. In that year, however, the alcohol 128 was isolated from male boll weevils (*Anthonomus*) in which it serves as a pheromone [2.104].

The term "pheromone" was introduced in 1959 by Karlson and Luscher to des-

[2.104]

CH_2OH

H

128

ignate substances excreted by insects in order to provoke in other insects of the same species, a specific reaction, a defined behavior or a developmental process. Hormones differ from pheromones in that their function is to coordinate the metabolism of tissues and organs of the same individual. Pheromones, on the other hand, constitute a chemical language which permits insects of the same species to communicate with each other. They should not be confused with the defensive substances such as irodomyrmecine mentioned earlier. They present an incontesible interest at the scientific, ecological and industrial levels.

- Scheme for the synthesis (1970) of the alcohol 128 [2.105].

[2.105]

hν

$PhN^{\oplus}(CH_3)_3Br^{\ominus}$

Br

Li_2CO_3

$CH_3CON(CH_3)_2$

CH_3Li

$(C_2H_5)_2O$

OH

1. OsO_4 / $NaIO_4$

2. Esterification

CO_2CH_3

$Ph_3P^{\oplus}-CH_2^{\ominus}$

(Wittig)

CO_2CH_3

1. $LiAlH_4$

2. H_2O

128

D. Monoterpenic derivatives with a cyclopropyl ring

The most characteristic representative of this group is trans-chrysanthemic acid 129 which possesses an irregular isoprenic structure [2.106].

It has been observed that several derivatives of chrysanthemic acid are

[2.106]

129

biodegradable insecticides. This presents an additional interest in their synthesis. Here are two examples which use very different strategies:

- Application of sulfone chemistry [2.107].

[2.107]

$ArSO_2Na$ CH_2Br CH_2-SO_2Ar $B^{\ominus}$ RO O SO_2Ar OR RO_2C H

R = H : 129

- Use of "carbene" chemistry [2.108].

Chrysanthemic acid or some of its derivatives are considered precursors of irregular terpenic compounds with artemisyl, lavandulyl or santolinyl skeletons [2.109].

Besides chrysanthemic acid, this group contains another important acid: pyrethric acid 131 (R=H) as well as its methyl ester 132 (R=CH_3) (in fact, it is the methyl ester which is called pyrethric acid) [2.110].

A strategy has been used which differs from those employed in the synthesis of chrysanthemic acid involving a 1,4-addition of a phosphonium ylid which permits the generation of either chrysanthemic acid or pyrethric acid depending on the starting substrate [2.111].

trans-Chrysanthemic acid exists in two enantiomeric forms: 129* (lR,2R), which possesses insecticidal activity and 129*A (lS,2S), which is devoid of this activity [2.112].

Pyrethric acid has been synthesized from these two mirror images as shown [2.113].

[2.108

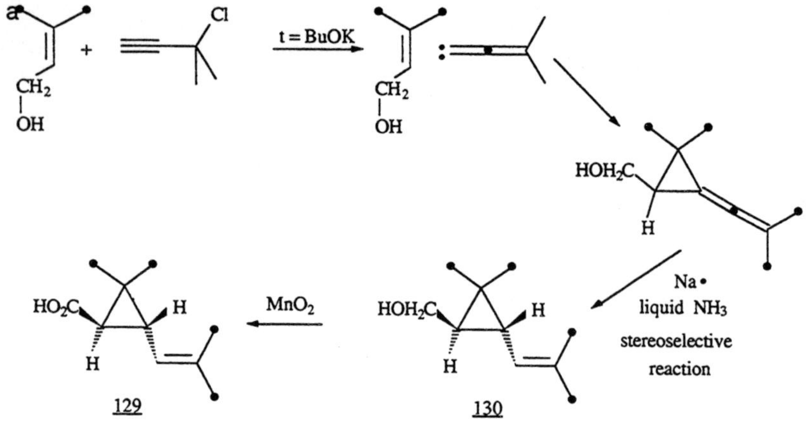

t = BuOK

$Na\bullet$
liquid NH_3
stereoselective
reaction

MnO_2

129 130

[2.109]

1 - oxalate
2 - Δ

artemisiatriene
skeleton
(scission A)

hν

santolinyl skeleton (scission C)

lavandulyl skeleton
(scission B)

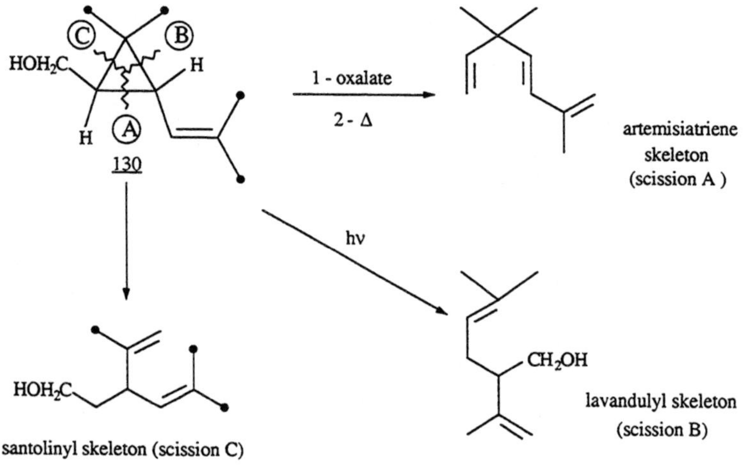

[2.110]

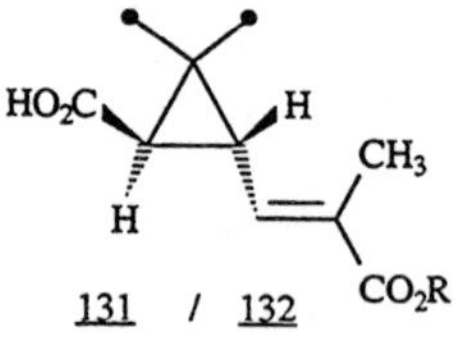

131 / 132

From the flowers of pyrethrum (*Chrysanthemum cinerariaefolium*) six insecticides derived from chrysanthemic and pyrethric acids can be extracted. These acids are esterified with three cyclopentenolones (rethrolones): cinerolone 133, jasmolone 134 and pyrethrolone 135 [2.114].

The six pyrethrum insecticidal esters are indicated below:

129 + 133 → Cinerine I; 132 + 133 → Cinerine II
129 + 134 → Jasmoline I; 132 + 134 → Jasmoline II
129 + 135 → Pyrethrine I; 132 + 135 → Pyrethrine II

[2.111]

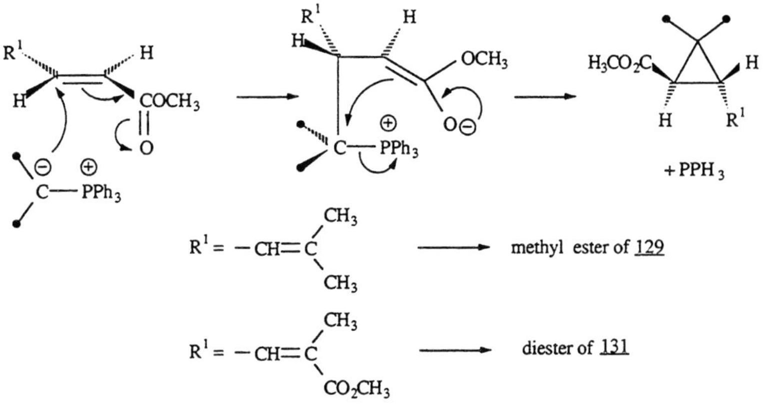

$R^1 = -CH{=}C(CH_3)_2$ → methyl ester of 129

$R^1 = -CH{=}C(CH_3)(CO_2CH_3)$ → diester of 131

+ PPH_3

[2.112]

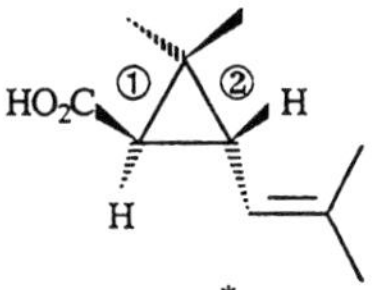

129* 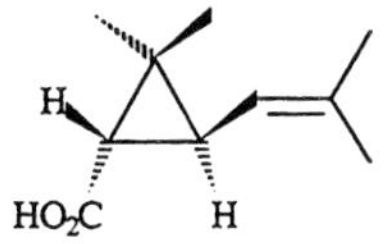 129*A

[2.113]

129* $\xrightarrow{O_3}$ + $(H_5C_2O)_2P(O)CH(CH_3)CO_2CH_3$ → B- Horner Reaction → 132

129*A $\xrightarrow{O_3}$ → $B^{\ominus}$ epimerization →

$SOCl_2$ epimerizaton at position 1

132 ←

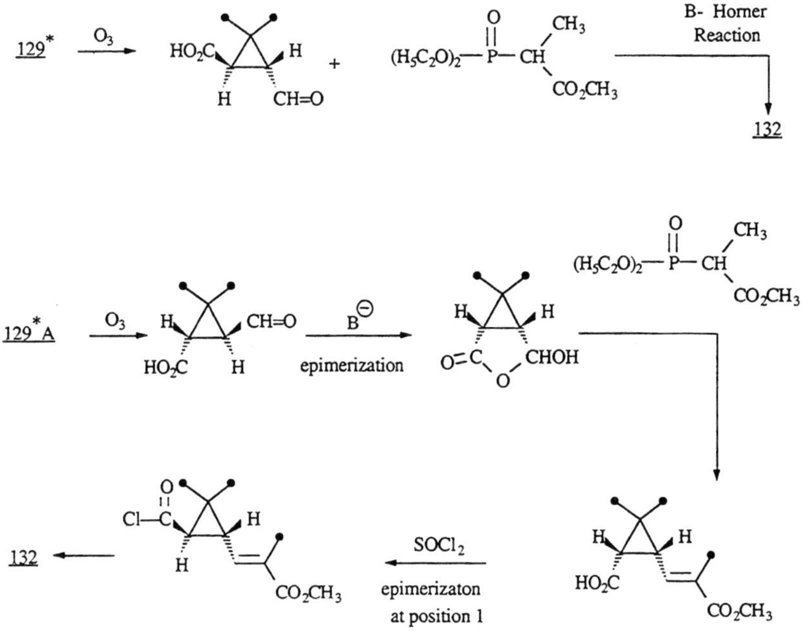

[2.114]

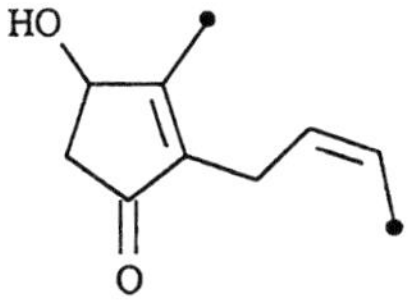

133 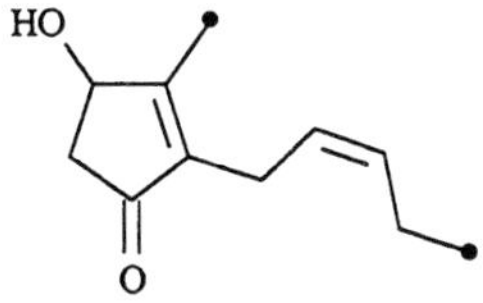 134 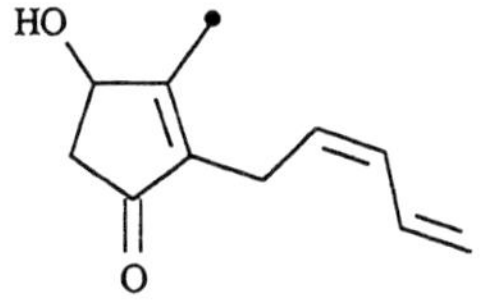 135

A synthetic pathway to pyrethrolone 135 is indicated schematically below. This type of synthesis, which is also valid for the two other rethrolones, uses a Wittig reaction in the absence of dissolved salts. As shown by Schlosser, these conditions favor the formation of the (Z) olefin from a non-stabilized ylid [2.115].

[2.115]

O CH$_2$OH multiple steps O O ⊕ PPh$_3$ ⊖ CH_2=CH-CH=O (absence of dissolved salts) O O H H

1 - deprotection
2 - base and OCH$_3$ O=C OCH$_3$

⊕ ⊖ K O O O H H Formation of the potassium salt ‡ CH$_3$O O O H H

B⊖ CH$_3$COCH=O HO O O ⊖ H H - H_2O HO O 135

Note: The condensation is carried out on the potassium salt of the β-ketoacid. It is interesting to note that the diketodienol resists intramolecular aldolization.

Bicyclic Monoterpenes

There are five principal types of bicyclic monoterpenic compounds which correspond to the following skeletons:

A. Thuyanes
B. Caranes
C. Pinanes
D. Camphanes-isocamphanes
E. Fenchanes [2.116]

It is also worthwhile to mention the existance of tricyclic monoterpenes such as tricyclene and cyclofenchene [2.117].

All of these derivatives belonging to these different groups do not have the same importance and only some will be discussed in detail here.

[2.116]

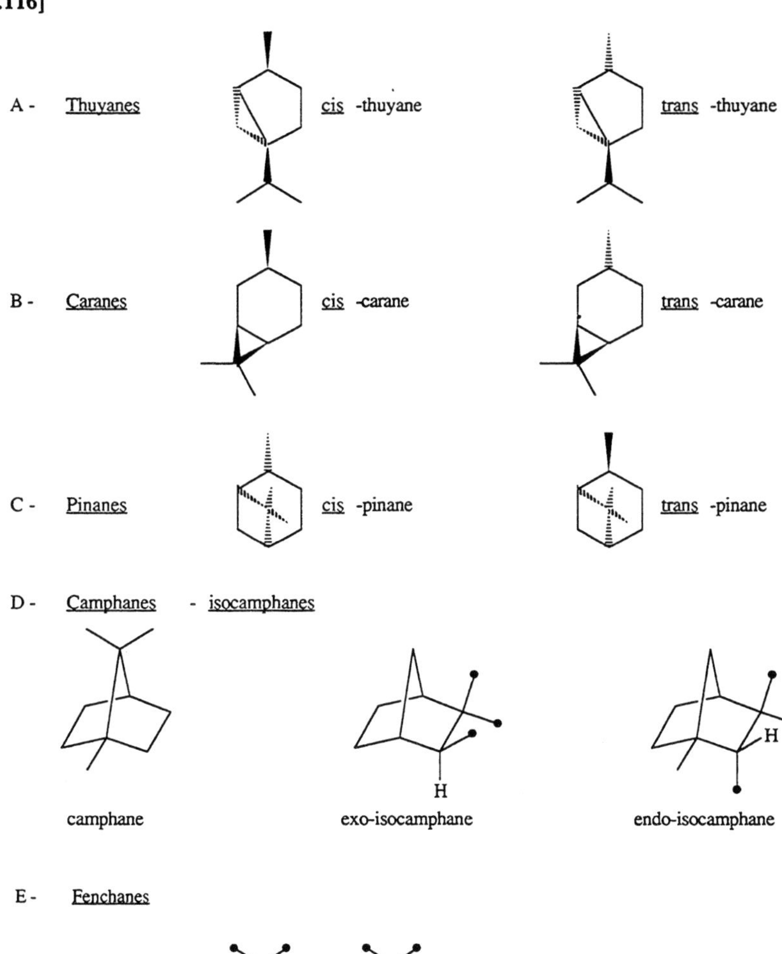
A - Thuyanes
cis -thuyane
trans -thuyane
B - Caranes
cis -carane
trans -carane
C - Pinanes
cis -pinane
trans -pinane
D - Camphanes - isocamphanes
H
H
camphane
exo-isocamphane
endo-isocamphane
E - Fenchanes

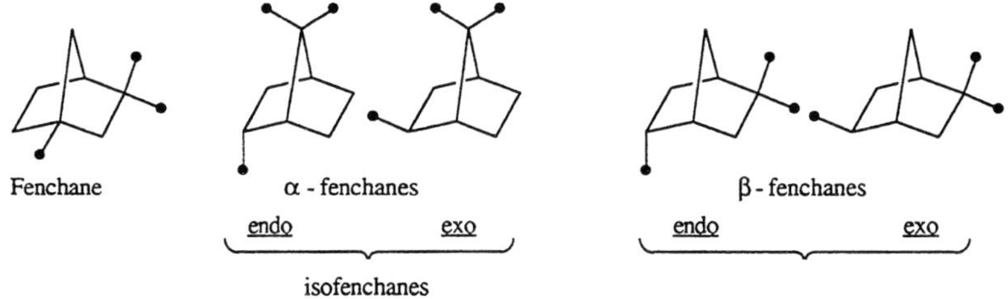
Fenchane
α - fenchanes
β - fenchanes
endo
exo
endo
exo
isofenchanes

[2.117]

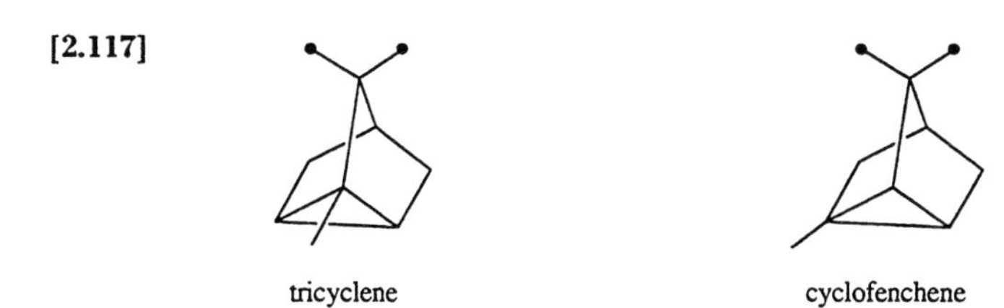
tricyclene
cyclofenchene

A. Derivatives of thuyane

1. The fact that these derivatives contain a cyclopropane ring renders them particularly fragile in acidic media.

Ring opening follows the rule of Markownikoff: protonation occurring on the least substituted carbon (in this case, the methylene) with the formation of two carbocations a and b being susceptible. Species b being the more stable, cyclopentanyl derivatives of the iridoid family are obtained [2.118].

[2.118]

a (secondary cation)   H⊕   H⊕   b (tertiary cation)   - H⊕

The principal natural products belonging to this group are:

- α-thuyene 136 found in its levo-rotatory form in the essential oils of eucalyptus (eucalyptus dives, in particular) and in its dextro-rotatory form in oregano oil and in some turpentines.
- sabinene 137 present in the essential oils of sabine, lavandin, *Juniperus horizontalis* leaves, some citrus fruit and in carrots. Ozonolysis of sabinene gives sabina ketone 138 which is present in the natural state in the essential oil of lavandin [2.119].

[2.119]

136   137 O_3 138   H⊕   139

In acidic media, sabina ketone rearranges to cryptone (4-isopropylcyclohexenone) 139. This can be explained in terms of the most basic site being the carbonyl oxygen.

>C=O

Protonation brings about the sequence shown below in which the three-membered ring is cleaved [2.120].

[2.120]

138 139

- The cis- and trans-hydrates of sabinene 140 and 141 are found in black pepper oil. The trans-isomer 141 has also been detected in American peppermint oil.

  The mixture of the two hydrates of sabinene can be obtained by the oxymercuration of sabinene (reaction dicovered by H. C. Brown) [2.121].

[2.121]

1 - $Hg(OAc)_2$, THF

2 - $NaBH_4$ then KOH

137 140 cis 141 trans

- In addition to the preceding compounds, (–) thuyone 142 and its epimer (+) isothuyone 149 are found in the essential oils of thuya, sage (*Salvia officinalis*), tansy and *Artimisia vulgaris* (common mugwort).

The four isomeric thuyols 144, 145, 146 and 147 are found in Artimisia cina (wormseed) oil while sabinol 148 is found in laurel leaf oil [2.122].

[2.122]

2/1

142 (-20°) 143 (+75°) 148 (+) 149 (-39.4°)

(-) thuyol 144

(-) neothuyol 145

(+) isothuyol 146

(+) neoisothuyol 147

The stereochemical relations which exist between the thuyones and the thuyols are in all respects similar to those found between the menthones and menthols. It is therefore not necessary to go into detail here.

Thuyones and thuyols can be obtained by the hydroboration-oxidation sequence of reactions (as described by Brown) carried out on α-thuyene. (This reaction will be dealt with later with regard to Δ–3-carene.) [2.123]

[2.123]

Stereochemical relations between the principal thuyane derivatives

(+) sabinol 148

(+) neoisothuyol 147

(β - thuyone) (+) isothuyone 143

(+) thuyane

(-) neothuyol 145

(+) isothuyol 146

(-) α-thuyene 136

(+) cis sabinene hydrate 140

(-) thuyol 144

(-) thuyone (α - thuyone) 142

(+) trans sabine hydrate 141

(-) sabinacetone 138

(-) cis-umbellulic acid

(+) cis-umbellulic acid

(+) sabinene 137

(-) umbellulone 149

2. A general scheme for the total synthesis of some thuyane derivatives can be described.

The starting point for this synthesis is methylheptenone 150, a compound of great importance in the synthesis of terpenoids. The hydroboration-oxidation of this compound gives the 1,4-diketone 151, which can also be obtained by ozonolysis of α-terpinene 63. When treated in refluxing alcohol containing 2% sodium hydroxide, this diketone gives the cyclopentenone 152 in a yield of 70% and this can be subsequently reduced with $LiAlH_4$ in etherial suspension to result in the cyclopentenol 153 (yield ~90%). The conversion of cyclopentenol into the bicyclic alcohol 140 makes use of a well known methylene transfer reaction first described by Simmons and Smith. The methylene-donating reagent is prepared from methylene iodide CH_2I_2 and a zinc-copper couple in ether and can be considered as an equilibrium between several species (in accord with the Schlenk equilibrium in organo-magnesium compounds).

$$CH_2I_2 + Zn/Cu \xrightarrow{(C_2H_5)_2O} 2\, I\, CH_2ZnI \longleftrightarrow (I\, CH_2)_2Zn + ZnI_2$$

This reactive species forms a complex with the hydroxyl group of the alcohol 153 and this explains both the stereoselectivity of this reaction (formation of the methylene bridge in a cis relation with the -OH group) and also the greater reactivity of allylic alcohols compared with isolated double bonds [2.124].

B. Derivatives of carane

1. Natural compounds

Only two natural products with this skeleton are known. One is Δ–3-carene 154 which is found principally in the dextro-rotatory form in turpentine oil, particularly in the oil from India (*Pinus longifola*). The other is Δ–4-carene 155 (*Pinus sylvestris*), sometimes known as Δ–2-carene. Anhydrous hydrochloric acid adds to Δ–3-carene to give two dichlorides, one of which is identical to the dichloride obtained from dipentene 71 and the other being identical to that of sylvestrene 156. By eliminating the two HCl molecules from 156 using aniline, two isomeric sylvestrenes 157 and 158 are obtained [2.125].

This mixture of hydrocarbons (157, 158) was initially described under the name "carvestrene" by Wallach who believed that they were natural components of certain pines. In fact, as their isolation had been made in the form of the dichloride 156, "carvestrene" was an artifact formed at the expense of Δ–3-carene.

In contrast to thuyane, no natural oxygenated derivatives of carane are known.

2. Oxygenated derivatives obtained by chemical means

Hydroboration and epoxidation reactions have been used to prepare a number of these substances.

The hydroboration of (+) Δ–3-carene can be achieved relatively economically by

[2.124]

150 —— 1 - B_2H_6, 2 - CrO_3, $H^{\oplus}$ (Brown & Garg) ——> 151 <—— ozonolysis —— 63

151 —— NaOH ethanolic ——> 152

152 —— $LiAlH_4$ / $(C_2H_5)_2O$ ——> 153 —— CH_2I_2 Zn / Cu (Simmons-Smith reaction) ——> 140 (HO H)

140 —— CrO_3 ——> (+) 138 (sabinacetone) and its mirror image —— Wittig $Ph_3P^{\oplus}$-$CH_2^{\ominus}$ ——> -) 137 (sabinene)

[2.125]

numbering according to IUPAC)

[(+) (1R,6S) –3-carene] in current nomenclature 154 —— HCl ——> 71 + 156

156 —— $C_6H_5NH_2$ ——> 157 + 158

155

a method perfected by the Research Center of Roure Bertrand Dupont in Grasse. It consists of using diisobutylaluminum hydride (Dibal) as a hydride source and methyl borate as a source of boron. The end product is a single crystalline alcohol (M.P. 27°C) called isocaranone 159 or (–) (lR,3R,4R,6S) cis–3-caranol. Oxidation of isocaranol gives very pure (–) isocaranone (cis-caranone) 160.

Reduction of isocaranone with triisobutylaluminum (Tibal) leads selectively to neoisocaranol 161 or (+) (lR,3S,4R,6S) cis–3-caranol. Base converts isocaranone 160 into an equilibrium mixture containing only 15–18% of caranone 162. It is therefore not possible to pass easily from the cis-series to the trans-series by this means.

When, however, (+) Δ–3-carene is treated with peracids (such as monoperphthalic acid, for example), the end product is cis,trans-epoxy (+) carane or α-epoxycarane 163. In the presence of Lewis acids such as $ZnBr_2$, this epoxide is converted into a mixture in which caranone 162 at 50% constitutes the major of the three principal components. Careful distillation permits its isolation in pure form [2.126].

[2.126]

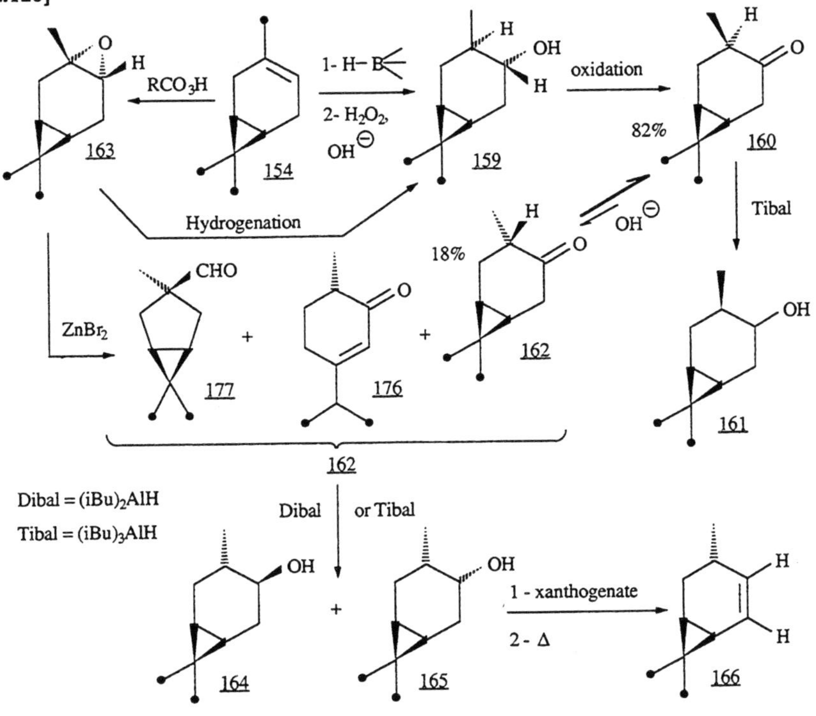

Dibal = $(iBu)_2AlH$

Tibal = $(iBu)_3AlH$

Reduction of caranone with "Dibal" or "Tibal" is not as selective as that for isocaranone. A mixture containing roughly equal quantities of caranol 164 and neocaranol 165 is formed. These two alcohols can be separated by careful distillation with neocaranol passing ahead. Pyrolysis of its xanthogenate permits veritable Δ–2-carene 166 to be obtained. Δ–3-Carene normally exists as a mixture of conformers such as 154a and 154b [2.127].

These conformations clearly demonstrate the steric crowding of the upper face (β-face), which explains why the hydroboration, epoxidation and reduction reactions occur stereospecifically on the lower face (α-face) [2.128].

[2.127]

154a ⇌ 154b

β–face

α–face

[2.128]

154a ⇌ 154b —(1- H–B< ; 2- H_2O_2, $OH^{\ominus}$)→ 159a —oxidation→ 160a

160a —Tibal→ 161a

160a (Pseudoequatorial) ⇌ 162a (Pseudoaxial)

These conformations permit a better understanding of why reduction of isocaranone 160a only occurs from the α-face giving rise stereospecifically to neoisocaranol 161a. The conformations attributed to the two caranones result from examination of their physicochemical properties. Whereas the β-face of isocaranone 160a is very crowded due to the presence of the pseudo-axial methyl, this is not the case for caranone 162a where approach to either face is hindered to roughly the same extent and reduction therefore gives a mixture of equal parts of the two caranols 164 and 165.

As shown earlier, epoxidation of Δ–3-carene is stereospecific with formation of the α-epoxide resulting from cis-addition of oxygen to the least crowded face.

Opening of this epoxide by nucleophilic reagents or aqueous mineral acids occurs with inversion of configuration at the carbon which is attacked. Generally, this type of cleavage proceeds in a way which places the two new substituents in trans-diaxial positions 167a rather than in trans-diequatorial positions 167b [2.129].

In this case, a careful physicochemical examination shows that the trans-diequatorial conformation is by far the more favored (half chair conformation of a six-membered ring).

Stereochemical studies in the corane series.

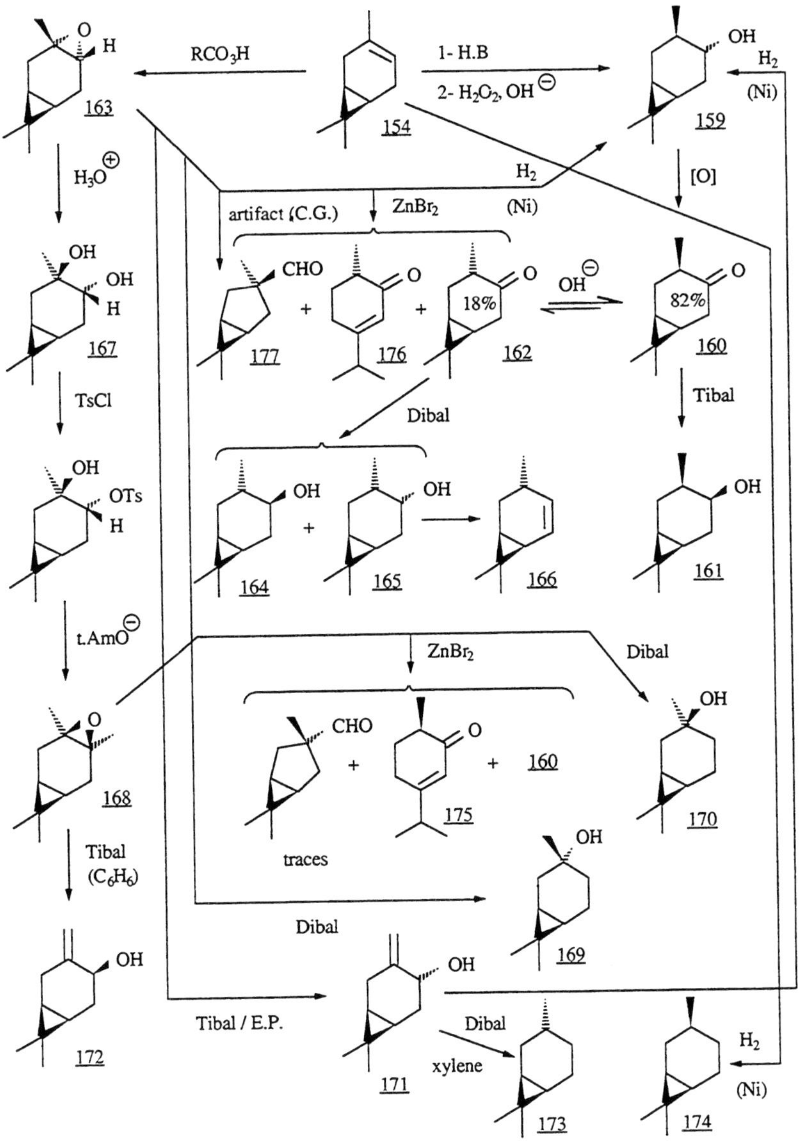

The monotosylate of the α-diol 167a (R=Tos), when treated with sodium tertiary amylate, affords stereospecifically β-epoxycarane 168 by trans-diaxial elimination with inversion of the configuration at carbon 3 [2.130].

The epoxides 163 and 168 can be reduced by $LiAlH_4$ or by "Dibal" to give the tertiary caranols 169 and 170, respectively [2.131].

In petroleum ether, however, "Tibal" reacts totally differently, with the reaction leading to the allylic alcohols 171 and 172, respectively [2.132].

[2.129]

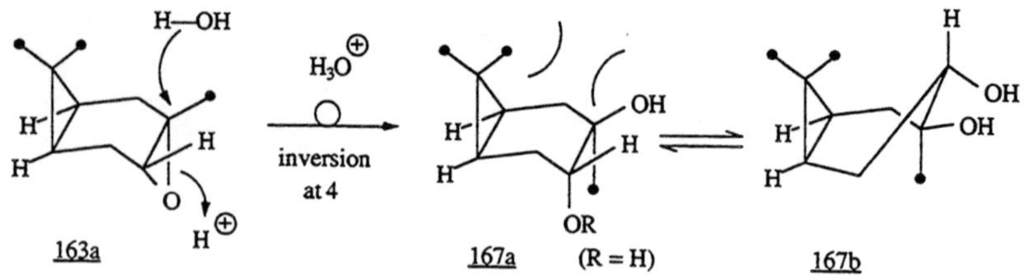

[2.130]

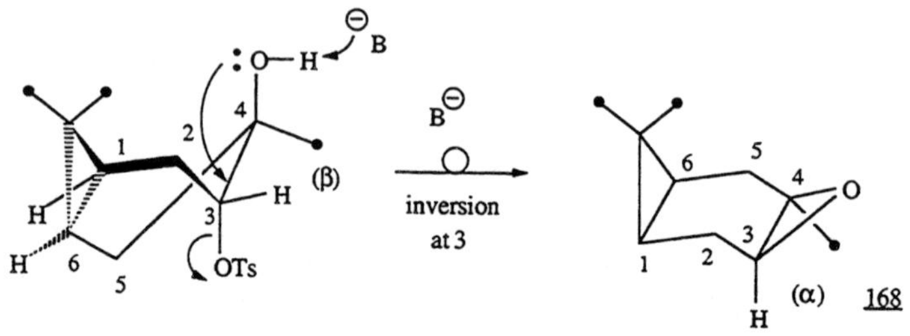

can also be represented as

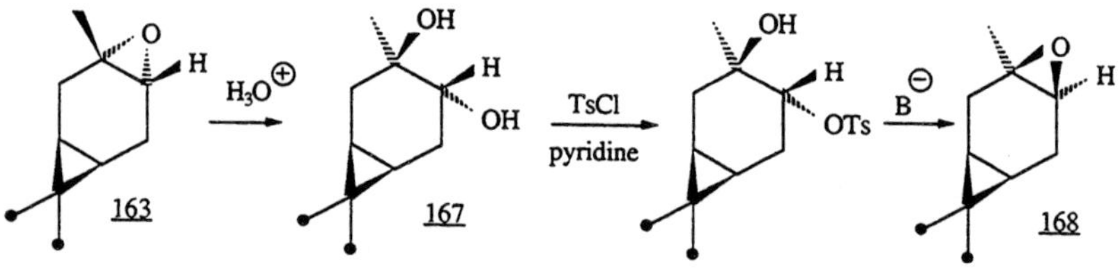

[2.131]

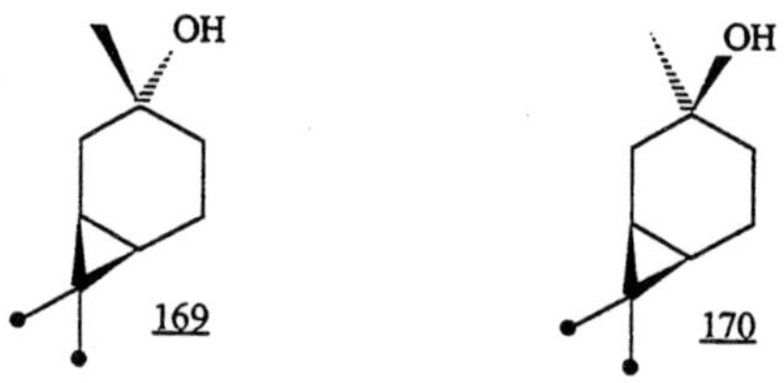

[2.132]

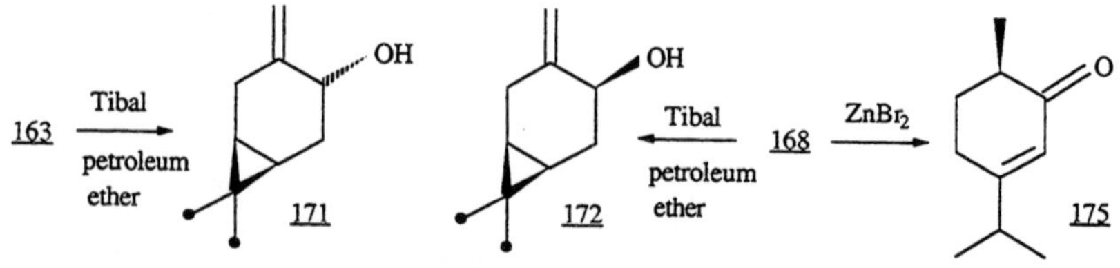

When treated with "Dibal" in refluxing xylene, the carenol 171 undergoes reductive hydrogenolysis with formation of trans-carane 173. cis-Carane 174 can be obtained directly from Δ–3-carene by catalytic hydrogenation (cis-addition of the H_2 molecule from the least hindered face) [2.133].

[2.133]

171 —Dibal, xylene→ 173   174 ←H_2, Raney Ni— 154

It should be noted that β-epoxycarane is much more stable than its α-isomer and, taken together, its reactions require more energetic reaction conditions. When treated with $ZnBr_2$, β-epoxycarane gives two products. The principal one (80%), just as for α-epoxycarane, is isocaranone while the minor product is (+) carvenone 175. Under the same conditions, α-epoxycarane gives (–) carvenone 176 accompanied by isocaranone. It is noteworthy in this regard that from the *single* substance (+) Δ–3-carene, it is possible to obtain both the (+) and the (–) carvenones. This can be interpreted as follows [2.134].

[2.134]

163 (β) —$ZnBr_2$→ A ($OZnBr_2$) —1,2-dihydride migration→ B ($OZnBr_2$) (α) → C ($OZnBr_2$) → 176

163 → 176

Inversion of configuration at carbon 4 can also be seen, particularly in the conversion of β-epoxycarane 168 into (+) carvenone 175. This can be interpreted as the result of a 1,2-hydride transfer from carbon 3 to the cationic carbon 4. If instead of this migration, a migration of the σ–2,3 doublet was to occur, a ring-contracted aldehyde 177 would be formed [2.135].

It is evident that this aldehyde with a contracted ring 177 possesses a plane of symmetry and would be devoid of optical activity (as confirmed by observation), whereas α-epoxycarane has a specific rotation of $[\alpha]_D^{20} = +13.4°$.

Note: Details of the physicochemical studies of the carane derivatives (IR, UV, NMR) can be found in the French language journal *Recherches,* 1967, **16,**

[2.135]

A OZnBr₂ (β) H (α) O—ZnBr₂ CH=O 177

133–57. The scheme represents the stereochemical transformations that have been carried out in the carane series.

C. Derivatives of pinane

The chemistry of α- and β-pinene has considerably influenced our understanding of the properties and reaction mechanisms of bridged bicyclic systems. In this regard, it has played an important role in the development of organic chemistry in general.

1. Introduction

Turpentine oil, which is the source of α- and β-pinenes 178 and 179, has been used since the end of the nineteenth century for the manufacture of derivatives for use in perfumery.

As early as 1820, Büchner reported the production of terpine 80, and in 1878, Tilden and later Wallach in 1885 were able to derive α-terpineol 19 from it. In France, the manufacture of terpineol resulted from the work of Bouchard and Bouari (1866) and Laporte (1889). Nevertheless, the structure of α-pinene was only suggested by Wagner in 1894 before being definitively established in 1896 by von Baeyer and it was not until 1907 that Wallach elucidated the structure of β-pinene [2.136].

[2.136]

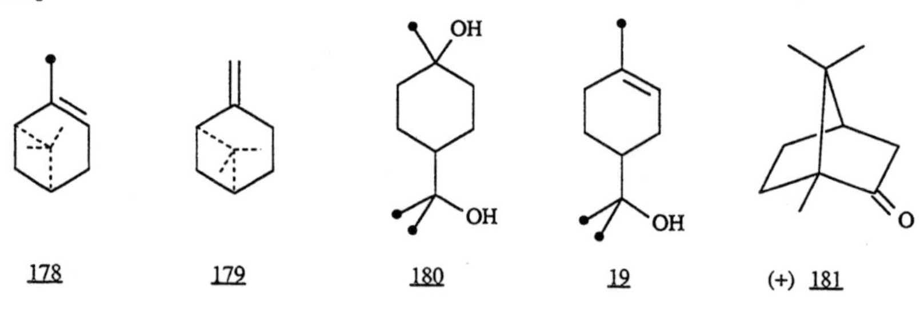

As for camphor 181, which has been known since ancient times, it was also only towards the end of the nineteenth century that it became an important raw material for the manufacture of celluloid and smokeless powders. Its structure was determined by Bredt in 1893 and its manufacture from turpentine oil was only industrialized from 1920 onwards. Nonetheless, the fundamental discovery at the chemical level occurred in 1899 when Wagner observed that the hydrochloride of camphene 182 rearranged into isobornyl chloride 183 in the presence of hydrochloric acid [2.137].

[2.137]

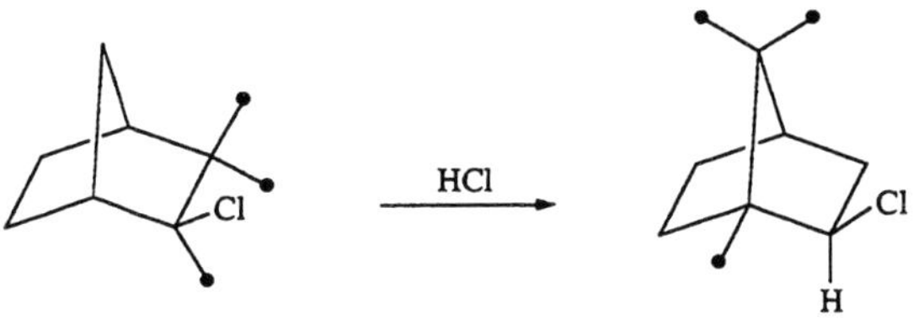

HCl

182 183

2. *Discussion of "reaction intermediate"*

To place this discussion in a historical context, it should be recalled that the first stable free radical, the triphenylmethyl radical 184, was discovered by Bomberg in 1900. Elsewhere, Norris in 1901 in the United States and Kehrmann and Wentzel in Germany in 1902 observed that the colorless derivatives of triphenylmethane such as triphenylcarbinol 185 or triphenylmethyl chloride 186 yielded strongly colored solutions when dissolved in concentrated H_2SO_4. They also noted that trityl chloride formed strongly colored complexes with Lewis acids such as $AlCl_3$ or $SnCl_4$ [2.138].

[2.138]

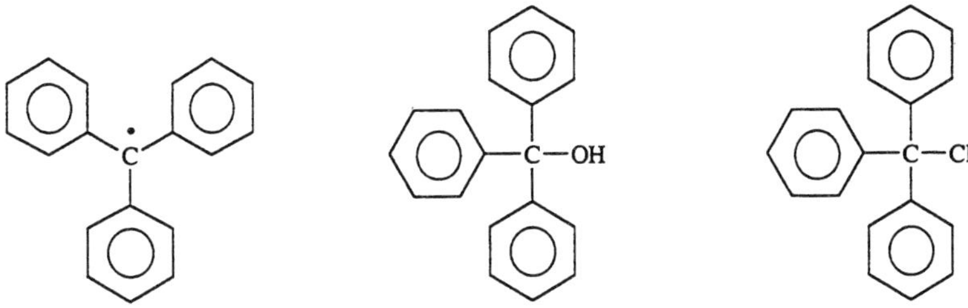

184 185 186

In 1902, von Baeyer reported that these compounds possessed characteristics of salts and called them "salts of carbonium." The first stable carbonium ion had must been discovered, but it is surprising to realize that carbonium ion chemistry was then confined for 20 years to the colorants of the triphenylmethane group.

In 1922, Meerwein and van Emster, while studying the rearrangement discovered by Wagner (see 182 → 183) and the reverse reaction from isobornyl chloride to camphene, made two important observations:

- The first concerns the effect of the solvent. While the reaction rate increases with the dielectric constant of the solvent, it is not exclusively dependent on it. Among the 13 solvents examined, there were two exceptions, liquid SO_2 and cresol. In fact, the rates measured in these solvents are much greater than would be predicted on the basis of their dielectric constants.

- The second concerns the accelerating effects of certain metal chlorides such as $SbCl_5$, $SnCl_4$, $FeCl_3$ and $SbCl_3$ which accelerate considerably the rearrangement of camphene hydrochloride.

Confronted with these phenomena, Meerwein concluded that "the rearrangement takes place only after the ionization of the substrate. The conversion of camphene hydrochloride into isobornyl chloride cannot be simply explained as the migration of a chlorine atom, but must involve the rearrangement of an intermediate carbocation." Thus was born the concept of carbonium ions as reaction intermediates.

In 1932, Whitmore generalized the theory of the Wagner-Meerwein rearrangement to include numerous other organic chemical reactions. Wilson re-examined this rearrangement in 1939 using radioactive isotope exchange techniques and came to the conclusion that the rearrangement proceeded in two stages. The first involved a rapid ionic equilibrium between an "ion pair" subsequent to separation of the chloride ion as concluded by Meerwein. The second was a slow biomolecular reaction between the ions thus produced and a molecule of hydrochloric acid [2.139].

[2.139]

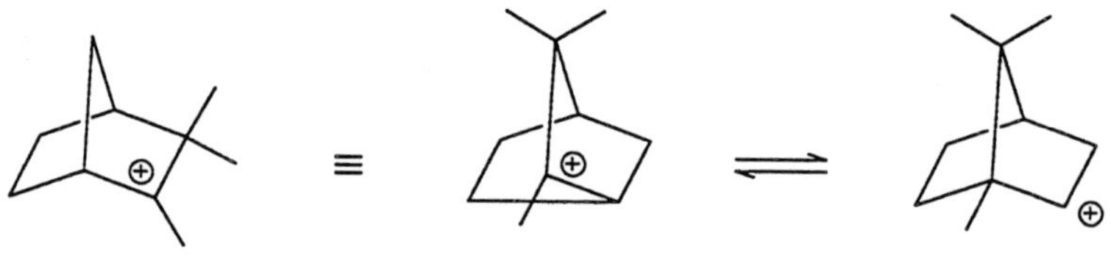

It should be mentioned that apart from the free radicals and carbocations discussed here, many other types of transient intermediates are currently being invoked. Some are electronically neutral like the arynes or the carbenes (bivalent carbon possessing in its "singlet" state one empty orbital and one containing a free pair of electrons) [2.140].

[2.140]

Others are charged: carbanions [2.141],

[2.141]

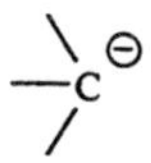

and radical ions such as the "cetyl" radical anion [2.142],

[2.142]

and radical cations such as [2.143]

[2.143]

3. *The problem of classic and non-classic cations*

The suggestion of Wilson excited the imagination of the physical organic chemists of whom some had a tendency to invoke the mesomeric effect rather than an equilibrium between transient species, as shown in this example [2.144].

[2.144]

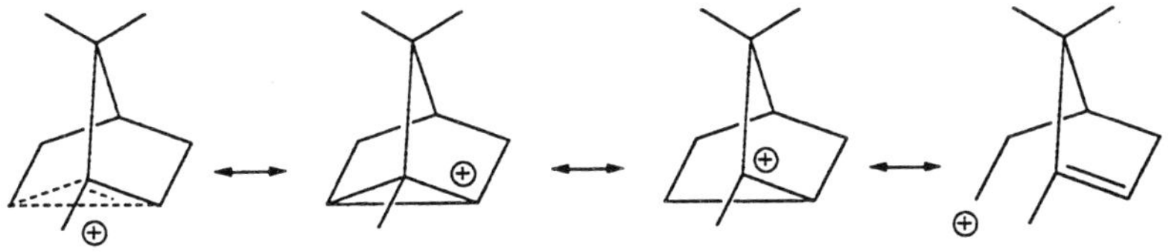

The concept of "bridged ions" or "non-classic ions" was thus created. A fundamental point should be made here, however. There is a great difference between the carbonium ions formed from triphenylmethane derivatives, which are stable in strongly acidic media and which have an appreciably long life span and the alkylcarbonium ions. These latter intermediates generally have an extremely short life span and can only be observed by traditional physical methods.

One physical method which provides direct proof of the existance of alkyl carbonium ions employs the phenomenon of electron impact, that is, organic cations are formed in mass spectrometric studies as the result of electron impact in the gas-phase. Carbonium ions in solution however, have been observed relatively recently—by Olah in 1965. His group obtained, for example, the tertiary-butyl cation [2.145]

[2.145]

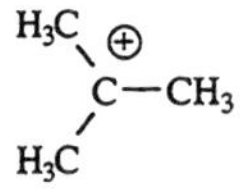

by dissolving tertiary-butyl fluoride in SbF_5 acting as both Lewis acid and solvent [2.146].

[2.146]

$$(CH_3)_3CF + (SbF_5)_2 \rightarrow (CH_3)_3\overset{\oplus}{C}\ \overset{\ominus}{Sb_2F_{11}}$$

SbF_5 can be diluted with SO_2 liquid or SO_2ClF. The fluorides of propyl, isopropyl, butyl and pentyl behave in the same fashion. Proton NMR spectroscopy in hyperacidic media at very low temperatures has provided the method of choice for studying carbonium ions in solution. The principal characteristic of the NMR spectra of these species is the considerable deshielding of the protons in the intermediates compared with those in the starting materials. The studies of Olah and his group have thus furnished decisive proof of the existance of alkylcarbonium ions in solution.

Returning to the Wagner-Meerwein rearrangement and the notion of non-classic carbonium ions, which is a consequence generalized from the propositions of Wilson, an enormous effort spanning 20 years was devoted to capturing the intimate structure of these cations. This problem has been particularly well studied using the norbornyl cation.

In 1949, Winstein suggested non-classic bridged structures a or b for this cation and this hypothesis was supported by Roberts and various other scientists on the basis of serious arguments. On the other hand, Brown defended vigorously the opposite point of view by affirming that the ion was classic with the "pseudo-symmetry" being the result of a very rapid Wagner-Meerwein rearrangement, as shown in c [2.147].

[2.147]

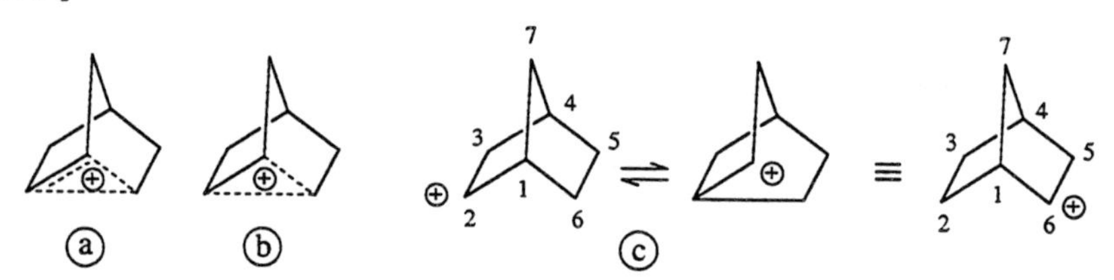

It was finally the group of G. A. Olah who resolved this controversy in 1970 thanks to their study of the structure of the norbornyl cation in superacidic media such as $SbCl_5$-SO_2, $SbCl_5$-SO_2ClF, $SbCl_5$-SbF_5, and FSO_3H-SbF_5-SbF_5-SO_2 (magic acid) using proton and ^{13}C NMR and Raman-laser spectroscopic techniques. In these media at low temperatures, the norbornyl cation has a long life span.

The work of Olah showed clearly that in SbF_5-SO_2ClF-SO_2F_2 at –154°C, the proton and ^{13}C NMR spectra were only compatible with the non-classic structure. A detailed examination of this work goes beyond the frame of this text, however, it should be pointed out that in the media used, the norbornyl cation is identical to nortricyclene "corner-protonated" at the bridgehead as shown in d, where a carbon atom is transitorily bound to five atoms [2.148].

This simplified representation corresponds to the formation of a tricentric molec-

[2.148]

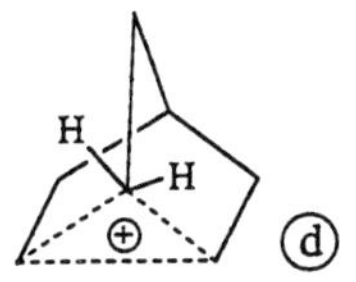

ular orbital by overlapping the sp3 orbital of the bridgehead carbon with two other orbitals from each of the two atoms which form the base of the triangular pseudo-cycle e. A cation such as this would be more correctly represented by f [2.149].

[2.149]

In order to designate carbocations such as these in which a carbon is bound to five other atoms, Olah has proposed the term "carbonium ions," whereas the term "carbenium ion" is reserved for classic carbocations that can be considered as having resulted from the coupling of a carbene with an electrophile [2.150].

[2.150]

$$\mathrm{(X)(Y)C}: + \mathrm{E}^{\oplus} \longrightarrow \mathrm{(X)(Y)C}^{\oplus}\mathrm{-E}$$

4. *Rearrangements of the carbon skeleton*

The Wagner-Meerwein rearrangement has already been widely discussed and will be further referred to in this text. It is characterized by geometrical considerations shown in the following scheme [2.151].

[2.151]

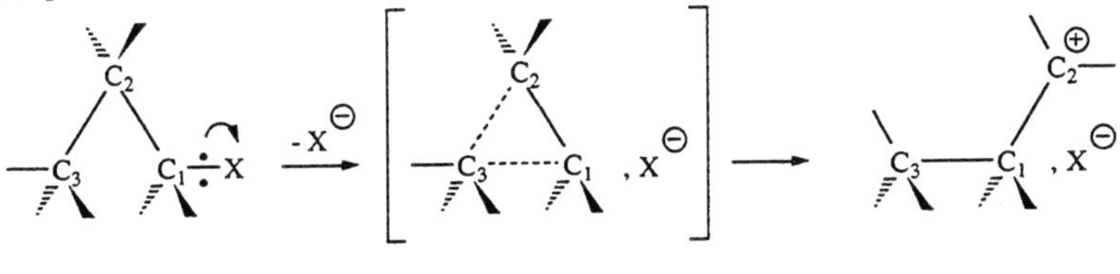

In order for this rearrangement to take place, it is necessary that the atoms X, C_1 and C_3 are significantly aligned and that the four atoms X, C_1, C_2 and C_3 are coplanar and have relative positions that are very close to those which they will occupy in the cyclic transition state.

In fact, this cyclic species will be stabilized by close association with the electron pair of the leaving group such as a hydroxyl or some other species with an electron pair. If the C_1-X bond was situated in an orthogonal plane to that of the three carbon atoms, no orbital overlap would occur and the cyclic transition state would not form due to lack of stability.

Examples: with rearrangement [2.152], without rearrangement [2.153]

[2.152]

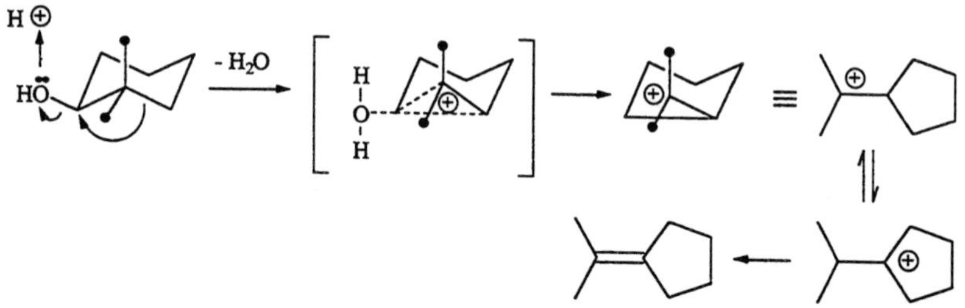

- H_2O

[2.153]

transelimination

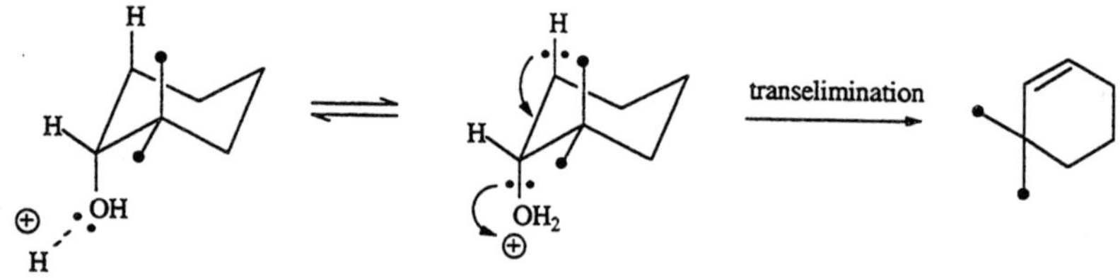

These conditions are found notably in the synthesis of borneols.

The so-called Nametkin rearrangement constitutes the second important case. It involves a 1,2 migration of a methyl group. The dehydration of camphenilol 187 in acidic media is a simple and often cited example of this rearrangement [2.154].

[2.154]

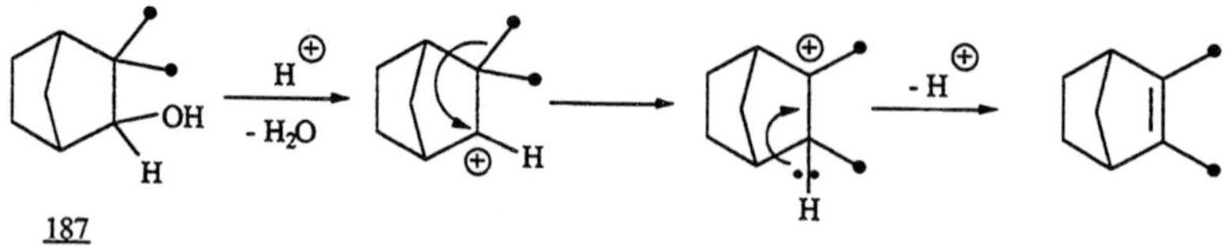

187

Later, an example of these two rearrangements will be seen in the conversion of (–) camphor into (+) camphor.

A third rearrangement is quite often observed in terpene chemistry and corresponds to a 1,3-hydride migration (a 1,3-shift). It has on occasion been designated by the name "6–2 Shift." The intermediate may be a carbene or a carbocation and examples of this rearrangement are the transformation of camphor 181 into tricyclene 188 when the hydrazone of camphor is treated with the yellow oxide of mercury (Meerwein reaction). In this case, the intermediate is believed to contain divalent carbon (i.e., a carbene) [2.155].

[2.155]

CHAPTER

3

Industrial Syntheses Starting from α- and β-Pinene

Until about 1940, pinenes were only used for the production of camphor, the fenchols and the free alcohols and esters of terpineol, borneol, and isoborneol. These compounds were obtained by submitting the pinenes to acid-catalyzed reactions. Towards 1950, two other types of reaction were employed: thermal reactions (pyrolyses) and oxidation reactions with the formation of hydroperoxides and epoxides. They will be discussed in this order.

Acid-Catalyzed Reactions

Either mineral acids or organic acids are used. The oldest methods are based on this type of reaction and these include the production of terpine, terpineol, camphor, and the borneols. They can be classed into two groups.

A. Reactions involving rupture of the cyclobutyl ring without rearrangement

Thanks to its ready crystallization, terpine 180 has been known for a very long time. Büchner reported it in 1820 and Deville obtained it by treating turpentine oil with a solution of nitric acid in 80% ethanol. Today, a solution of sulfuric acid concentrated to between 20 and 40% is used with the reaction taking place around 30°C in the absence of air. The isolated end product is the monohydrate of cis-terpine (MP = 116°–117°C) [3.01]

[3.01]

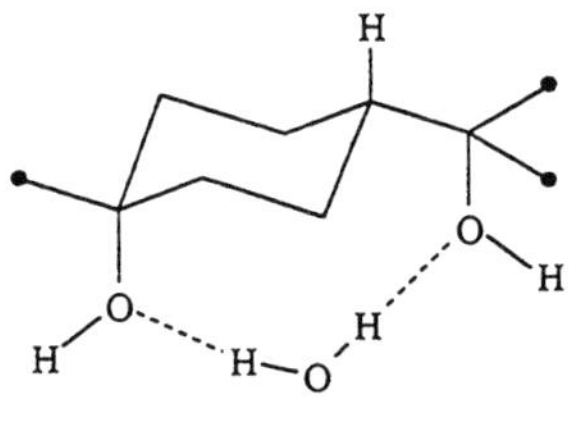

H O H H O H H H O H

The formation of terpine involves the mediation of a doubly charged cation which is subsequently neutralized by reaction with water [3.02].

[3.02]

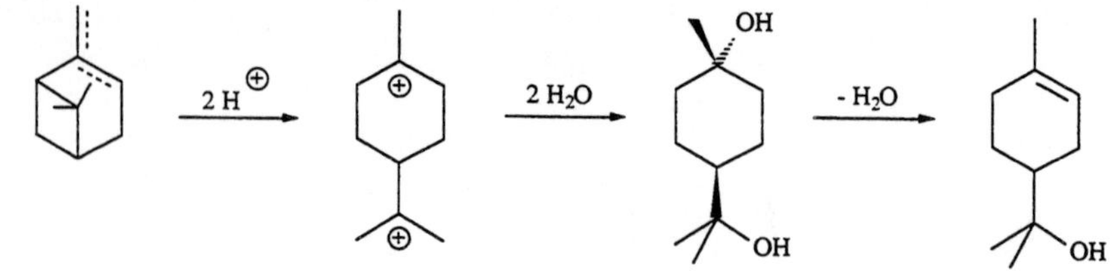

2 H$^{\oplus}$ → 2 H_2O → - H_2O →

178 + 179 180 19

In its own right, terpine has especially found use as a pharmaceutical. It is also used as an intermediate in the production of α-terpineol 19 of high quality. This process requires extremely strict dehydration conditions with regard to the nature of the dehydrating agent as well as the pH, the temperature and the duration of the reaction. Under specially selected conditions, a product containing approximately 95% α-terpineol can be obtained with a total yield of 99.5%.

Yet, the dehydration of terpine under uncontrolled conditions gives a complex mixture containing the α, β and γ-terpineols [3.03].

[3.03]

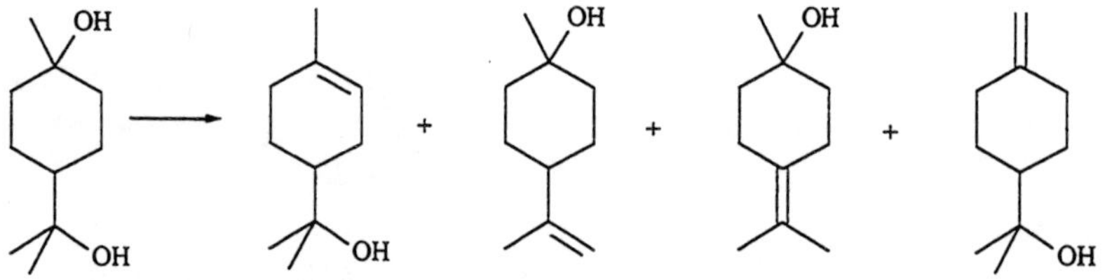

180 → 19 (α) + 68 (β : cis + trans) + + γ

By far the largest quantities of terpineol are manufactured directly from turpentine oil. This oil is treated with an organic acid such as acetic acid in the presence of small quantities of mineral acid as catalyst (Bertram-Walbaum reaction). Crude terpinyl acetate 189 is thereby obtained and this can be saponified and rectified to give terpineol [3.04].

[3.04]

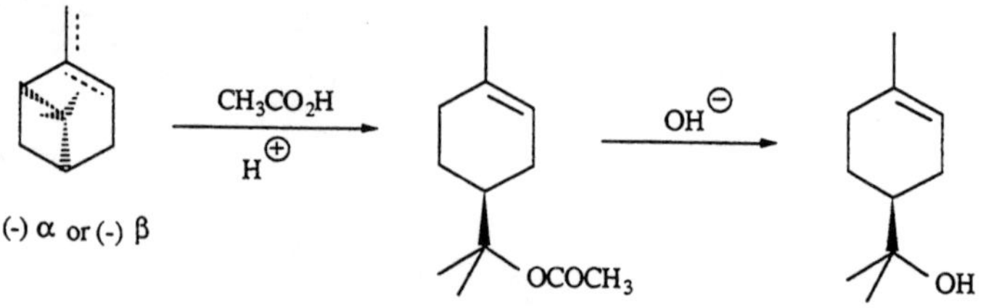

(-) α or (-) β —CH_3CO_2H, H$^{\oplus}$→ (-) 189 —OH$^{\ominus}$→ (-) 19

It should be noted that terpinyl acetate as well as other esters of α-terpineol are also employed in large quantities by perfumers. For instance, even in 1964 direct perfumery use in America alone was of the order of 2,000 metric tons. It is understandable that there are numerous patents covering methods of producing α-terpineol.

In all of these processes, a certain number of valuable terpenic by-products can be obtained in addition to α-terpineol. The different processes involving rupture of the cyclobutyl ring without rearrangement are shown here schematically [3.05].

[3.05]

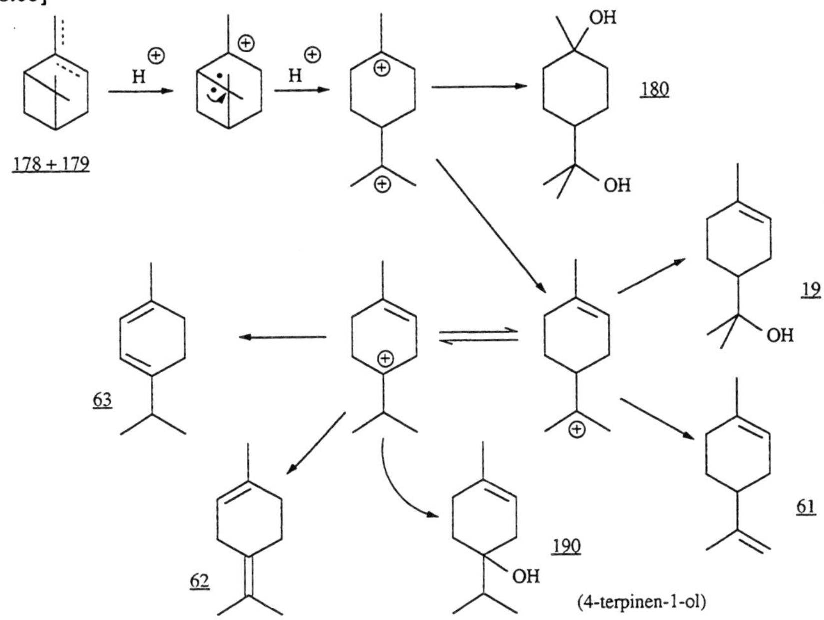

Although the rupture of the four-membered ring has been recognized for a long time, it was only fully interpreted in 1947 by Moscher on the basis of a fragmentation reaction discovered by Whitmore and Stahly in 1933.

By treating methylisopropylcarbinol 191 with 75% H_2SO_4 at 80°C, Kline and co-workers obtained 3,4,5,5-tetramethyl–2-hexene 192 as the result of dimerization.

Whitmore and Moscher observed that when they heated this hydrocarbon under reflux in the presence of naphthalene 1-chlorosulfonic acid, it fragmented to obtain a mixture of isobutylene and 3-methyl–2-pentene. The sequence shown below offers a possible explanation for this result [3.06]. It is possible to see, as Moscher noted in 1947, the analogy that exists between the fragmentation reaction of tetramethylhexene and that of the pinenes under the same conditions [3.07].

These openings of the cyclobutyl ring without rearrangement occur in aqueous mineral acids or in anhydrous organic acids. If, however, anhydrous mineral acids (particularly HCl) are used, the reactions are radically different and

[3.06]

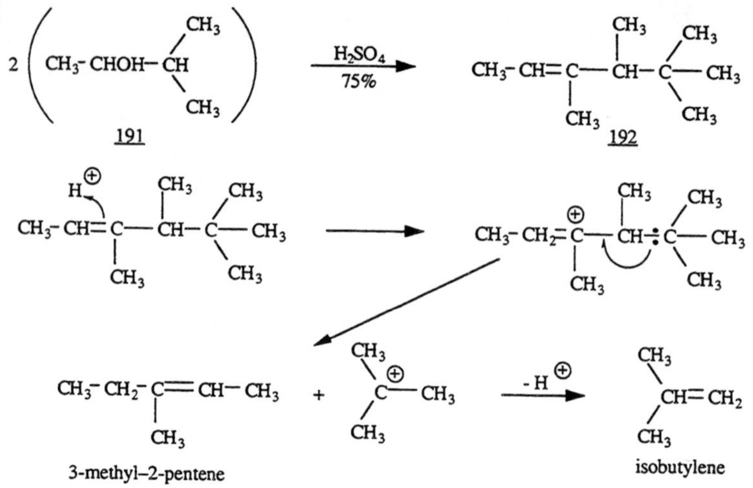

$$2\,(CH_3\text{-}CHOH\text{-}CH(CH_3)_2)\ \underline{191} \xrightarrow[75\%]{H_2SO_4} CH_3\text{-}CH{=}C(CH_3)\text{-}CH(CH_3)\text{-}C(CH_3)_3\ \underline{192}$$

$$CH_3\text{-}CH{=}C(CH_3)\text{-}CH(CH_3)\text{-}C(CH_3)_3 + H^{\oplus} \longrightarrow CH_3\text{-}CH_2\text{-}C^{\oplus}(CH_3)\text{-}CH(CH_3)\text{-}C(CH_3)_3$$

$$\longrightarrow CH_3\text{-}CH_2\text{-}C(CH_3){=}CH\text{-}CH_3 + (CH_3)_2C^{\oplus}\text{-}CH_3 \xrightarrow{-H^{\oplus}} (CH_3)_2C{=}CH_2$$

3-methyl-2-pentene

isobutylene

[3.07]

H⊕ 178

192 → H⊕ see above

dipentene 61

terpinolene 62

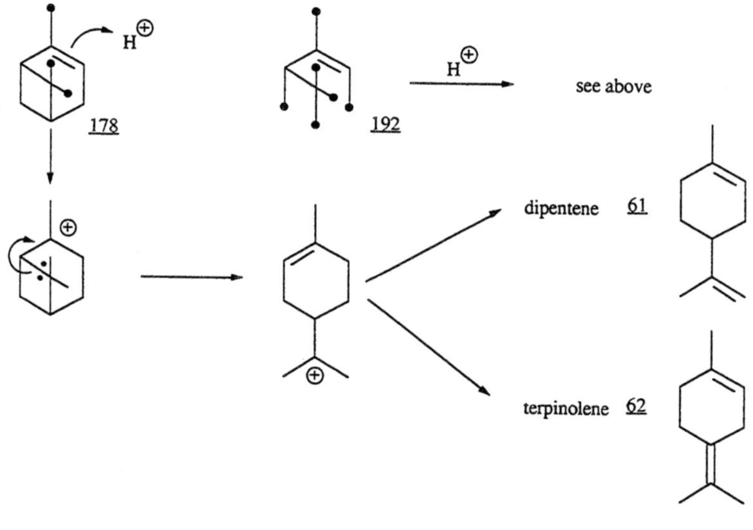

rearrangements are observed which give access to derivatives of other series: the camphanes and fenchanes. Reactions of this type are, in fact, actually used in the production of borneols, fenchols, camphor, and fenchone.

B. Reactions involving rupture of the cyclobutyl ring accompanied by rearrangement

The most important application of this type of reaction is the industrial synthesis of camphor. The procedures used can be classed into two groups.

First group: Involving a five-step synthesis

Dry HCl adds to α- and β-pinene with simultaneous isomerization to form bornyl chloride 193a and isobornyl chloride 193. This stereoselective formation of bornyl chloride constitutes the first step. The second step consists of eliminating the chloride using a variety of basic reagents such as the alkaline or lead phenate salts, aniline, etc., with formation of camphene 194 along with a small quantity of tricyclene 188. In the third step, camphene combines with organic acids almost quantitatively and stereoselectively to form the corresponding esters of isoborneol. Catalysts such as 50% sulfuric acid or, better still, para-toluenesulfonic acid are used as in the procedure of Barbier and Grignard. The saponification of these esters constitutes the fourth step, giving a mixture of isoborneol 195 with a small amount of borneol 196. It is possible, however, before reaching this step, to carry out a purification in order to have directly usable isobornyl esters only. In the fifth step, the mixture of borneols is oxidized or dehydrogenated to finally obtain camphor 181. The following scheme summarizes these five steps [3.08].

It can be noted that each of the first three steps involves a Wagner-Meerwein rearrangement. The industrial importance of this process should now be evident.

Second group: A more direct method

The preceding strategy requires a long series of reactions. Although each is carried out with a yield of around 90%, the overall result is a mediocre yield ($0.9^5 \sim 0.6 \rightarrow$ 60%). A means of directly converting the pinenes into bornyl esters by treatment with organic acids has therefore been found. Although these acids generally give monocyclic derivatives, it is possible, by modifying the conditions and by using special organic acids like oxalic, ortho-chloro- benzoic or tetrachlorophthalic acids, to obtain the corresponding bornyl esters. Unfortunately, despite their apparent elegance, the various methods of this group give rather dissatisfactory yields and the final product has a cost price above that of the preceding methods.

Returning to the schematic representation of the reactions in the first group, it is possible to note that simultaneously with the Wagner-Meerwein rearrangement which involves migration of the $\sigma(C_1\text{-}C_6)$ electron pair, there is also, to a lesser degree, migration of the $\sigma(C_1\text{-}C_7)$ pair. This latter migration gives rise to fenchane derivatives. Two groups of products that subsequently need to be separated are therefore obtained. Any realization of these procedures consequently requires the use of powerful distillation columns (generally under reduced pressure). The following scheme shows how the fenchyl derivatives are formed [3.09].

Dehydration of the α- and β-fenchols (197 and 198, respectively) provides a complex mixture of fenchenes: α-fenchene 199, β-fenchene 201, γ-fenchene 202 and cyclofenchene 203. It can be seen that α-fenchene results from a Wagner-Meerwein rearrangement, whereas both β-fenchene and γ-fenchene arise from a 6–2 displacement of hydride followed by a Wagner-Meerwein rearrangement. As for cyclofenchene 203, it is formed by the 1,3-shift of a σC-H pair with elimination of a proton [3.10].

[3.08]

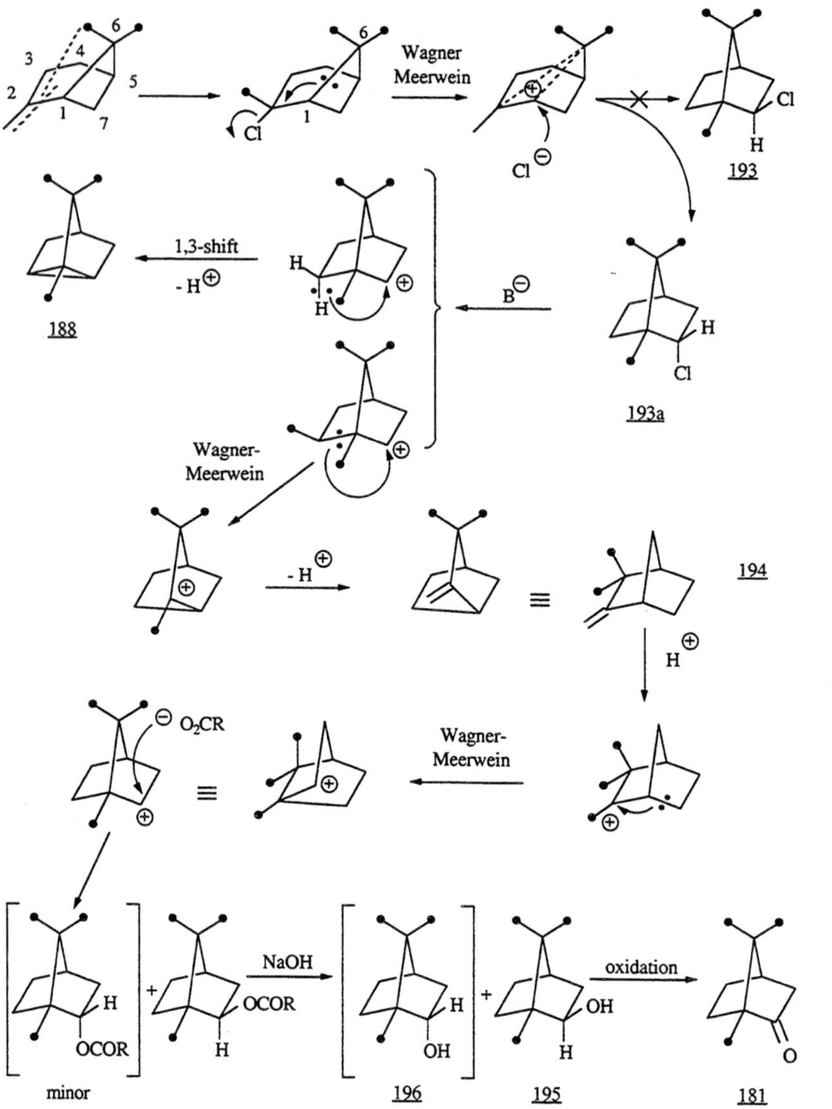

The mechanism of these reactions has been demonstrated beyond doubt by treating the O-deuterated fenchol (-O-D) with $DNaSO_4$.

With regard to fenchone, it is probably useful to discuss here the so-called Haller and Bauer reaction. This consists of treating a ketone with sodium amide in refluxing toluene to form the amide.

$$R\text{-}CO\text{-}CH_2R' + NaNH_2 \rightarrow RCONH_2 + R'\text{-}CH_3$$

[3.09]

[3.10]

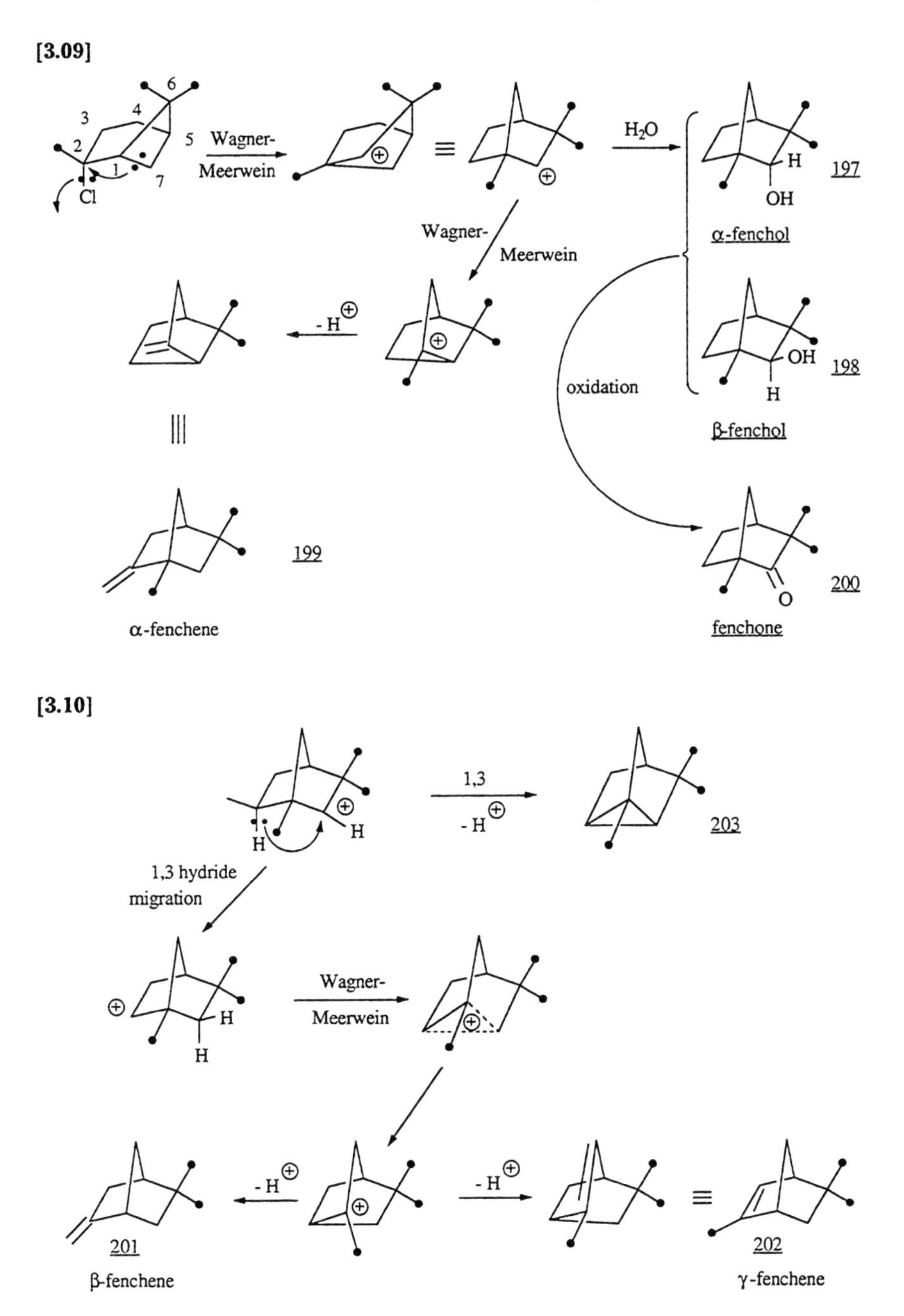

The amide can then be hydrolyzed to the corresponding acid. When applied to fenchone, this reaction should in theory afford two separate products; however, in practice only a single product is obtained. This result can be explained by examining the structures of the transition states corresponding to the two possible modes of cleavage [3.11].

[3.11]

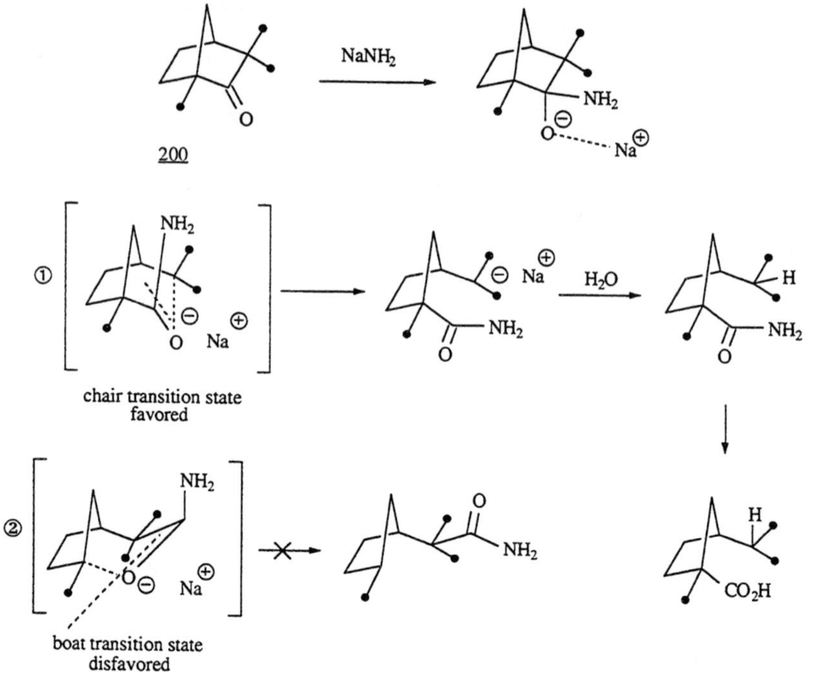

chair transition state favored

boat transition state disfavored

Finally, it is possible to cite a series of reactions having purely academic interest. These permit the conversion of (–) camphor obtained from (–) pinene into (+) camphor [3.12].

Pyrolysis Reactions of the Pinenes

A. Introduction

Although the use of acid-catalyzed reactions goes back more than a century, it was only from 1945 onwards that the pyrolysis reactions of the pinenes were developed.

It should be noted from the outset that in contrast with the acid-catalyzed reactions where both pinenes pass through the same carbocation, pyrolysis of α-pinene on one hand and β-pinene on the other does not lead to the same products.

Thermolyis of these hydrocarbons will obviously only give rise to other hydrocarbons. Among these are limonene 61*, dipentene or (±) limonene 61, 1-myrcene and the 2,3-ocimenes and the allo-ocimenes 6–9.

These results are interesting in that some of the hydrocarbons thus obtained can serve as starting materials giving access to a large number of compounds essential to the perfumery industry.

The reactions observed in the course of pyrolysis are shown here [3.13].

[3.12]

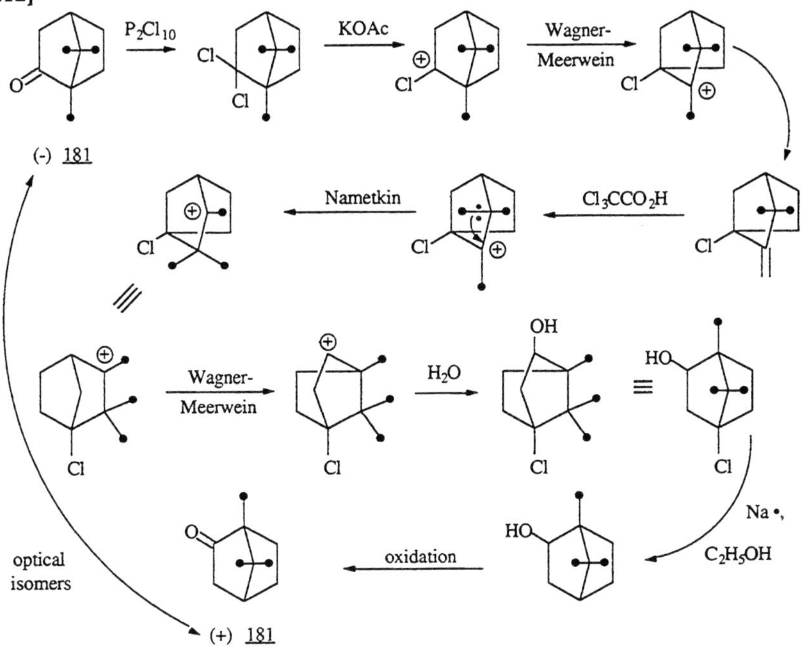

The reactions discussed here are no longer ionic but instead involve free radicals, and it is possible to see that each type of pinene gives different products.

In the case of α-pinene, the first homolytic bond scission occurs between 400° and 450°C and gives a biradical for which the two mesomeric reflect symmetrically across a plane. As a result, all of the products obtained from the pyrolysis of α-pinene are devoid of optical activity as seen in the formation of dipentene (+) 61.

Above 500°C, a double rupture occurs with intermediate formation of cis-ocimene 2. This substance has a life span at this temperature of a few millionths of a second and it isomerizes by [1,5] hydrogen migration to allo-ocimenes. It is nevertheless possible, with special apparatus, to "quench" the cis-ocimene and to avoid its isomerization to allo-ocimenes.

For β-pinene, the two mesomeric forms of the initially formed biradical cannot be reflected across a plane of symmetry. Consequently, the original asymmetry is conserved and the final products are optically active. This is the case for the formation of limonene 61* and of 1-(7),8(9)-para-menthadiene 204. Of all the various products obtained from the pyrolysis of β-pinene, however, myrcene 1 presents the biggest industrial interest.

Under the action of dry or aqueous HCl, myrcene gives a complex mixture of chlorides which can then be converted into the corresponding alcohols. In fact, it is necessary to distinguish between the action of dry and aqueous HCl. The action of aqueous HCl gives essentially linalyl chloride, however, with anhydrous HCl, one can distinguish between reactions which take place in the presence or in

[3.13]

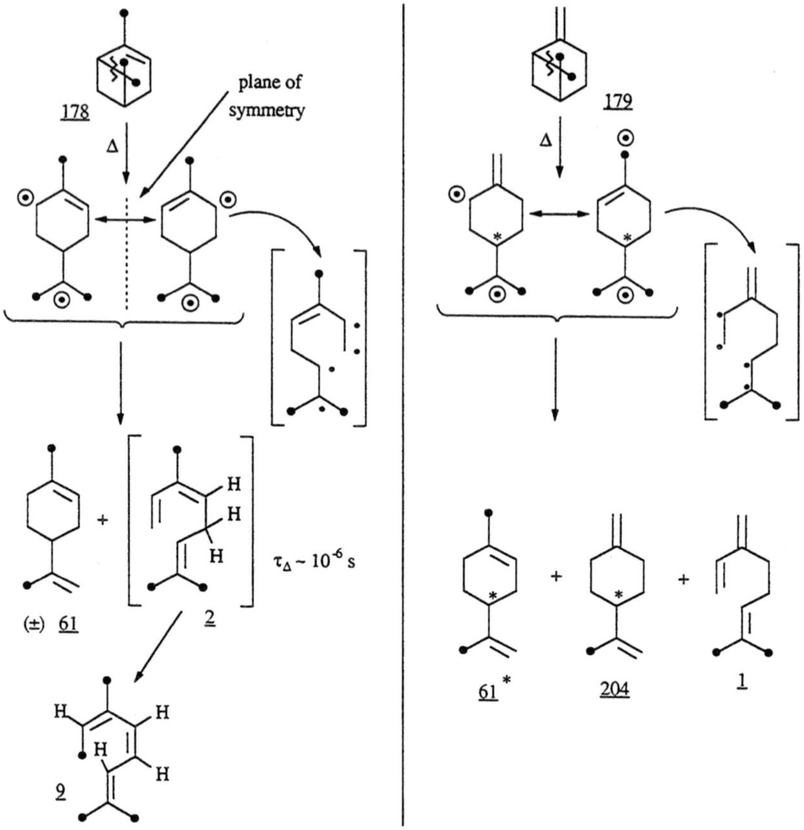

the absence of catalysts (of which cuprous chloride Cu_2Cl_2 is the most important). From the industrial point of view, the more interesting processes are the uncatalyzed reactions and these will be examined here in detail.

B. Myrcene as substrate

When myrcene is treated at 0°C and in the absence of catalyst with a stream of dry HCl, a mixture containing geranyl chloride 205, neryl chloride 206, myrcenyl chloride 207 and α-terpenyl chloride 208 is obtained [3.14].

This mixture forms another mixture of the corresponding alcohols when treated with an aqueous solution of sodium carbonate. When cuprous chloride is added to the reaction mixture, this process evolves completely differently, as shown in the following scheme [3.15].

Under these conditions, the equilibrium is greatly displaced in favor of the formation of the two isomeric chlorides: geranyl 205 and neryl 206. Less than 1% myrcenyl chloride 207 is formed. The chlorides obtained can be transformed into

[3.14]

HCl gas, dry, 0°C

1 → 206 + 205 (45%) + 207 (25%) + 208 (30%)

[3.15]

does not participate in the reaction

R = CH_2

the acetates by treatment with sodium acetate. Two distinct mechanisms may be operating in this substitution reaction.

a) *A monomolecular solvolytic reaction* (reaction with the solvent) also catalyzed by Cu_2Cl_2. The end product is either linalyl acetate 209 or linalool 13, depending on whether acetic acid or water is used as solvent.
b) *A bimolecular SN_2 reaction.* In this case, the acetates of nerol 210 and geranyl 211 are formed exclusively [3.16].

It is evident that the economic viability of these processes depends on the use of

[3.16]

S_N1' HOAc - NaOAc or $NaHCO_3$

Cl

S_N2 NaOAc / Et_3N

OR

R = H 13
R = CH_3CO 209

H
OAc
210

OAc
+
211

corresponding alcohols nerol and geraniol ← saponification or hydrolysis

highly efficient distillation equipment. This is true both for the purification of myrcene used as a starting material and for the separation of the alcohols and esters which are formed. It is possible to convert the geraniol-nerol mixture into citronellol, citral and citronellal; however, these processes would not be economical. Furthermore, in order to obtain these three compounds and their derivatives, it is necessary to have access to pure β-pinene and this necessitates carrying out an extremely sophisticated fractionation of turpentine oil with α-pinene as a by-product. Indeed, a good deal of research has been incited into ways of isomerizing α-pinene into β-pinene. At first, access to citronellal and citronellal was dependent on this approach, but, as will be shown later, this problem no longer exists for citral.

C. Generation of derivatives of citronellol and citronellal

Hydrogenation of α- and β-pinenes leads to cis-pinane, which when pyrolyzed gives dihydromyrcene 212 along with cyclopentene derivatives 212'. The conditions for this pyrolysis must be rigorously controlled in order to diminish the proportion of these 1,2-dimethyl–3-isopropenylcyclopentanes which are difficult to eliminate when formed as impurities. This pyrolysis occurs with retention of optical activity and leads ultimately to different optically active derivatives [3.17].

As shown earlier, dihydromyrcene 212 can be treated in three different ways. It is known that diisobutylaluminum hydride (Dibal) attaches itself to vinyl groups at temperatures betwen 100° and 120°C to give a mixed trialkylaluminum.

$$\text{-CH=CH}_2 + \text{H-Al(iBu)}_2 \xrightarrow{\Delta} \text{-CH}_2\text{-CH}_2\text{-Al(iBu)}_2$$

This aluminum compound can be directly oxidized by atmospheric oxygen in mixed alcoholate solutions to ultimately convert into a primary alcohol. The

[3.17]

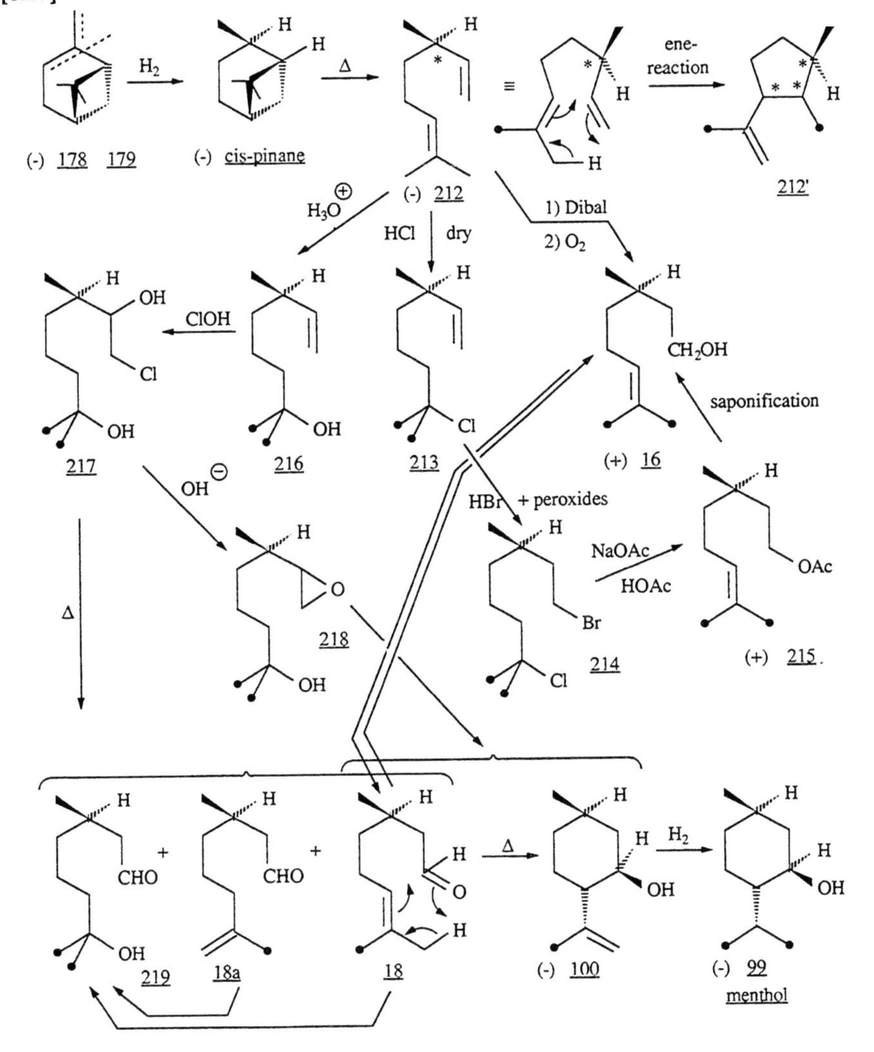

overall result is "formally" equivalent to an "anti-Markownikoff" hydration of the vinyl group.

$$\text{-CH-CH}_2\text{-Al(iBu)}_2 + 1.5\,O_2 \rightarrow \text{-CH}_2\text{-CH}_2\text{-O-Al(iBu)}_2 \xrightarrow{H_3O^+} \text{-CH}_2\text{-CH}_2\text{-OH}$$

This reaction (initially developed by Ziegler), when applied to dihydromyrcene 212, gives (+) citronellol 16. This alcohol can be dehydrogenated in the gas phase (in the presence of finely powdered copper, for example) to give (+) citronellal 18.

When treated with dry HCl, dihydromyrcene is converted into dihydromyrcenyl chloride 213 which can be treated with dry hydrobromic acid in the presence of peroxides (Karasch effect) to form 3,7-dimethyl 1-bromo–7-chlorooc-

[3.18]

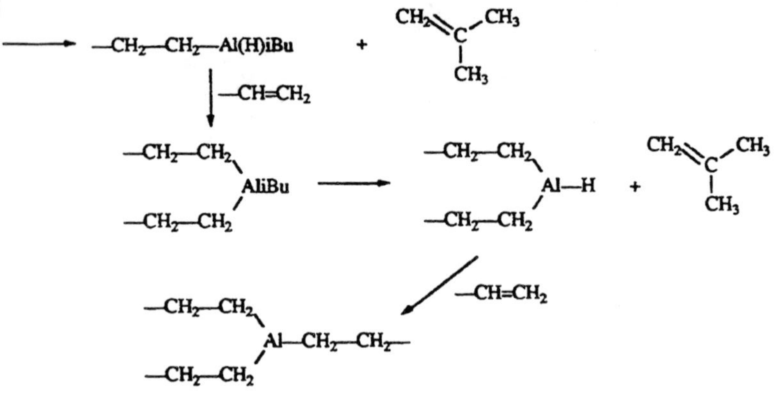

tane 214. On further treatment with NaOAc-HOAc, this compound leads to (+) citronellyl acetate 215. In the presence of mineral acids, dihydromyrcene furnishes dihydromyrcenol 216. If treated with calcium hypochlorite in the presence of CO_2, dihydromyrcenol gives the hydroxychlorohydrin 217 which can be transformed into the hydroxy-epoxide 218 in the presence of base.

Pyrolysis of the hydroxychlorhydrin 217 results in a mixture of three products: α-citronellal 18a, β-citronellal 18 and hydroxycitronellal 219. In acidic media, the hydroxy-epoxide 218 leads to two major products: β-citronellal 18 and (–) isopulegol 100 which can be hydrogenated to give (–) menthol 99 [3.18].

Hydroxycitronellal 219 is an extremely important raw material for the perfumery industry. It has an agreeable floral, lily-of-the-valley odor, but is not a natural product. The traditional procedure used for its production involves prior protection of the aldehyde function, followed by hydration of the residual double bond using 50% H_2SO_4 at around 0°C [3.19].

[3.19]

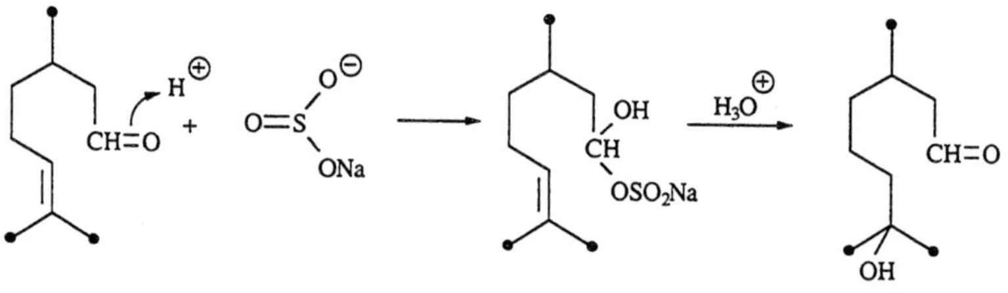

The aldehyde function is usually masked by using sodium bisulfite and the derivative thus formed can be hydrated to give the bisulfite derivative of hydroxy-dihydrocitronellal (correct name for hydroxycitronellal). Treatment with Na_2CO_3 regenerates the target product, hydroxy-aldehyde. The interest presented by the conversion of citronellal into isopulegol in the synthesis of (–) menthol has already been discussed.

In summary, it is possible to conclude that α-pinene and β-pinene, through the mediation of dihydromyrcene, open a way of access to citronellol and its esters, to citronellal and hence to hydroxycitronellal and to the menthols via isopulegol.

At the technical level, the development of these processes has required the use of extremely powerful distillation equipment. J. P. Bain of the Glidden Company has been one of the principal originators of these processes.

Hydroperoxidation Reactions

A. Introduction

Although both α- and β-pinene are capable of reacting with oxygen, the most interesting industrial products are obtained from α-pinene. Therefore, the discussion here will be limited to the hydroperoxidation of α-pinene and (–) α-pinene in particular. These reactions are caused either by the action of air or by the action of oxygen with or without catalysts.

First, it can be recalled that the autoxidation of cumene has been used on a grand scale for the production of phenol and acetone [3.20].

[3.20]

H $\xrightarrow{O_2}$ O—OH $\xrightarrow{\Delta}$ OH + CH_3—C(=O)—CH_3

The oxygen molecule can be considered as constituting a biradical:

$$\cdot\,\overline{\underline{O}}-\overline{\underline{O}}\,\cdot$$

and reacts as if it was a biradical; this is the phenomenon of autoxidation, well known since the celebrated work of Moureu and Dufraisse. The most vulnerable sites of attack correspond to allylic C-H bonds with the following free radical mechanism involving chain reactions [3.21].

The final chain reaction product is a hydroperoxide.

B. Hydroperoxidation of α-pinene

1. Products obtained

The preceding sequence applied to α-pinene leads to the formation of primary reaction products, verbenyl hydroperoxide 220 in addition to a small quantity of myrtenyl hydroperoxide 221 [3.22].

The oxidation reaction is carried out at a relatively elevated temperature (roughly 70°–80°C) to the extent that a part of the hydroperoxides decompose as they are formed. This approach permits the maintenance of a stationary concentration of hydroperoxides and thus limits the risk of any violent decomposition.

[3.21]

R—CH=CH—CH(H)$_2$ + •O–O• ⟶ [R—CH=CH—ĊH$_2$ ⟷ R—ĊH—CH=CH$_2$] + HO–O•

R—CH=CH—ĊH$_2$ + •O–O• ⟶ R—CH=CH—CH$_2$—O–O•

R—CH=CH—CH$_2$—O–O• + H$_2$C(H)—CH=CH—R ⟶ [H$_2$Ċ—CH=CH—R] + R—CH=CH—CH$_2$—O–OH

[3.22]

O–OH

220

CH$_2$–O–OH

221

O–OH

221a

R–CH=CH–CH$_2$–O–OH —Δ⟶ R–CH=CH–CH$_2$–O• + •OH

Hydroperoxides formed in this way function as epoxidizing agents with the result that (–) α-pinene is partially transformed into trans-pinane oxide 222.

As soon as this reaction is terminated, the contents of the reactor are cooled and the products of hydroperoxidation are reduced using sodium sulfite or, more simply, by boiling in the presence of sodium hydroxide. A complex mixture of alcohols is thereby obtained as well as a small quantity of trans-pinane oxide 222, verbenone 223 and myrtenal 224.

When the reducing agent is $NaHSO_3$, the alcohols are made up of (–) trans verbenol 225 along with traces of (+) cis verbenol 226, (+) cis–1-pinene–2-ol 227 and small quantities of myrtenol 228.

When, however, the reduction is effected with aqueous sodium hydroxide, no pinenol is formed. This substance, in fact, isomerizes into the verbenols under the action of the boiling water. In this way, a very high yield of verbenol is obtained and can be purified either by distillation or by the action of boric acid.

After purification, (–) trans verbenol 225 is submitted in turn to two types of reaction: an acid-catalyzed process and pyrolysis [3.23].

In fact, during the hydroperoxidation reaction, there is a very complex mixture of α-pinene, oxygen, the free radicals r^1, r^2, r^3, ·OH, ·H and the reaction products. The radicals r^2 and r^3 are likely to give coupling reactions (dimeriza-

[3.23]

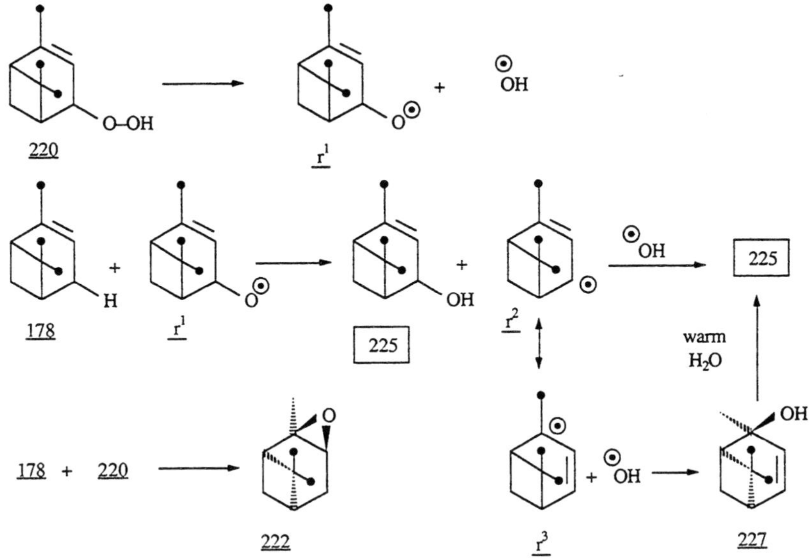

220 → r[1] + ·OH

178 + r[1] → 225 + r[2]; r[2] + ·OH → 225

r[2] ↕ r[3]; r[3] + ·OH → 227 → (warm H_2O) 225

178 + 220 → 222

tions) just as ·H and ·OH can associate to give rise to the traces of water that are always observed in the reaction mixture. In addition, the free radicals r[2] and r[3] can also react with α-pinene to form polymers which are also detectable [3.24].

[3.24]

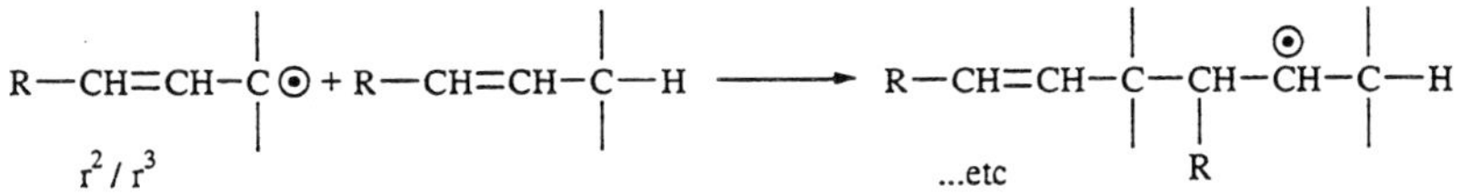

R—CH=CH—C· + R—CH=CH—C—H ⟶ R—CH=CH—C—CH—ĊH—C—H (R) ...etc

r^2 / r^3

As for the hydroperoxide 221, it gives myrtenol 228 [3.25].

[3.25]

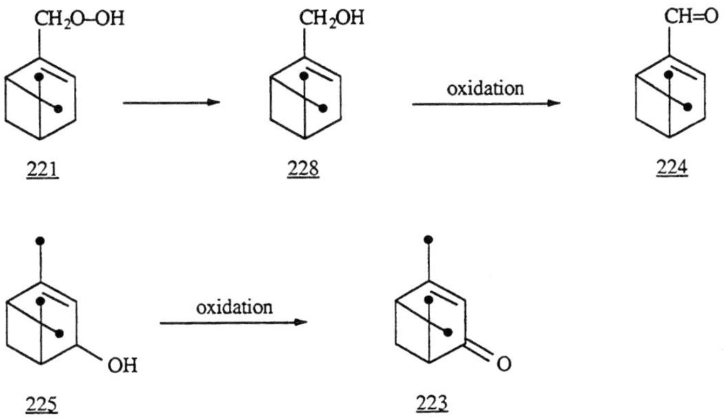

221 (CH_2O–OH) → 228 (CH_2OH) —oxidation→ 224 (CH=O)

225 (OH) —oxidation→ 223 (O)

In a particularly oxidizing reaction medium, verbenol and myrtenol are partially oxidized into verbenone 223 and myrtenal 224.

To make this process more worthwhile economically, it is necessary to recuperate the by-products. For this reason, trans-pinane oxide 222, in itself rather uninteresting, can be isolated and transformed into carvone 84 by the following sequence: it is converted by the action of hot water saturated with CO_2 into the 1,2-ethylenic diol 229 which is then oxidized into 8-hydroxy carvone 230. This is then dehydrated to give carvone 84 [3.26].

[3.26]

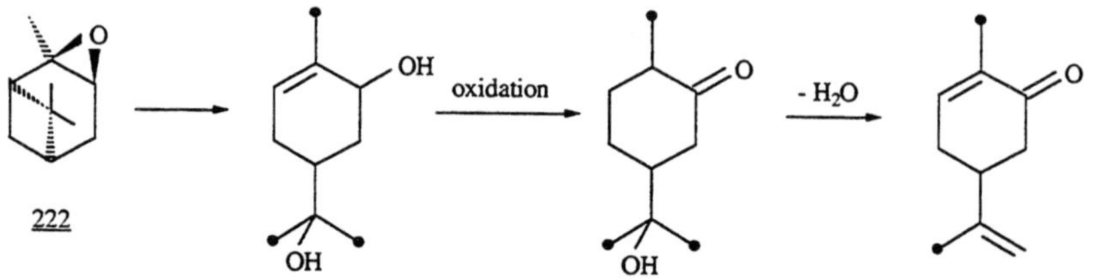

222 → 229 —oxidation→ 230 —- H_2O→ 84

An important product of the hydroperoxidation of α-pinene, verbenol 225, can be used as a starting point for further syntheses.

2. *Transformation of verbenol 225*

Acid-catalyzed reactions

The following scheme summarizes the different possibilities offered by this approach [3.27].

When treated in acidic media, (–) verbenol 225 immediately gives the cation 231 which then undergoes cleavage of the cyclobutyl ring to form the doubly charged ion 231a by a mechanism already discussed [3.28].

This species can evolve in a number of different ways to generate limonenol 232, isopiperitol 233, and the corresponding glycols 234 and 235. These two glycols, which are formed in minor proportions, can be dehydrated and converted into the two preceding alcohols 232 and 233. If these are hydrogenated in the presence of a nickel catalyst, a mixture of menthols is obtained in which (–) menthol 99 dominates. In this way, a new reaction sequence leading to menthol is available.

Pyrolysis of (–) verbenol

As in the case of the pinenes, the pyrolysis of verbenol proceeds by a radical mechanism via a biradical for which the two mesomeric forms 236 and 237 are not symmetrical. Each of these evolves into different products—respectively, isopiperitol 233 and limonenol 232; however, they are not stable under the conditions of pyrolysis. Both undergo ring-opening reactions giving, respectively, cit-

[3.27]

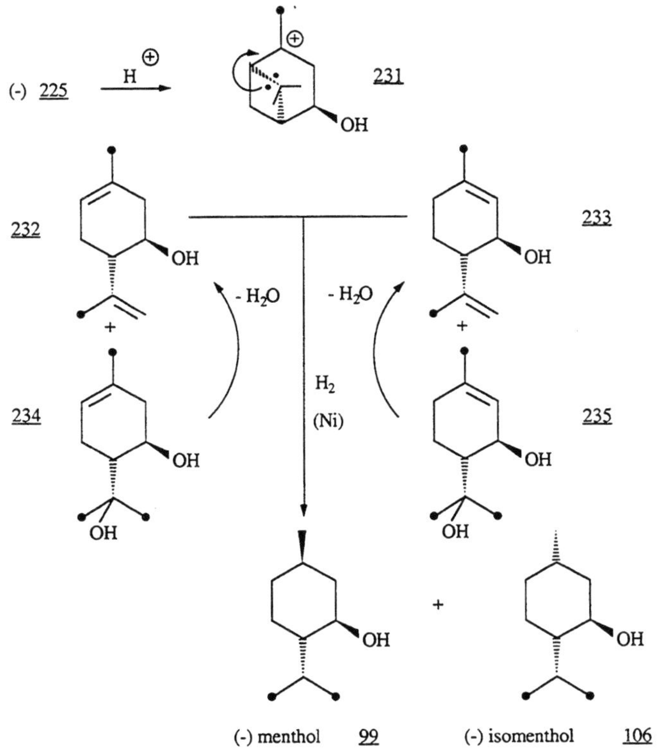

(-) 225 → ($H^{\oplus}$) 231

232 + 234 (- H_2O) ← 231 → (- H_2O) 233 + 235

H_2 (Ni)

(-) menthol 99 + (-) isomenthol 106

[3.28]

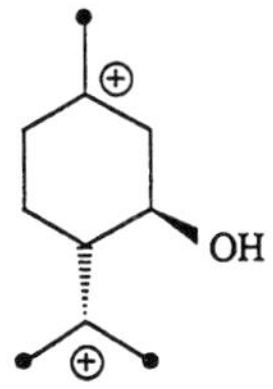

231a

ral 17 and isocitral 17a. These two compounds react with acetone in the presence of base to yield pseudoionone 22. This substance can be cyclized in acidic media to obtain α- and β-ionones. Replacement of acetone with methylethylketone leads to the methylionones 26, 27, 28, and 29 [3.29].

If the conditions of pyrolysis are modified and strictly controlled, verbenol undergoes another type of ring-opening process which gives principally pseudocyclocitral 238. Application of the Wolff-Kischner reaction to this substance leads to 1,1,3,6-tetramethyl–2-cyclohexene 239 and this unsaturated compound undergoes a Prins reaction to give a mixture of the tetramethylcyclohexenylcarbinols 240 which can in turn be oxidized to formyltetramethylcyclohexenes 241. The isomeric aldehydes condense with acetone in alkaline media to give a mixture of the α-, β- and γ-irones 242. These ketones, which are known to be the

[3.29]

225 OH Δ 236 OH 237 OH 233 O H 232 O H 17 CH=O Na_2CO_3 CH=O 17a $^{\ominus}$OH CH_3COCH_3 22 O CH_3 O CH_3 + O CH_3 24 25

characteristic constituents of the essential oil of iris, will be discussed later [3.30].

Note: Another synthesis of the irones from α-pinene can be mentioned. It consists of the following sequence [3.31].

The ozonolysis of α-pinene 178 to obtain the keto-aldehyde 243, is a well known process. On heating in the presence of palladium, this keto-aldehyde subse-

[3.30]

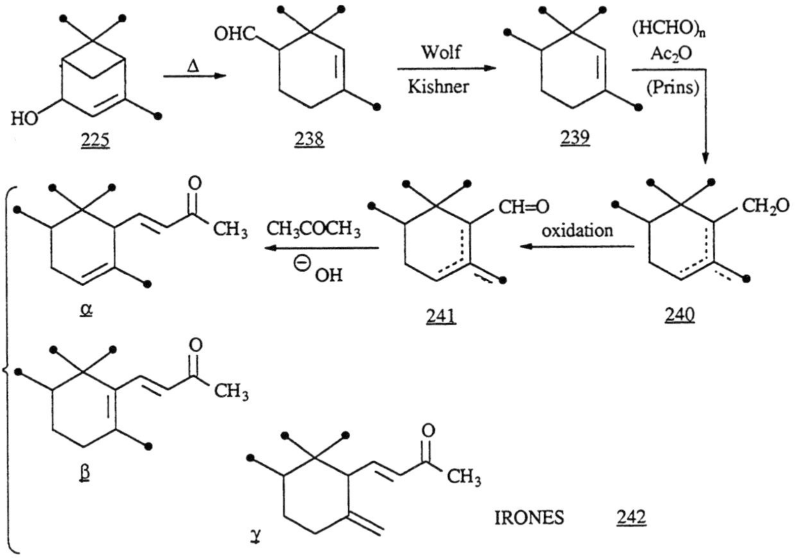

[3.31]

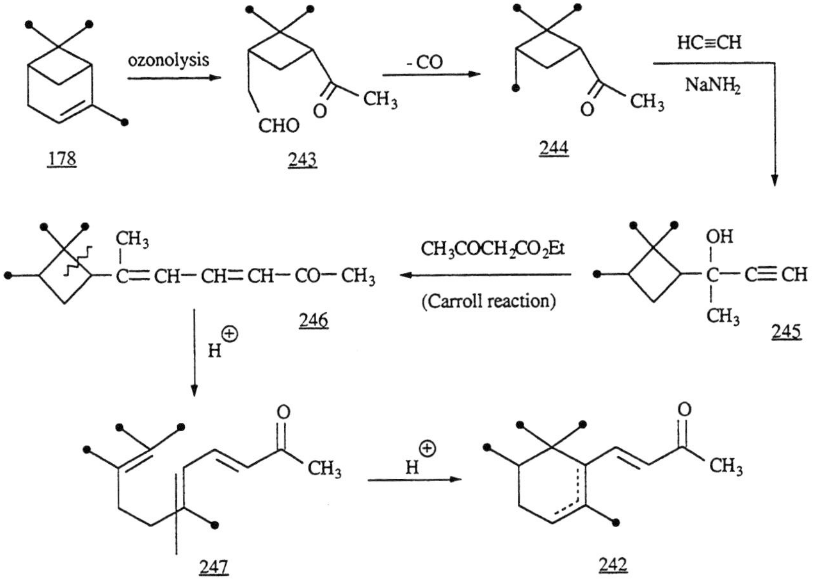

quently can decarbonylate to give the tetramethylacetylcyclobutane 244. In the presence of acetylene and sodium amide, this compound gives a tertiary acetylenic alcohol 245 which can be transformed into a cyclobutyl dienone 246 by reaction with ethyl acetoacetate (generalized Caroll reaction, discussed in the appendices).

The dienone 246 is transformed in acidic media into a mixture of pseudoirones

247 which can be cyclized in concentrated sulfuric or phosphoric acids to form a mixture of α- and β-irones.

The carboxylic acid corresponding to the keto-aldehyde 243 can be obtained in high yield by another route using a "crown ether." The "crown ethers" as well as other macrocyclic polyethers having 16–20 units can complex with alkali metal cations. The associated anion, considered "naked" because there is no specific solvation by the cation, possesses full reactivity and can be dissolved. Thus, $KMnO_4$, in the presence of crown ether A, capable of "crypting" the potassium ion, becomes soluble in benzene (to around 0.1 M). The resulting solution (purple benzene) shows a remarkable oxidizing power. The reaction with α-pinene takes place under homogeneous conditions and gives the corresponding cyclobutyl derivative [3.32].

[3.32]

$K^{\oplus}$ $MnO_4^{\ominus}$

A

(18 units)

(-) 178 —$KMnO_4$ / A / benzene→ [CO_2H] ≡ [CO_2H]

C. Optically active linalools via hydroperoxidation of the pinanes

A synthesis of inactive linalool using myrcene obtained by pyrolysis of pinene has been described earlier.

It will be shown that the pinenes can also be used in the synthesis of optically active linalool. cis-Pinane 248 is obtained from the hydrogenation of α-pinene and, in contact with air, this product gives the hydroperoxide 249. When treated with sodium hydroxide in solution or with sodium methylate, cis–2-pinanol 250 is obtained and this is transformed into linalool 13 with a yield near to 95% on pyrolysis at around 600°C. The following scheme summarizes the different possibilities [3.33].

It should be noted that the hydroperoxide of pinane 249 is one of the best catalysts known for the polymerization of "GR-S" rubber.

[3.33]

(-) α-pinene 178

H_2, catalyst

(-) cis-pinane 248

1) O_2
2) NaOH

(2S)(-) cis–2-pinanol 250

Δ

(3S)(+) linalool 13

(2S)(+) trans–2-pinanol

(+) α-pinene 178

H_2, catalyst

(+) cis-pinane 248

1) O_2
2) NaOH

(2R)(+) cis–2-pinanol 250

Δ

(3R)(-) linalool 13

(2R)(-) trans–2-pinanol

O–OH

249

yield = 95%

yield = 75%

CHAPTER

4

Total Synthesis of Aliphatic Terpenoids of Industrial Importance

Introduction

At the end of World War II, the market for natural perfumery raw materials was largely anarchic giving free reign to many speculative practices. In any case, wide fluctuations in the value of these raw materials have always existed for climatic, meteorological, and various other reasons.

This is why, from the years 1948–1950 onwards, numerous researchers have been trying to find strategies of economic synthesis that would liberate the industry from its dependance on natural products and that would provide thresholds which permit speculation to be controlled.

The previously discussed syntheses using the α- and β-pinenes were developed with this in mind. They certainly took account of the enormous world production of turpentine oil and "pine oil" which was of the order of 300 000 metric tons and which represented alone more than 80% of the total worldwide production of all essential oils (see Table I).

The principal substrates indispensible to the perfumery industry are:

- linalool drawn from the oil of Bois de rose produced principally in Brazil and "Shiu" or Ho oil from Taiwan;
- citral extracted from Lemongrass oil from the East Indies and from Guatemala. A more recently exploited source, *Litsea cubeba* or "May-Chang" of Chinese origin, made its appearance on the world market around 1950, although its production is limited to some 50 metric tons per year;
- geraniol and nerol obtained from the oils of citronella of Asian (Java, China) or South American (Guatemala) origin or in lesser amounts from palmarosa oil (Indies);
- citronellal also furnished by citronella oil or from the oil of *Eucalyptus citriodora* of Australian origin;
- menthol coming principally from cornmint (*Mentha arvensis*) of Chinese, Japanese or Brazilian origin.

In addition, other criteria have motivated the research of the Société Roure Bertrand Dupont. A number of essential oils which have a fundamental impor-

Table I

Principal essential oils	metric tons/year
Lemon	2,300
Citronella	2,300
Camphor oil	250
Dementholated *Mentha arvensis*	2,100
Mentha piperita	2,200
Lemongrass	310
Bois de rose	160
Spearmint	1,400
Lavandin	750
Lavender	200
Patchouyl	500
Vetyver	260
Sandalwood	70
Orange	12,000

Figures in 1984: Brian M. Lawrence, *Perfumer & Flavorist*, 10, October–November 1985, pp 2–16.

tance for the perfumery industry—such as the oils of lavender, lavandin, spike lavender, bergamot, petitgrain and clary sage—are principally composed of a mixture of linalool and its acetate. In fact, the total of these two constituents often exceeds 80% of the essential oil. In order to master the artificial reconstitution of these oils, it was necessary at all costs to have access to these two products.

At the same period and for the same economic reasons, an analogous problem was encountered by the Swiss company, Hoffmann-La Rôche (large producer of vitamins A, E and K) with regard to citral, a raw material used in the synthesis of vitamin A.

The total value of essential oils produced around 1960 exceeded $250 million. In addition to the essential oils mentioned here, it can be noted that:

- production of the essential oil of sclary sage could exceed 100 metric tons, but with a large contribution from production in Russia, accurate statistics are difficult to obtain;
- the annual production of the oils of geranium would be of the same order of greatness;
- the oils of bergamot, petitgrain, orange, lemon are larger still.

General scheme of these syntheses

The syntheses developed in Grasse at the Research Center of Roure Bertrand Dupont, at Hoffmann-La Rôche and later at BASF and Rhône-Poulenc, all use acetone and acetylene. For this reason, they are commonly called the "acetylenic syntheses of the terpenoids." The various possibilities are summarized in Table II.

It is evident from Table II that the "key" step in all of these syntheses is the preparation of 2-methyl–2-hepten–6-one 253 or methylheptenone, with the whole network of reactions hinging on dehydrolinalool 254. The general scheme also shows that acetylene is used straight away in the first step, in the preparation of 2-

Table II

$(H_3C)_2C{=}O$ + HC≡CH ⟶ $(H_3C)_2C(OH){-}C{\equiv}CH$ 251 ⟶ $(H_3C)_2C(OH){-}C{=}CH_2$ 252

252 + $H_3C{-}CO{-}CH_2{-}CO_2C_2H_5$ ⟶ 253 ⟵ other types of synthesis

253 + HC≡CH ⟶ 254 ⟶ linalool 13 ⟶ vitamins E and K

254 ⟶ pseudoionones ⟶ α-ionone (perfumery)

pseudoionones 22 ⟶ β-ionone 25 ⟶ vitamin A

254 ⟶ citral 17 ⟶ citronellal 18 ⟶ citronellol 16

citral 17 ⟶ geraniol / nerol 14 15

citronellal 18 ⟶ menthols 99

citronellal 18 ⟶ hydroxycitronellal 219

methyl–3-butyn–2-ol 251, then again in the production of dehydrolinalool 254 from methylheptenone. This condensation of acetylene on carbonyl derivatives or the "ethynylation reaction" is of major importance and deserves detailed discussion.

Subsequent to this reaction, semi-hydrogenation intervenes with 2-methyl–3-butyn–2-ol being converted into 2-methyl–3-buten–2-ol while dehydrolinalool is converted into linalool. In general terms, these reactions permit the selective reduction of triple bonds to double bonds with stereochemical control. These reactions will also be discussed. The different methods of synthesizing methylheptenone will also be described along with the various reactions based on the use of dehydrolinalool.

Ethynylation Reactions

In 1900, Favorskii reported that aldehydes and ketones condensed with acetylene in the presence of a large excess of pulverized potassium hydroxide. Much later, Reppe introduced the term "ethynylation" (*Aethinierung*) to designate the addition of acetylene to the cabonyl group of aldehydes and ketones as well as to vinylamines (imines). This type of reaction can be represented as follows [4.01].

[4.01]

$$R{-}C{\equiv}C{-}Y \; + \; {>}C{=}Z \longrightarrow R{-}C{\equiv}C{-}\overset{|}{\underset{|}{C}}{-}Z{-}Y$$

If R = H → ethynylation; R ≠ H → alkynylation

Y = H or metal
Z = O or = NH

The ethynylation reactions can be:

- stoichiometrical
- semi-catalyzed
- catalyzed.

The most important industrial procedures use acetylene and potassium hydroxide, acetylene and alkali metals, alkaline earths or copper. In the laboratory, magnesium acetylenes are also used, particularly mono-magnesium acetylide which can be prepared in tetrahydrofuran. As a general rule, the role of the solvent is of fundamental importance.

A. Stoichiometric ethynylation

Since the first work of Favorskii, the main modifications made to this method have been localized on the choice of solvent such as ethers, acetals or amines. The necessity for using enormous excesses of potassium hydroxide raises a fundamental technical problem. The product most frequently used is industrial potassium hydroxide which titrates to the equivalent of 85–87% KOH. The remainder is constituted by H_2O and traces of K_2CO_3. This necessitates studying the equilibria of the binary liquid-solid system H_2O-KOH which was more or less well known when research workers in Grasse started their studies on this subject. It can be seen in Figure 1, showing the liquid-solid equilibria of the H_2O-KOH system, that industrial potassium hydroxide corresponds to eutectic KOH, H_2O-KOH, for which the melting point is nearly 100°C. In the course of ethynylation, only anhydrous potassium hydroxide intervenes and the reaction produces a molecule of water, which must be fixed by a second molar equivalent of the eutectic [4.02].

[4.02]

$$R^1R^2C{=}O + HC{\equiv}CH + KOH \rightleftharpoons R^1R^2C(O^{\ominus}K^{\oplus})C{\equiv}CH + H_2O \;(\rightarrow KOH)$$

Stoichiometrically, it is necessary to bring into action two "molar equivalents" of the eutectic for each mole of the carbonyl derivative, that is, four mole equivalents of KOH [4.03].

[4.03]

$$R^1R^2C{=}O + 2[KOH, H_2O\text{-}KOH] + HC{\equiv}CH \longrightarrow R^1R^2C(O^{\ominus}K^{\oplus})C{\equiv}CH + 3(KOH - H_2O)$$

Although a minimum of four moles of potassium hydroxide is necessary, six are

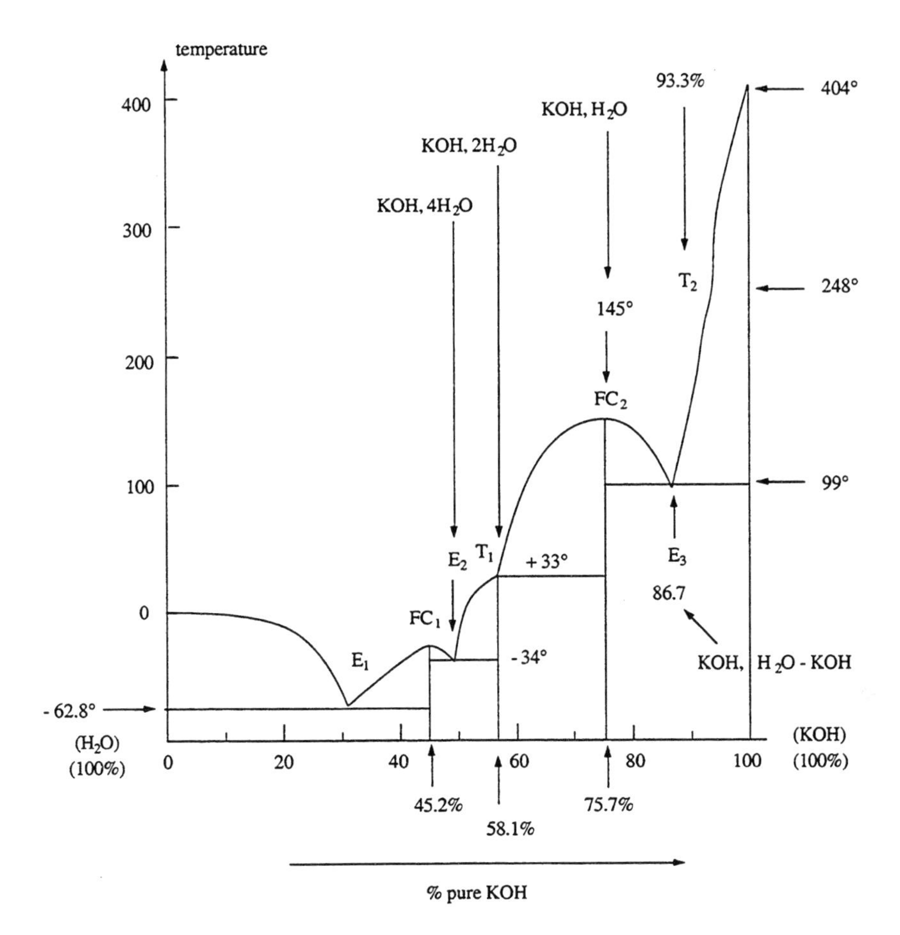

Figure 1. Equilibrium of the binary system H²O-KOH

generally used (three mole equivalents of the eutectic) and these are dispersed by heating in a convenient solvent (which can then be removed and replaced by another solvent). The suspension is saturated on cooling with acetylene and, while continuing this saturation process, the carbonyl derivative is slowly added. At the end of the reaction, the product is hydrolyzed and the acetylenic alcohol is extracted and then isolated by distillation. Methylbutynol and dehydrolinalool are thus obtained from acetone and methylheptenone respectively with yields of over 90%.

The reaction can also be carried out under moderate pressure ($\leq$ 10 bars). This does not modify the yields, but the reaction rates are increased. In addition, the small amount of diol that is formed when the reaction is carried out at normal pressure becomes totally inexistant when the reaction operates under pressure. This diol arises from the condensation of a second molecule of carbonyl derivative with the acetylenic alcohol just formed [4.04].

The main inconvenience of this process is the consumption of a large excess of

[4.04]

$$R^1R^2C(OH)-C\equiv CH + R^1R^2C=O \xrightleftharpoons{KOH} R^1R^2C(OH)-C\equiv C-C(OH)R^1R^2$$

potassium hydroxide. This is the origin of research into other more economic conditions of ethynylation.

It is possible to use quasi-stoichiometric quantities of sodium amide $NaNH_2$, but this reagent is costly.

B. Semi-catalytic ethynylation

1. By potassium

A better approach would be to find an economical system in which a base could function semi-catalytically so that a molar ratio (acetylenic alcohol to base) greater than 1 could be obtained. The use of polar organic solvents and specially designed reactors which allow liquid NH_3 to be used constitute some of the possible solutions that have been explored.

Blumenthal demonstrated that it is possible to use semi-catalytic quantities of potassium hydroxide in solvents such as the sulfoxides and ethylenediamine. For each liter of solvent, 0.38–0.5 moles of KOH and 5.0–6.25 moles of the carbonyl substrate are used. If the reaction is carried out at a temperature below 40°C, the ethynylcarbinol is the principal product, whereas between 40° and 70°C, the acetylenic diol becomes preponderant. Although sodium hydroxide is completely inactive in the Favorskii procedure, it can be used in the Blumenthal system.

Under the preceding conditions, when the temperature is below 40°C, the yield of conversion of acetone into methylbutynol reaches 71%. When potassium hydroxide titrating to 90% KOH is used, the yield relative to potassium hydroxide is 647%.

With sodium hydroxide titrating to 95%, the yield of conversion is still 47% relative to acetone and 470% relative to sodium hydroxide.

In his studies, Tedeschi used liquid ammonia as solvant. NH_3 boils at –33°C, hence its easy separation once the reaction is complete. The solubility of acetylene in NH_3 is equimolar and remains practically constant at between –40° and +30°C. This phenomenon is due to the formation of a complex by hydrogen bonding between NH_3 and $HC\equiv CH$ which can be written as $HC\equiv C\text{-}H \leftarrow NH_3$.

Examination of Table III shows that there is no advantage in charging 500 ml of liquid NH_3 with more than 18 moles of acetone.

The mechanism of ethynylation of acetone by KOH in liquid NH_3 is relatively complicated, but can now be correctly analyzed. For a long time, it was thought that potassium acetylide was generated and this then reacted with acetone [4.05].

In fact, the reaction is completely different, at least in the semi-catalytic reactions undertaken in liquid NH_3. Serious arguments (which will not be dealt with

Table III. Moles Acetone/500*ml* liquid NH_3

Degree of cenversion = moles obtained/moles of acetone (or of KOH)			
Moles acetone/ 500*ml* liquid NH_3	To acetone	To KOH	g Carbinol/500*ml* liquid NH_3
1	95	95	80
12 ↗	82 ↘	660 ↗	830 ↗
18 ↗	75 ↘	902 ↗	1 138 ↗
24 ↗	52 ↘	835 ↘	1 050 ↘
12 NaOH	81	647	815

[4.05]

$$HC{\equiv}CH + KOH \rightleftharpoons KC{\equiv}CH + H_2O$$

$$H_2O + KOH \rightleftharpoons KOH \cdot H_2O$$

$$(CH_3)_2{-}C{=}O + KC{\equiv}CH \rightleftharpoons (CH_3)_2{-}C(O^{\ominus}K^{\oplus}){-}C{\equiv}CH$$

$$[\,H_3N \rightarrow HC{\equiv}C^{\ominus}\,],\ KOH$$

here) permit the conclusion that the real intermediate that intervenes in the ethynylation reactions carried out in donor solvents such as NH_3, is the complex $HC{\equiv}C\text{-}H \leftarrow NH_3$ which gives an ionic combination with KOH.

The mechanism can therefore be represented by the reactions shown below [4.06].

Reaction 3 is almost irreversible at 25°C and becomes completely so at temperatures over 60°C. Reactions 3, 4 and 5 constitute the catalytic cycle.

2. *By metals*

In this case, liquid NH_3 remains a much used solvent in which veritable metal acetylides (HC≡C-Metal) are formed.

An example of this process carried out with sodium is described here.

In 10 kg of anhydrous liquid ammonium, three portions of 5 g of ferric chloride $FeCl_3$ (rather violent reaction) are added. Following this, 500 g of finely cut sodium are introduced and this provokes an abundant emanation of hydrogen. Finally, a current of acetylene is introduced, thereby obtaining HC≡C Na in solution.

The acetylides of potassium, rubidium and cesium can be prepared in the same way. These (including sodium acetylide) are not solvated. The acetylides of lithium, calcium, strontium and barium, when prepared in the same manner, however, exist as ammoniates Metal-C≡C-H, x NH_3.

Solvated sodium and lithium acetylides are strongly ionized in liquid ammonia. They can be electrolyzed in this solvent with deposition of carbon at the anode and of the metal at the cathode. It is also possible to use other organic solvents for forming metallic acetylides. Table IV shows the reactions with acetylene of sodium in different solvents.

[4.06]

$$HC{\equiv}CH + NH_3 \rightleftharpoons H_3N \rightarrow HC{\equiv}CH$$

$$\rightleftharpoons NH_3 + H\cdots C{\equiv}C\text{-}H \leftarrow NH_3 \rightleftharpoons H_3N \rightarrow HC{\equiv}C^{\ominus}\ \overset{\oplus}{NH_4} \quad (1)$$

$$KOH + H_3N \rightarrow HC{\equiv}C^{\ominus} \rightleftharpoons [H_3N \rightarrow HC{\equiv}C^{\ominus}]\ KOH \quad (2)$$

$$t° > 60°C \quad \rightleftharpoons O{=}C(CH_3)_2 \quad t° = 25°C$$

$$\left[H_3N \rightarrow HC{\equiv}C{-}C(CH_3)_2O^{\ominus} \right] KOH \quad (3)$$

$$\rightleftharpoons + [H_3N \rightarrow HC{\equiv}C^{\ominus}]$$

$$[H_3N \rightarrow HC{\equiv}C^{\ominus}], KOH + (CH_3)_2C(O^{\ominus})C{\equiv}C\text{-}H \leftarrow NH_3 \quad (4)$$

$$\rightleftharpoons \overset{\oplus}{NH_4}$$

$$NH_3 + (CH_3)_2C(OH)C{\equiv}C\text{-}H \leftarrow NH_3 \quad (5)$$

The dispersions of acetylides formed in this way can of course be used for other reactions than ethynylation. For instance, methyl sulfate $SO_4(CH_3)_2$ reacts with sodium acetylide prepared in xylene to give propyne with a yield above 85%. With ethyl sulfate, 1-butyne ($HC{\equiv}C\text{-}C_2H_5$) is obtained with a similar yield. These terminal acetylenic hydrocarbons can in turn be "sodiated" and so on.

Ethynylation using sodium or lithium acetylides in NH_3 has been known for a long time. It can be applied just as well to saturated aldehydes, unsaturated

Table IV

Solvents	t(°C)	time (hours)	Yield of HC≡CNa
xylene	100	2.5	>99%
	120	2.75	71.5%
	130	3	7.3%
di n-butyl "carbitol"	65	2.5	>99%
	95	1	>99%
	130	3.3	59%
	150	1.5	49%
	175	1.25	31%
di n-butyl ether	80	2	>99%
	100	1	>99%
dioxane	65–70	2	>99%

aldehydes and ketones. When carried out at normal pressure, this reaction is not catalytic. Nonetheless, De Malde and collaborators (E.N.I. = Ente Naz. Idrocarburi) claimed in 1962 a catalytic intervention of sodium acetylide in liquid NH_3 under pressure for the ethynylation of acetone.

First, sodium or lithium acetylide is prepared in ammonia at –40°C and then acetylene is introduced at the same temperature. Temperature is then raised to 0°C while continuing the introduction of acetylene until a pressure of 6 bars is obtained. At this point, acetone is slowly introduced. The yield relative to acetone is good (71%) and 1 mole of catalyst furnishes 28 moles of methylbutynol.

Cesium gives even better results with the yield relative to acetone being quantitative and 1 mole of catalyst giving 79 moles of methylbutynol.

C. Catalytic ethynylation

At the industrial and commercial level, by far the most important use of this type of ethynylation is the production of propargyl alcohol and butyndiol by the reaction of acetylene and formaldehyde in the presence of cuprous acetylide.

A mixture of nitrogen, acetylene and aqueous formaldehyde is streamed in the gas phase over a catalyst based on cuprous acetylide at a temperature of around 100°C and under a pressure increasing from 5 to 50 bars. Both butyndiol and propargyl alcohol are obtained simultaneously with the former predominating.

This type of catalytic ethynylation was discovered by Reppe. In addition to demonstrating the fact that acetylene could be used under pressure even on a large scale, he created a reaction medium by which products of considerable economic importance could be obtained. Table V gives some of the most salient examples of this.

The polymers derived in this way have a very large utility:

- Polyamide–4 is an analogue of Nylon;
- Polyurethane is widely used as foam for the manufacture of filling materials for cushions, etc.;
- The polyvinylpyrrolidones were a spectacular success during the second World War II as a substitute for blood plasma;
- In the form of a co-polymer with vinyl acetate, vinylpyrrolidone possesses film-forming properties that have been much appreciated by cosmetologists, particularly in hair lacquers, dyes and tints, shaving creams, lipsticks, skin creams, toothpastes, etc. A detailed discussion of these products does not fall within the scope of this work, however.

In 1965, Nogaideli reported that methylbutynol could be prepared by streaming the vapors of acetone and acetylene over a solid catalyst carrying sodium hydroxide. The catalyst is prepared by mixing 25 g of clay and 75 g of molten sodium hydroxide to give spheres of 5 mm diameter that become activated after drying at 130°C for 3 hours.

The best results are obtained at 120–125°C with a yield of 20–25%. The

Table V

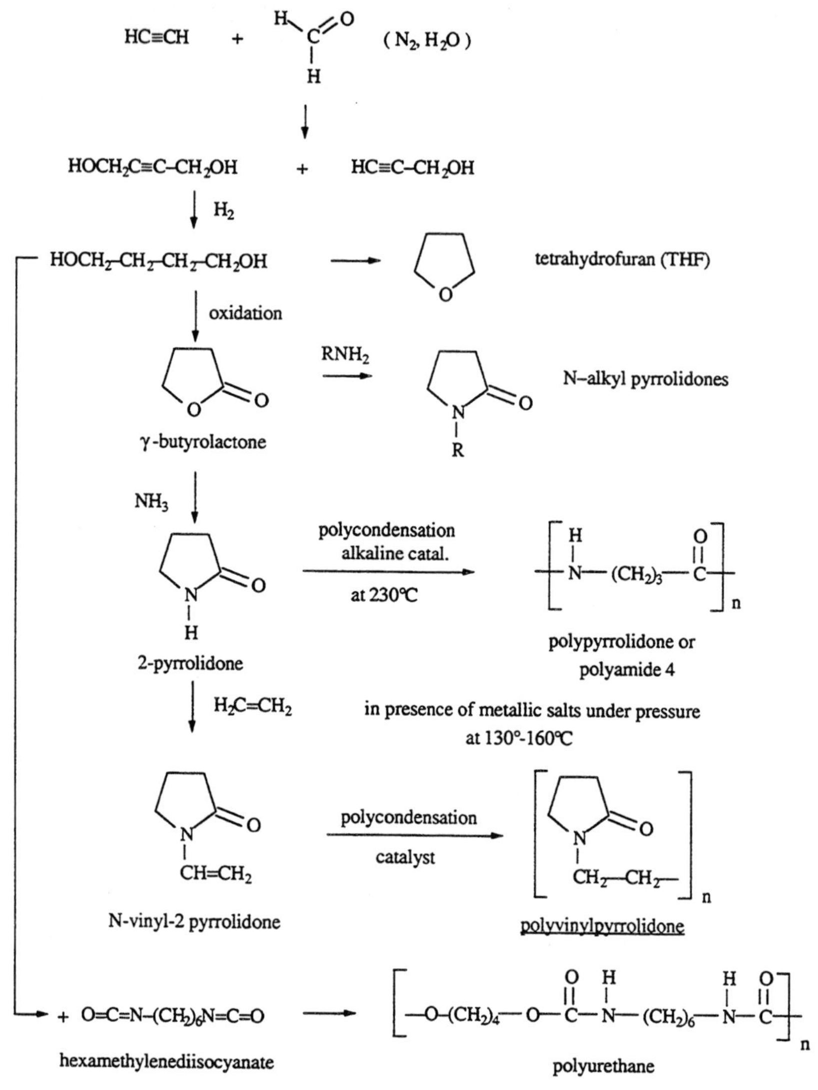

$HC \equiv CH$ + $H-CHO$ (N_2, H_2O)

↓

$HOCH_2C \equiv C-CH_2OH$ + $HC \equiv C-CH_2OH$

↓ H_2

$HOCH_2-CH_2-CH_2-CH_2OH$ → tetrahydrofuran (THF)

↓ oxidation

γ-butyrolactone → (RNH_2) N-alkyl pyrrolidones

↓ NH_3

2-pyrrolidone → polycondensation, alkaline catal., at 230°C → polypyrrolidone or polyamide 4: $[-NH-(CH_2)_3-CO-]_n$

↓ $H_2C=CH_2$ in presence of metallic salts under pressure at 130°-160°C

N-vinyl-2 pyrrolidone → polycondensation, catalyst → polyvinylpyrrolidone

+ $O=C=N-(CH_2)_6N=C=O$ (hexamethylenediisocyanate) → polyurethane: $[-O-(CH_2)_4-O-CO-NH-(CH_2)_6-NH-CO-]_n$

reagents which have not reacted can be recuperated and recycled. This process has the following advantages:

- absence of by-products
- sodium hydroxide costs less than potassium hydroxide
- no hydrolysis step.

Crude methylbutynol is simply isolated by distillation.

The Frantz (air reduction) method uses an ion-exchange resin. A current of acetylene in acetone under a pressure of 10 bars, to saturate the acetylene current with acetone, is sent over an "Amberlite IRA 400" catalyst (in the form of CN^-) at 30°–35°C. The yield of methylbutynol is between 85 and 95% when calculated relative to the amount of acetone actually consumed.

The majority of the ethynylation procedures described here for use with acetone can also be applied to methylheptenone for the production of dehydrolinalool 254.

D. Semi-hydrogenation reaction

When hydrogenated without adequate precaution and in the absence of catalysts, the triple bond rapidly combines with two equivalents of hydrogen [4.07].

[4.07]

$$—C{\equiv}C— \quad + \quad 2\,H_2 \xrightarrow{\text{catalyst}} \diagdown CH_2—CH_2 \diagdown$$

Thanks to the use of appropriate techniques, it is possible to stop hydrogenation so as to preserve the double bond. Lindlar of Hoffmann-La Rôche was the first to describe the precise conditions that permit the selective semi-hydrogenation of triple bonds. A satisfactory catalyst can be prepared from palladium deposited on $CaCO_3$ and poisoned by the addition of a small quantity of quinoleine and traces of lead. $BaSO_4$ can also be used as a support for the palladium to obtain a catalyst generally superior to the one using $CaCO_3$.

A particularly useful catalyst for the selective hydrogenation of propargyl carbinols such as methylbutynol and dehydrolinalool is prepared by the deactivation of palladium by addition of an equal weight of pulverized potassium hydroxide.

These catalysts allow the hydrogenation to be stopped at the ethylenic stage [4.08].

[4.08]

$$(CH_3)_2C(OH)–C{\equiv}CH \xrightarrow[\text{Lindlar}]{\text{Pd}} (CH_3)_2C(OH)–C{=}CH_2$$

When bisubstituted acetylenes are reduced in this way, ethylenic compounds are obtained with the cis configuration [4.09].

[4.09]

$$R–C{\equiv}C–R' \longrightarrow \text{(cis) } RHC{=}CHR'$$

It is also possible to make use of metals for the reduction. The stereochemistry of the double bond depends in these cases on the nature of the metal and the reaction conditions [4.10].

[4.10]

$$R{-}C{\equiv}C{-}R' + 2\,Na + 2\,NH_3 \longrightarrow \underset{trans}{(R)(H)C{=}C(H)(R')} + 2\,NaNH_2$$

$$R{-}C{\equiv}C{-}R' + Zn \xrightarrow[95°C,\ 24\ h]{HOAc,\ 99\%} \underset{cis}{(R)(H)C{=}C(R')(H)} + Zn(OAc)_2$$

It should be noted that the selectivity of the semi-hydrogenation of acetylenes does not result from differences between the rates of hydrogenating the triple and the double bonds. It comes from the fact that the triple bond adsorbs more energetically to the catalyst and this permits unreduced acetylenes to displace the other functional groups from the surface of the catalyst.

Methods for the Synthesis of Methylheptenone

Methylheptenone is at the crossroads of a number of very economically important syntheses. It can be prepared by following three different approaches depending on the starting material used:

- 2-methyl–3-buten–2-ol (methylbutenol)
- 2-methylpropene (isobutylene)
- isoprene.

A. Syntheses starting from methylbutenol

This synthesis was the subject of one of the first endeavors of the Research Center of Roure Bertrand Dupont in Grasse.

In 1940, Carroll published work on the action of α-ethylenic alcohols on compounds with active methylene groups in the presence of alkaline ethylates or acetates. The first two publications were consacrated to the action of ethyl acetoacetate on linalool and on geraniol as well as on cinnamic alcohol and phenylvinylcarbinol.

Carroll was trying to obtain by that route the corresponding acetates by using what could be considered an inverse Claisen reaction [4.11].

[4.11]

$$CH_3COCH_2CO_2C_2H_5 + C(R^1)(R^2)(OH)(CH{=}CH_2) \longrightarrow CH_3{-}C({=}O){-}O{-}C(R^1)(R^2)(CH{=}CH_2) + CH_3CO_2C_2H_5$$

In fact, Carroll did indeed observe the formation of these acetates, but the yield did not exceed 20%. He noted the simultaneous formation of a ketone with a yield in the range of 40%. It was this result that constituted the fundamental point.

In his third publication he tried to explain the formation of this ketone through the intervention of a reaction involving a Michael addition followed by cleavage of the intermediate β-ketoester to give the observed γ-ethylenic ketone [4.12].

[4.12]

$$CH_3{-}CO{-}\overset{CO_2C_2H_5}{\underset{\ominus}{CH}} + CH_2{=}CH{-}\overset{R^1}{\underset{R^2}{C}}{-}OH \longrightarrow CH_3{-}CO{-}\overset{CO_2C_2H_5}{CH}{-}CH_2{-}CH{=}C\langle^{R^1}_{R^2} + {}^{\ominus}OH$$

$$CH_3{-}CO{-}CH_2{-}CH_2{-}CH{=}C\langle^{R^1}_{R^2} + C_2H_5OH + CO_2$$

γ- ethylenic ketone

Precise kinetic study of this reaction shows that, first, ethanol and a gas which is liberated at high temperature are formed. The gas is carbon dioxide CO_2. It is therefore necessary to propose an exchange reaction between the ethylenic alcohol present in the reaction mixture and the C_2H_5O- of ethyl acetoacetate. From this observation came the impetus for the subsequent search for exchange catalysts which are more efficient than sodium ethylate (such as magnesium and aluminium alcoholates). The use of catalysts such as these has permitted us to attain yields over 80% thereby suggesting the possibilitiy of using this reaction, which we have called the "Carroll Reaction," on an industrial scale (Recherches, 1955, 5, 3).

When applied to methylbutenol, this reaction leads to methylheptynone for which the following mechanism has been proposed [4.13].

The exchange reaction, which constitutes the first phase of this reaction, starts at temperatures over 110°–120°C. The second phase can be considered as a sigmatropic reaction involving 10 electrons (4n + 2), which is thermally permitted. This requires a higher temperature and gives intermediateB. B, which as a β-keto acid, finally decarboxylates by a thermal route and liberates methylheptenone 253.

Researchers at Hoffmann-La Rôche (Basel) have used a variant of this reaction, which consists of preparing methylbutenyl acetoacetate A by the action of diketene on methylbutenol in the presence of catalysts [4.14].

The acetoacetic ester thus formed is then pyrolyzed to obtain methylheptenone 253. These results constitute a confirmation of the mechanism proposed for the reaction of Carroll. The pyrolysis can also be catalyzed by aluminum alcoholates.

Another variant devised by workers at BASF uses the following sequence [4.15].

The dimethylacetal 255 can be prepared from three rather inexpensive raw materials: acetylene, acetaldehyde and methanol [4.16]. This compound 255 is implicated in the industrial synthesis of vitamin A by Eastman-Kodak.

A third synthesis of methylheptenone, also realized by workers at Hoffmann-La Rôche, is based on the chemistry of enol ethers. It consists of condensing

[4.13]

① $CH_3COCH_2-\overset{O}{\overset{||}{C}}-OC_2H_5$ + $(CH_3)_2C(OH)CH=CH_2$ (252) $\xrightarrow{(iPrO)_3Al}$ $CH_3COCH_2COO-C(CH_3)_2-CH=CH_2$ Ⓐ + C_2H_5OH

② CH_3 … O … H … O ; CH_3, CH_3 $\xrightarrow{\Delta}$ Ⓑ CH_3 … O … H … O ; CH_3, CH_3 $\xrightarrow{\Delta}$ enol (CH_3 … O–H ; CH_3, CH_3) + CO_2

ketone ⇌ enol

253 (methylheptenone)

[4.14]

$H_2C=$ (diketene: CH_2, O, O) + $H-\ddot{O}$–$C(CH_3)(H_3C)$–$CH=CH_2$ ⟶ $CH_2=C(HO)-CH_2-\overset{O}{\overset{||}{C}}-O-C(CH_3)_2-CH=CH_2$

↓

$CH_3-\underset{O}{\underset{||}{C}}-CH_2-\overset{O}{\overset{||}{C}}-O-C(CH_3)_2-CH=CH_2$ Ⓐ

[4.15]

252 (OH) + 255 (OCH_3, CH, CH_2, OCH_3, C, O, CH_3) $\xrightarrow[-CH_3OH]{\Delta}$ (O, HC, OCH_3, H, O) $\xrightarrow{\Delta}$ CH_3OH + [(O, H, O)]

CLAISEN rearrangement ⟶ [(CH=O, O)]

CH₃OH / retro-Claisen ⟶ 253 (O) + $H-\overset{O}{\overset{||}{C}}-OCH_3$

[4.16]

$$CH_3CH{=}O + HC{\equiv}CH \longrightarrow CH_3{-}CH(OH){-}C{\equiv}CH \xrightarrow[KOH]{CH_3OH} CH_3{-}CH(OH){-}CH{=}CH{-}O{-}CH_3$$

$$\xrightarrow[H^{\oplus}]{CH_3OH} CH_3{-}CH(OH){-}CH_2{-}CH(OCH_3)_2 \xrightarrow[\text{catalytic}]{\text{dehydrogenation}} CH_3{-}C(=O){-}CH_2{-}CH(OCH_3)_2$$

255

methylbutenol with the methylpropenyl ether 256 and pyrolyzing the mixed acetal 257 thus obtained. The vinylallylic ether formed as an intermediate in this reaction undergoes a Claisen rearrangement to give methylheptenone 253 [4.17].

[4.17]

OH + OCH_3, H_3C–C=CH_2 (256) $\xrightarrow{H^{\oplus}}$ O, OCH_3 $\xrightarrow[- CH_3OH]{\Delta}$ [O] ⟶ O (253)

A fourth synthetic route has been elaborated by Japanese researchers. It is based on the radical reaction of acetone with methylbutenol in the presence of peroxides—however, little information on this process is available [4.18].

[4.18]

OH + $CH_3{-}CO{-}CH_3$ ⟶ OH, O (258) + O (253) + O (253a)

This reaction requires heating in an autoclave at 130°C for 24 hours under a nitrogen atmosphere. A mixture of methylheptenone 253, its isomer 253a and hydroxymethylheptanone is obtained. The last of these must be dehydrated.

B. Syntheses starting from isobutylene

This discussion will deal essentially with the industrial synthesis of methylheptenone used by BASF. Acetone, isobutylene and formaldehyde are condensed at high temperatures and under high pressure to give principally 2-methyl–1-hepten–6-one 253a, which must then be isomerized to β-methylheptenone by treatment with special catalysts [4.19].

[4.19]

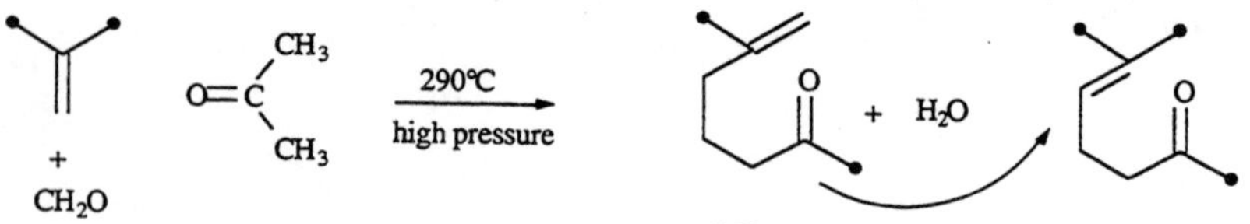

290°C high pressure; + CH_2O; + H_2O; catalysts; 253a; 253

In the unlikelihood that this reaction could be termolecular, a mechanism involving two bimolecular steps can be envisaged, starting in two different ways [4.20].

[4.20]

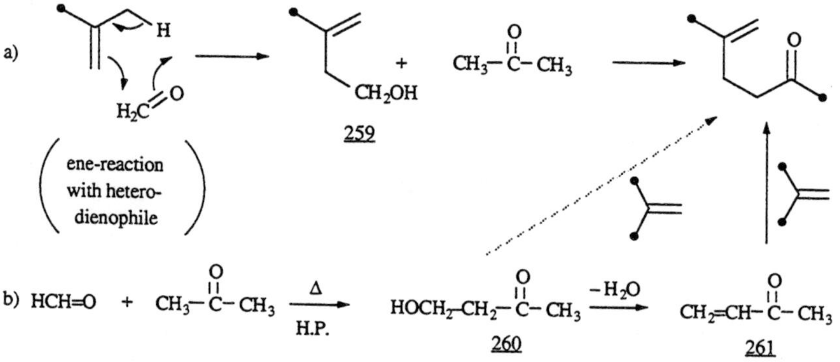

a) (ene-reaction with hetero-dienophile) → 259 (CH_2OH) + $CH_3-CO-CH_3$ →

b) $HCH{=}O + CH_3-CO-CH_3 \xrightarrow[H.P.]{\Delta} HOCH_2-CH_2-CO-CH_3$ (260) $\xrightarrow{-H_2O} CH_2{=}CH-CO-CH_3$ (261)

If 3-oxobutanol 260 is prepared separately and then condensed with isobutylene at 270°C under high pressure, a mixture of 253 and 253a is indeed obtained, although the β isomer 253 predominates.

Methylvinylketone 261 also reacts with an excess of isobutylene at 270°C under pressure. The results vary depending on whether there is water in the reaction mixture [4.21].

[4.21]

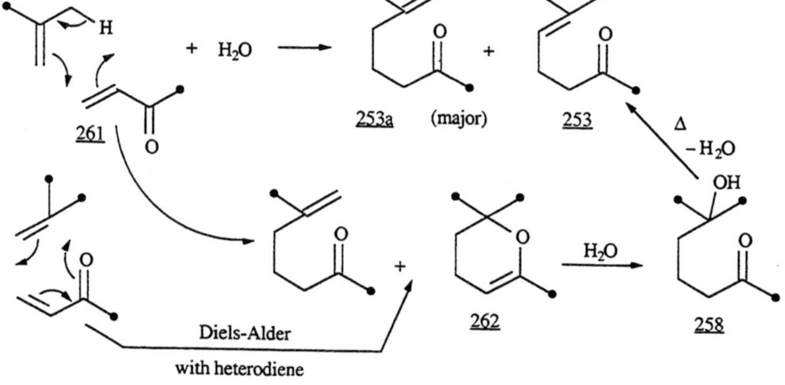

261 + H_2O → 253a (major) + 253

Diels-Alder with heterodiene → + 262 $\xrightarrow{H_2O}$ 258 $\xrightarrow[-H_2O]{\Delta}$ 253

C. Syntheses starting from isoprene

This essentially concerns the work carried out by Rhône-Poulenc. The key material is 2-methyl–4-chloro–2-butene 263 [4.22].

[4.22]

+ HCl at 33° Be → 263 (Cl)

Methylchlorobutene is condensed with acetone in the presence of sodium hydroxide and catalytic quantities of iodine or amines. An iodine-chlorine exchange takes place, thereby accelerating the reaction because the I- ion is a better leaving group than the Cl- ion [4.23].

[4.23]

263 + $^{\ominus}CH_2$–$COCH_3$ —NaOH, catalysts→ 253

This approach requires extremely pure isoprene. It concerns more the polymer and rubber industries than the perfumery industry. It suffices only to say that there are several industrial processes for the production of isoprene. Discussion of these, however, is outside the scope of this text.

Industrial Chemistry of Dehydrolinalool

A. Preparation of linalool and its derivatives

The methods of selective semi-hydrogenation just discussed permit the conversion of dehydrolinalool into linalool. The esters of linalool can then be prepared by direct acylation. The most important of these, linalyl acetate, can be prepared using acetic anhydride in the presence of a mineral acid or acetyl chloride associated with an organic base such as dimethylamine. The direct acetylation of linalool in this way does, however, give rise to the formation of by-products: terpinyl acetate (by cyclization), geranyl acetate (by allylic rearrangement) and terpenic hydrocarbons (by elimination) in the first case; chlorinated derivatives in the second case.

When solely natural linalool was available, these processes were the only ones which could be used. They also permit the acetylation of synthestic linalool prepared from the pinenes.

A far better method consists of treating dehydrolinalool with acetic anhydride

[4.24]

254 (OH) —H_2→ 13 (OH) —Ac_2O / H_3PO_4→ 209 (OAc) + 189 (OAc) + 210 + 211 (OAc)

13 —CH_3COCl / $N(CH_3)_2$-C6H5→ 209 (OAc) + $\overset{\oplus}{HN}(CH_3)_2$-C6H5 $\overset{\ominus}{Cl}$

254 —Ac_2O / H_3PO_4→ 264 ($OCOCH_3$) —H_2 / Pd / $CaCO_3$→ 209 (OAc)

[4.25]

CH_3, O, H, O, O, CH_3 —Δ→ $CH_3-\overset{O}{\overset{\|}{C}}-CH_2CH_2-CH{=}\underset{CH_3}{C}-CH_2CH_2-CH{=}C(CH_3)_2$

265

[4.26]

265 —1) HC≡CH; 2) semi-hydrogenation→ 267a (trans) (H, OH) + 267 (cis) (H, OH)

HOAc / Ac_2O

268 (4 isomers)

hydrolysis of esters

269' : (CH_2OH, H, H) + (H, H, CH_2OH)

269 : (H, CH_2OH, H) + (H, H, CH_2OH)

in the presence of traces of phosphoric acid. Dehydolinalyl acetate 264 prepared in this way is extremely pure and its selective semi-hydrogenation leads to linalyl acetate 209. It should be noted that this semi-hydrogenation is even more selective than that of dehydrolinalool itself [4.24].

When applied to linalool, the Carroll reaction (of its diketene variant) allows the preparation of geranylacetone 265 in very high yields [4.25].

In fact, a mixture of cis and trans geranylacetones (isomeric at the $\Delta^{6,7}$ bond) is obtained. Ethynylation of geranylacetone followed by semi-hydrogenation gives the nerolidols 267 and 267a. Allylic rearrangement of these isomeric nerolidols in HOAc-Ac_2O leads to four isomeric farnesyl acetates 268 which are converted by saponification into a mixture of four corresponding farnesols 269.

Careful distillation under reduced pressure does, however, permit the separation of two cis and trans nerolidols. The latter is generally the natural compound that is found, for example, in cabreuva oil which is obtained from the distillation of the wood of *Myroxylon pereire* of Brazilian origin [4.26].

"All trans" farnesol 269 is also found in nature. It is a major constituent of the essential oil of *Melaleuca viridiflora.*

The discovery by Normant of the vinyl-magnesium species has made possible a very direct synthesis of linalool [4.27].

[4.27]

253 + CH_2=CH–MgBr (V) → THF, then hydrolysis → 13

The species V are obtained by the action of vinyl bromide (or chloride) on magnesium in the presence of tetrahydrofuran (the preparation of these magnesium complexes cannot be done in diethyl ether). Even though this procedure transforms methylheptenone into linalool in a single step, it is not used industrially.

Another particularly elegant although nonindustrial variant has been developed by Julia. It uses the cyclopropyl ketone 270* and methylvinylketone 261 as shown in the following scheme [4.28].

[4.28]

270 ($COCH_3$) → CH_3MgI / $C_2H_5OC_2H_5$ → 271 (CH_3, OH, CH_3) → HCl at 22° Be → 272 (Cl, CH_3, CH_3)

272 → 1°) Mg, 2°) methyl vinyl ketone → 13 (OH)

* Note on the preparation of methylcyclopropylketone [4.29].

[4.29]

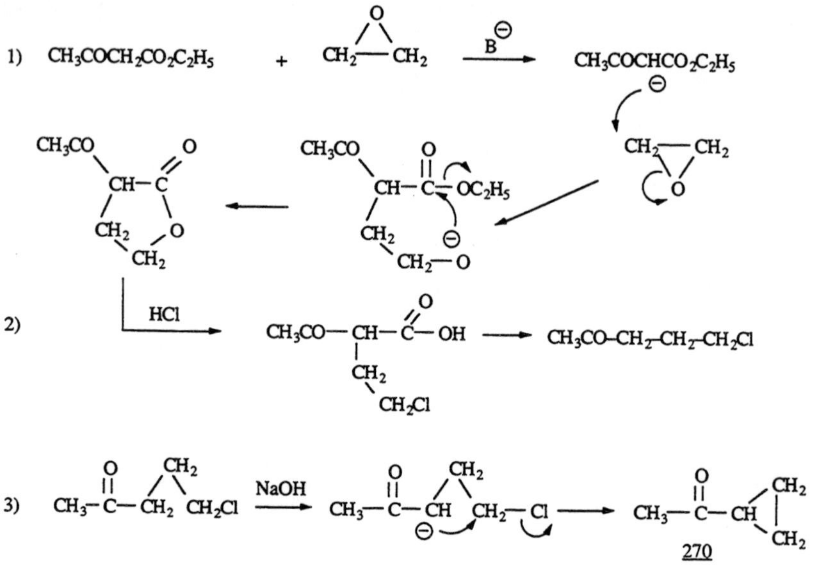

1) $CH_3COCH_2CO_2C_2H_5$ + CH_2—CH_2 (O) $\xrightarrow{B^\ominus}$ $CH_3COC^\ominus HCO_2C_2H_5$

2) $\xrightarrow{HCl}$ CH_3CO—CH(CH_2CH_2Cl)—C(=O)—OH ⟶ CH_3CO-CH_2-CH_2-CH_2Cl

3) CH_3—C(=O)—CH_2 CH_2 CH_2Cl $\xrightarrow{NaOH}$ CH_3—C(=O)—$CH^\ominus$ CH_2—Cl ⟶ CH_3—C(=O)—CH(CH_2)(CH_2) 270

This strategy can be applied to the preparation of nerolidol. In this case, the homoallylic halide 272 is condensed once again (via its magnesium derivative), with methylcyclopropylketone to form the tertiary cyclopanyl alcohol 273. Opening of the small ring by hydrochloric acid, or better still, with the HBr-H_2O azeotrope, gives specifically the bromide 274 (entirely trans). The magnesium derivative of this bromide reacts with methylvinylketone 261 to obtain, after hydolysis, trans-nerolidol 267a [4.30].

[4.30]

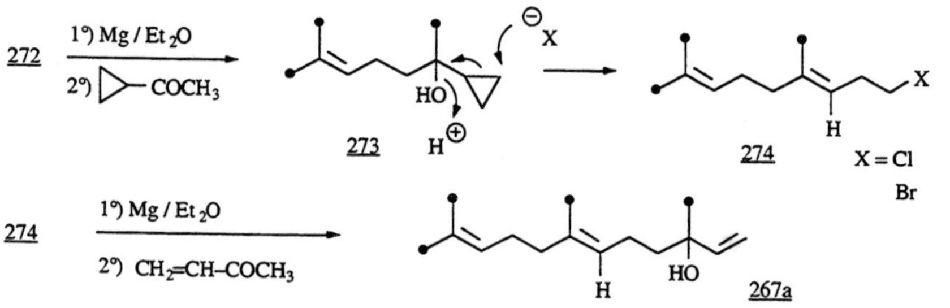

272 $\xrightarrow[2^\circ)\ \triangleright\text{—}COCH_3]{1^\circ)\ Mg / Et_2O}$ 273 (HO, $H^\oplus$, $X^\ominus$) ⟶ 274 (H; X = Cl, Br)

274 $\xrightarrow[2^\circ)\ CH_2{=}CH\text{-}COCH_3]{1^\circ)\ Mg / Et_2O}$ 267a (H, HO)

The reaction sequence is as follows: Carroll-ethynylation followed by semi-hydrogenation, applied to nerolidol, resulting in the geranyllinalools 275 which can be converted into geranylgeraniols 276 by allylic rearrangement. (The key intermediate is farnesylacetone 277 which issues from the Carroll reaction.) These two alcohols in their all-trans forms are major constituents (although without odor) of jasmin concrete [4.31].

[4.31]

267 Carroll 277 278 OH H2, catal. 279 275 OH 280 OH 276 CH2OH H 281 OH 282 CH2OH

Total hydrogenation of farnesylacetone 277 permits the generation of hexahydrofarnesylacetone 279 which can be converted by ethynylation into dehydroisophytol 280. This can then be semi-hydrogenated to give isophytol 281 which leads to phytol 282 by an allylic rearrangement.

In fact, the hydrogenation of farnesylacetone introduces two asymmmetric carbon atoms. Ethynylation creates a third. This third center of asymmetry disappears in the course of the allylic rearrangement, but this still results in the formation of two geometric isomers E and Z in phytol. A mixture of diastereoisomers and geometric isomers is thus obtained. When the starting materials are racemic, a number of different products are formed:

- for 279: $2^2/2 = 2$ products
- for 280 and 281: $2^3/2 = 4$ products
- for 282: $2^2/2 \times 2 = 4$ products.

B. Direct synthesis of the pseudoionones

1. Generalized Carroll reaction

Very early on, workers at the Research Center of Roure Bertrand Dupont in Grasse attempted to extend the Carroll reaction to α-acetylenic alcohols. This is what they have called the "generalized Carroll reaction." When applied to dehydrolinalool, two "apparently competing" reactions are observed.

a. Principal reaction

[4.32]

Dehydrolinalool 254 + $CH_3COCH_2CO_2C_2H_5$ $\xrightarrow{(iPrO)_3Al}$ C_2H_5OH +

Δ

283 + CO_2 $\xrightarrow[(iPrO)_3Al]{\Delta}$ 22

The reaction proceeds as in the synthesis of methylheptenone. In this case, however, a β-allenic ketone 283 is formed and this then isomerizes under the reaction conditions to give pseudoionone 22 essentially in its trans form.

b. Secondary reaction

Another ketone 284 is obtained along with pseudoionone. The exact mechanism of its formation has not been totally explained, but two distinct routes can be envisaged:

- In the first, the β-allenic ketone evolves by an intramolecular ene-reaction involving one of the two double bonds of the cummulated diene [4.33].

[4.33]

283 $\xrightarrow{\Delta}$ 284

Some reactions of this type are known.

- Before discussing the second possibility, it should be noted that the pyrolysis of dehydrolinalool gives a tertiary α,β-ethylenic alcohol with an iridoid structure 285. It is therefore possible that ethyl acetoacetate reacts with this alcohol according to the normal Carroll process to give 284 [4.34].

2. *Reaction of Saucy-Marbet II (Hoffmann-La Rôche)*

Before going into the details of this reaction, it should be recalled that alcohols react with enol ethers in acidic media to give mixed acetals [4.35].

[4.34]

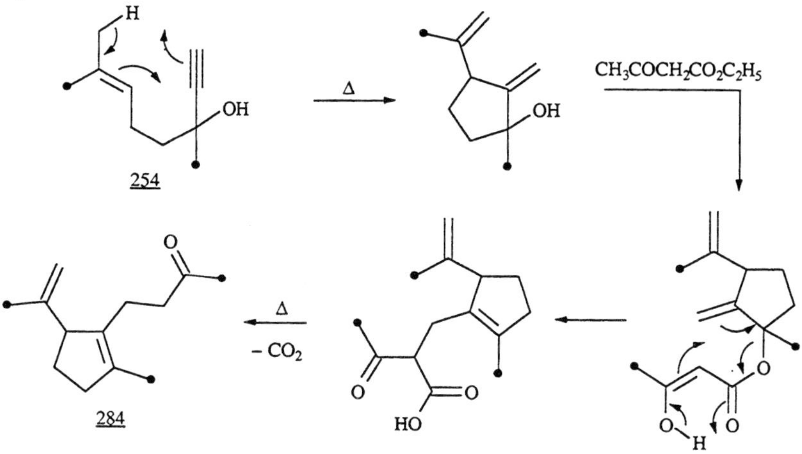

[4.35]

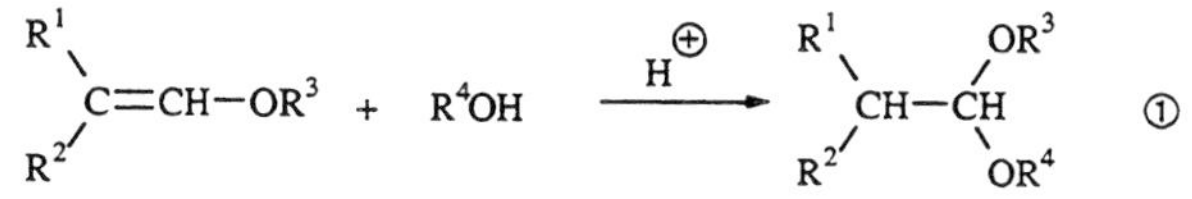

It is also known that the pyrolysis of acetals in neutral media furnishes the enol ether in an irreversible reaction. For example [4.36].

[4.36]

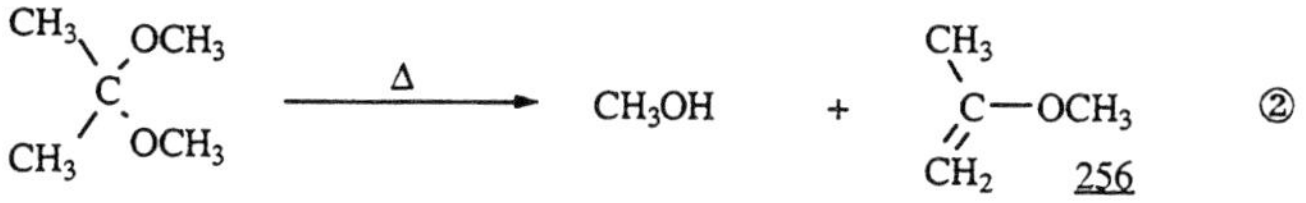

If the reaction is conducted at a relatively high temperature, the formation of mixed acetals (1) becomes reversible and it can be written, taking account of reaction 2 [4.37].

[4.37]

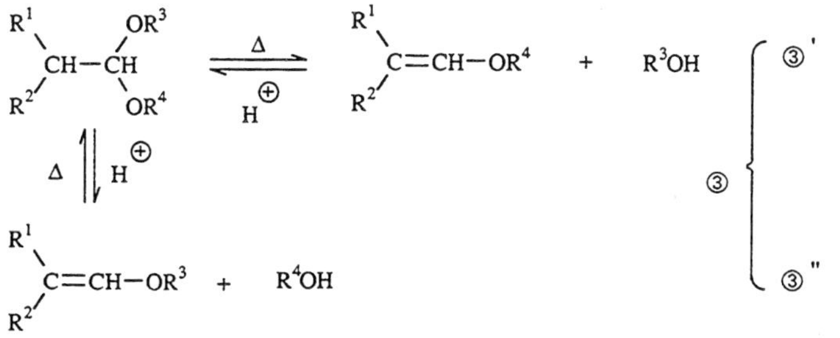

Thus, the condensation of dehydrolinalool with methylisopropenyl ether 256 under acidic conditions gives rise to the formation of the mixed acetal 286. The pyrolysis of 286 can either regenerate dehydrolinalool and the enol ether 256, or convert 286 into the enol ether 287 by loss of methanol [4.38].

[4.38]

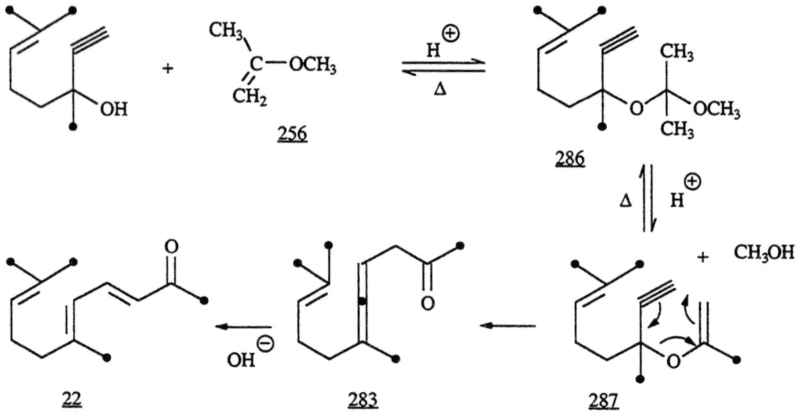

Under these reaction conditions, compound 287 undergoes a thermal vinyl-propargyl migration (related to the Claisen rearrangement of allylic ethers of enols). The β-allenic ketone already encountered is thus formed. As there is an excess of methylisopropenyl ether which fixes methanol, the equilibrium is displaced in favor of the propargyl enol ether 287 which rearranges to 283. This then contributes definitively to displacing the equilibrium just mentioned. In the presence of a small quantity of base, the β-allenic ketone 283 progresses to the more stable conjugated structure of pseudoionone 22.

This sequence gives very high yields of pseudoionone and the procedure is used by Hoffmann-La Rôche in the production of β-ionone required for the manufacture of vitamin A.

If this sequence is applied to 6-methyldehydrolinalool 288, a pseudoirone 289 is obtained. It can be noted in passing that the application of the generalized Carroll reaction to the same compound, 6-methyldehydrolinalool, also gives the same pseudoirone 289, essentially in the trans form. This is an important fact, because in the course of the cyclization, this geometry is responsible for the preferential formation of the isomer of α-irone which is the most olfactively interesting among the 10 possibilities. Unfortunately, the generalized Carroll reaction always gives a cyclopentenyl ketone as a by-product [4.39].

[4.39]

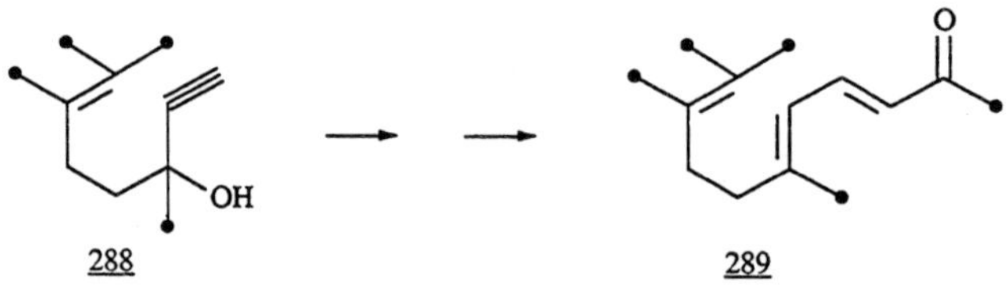

In the same way, methylbutynol 251 results initially in the formation of the β-allenic ketone 290 which can then be isomerized in alkaline media to the dienone

291. The β-allenic ketone 290 can, with some difficulty, be selectively hydrogenated to methylheptenone 253 [4.40].

[4.40]

OH

251

290

O

$OH^{\ominus}$

O

291

H_2

O

253

Methylheptenone can also be obtained by the reaction of Saucy-Marbet II, described above [4.41].

[4.41]

OH + CH_3 OCH₃ CH_2 $H^{\oplus}$ O OCH_3 Δ O + CH_3OH

253

When applied to linalool, this reaction affords geranylacetone 265 [4.42].

[4.42]

OH

O

265

C. Synthesis of citral

Indirect synthesis of pseudoionones

In itself, citral only represents a minor interest. Its real importance is linked to its facile conversion into pseudoionones. For this reason, much work has been devoted to its total synthesis. This has been done in parallel with the syntheses carried out starting from the pinenes—which, as we have seen, always give complex mixtures.

World production of the essential oil of lemongrass is around 2,000 metric tons per year and from this, it is possible to extract between 1,500 and 1,550 metric tons of natural citral. Yet the annual consumption in the United States alone is close to 4,350 metric tons—broken down as follows:

citral (perfumes and flavors)	150 to.	corresponding to 150 to. of citral
ionones (perfumes)	425 to.	corresponding to 600 to. of citral
vitamin A	825 to.	corresponding to 2,800 to. of citral
vitamin E	320 to.	corresponding to 800 to. of citral

1. Reaction of Saucy-Marbet I

Treatment of dehydrolinalyl acetate 264 with a mixture of acetic anhydride/acetic acid in the presence of salts of silver or copper gives three principal products: the allenyl acetate 292 (major product), the dienyl acetate 293 and the diacetyl ester of citral hydrate 294. This reaction is carried out by heating the mixture to 100°–120°C and then treating the product mixture with sodium carbonate in aqueous or hydro-alcoholic solution to give citral. Treatment with alkali in the presence of acetone gives pseudoionone [4.43].

[4.43]

264 → 292 + 293 ($OCOCH_3$) + 294 ($OCOCH_3$, $OCOCH_3$)

$^{\ominus}OH$, CH_3COCH_3 ; H_2O / Na_2CO_3 → CHO

2. Direct transformation of dehydrolinalool into citral

It can be noted that these two compounds are structural isomers of each other. After numerous attempts, the Rhône-Poulenc company was the first to propose a method that permitted the direct isomerization of dehydrolinalool into citral. Dehydrolinalool is heated in the presence of alkyl vanadates as catalysts. Other metals such as molybdenum and zirconium can also be used. The isomerization takes place in the liquid phase.

The first step of the reaction apparently corresponds to an exchange between alkoxy moieties with the formation of dehydrolinalyl orthovanadate. Although the patents of Rhône-Poulenc make no mention of the mechanism, it can be reasonably envisaged that the real reaction intermediate is the orthovanadate of dehydrolinalool. According to this hypothesis, the following sequence can be realized for this elegant transformation [4.44].

[4.44]

1st step

+ O=V(OR)$_3$ ⟶ 3 ROH +

DHL = dehydrolinalool

(the process repeats for the two other DHL groups)

2nd step

It should, however, be pointed out that this reaction furnishes very low r-transformation levels if high yields are sought. The optimum yield reaches 75% for a τ-transformation level of around 20% [4.45].

[4.45]

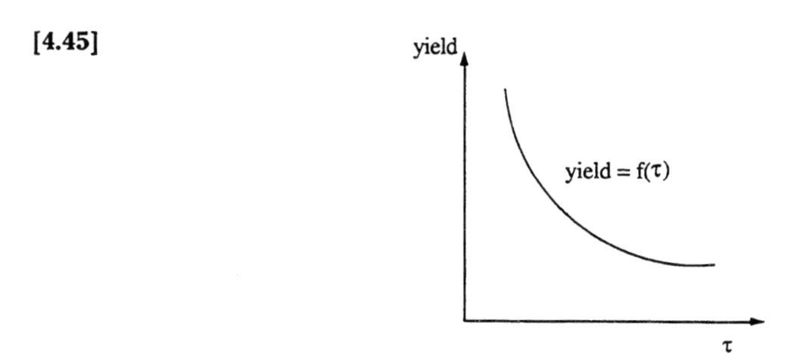

This methodology has been noticably improved by Hoffmann-La Rôche thanks to the use of complex catalysts such as [4.46].

[4.46]

$$(R_3SiO)_n\!-\!\underset{\underset{\displaystyle O}{\|}}{V}\!-\!(OR')_m \qquad \text{with } m+n=3$$

$$(R_3SiO)_m\!-\!\underset{\underset{\displaystyle O}{\|}}{V}\!-\!(OSiR'_3)_n \qquad \text{- id -}$$

An example of this type of catalyst is bis-(trimethylsilyloxy) triphenylsilyloxy-vanadiumoxide [4.47].

[4.47]

$$[(CH_3)_3SiO]_2\!-\!\underset{\underset{\displaystyle O}{\|}}{V}\!-\!OSi(C_6H_5)_3$$

In contrast with the alkyl vanadates, the catalysts shown above remain stable and conserve their activity over a long period. Moreover, practically no decomposition products are observed.

The reaction is carried out in a solvent such as nitrobenzene at a temperature near to 120°–125°C. Yields are about 80%, but recycling the dehydrolinalool raises this to around 90%.

It can be noted that the patent of Rhône-Poulenc was applied for in 1968 and that of Hoffmann-La Rôche in 1971.

Syntheses of Terpenoids

A. Syntheses of terpenoids by telomerization

A second method of synthesis for obtaining aliphatic terpenoids has been developed in the Soviet Union by Leets and Petrov from 1956 onwards. The strategy used in this approach is based on telomerization reactions in which the diene is isoprene and the telogen is chloromethylbutene (predominantly 1-chloro–3-methyl–2-butene obtained by the addition of hydrogen chloride to isoprene). Very little information exists concerning the industrial applications of this process. If it is assumed that the mechanism is ionic, the following sequence can be envisaged [4.48]. This process could then repeat itself.

It is, however, uncertain whether the reaction is indeed ionic. Furthermore, it should not be forgotten that in the course of the telomerization process, the addition of the telogen can give four distinct types of coupling. Consider the molecule isoprene [4.49].

In this case, the addition can take place across the following bonds: 1,2; 3,4; 1,4 (cis); and 1,4 (trans). In each case there are two possibilities because the molecule is not symmetrical, thus forming eight chlorinated derivatives of which

[4.48]

CH_2Cl + $AlCl_3$ ⟶ $\overset{\oplus}{CH_2}\cdots\overset{\ominus}{AlCl_4}$

Cl + $AlCl_3$ ⟵ $\ominus AlCl_4$

[4.49]

CH_3

CH_2 C 2 CH CH_2

1 3 4

only three possess the regular isoprene coupling. These are linalyl, neryl, and geranyl chlorides. A fourth, corresponding to lavandulyl chloride, has an irregular coupling. The remaining four correspond to isoprene linkages having no interest. The following table summarizes the different possibilities [4.50].

Furthermore, if 3-chloro–3-methyl–1-butene, which exists in small quantities in the crude telogen (chloromethylbutene), is allowed to intervene, eight supplementary chlorinated derivatives having irregular isoprene linkages and also having no known utility are encountered.

The complexity of this problem is evident. Nevertheless, it has been made controllable in the polymerization of isoprene thanks to catalysts of the Ziegler-Natta type. Thus, isoprene can be converted into a stereoregular homopolymer entirely cis–1,4 and having a structure resembling that of natural rubber.

Several modifications have been made to the initial binary system (organo-aluminum + titanium salt) by the coupled use of electron-donating transition metals and Lewis acids as interesting complexing agents. By using tertiary systems—$Al(C_2H_5)_3$, $TiCl_4$, amines—and by controlling the quantity of amines introduced, it is in fact possible to pass from one stereoregular structure (cis–1,4) to another (trans–1,4).

Obviously the aim of researchers in the perfume and pharmaceutical fields is not the synthesis of rubber. This does not mean, however, that this type of approach will not bring, over the course of the next few decades, original solutions using functionalized isoprene units in analogy with biosynthetic processes.

With this in mind, the Chuisoli group at Montecatini has recently developed a reaction involving chloromethylbutene and isopentenyl acetate dissolved in a mixture of benzene and methanol in the presence of nickel complexes [4.51].

The end product is a mixture of geranyl and neryl acetates in the proportion 85:15.

[4.50]

(I) 1,2-addition

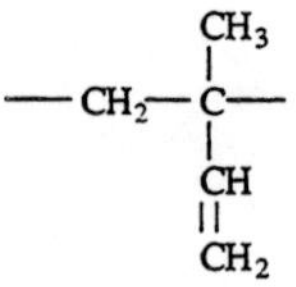

$$-CH_2-C(CH_3)(CH=CH_2)-$$

a) Cl (linalyl)

b) $ClCH_2$ CH_3

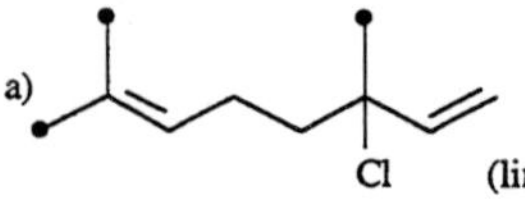

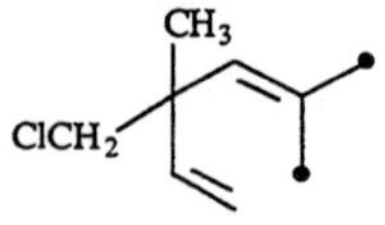

(II) 3,4-addition

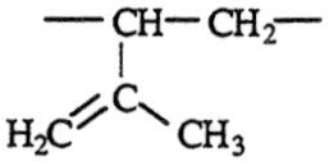

$$-CH(C(CH_3)=CH_2)-CH_2-$$

a) CH_2Cl (lavandulyl)

b) $ClCH_2-CH-CH_2-CH=C(CH_3)_2$

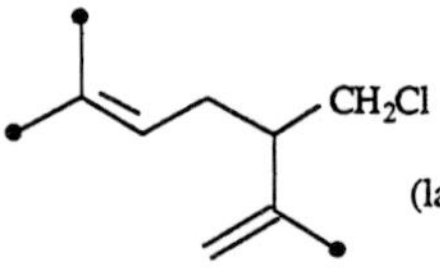

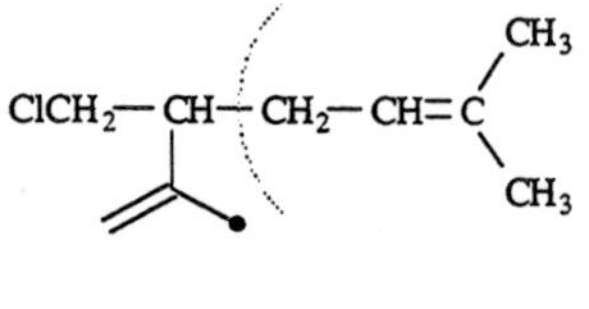

(III) cis–1,4-addition

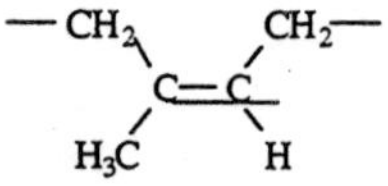

$$-CH_2(H_3C)C=C(H)CH_2-$$

a) H (neryl) CH_2Cl

b) $ClCH_2(H_3C)C=C(H)CH_2-CH_2-CH=C(CH_3)_2$

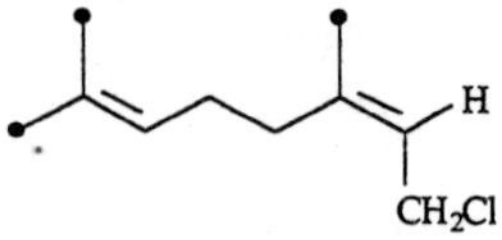

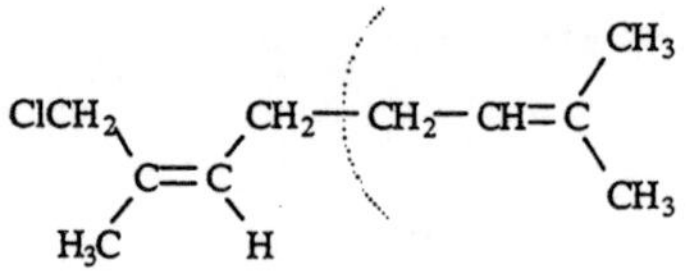

(IV) trans–1,4-addition

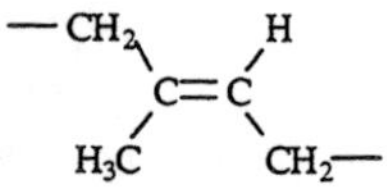

$$-CH_2(H_3C)C=C(H)CH_2-$$

a) CH_2Cl H (geranyl)

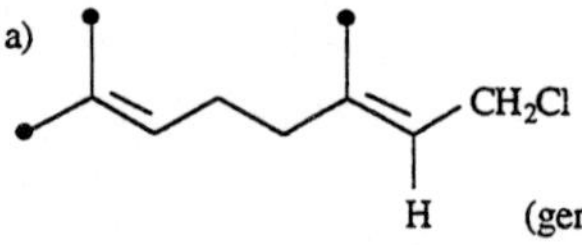

b) $ClCH_2(H_3C)C=C(H)CH_2-CH_2-CH=C(CH_3)_2$

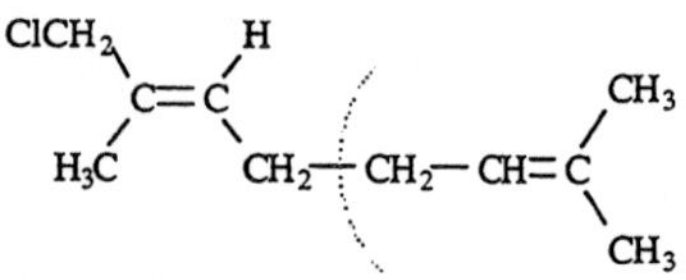

[4.51]

CH_3 / CH_3 > C=CH−CH_2Cl + (Ni complex) CH_2OCOCH_3, X, Ni, thiourea

C_6H_6 | CH_3OH

CH_2OCOCH_3, H + H, CH_2OCOCH_3 + $NiCl_2$

All of these syntheses of aliphatic terpenoids just discussed are of interest not only in the domain of perfumery. They have reached a high degree of industrialization because of their importance in the manufacture of compounds of great biological interest such as vitamins A, E, and K and the carotenoids.

B. Recent syntheses

As discussed in Chapter 2, the near totality of the aliphatic monoterpenoids is obtained from a mixture of geraniol-nerol. Dehydrogenation gives citral (mixture of geranial and neral) which can be transformed by selective hydrogenation into citronellal. Citral condenses with acetone or methylethylketone to give the ionones and the methylionones. Citronellol is obtained from selective hydrogenation of the geraniol-nerol mixture. These possibilities are depicted in Figure 1.

Some more recent reactions, involving myrcene and having great industrial interest, should also be considered.

1) The addition of secondary amines to these conjugated dienes in the presence of sodium or lithium leads to allylic amines. With diethylamine, the product obtained is diethylgeranylamine, an intermediate in the synthesis of Lyral (IFF).

The oxide of this same amine can undergo a [2,3]-sigmatropic rearrangement resulting in an intermediate that can be reduced to linalool by the action of zinc in acetic acid (Figure 2).

Deamination of 7-hydroxydiethylgeranylamine can be achieved in a single step in the presence of an organometallic catalyst derived from palladium. In this case, passage through the quaternary ammonium salt is avoided, but it also gives (along with the hydroxydiene myrcenol necessary for the synthesis of "Lyral") two isomeric ocimenols (Figure 3). Myrcenol can be separated from these two ocimenols by careful distillation.

2) Another very important application involves the use of optically active organometallic catalysts. Pure diethylgeranylamine or pure diethylnerylamine are used. The former can be obtained by adding diethylamine to myrcene in the presence of lithium. The latter comes from the telomerization of isoprene in the presence of the same amine and the same metal. Figure 4 indicates the composition of

Figure 1

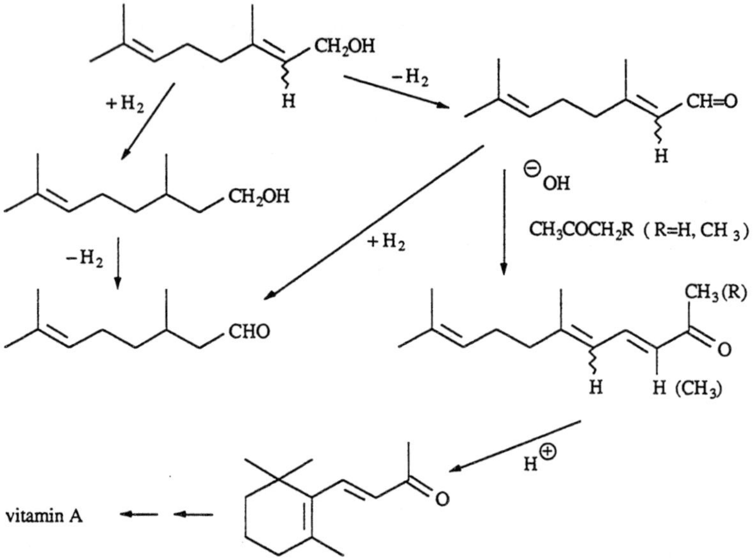

the two amines obtained in this way. It can be noted that in the second case, a small quantity of the compound resulting from tail-tail (or Z-Z) coupling is obtained.

These allylic amines can be isomerized to enamines. Use of a rhodium complex containing an atropimer with axial chirality as a ligand ("BINAP") (Figure 5), gives the enamine "E." It is obtained in almost quantitative yield and with a 95% enantiomeric excess, regardless of which enamine (geranyl or neryl) was used at the outset.

Depending on the configuration of the "BINAP" used in this process, either diethylgeranylamine or diethylnerylamine will give the levorotatory or the dextro-rotatory enamine. On hydrolysis, these give, respectively, the 3R- (+) or the 3S (–) citronellal (Figure 6).

As shown in Figure 7, it is possible to obtain (–) menthol from (3R) (+) cit-ronellal. These last two series of syntheses were realized by Noyori and Takaya (University of Nagoya) and Otsuka and colleagues (University of Osaka).

Figure 2

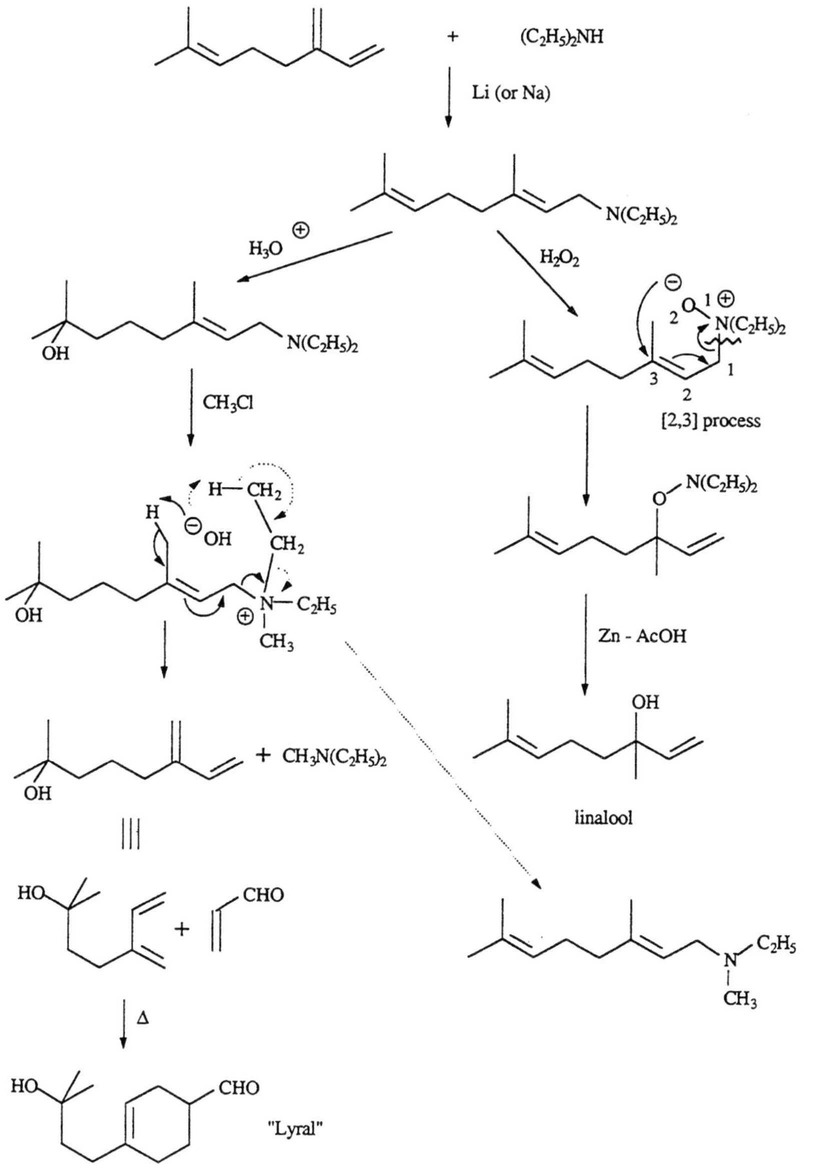

+ $(C_2H_5)_2NH$

Li (or Na)

$N(C_2H_5)_2$

$H_3O^{\oplus}$

H_2O_2

OH

$N(C_2H_5)_2$

CH_3Cl

$\ominus$ 2 O 1 $\oplus$ $N(C_2H_5)_2$

3 2 1

[2,3] process

H — CH_2

H $\ominus$OH CH_2

OH $\oplus$ N — C_2H_5

CH_3

O — $N(C_2H_5)_2$

Zn - AcOH

OH

OH + $CH_3N(C_2H_5)_2$

linalool

|||

HO + CHO

N — C_2H_5

CH_3

Δ

HO CHO

"Lyral"

Figure 3

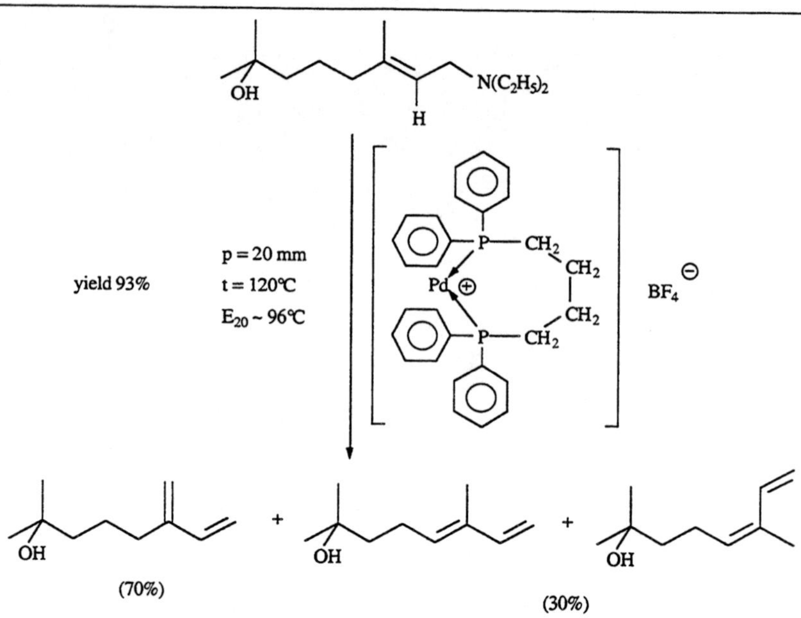

$N(C_2H_5)_2$

OH

H

yield 93%

p = 20 mm

t = 120°C

E_{20} ~ 96°C

$BF_4^{\ominus}$

P — CH_2

CH_2

Pd ⊕

CH_2

P — CH_2

OH + OH + OH

(70%)

(30%)

Figure 4

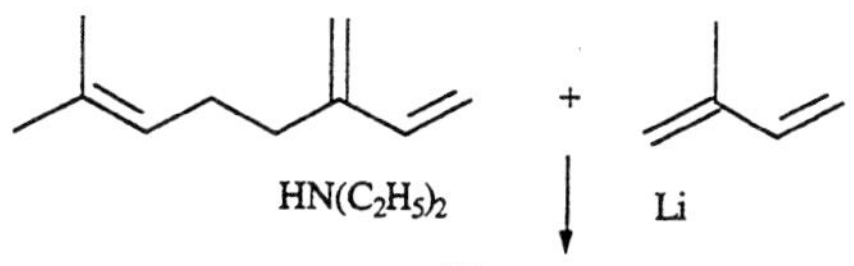

Myrcene + Isoprene, $HN(C_2H_5)_2$, Li (%)		Compounds obtained $R = -N(C_2H_5)_2$
93.6	0	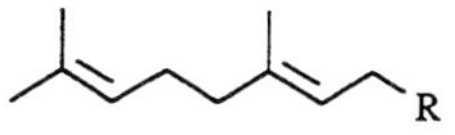
1	94.9	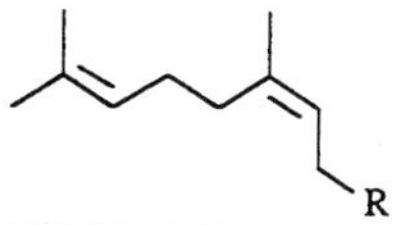
.6	.1	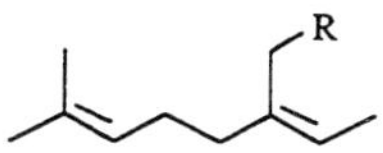
1.3	.1	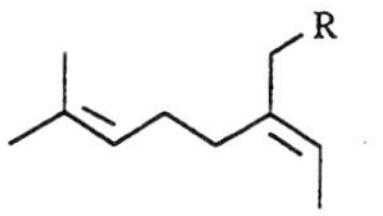
2.3	2.1	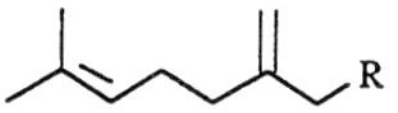
0	1.5	Z - Z 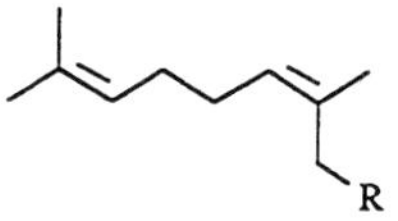

Figure 5

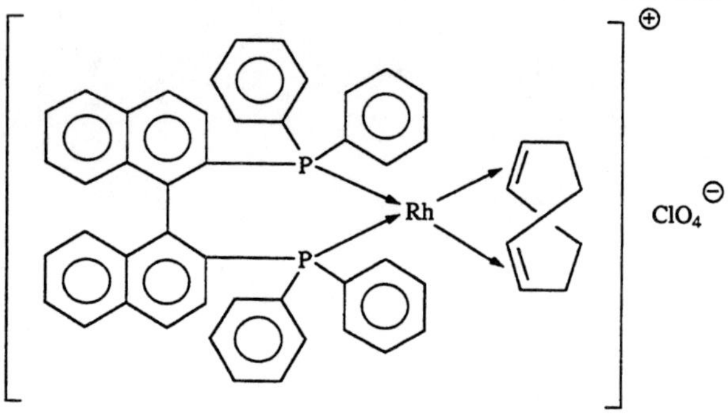

$ClO_4^{\ominus}$

BINAP

Figure 6

$[(DCO)R(+)PhBinap]^{\oplus}$

H_R H_S $N(C_2H_5)_2$ H

S (-)

e.e. = 96%

H H_S $N(C_2H_5)_2$ H_R

R (+)

THF

100°C, 3h

yield 100%

e.e. = 94%

$H_3O^{\oplus}$

CHO

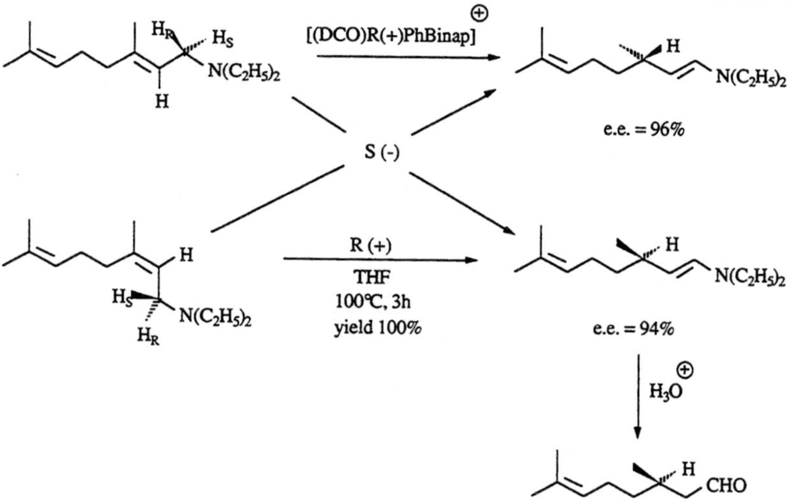

(3R) (+) citronellal

Figure 7

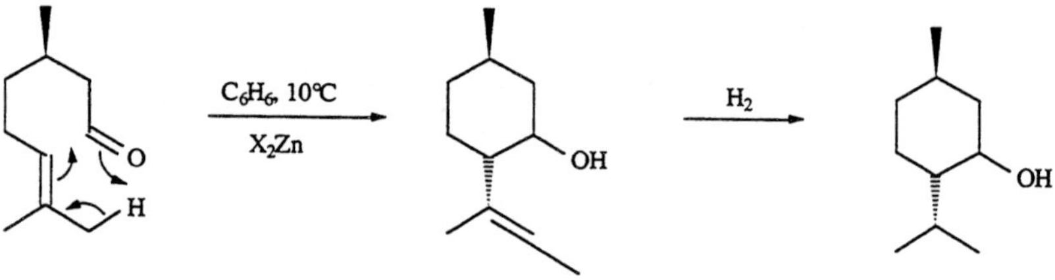

C_6H_6, 10°C

X_2Zn

H_2

OH

X	HO (wedge)	HO (hashed)	yield
Cl	88%	12%	60%
Br	94%	6%	>70%

CHAPTER

5

Biogenesis of Natural Substances

Introduction

The question of how natural substances are elaborated by vegetable and animal species has intrigued researchers from the start. The answers and the conclusions that can be drawn from them, will of course permit a more rational study of them.

Near the end of the last century, Wallach first noticed the regular construction of these substances and postulated the role of isoprene as the corner-stone of these edifices. Thirty years later, Ruzicka and his school established from this observation the "isoprene structure rule," which has contributed much to the often complicated determination of structures (for example, with the tetra- and pentacyclic triterpenes).

During more than 15 years, any substance that did not obey this rule was immediately considered suspect. Yet now,. numerous exceptions to this rule are known and these must be accomodated by the theories of biogenesis.

The first exception concerned the structure of eremophilone 1, observed in 1937 by Simonsen and his co-workers. Other exceptions can be found in the sesquiterpene series and some abnormal skeletons have been reported in the monoterpenoids (lavandulol, artemisiaketone, lyratol, ferulol . . .) [5.00].

[5.00]

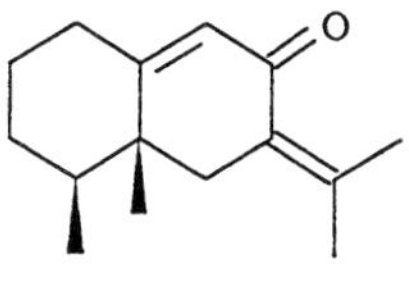

O

1

In fact, it seems that in the back of Ruzika's mind, the isoprene rule represented a "biogenetic idea." In this spirit, he advanced the theory in 1923 that plants would have the capacity to cyclize nerolidol (which he had just succeeded in synthesizing) or farnesol to give bicyclic sesquiterpenes with eudesmane or cadalane skeletons. Furthermore, he had just determined their skeletons by dehydrogenating numerous sesquiterpenes in the presence of sulfur.

The remarkable way in which this hypothesis was verified will be seen from the ensuing discussion. As will be seen in Chapter 7, Ruzika also proposed in 1926 that macrocyclic substances such as civettone and muscone were derived from oleic acid and palmitic acid, respectively. Although no experiment has yet confirmed this hypothesis, there are a number of findings that corroborate it indirectly.

While acceptance of the isoprene hypothesis gradually gained ground among organic chemists, there was increasing reflection over the possible biological origin of this "isoprene unit." For some time, another hypothesis had been envisaged in which the biogenesis of numerous natural substances could result from the condensation of units which were "acetic acid derivatives." Advanced by Collie between 1893 and 1907, this theory notably concerned natural products with aromatic rings, and it has been verified for all substances which today are termed "acetogenic substances"—so-called because they are formed by condensation of acetic acid units to which fatty acids can also be attached. It will be shown how the "isoprenic" and "acetic" hypotheses were eventually reconciled. Empirical support for the hypothesis advanced by Collie was, at that era, extremely fragile. He had, in fact, observed that diacetylacetone, when treated in weak alkaline media, "dimerized" by aldolization and crotonization to give a naphthalenic derivative 2, a close analogue of musizine 3 [5.01].

[5.01]

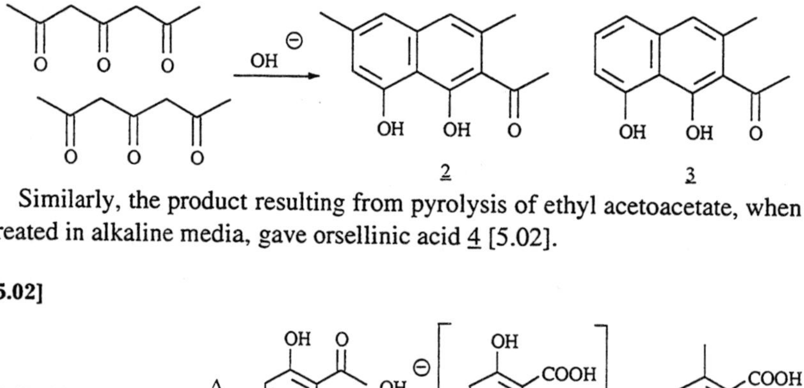

Similarly, the product resulting from pyrolysis of ethyl acetoacetate, when also treated in alkaline media, gave orsellinic acid 4 [5.02].

[5.02]

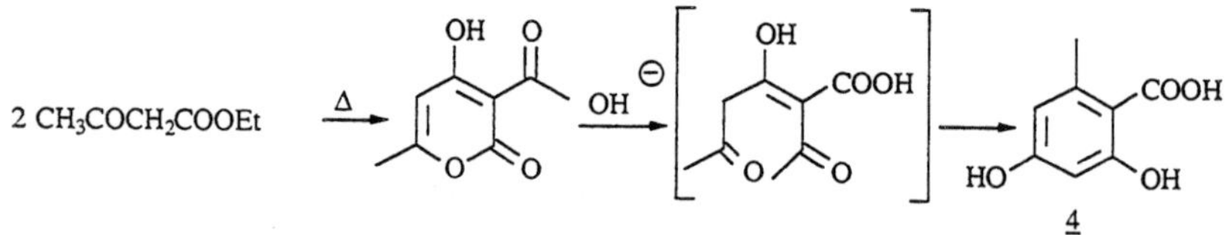

2 CH_3COCH_2COOEt

Collie was struck by the similarity of 2 with certain natural naphthalenic compounds and by the existance of orsellinic acid in natural products derived from lichens. He therefore advanced the hypothesis that such compounds could be formed in nature from "acetic units."

After having fallen into neglect for nearly half a century, this hypothesis was finally revived in 1953 by Birch and co-workers who, after first verifying and developing Collie's original studies, conducted their own experimental studies on the biosynthesis of natural aromatic products with the aid of sodium acetate labeled with ^{14}C. The biosynthesis of methylsalicylic acid metabolized by *Penicillium griseofulvum* affords the following labeling pattern in which the asterisks show the position of the incorporated ^{14}C [5.03].

[5.03]

$$4\ CH_3\text{-}^{14}COOH \longrightarrow$$

Meanwhile, the Swedish school of Gatenbeck examined in detail the biosynthesis of orsellinic acid 4 in haetomium cochliodes starting from doubly labeled acetic acid: $CH_3\text{-}^{14}C^{18}OOH$ [5.04].

[5.04]

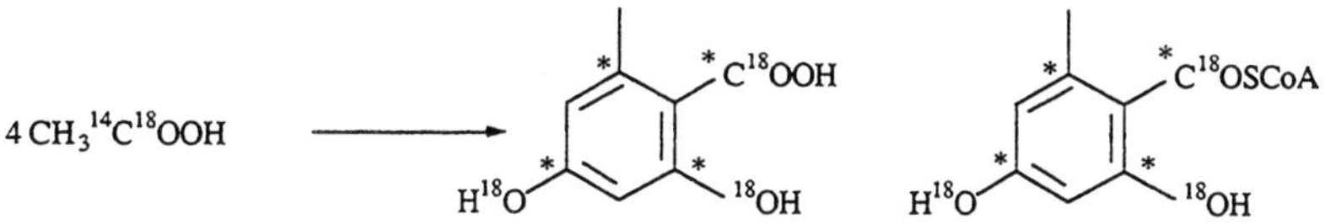

$$4\ CH_3{}^{14}C^{18}OOH \longrightarrow$$

Orsellinic acid was not isolated in its free state, but rather as 4a in which it is combined with coenzyme A (HSCoA). Hydrolysis of 4a gives the free acid 4 which is then decarboxylated. It was noted that the quantity of ^{18}O in the total CO_2 obtained from the group $-C^{18}OOH$ was found to be half that coming from the hydroxyl groups. Furthermore, the total degradation of the molecule showed that the distribution of radiocarbon was as expected.

These results taken together confirm the postulated condensation of active acetate units to give polyketomethylene acids as intermediates. This is known as the "polyacetic" rule of Birch.

Additionally, the fact that orsellinic acid was isolated as a complex with coenzyme A demonstrates that the latter substance intervenes in the biosynthesis. Before entering into the details of this biosynthesis, however, it is necessary to return to some fundamental problems of biochemistry.

Reminder of some notions of biochemistry

Recent studies have been carried out on the chemical composition of the simplest cells, those of the mycoplasms which constitute a very important family in the realm of bacteria. Roux and Nocard first identified a species of this genus (*M. mycoides*) as being the agent of cattle peripneumonia. French phytopathologists also discovered mycoplasms which were pathogenic to plants. Yellow perishing of lavandin and an affliction of tomatoes are also due to a new species of mycoplasm; insects are generally the vectors of these mycoplasmoses.

These studies have also suggested that the first cell that appeared on Earth was probably made up of a combination of only 30 different organic molecules. This group of 30 primordial biomolecules would have comprised 20 amino acids, 5 nitrogenous aromatic bases, 1 fatty acid (palmitic acid), 2 sugars (α-D-glucose and α-D-ribose), an alcohol (glycerol) and an amine (choline). In fact, this list can be shortened to 25 because recent research relating to the genetic code suggests that the first living cell needed only 16 amino acids of the 20 known to be present at this time in proteins. Whatever their precise number, the essential biomolecules must be considered the ancestors of all other biomolecules. They represent the first alphabet of living matter. Phosphoric acid should also be added to these.

At first appearance, these different biomolecules do not seem to manifest any underlying chemical interrelationship. In fact, however, they are linked by the enzymatic metabolic pathways through the successive reactions that share common intermediates. For example, although glucose, palmitic acid and alanine appear to be entirely different, studies using isotopic labeling show that in the course of metabolism, all of the carbons of glucose can be used by the living cell to form the carbon skeleton of alanine [$CH_3CH(NH_2)$-COOH] and that four of its six carbon atoms form the carbon skeleton of palmitic acid through the mediation of acetic acid.

Living organisms have evolved into extremely differentiated forms with new, more complex and varied biomolecules being formed, however, these are still structural and metabolic derivatives of the 30 primordial biomolecules.

Numerous specialized biomolecules are extremely complex and appear to have little similarity with the 30 primordial biomolecules. (Among these are the pigments, the constituents of waxes and essential oils, the hormones, vitamins, antibiotics, alkaloids, etc.) Nevertheless, recent research into the biogenesis of several of these substances shows that they can be classed into a few categories which are all derived from the primordial biomolecules or their transformation products: acetogines, terpenes, alkaloids, . . . Most of the alkaloids for example, come from the primordial amino acids.

While avoiding all metaphysical considerations, one can still ask what the origin of these biomolecules might be.

Various studies suggest that at the dawn of the Earth's history, particular conditions favored the existence at high concentrations of numerous different organic compounds present on the surface of the ocean waters. The first living cell would have originated in this sort of "soup" which was rich in organic compounds. It now seems to be accepted that the age of the Earth is around 4.5 billion years: This is the date given by geologists to estimate the date at which the superficial crust of the Earth stabilized. Study of fossils shows that bacteria similar to those known today existed about 3 billion years ago.

Around 1920, a Russian biochemist, Oparin, and later an English biochemist, Haldane, suggested that the physical and chemical processes which occurred in the primitive atmosphere resulted in the formation of simple organic compounds such as the amino acids and the sugars. These compounds may have formed from methane, ammonia and water vapor, which were supposedly present in the primitive atmosphere. According to this theory, the gases were activated by radiant solar energy and electrical discharges (lightening) and reacted with each other. The simple organic molecules thus formed, condensed and dissolved in the primitive ocean. Oparin supposed that the first living cells appeared in that warm solution of organic products.

It is obvious that a hypothesis as revolutionary as this was not immediately accepted because of the lack of empirical proof and knowledge of the constitution of the atmosphere during the period corresponding to the appearance of life. The most recent research indicates that 3 to 3.5 billion years ago, the atmosphere was rich in nitrogen, hydrogen, CO_2 and CO but its ammonia and methane content was probably lower. There was no free oxygen! Furthermore, it is believed that the temperature of the Earth's crust was around 70°–80°C during that period and that a large part of its surface was covered with water.

The first laboratory experiments to address this were undertaken around 1950 by, essentially and independently, Miller and Fox, and Pavlovskaya and Matheus and were aimed at examining gaseous mixtures which simulated the composition of the primitive atmosphere. The first pair of researchers submitted these mixtures to electrical discharges and was able to show evidence for the generation of amino acids. The other two used electrical discharges and high energy irradiation. They were able to observe the formation of elementary sugars. This series of studies continued until nearly 1960. Since then, it has become certain that almost any atmosphere *without any excess of oxygen* (a situation in which the Earth necessarily found itself) and using just about *any source of energy* ($h\upsilon$, electrical sparks, radioactivity, etc.) was sufficient to generate what modern analytical techniques showed to be the majority of the elementary biomolecules.

In this vein, Ponnamperuma and his colleagues and Calvin and collaborators were able, in 1963, to obtain adenine by irradiating a mixture of CH_4, NH_3 and water. Adenine accompanied by a little guanine was also obtained by irradiating an aqueous solution of hydrocyanic acid with ultraviolet light [5.05].

[5.05]

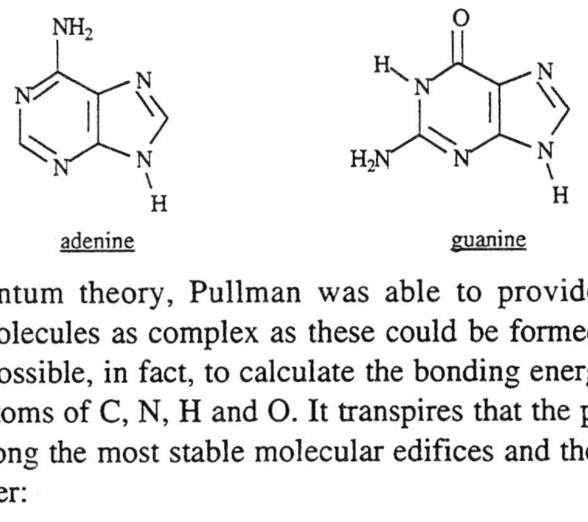

adenine guanine

Using the quantum theory, Pullman was able to provide some basis for explaining how molecules as complex as these could be formed under the above conditions. It is possible, in fact, to calculate the bonding energy of the different combinations of atoms of C, N, H and O. It transpires that the purine and pyrimidine bases are among the most stable molecular edifices and their relative stabilities follow the order:

adenine>guanine>cytosine>thymine.

One can therefore see the feasibility of this phenomenon by imagining that irradiation will fragment the molecules and these will then reorganize to give ultimately the most stable combinations.

One conclusion of these brief considerations could well be the absence of continuity between what is living and what is not. One problem is still outstanding and that is the origin of optical activity. Some extremely original explanations have recently been proposed, but they will not be discussed here.

Some Aspects of Enzymology and Biochemistry

Mechanisms of biosynthesis

Despite the fascinating nature of these studies, it is difficult to imagine how substances as complex as proteins (which constitute 50% by weight of cells) and enzymes could have developed. The latter form a class of proteins that is both vast and highly specialized. They represent the primary element in expressing the action of genes by catalyzing thousands of chemical reactions that are involved in the metabolism of cells.

The history of enzymology is in large part the history of biochemistry itself. The discovery and the first studies of enzymatic digestion in the stomach, between 1760 and 1825, constituted the first major experiments in chemical catalysis. Although Pasteur had recognized that fermentation is catalyzed by enzymes, he affirmed in 1860 that these enzymes are inextricably tied to the structure and the life of yeast cells. Thus, it was an event of prime importance in the history of enzymology when, in 1897, Buchner succeeded in extracting from yeast the enzymes responsible for alcoholic fermentation. This showed clearly that these enzymes could function independently of all cellular structures. It was, however, a long time before an enzyme was crystallized and purified. This was achieved by Summer in 1926 with the Jack bean urease enzyme. Summer proved that the crystals he had obtained were of proteic nature and he concluded that all enzymes were proteins, in contradiction with the generally held opinion of that time. His ideas were not immediately accepted and it was only between 1930 and 1936, when Northrop isolated pepsin, trypsin and chymotrypsin in crystalline form, that the proteic nature of enzymes was finally accepted.

Presently, more than 1,000 different enzymes have been identified. Many have been completely purified and more than 150 have been crystallized. Although the majority of the enzymes responsible for the fundamental metabolic routes and for cellular regulation are now well known, there are good reasons—genetic in particular—for believing that many remain to be identified. Over recent years, extremely important discoveries have opened new perspectives in enzymological research and have contributed to a better comprehension of the role of enzymes in cellular biology. These discoveries involve the genetic control of enzyme synthesis, the mechanisms of autoregulation to which numerous enzymes are submitted and the role of enzymes in the development and differentiation of cells. Without going into details on the classification of enzymes, it should be simply noted that an international system of nomenclature covers the chemical reactions catalyzed by enzymes and is based on the nature of the reactions themselves. These systematic names are used each time precise identification of the enzyme is necessary. As some of these systematic names are quite long, however, common (trivial) names are still currently used, as shown in this example:

Common name: hexokinase
Systematic name: ATP, hexose phosphotransferase enzyme
Reaction catalyzed: ATP + glucose → glucose–6-phosphate + ADP

Like other proteins, enzymes can also be classified according to their chemical composition, into holo- and heteroenzymes.

Holoproteins, for which hydrolysis yields only amino acids.

Heteroproteins, for which hydrolysis yields amino acids plus other organic and inorganic compounds. These are called "prosthetic groups" (nucleoproteins, lipoproteins, phosphoproteins, metalloproteins, glycoproteins).

There does exist, however, a more functionally significant way of classifying enzymes. Some enzymes owe their activity only to their proteic structure, while others require, in addition, nonproteic groups or "cofactors." A cofactor can either be a complex organic molecule called a "coenzyme" or a metal ion, both of which are sometimes necessary. The cofactors are usually thermostable while the majority of proteins are thermolabile. Enzymes that require a cofactor will bind it with variable affinities. In the majority of cases, the cofactor that is essential for activity can be separated from the enzyme protein by dialysis or by other means, although some cofactors are bound to the protein molecule by covalent bonds. The intact enzyme-cofactor complex is called a "holoenzyme." After elimination of the cofactor, the remaining protein is called the "apoenzyme."

When enzymes use metal ions as cofactors, these ions can play one of two possible roles: either they act by establishing a bridge between the enzyme and its substrate by formation of a coordination complex or they themselves exert some catalytic activity. It is known, for example, that the iron atoms of the catalases, which are capable of decomposing hydrogen peroxide, are the catalytic centers of the enzyme. Simple iron salts display a certain potential for decomposing hydrogen peroxide, but this potential is greatly enhanced by the enzyme protein.

Coenzymes usually function as intermediates in the transport of electrons, of atoms or of functional groups which are transferred in the overall reaction that is being catalyzed. Some coenzymes are very strongly bound to the enzyme protein. These are usually called "prosthetic groups," as in the case of ordinary heteroproteins.

The principal coenzymes encountered in the biosynthesis of natural products are the following:

- Nicotinamide adenine dinucleotide (NAD): transfers hydrogen atoms and electrons;
- Nicotinamide adenine dinucleotide phosphate (NADPH): idem;
- Coenzyme Q: idem;
- Coenzyme A: transfer of acyl groups (particularly acetyl);
- Biocytin: transfer of carbon dioxide.

The synthesis of one of these coenzymes, coenzyme A, will be discussed in detail, however, before doing so, it may be useful to recall the structures of a certain number of fundamental substances, starting with D-ribose.

This substance can be represented by the conventional Fischer projection by the structure a (deriving from D-glyceraldehyde) which indicates three asymmetric carbons. The phenomenon of mutarotation has shown that the numerous "oses" behave in solution as if they possessed a supplementary asymmetric carbon atom compared with those indicated for the linear Fischer projections. The corresponding structures result from the formation of an internal hemiacetal for which the cyclic structure is a furan or pyran derivative. Thus, D-ribose can exist in the form b (β-D-ribofuranose) or in the form c (β-D-ribopyranose). Forms b and c are Haworth projections in which the thickness of the lines indicates three-dimensional proximity to the reader. Even though these planar representations do not reflect the real conformations of these substances, they remain in common use because of their simplicity [5.06].

[5.06]

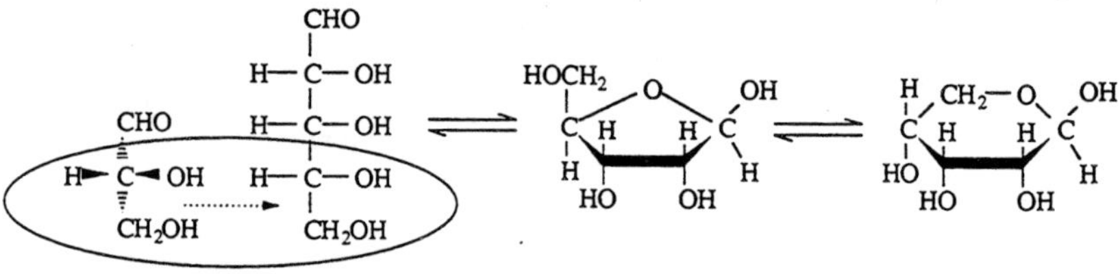

D-glyceraldehyde a b c

Another biologically important substance is the purine d, the result of fusing pyrimidine with imidazole, an example being the fundamental substance, 6-aminopurine or adenine e [5.07].

[5.07]

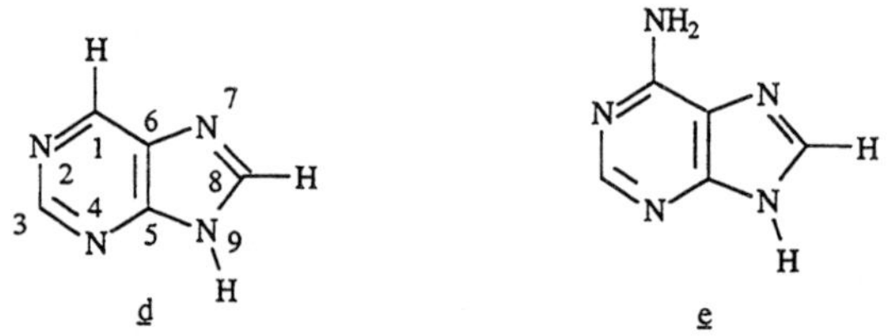

d e

The condensation of an "ose" with a purine base gives a nucleoside. In the case of β-D-ribofuranose b and adenine e the nucleoside is adenoside f (9-β-ribofuranosyladenine) often represented more simply by structure f' [5.08].

The total synthesis of adenosine was described by Lord Todd and his co-workers in 1948 starting from uric acid g. When a nucleoside such as adenosine is esterified with phosphoric acid, it gives a mononucleotide; in this case, 3'-adenylic acid or 3'-adenosine phosphoric acid h. This acid results principally from hydrolysis of RNA, particularly when catalyzed enzymatically by nucleases [5.09].

Esterification of the 5'-hydroxyl with phosphoric acid leads to another

[5.08]

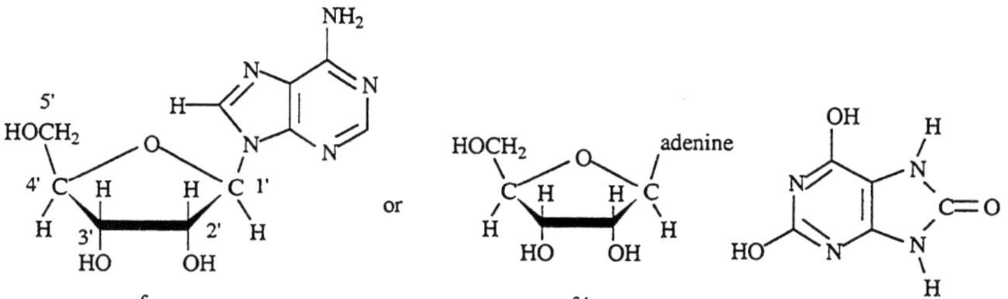

f or f'   g

[5.09]

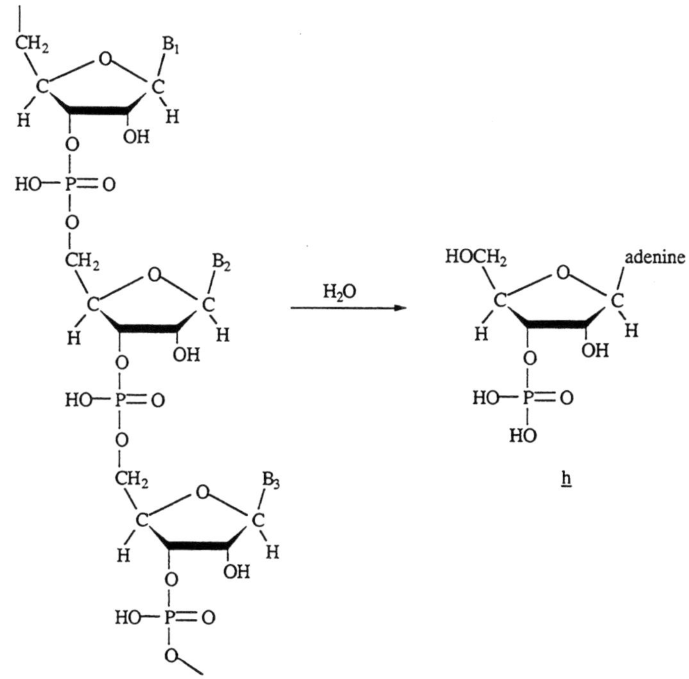

part of RNA $\xrightarrow{H_2O}$ h

mononucleotide called adenosine monophosphate or AMP. This can be transformed into a dinucleotide, adenosine diphosphate or ADP and into a trinucleotide, adenosine triphosphate or ATP.

ATP was isolated from extracts of muscle tissue in 1929 by Fiske and Subbarov. In 1937, Warburg showed that the formation of ATP is coupled to the dehydrogenation of 3-phosphoglyceraldehyde and between 1939 and 1941, Lipman demonstrated the essential role of ATP in the transfer of energy [5.10].

There are numerous important functions performed by ATP (as there are also for its analogues in which the adenine base has been replaced by other bases such as guanine, cytosine and uracil). ATP is the principal chemical reservoir of energy in cells. It is also a transporter of the phosphoryl group for which the bonds

[5.10]

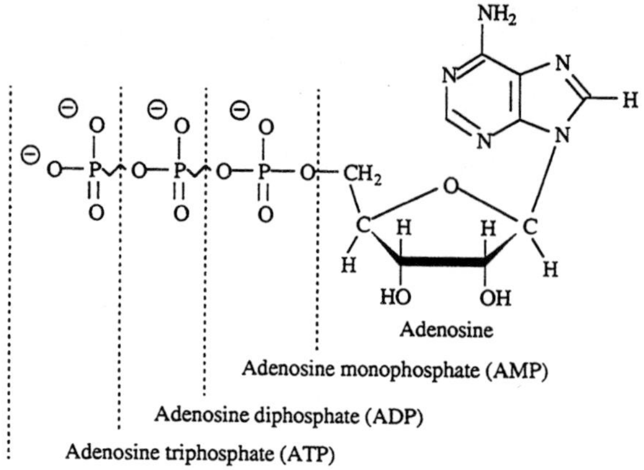

Adenosine

Adenosine monophosphate (AMP)

Adenosine diphosphate (ADP)

Adenosine triphosphate (ATP)

indicated by (~) are rich in energy. After dephosphorylation, the resulting ADP and AMP are once again phosphorylated in cellular respiration to regenerate ATP.

A second important function of ATP and of ADP corresponds to their intervention as coenzymes in biosynthetic transfer reactions. ATP acts as an energy rich precursor for mononucleotide units and nucleic acids. This is its third important function. During the biosynthesis of DNA and RNA, ATP, like its analogues, loses its terminal pyrophosphate group to become a mononucleotide unit of the nucleic acids.

At pH 7, ATP and ADP are both strongly charged anions. ATP has four ionizable hydrogens and three of these have pK values between 2 and 3. They are consequently completely dissociated at pH 7. The other proton, having a pK_a of 6.5, is 75% dissociated at pH 7. ADP has three ionizable hydrogens of which two are totally dissociated at pH 7, while the third, with a pK_a of 7.2, is 39% dissociated at this pH.

In intact cells, there are only traces of free ionized ATP and ADP. They are present as magnesium salts due to the high concentration of Mg^{2+} in the intracellular liquid. In numerous enzymatic reactions in which ATP intervenes as a donor of the phosphoryl group, its active form is $MgATP^{2-}$. In other reactions, the most active form is MnATP.

Having reviewed these fundamental facts, it is now possible to return to the structure of coenzyme A and its synthesis. This substance was isolated during the years 1947 and 1948 by the Lipman and Kaplan groups. They, as well as Lynen's group, demonstrated its precise role in transporting acetyl groups (acyl groups, in general). It has the following structure [5.11].

In the course of their work, Lipman and Kaplan and their co-workers showed that numerous reactions depended on the transfer of acetyl groups. It was necessary for the coenzyme to be thermostable and to combine with acetic acid to generate a donor of the activated acetyl group. They isolated coenzyme A (abbreviated to HSCoA where A is for acetyl). The acetylated form of coenzyme A (Acetyl CoA or Acetyl SCoA) is the thioester of acetic acid with the thiol group at the

[5.11]

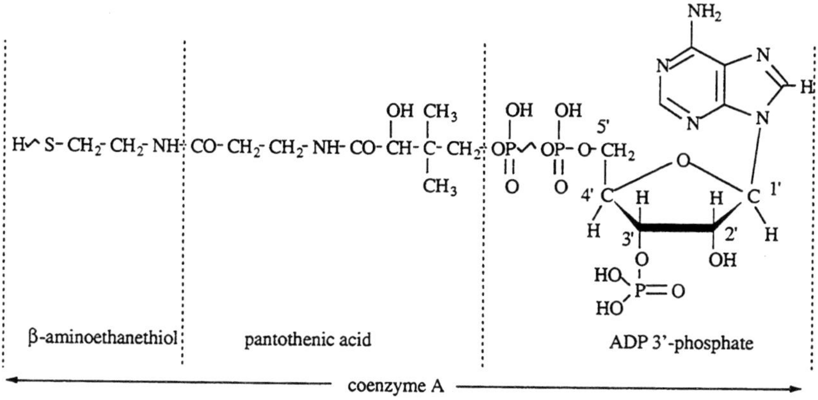

extremity of the coenzyme A molecule. This thioester bond is rich in energy (~), which means that its hydrolysis is accompanied by a strongly negative free standard enthalpy.

$$\text{Acetyl SCoA} + H_2O \rightarrow CH_3COOH + \text{HSCoA} \; \Delta G'_0 = -7.5 \text{ Kcal.mole}^{-1}$$

Examination of the structure of coenzyme A suggests that its synthesis should involve on the one hand a synthesis of pantothenic acid and its condensation with β-aminoethanethiol and on the other hand, the synthesis of adenosine. Finally, as a last step, it would be necessary to introduce the pyrophosphate moiety onto one of these two units and to weld the two sections together. This total synthesis was performed by Khorana and Moffat after seven years of research developing methods for the synthesis of dissymmetric pyrophosphates. The principal steps of this total synthesis will be discussed briefly.

Synthesis of pantothenic acid (vitamin B3)

Following are the principal steps of the synthesis of pantothenic acid (vitamin B3) [5.12]:

[5.12]

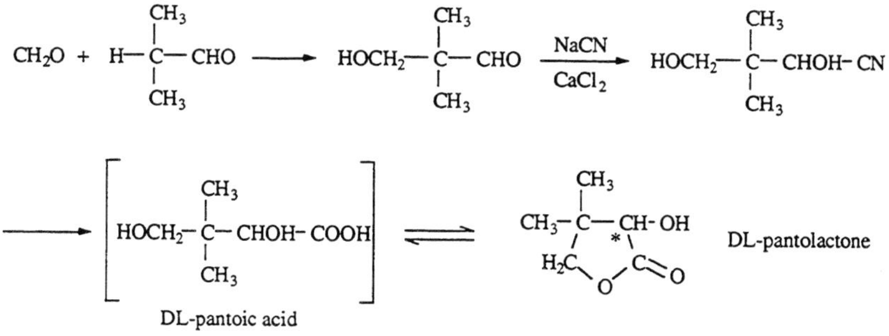

Vitamin B_3 is D(+) pantothenic acid. It can be obtained industrially by separating the sodium salt of DL-pantoic acid into its enantiomers with the aid of quinine hydrochloride. Pantolactone D(–) is then condensed with β-alanine in methanolic solution to obtain D(+) pantothenic acid [5.13].

[5.13]

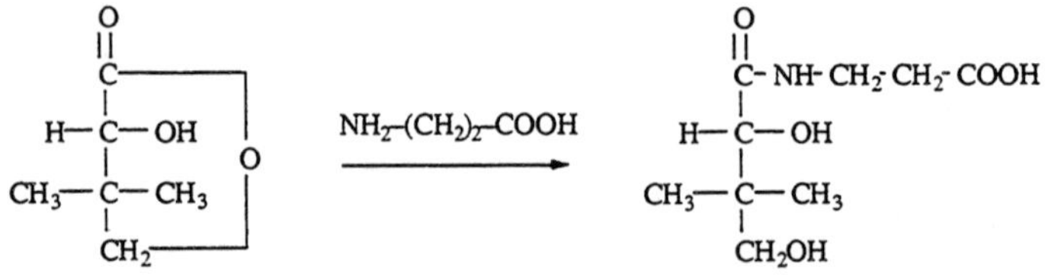

$$\underset{\text{D(-)-pantolactone}}{\text{H–C(OH)(C=O)–C}(CH_3)_2\text{–}CH_2\text{–O (lactone)}} \xrightarrow{NH_2\text{–}(CH_2)_2\text{–}COOH} \underset{\text{D(+) pantothenic acid}}{\text{O=C(NH–}CH_2\text{–}CH_2\text{–COOH)–CH(OH)–C}(CH_3)_2\text{–}CH_2OH}$$

It is necessary beforehand to protect the secondary hydroxyl in the form of its benzyl ether for the rest of the synthesis [5.14].

[5.14]

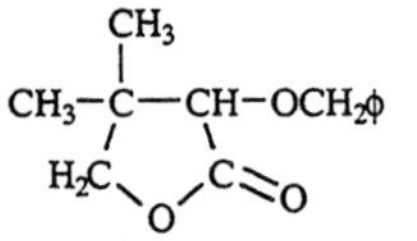

$$CH_3\text{–}C(CH_3)\text{–}CH\text{–}OCH_2\phi \text{ (ring: } H_2C\text{–O–C=O)}$$

This furnishes pantothenic acid with its secondary hydroxyl still blocked [5.15].

[5.15]

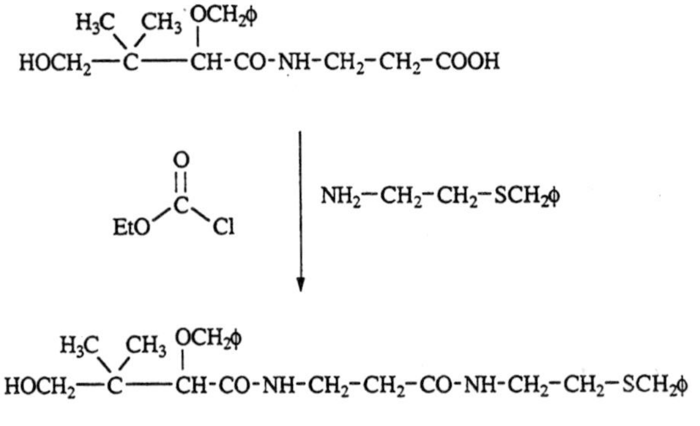

$$HOCH_2\text{–}C(CH_3)_2\text{–}CH(OCH_2\phi)\text{-CO-NH-}CH_2\text{–}CH_2\text{–}COOH$$

$$\downarrow \quad EtO\text{–C(=O)–Cl}, \quad NH_2\text{–}CH_2\text{–}CH_2\text{–}SCH_2\phi$$

$$HOCH_2\text{–}C(CH_3)_2\text{–}CH(OCH_2\phi)\text{-CO-NH-}CH_2\text{–}CH_2\text{-CO-NH-}CH_2\text{–}CH_2\text{–}SCH_2\phi$$

This benzyl ether of pantothenic acid is then treated with the benzyl ether of β-aminoethanethiol in the presence of chloroethoxycarbonyl to give a diether which, under the action of an excess of dibenzylchlorophosphoridate at –20°C, gives the compound shown below [5.16].

[5.16]

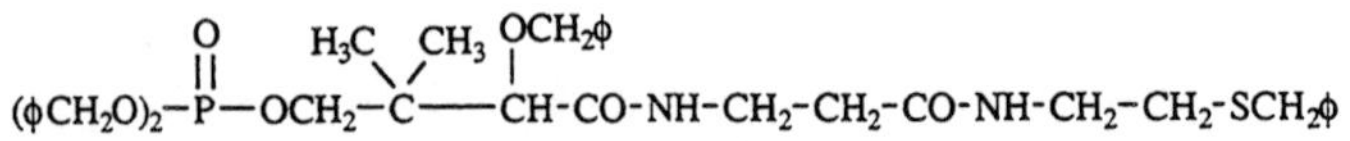

$$(\phi CH_2O)_2\text{–P(=O)–}OCH_2\text{–}C(CH_3)_2\text{–}CH(OCH_2\phi)\text{-CO-NH-}CH_2\text{–}CH_2\text{-CO-NH-}CH_2\text{–}CH_2\text{-}SCH_2\phi$$

Hydrogenolysis of the four benzyl ether functions using sodium in ammonia finally results in DL-pantetheine–4'-phosphate [5.17].

[5.17]

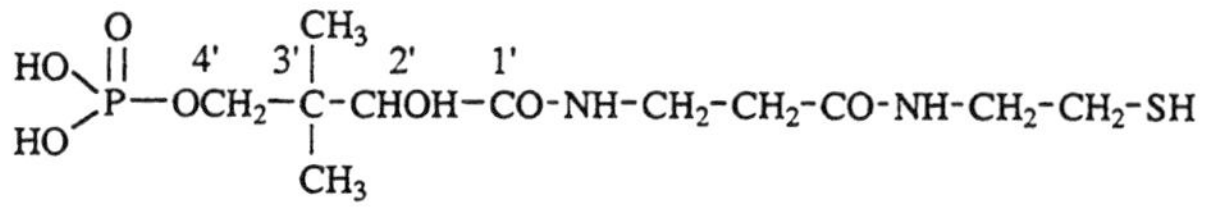

$$HO\text{-}P(=O)(OH)\text{-}OCH_2\text{-}C(CH_3)_2\text{-}CHOH\text{-}CO\text{-}NH\text{-}CH_2\text{-}CH_2\text{-}CO\text{-}NH\text{-}CH_2\text{-}CH_2\text{-}SH$$

(positions 4', 3', 2', 1' marked on the chain)

Synthesis of adenosine (Lord Todd and collaborators, 1948)

Uric acid, or more exactly its dipotassium salt, is the starting material. When treated with phosphorus oxychloride, it gives 1,3-dichloro–8-hydroxypurine which is prepared as its ammonium salt [5.18].

[5.18]

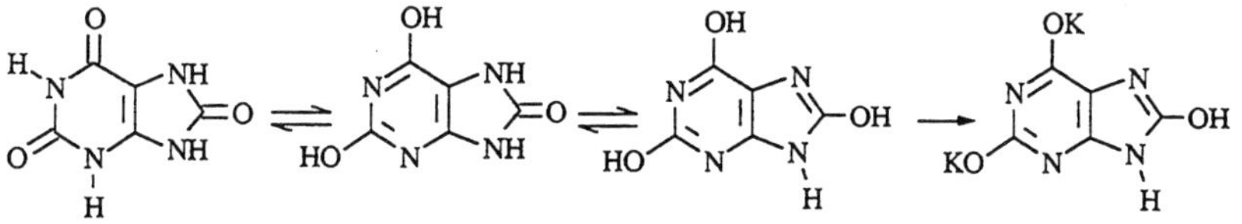

This salt (dry), in refluxing dimethylaniline and phosphorus oxychloride, is transformed into 1,3,8-trichloropurine, which when treated at 100°C with a saturated aqueous solution of ammonia, undergoes quantitative transformation into 3,8-dichloroadenine which subsequently affords a silver derivative under the action of silver nitrate in ammonia solution [5.19].

[5.19]

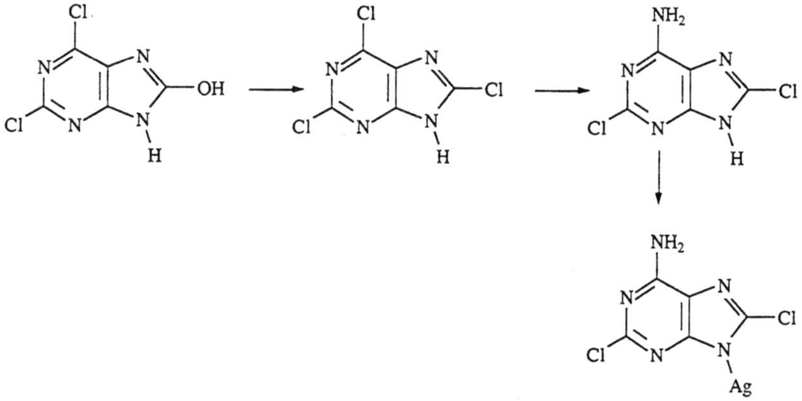

This derivative is then condensed with chlorodiaceto-D-ribofuranose which is prepared in the following manner.

It should first be recalled that triphenylmethyl chloride, more commonly called trityl chloride, reacts specifically with primary alcohols to give the corresponding trityl ethers. Thus, when D-ribose is treated with trityl chloride, its trityl ether is obtained. Under the action of acetic anhydride in pyridine, this gives a triacetylated trityl derivative. It is necessary to block the primary alcohol function to prevent the formation of ribopyranose in the course of this acetylation. The primary alcohol function must then be regenerated to continue this acetylation to the tetraacetylated stage. Once this has been done, treatment with hydrochloric acid in dietyl ether forms triacetylchloro-D-ribofuranose because only the hemi-acetal hydroxyl is transformed into its chloro derivative [5.20].

[5.20]

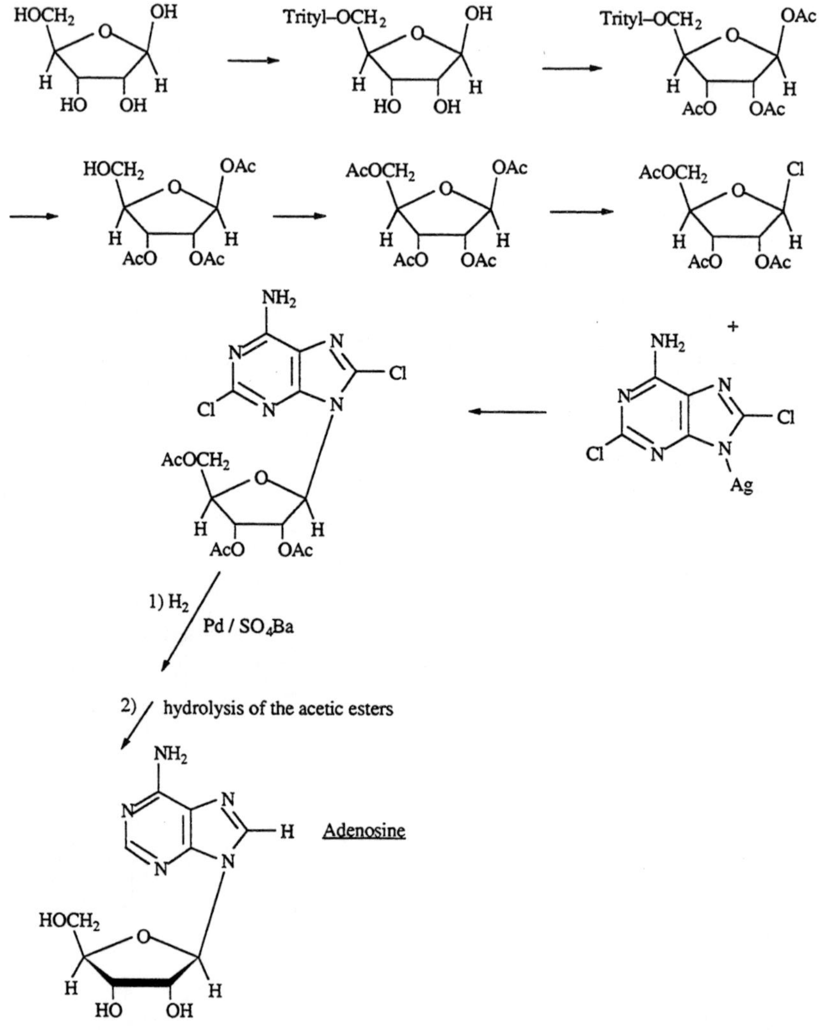

1) H_2 Pd / SO_4Ba

2) hydrolysis of the acetic esters

Adenosine

Adenosine

Coupling of pantetheine phosphate and adenosine, generation of coenzyme A

Adenosine is first treated with dibenzylchlorophosphoridate which leads to a mixture of three derivatives A, B and C. These are then hydrolyzed and hydrogenolyzed to give a mixture of equal quantities of the two derivatives D and E [5.21].

[5.21]

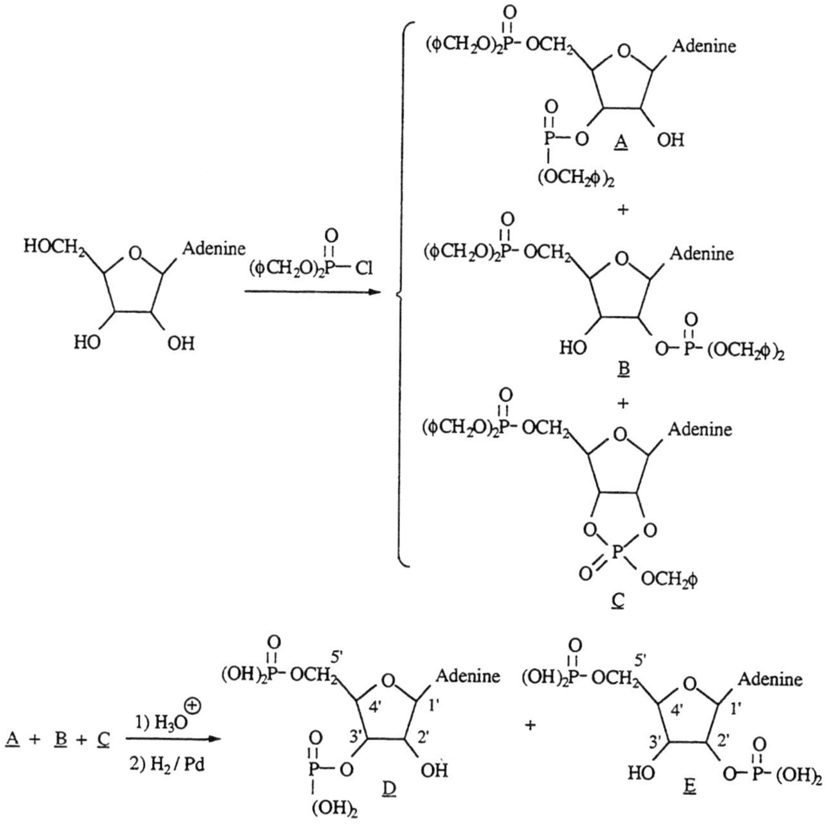

A + B + C → 1) $H_3O^{\oplus}$ 2) H_2 / Pd → D + E

The Khorana method intervenes essentially at this stage. It should be recalled that N,N'-dicyclohexylcarbodiimide, or DCC, is a powerful dehydrating agent. Using DCC in the presence of triethylamine, the mixture of D + E gives the derivative F [5.22].

[5.22]

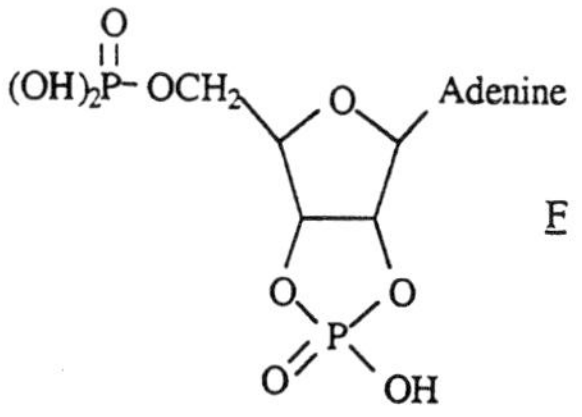

F

It can be noted that morpholine and DCC react to give 4-morpholino-N,N'-dicyclohexylcarboxyamidine (MDCCA). Under the action of this amidine, compound F is transformed into G [5.23].

[5.23]

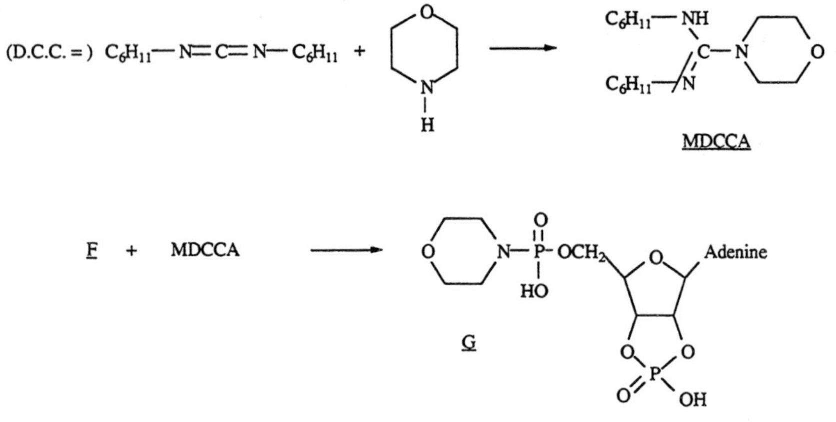

(D.C.C. =) $C_6H_{11}—N=C=N—C_6H_{11}$ + morpholine ⟶ MDCCA

F + MDCCA ⟶ G

Condensation of pantetheine phosphate with this product (G) in the presence of anhydrous pyridine obtains the derivative H, which is finally hydrolyzed to a mixture containing nearly equal quantities of coenzyme A and isocoenzyme A [5.24].

[5.24]

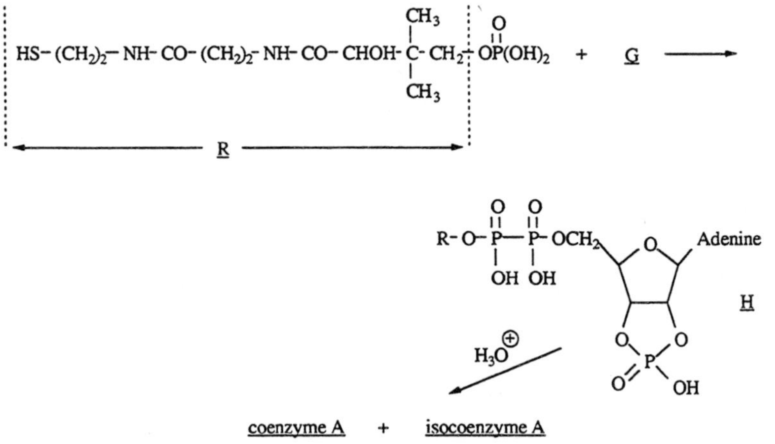

$HS-(CH_2)_2-NH-CO-(CH_2)_2-NH-CO-CHOH-C(CH_3)_2-CH_2-OP(O)(OH)_2$ + G ⟶ H

R

H ⟶ ($H_3O^{\oplus}$) coenzyme A + isocoenzyme A

These two coenzymes can then be separated and purified by ion-exchange chromatography (cellulose DEAE in its chloride form).

This constitutes the total synthesis *"in vitro"* of coenzyme A. This leads to the question of how it is generated *"in vivo,"* or in other words, what is its biosynthesis. In 1959, G. M. Brown demonstrated the series of processes summarized in the following scheme [5.25].

Reaction 1 is catalyzed by an enzyme, pantothenic acid kinase; reaction 2 by a coupling enzyme; and reaction 3 by phosphopantothenylcysteinedecarboxylase.

Two other extremely important coenzymes that also intervene in biosynthetic processes are NAD and NADP. These are coenzymes associated with a class of enzymes called "pyridine nucleotide dehydrogenases" and which are involved

[5.25]

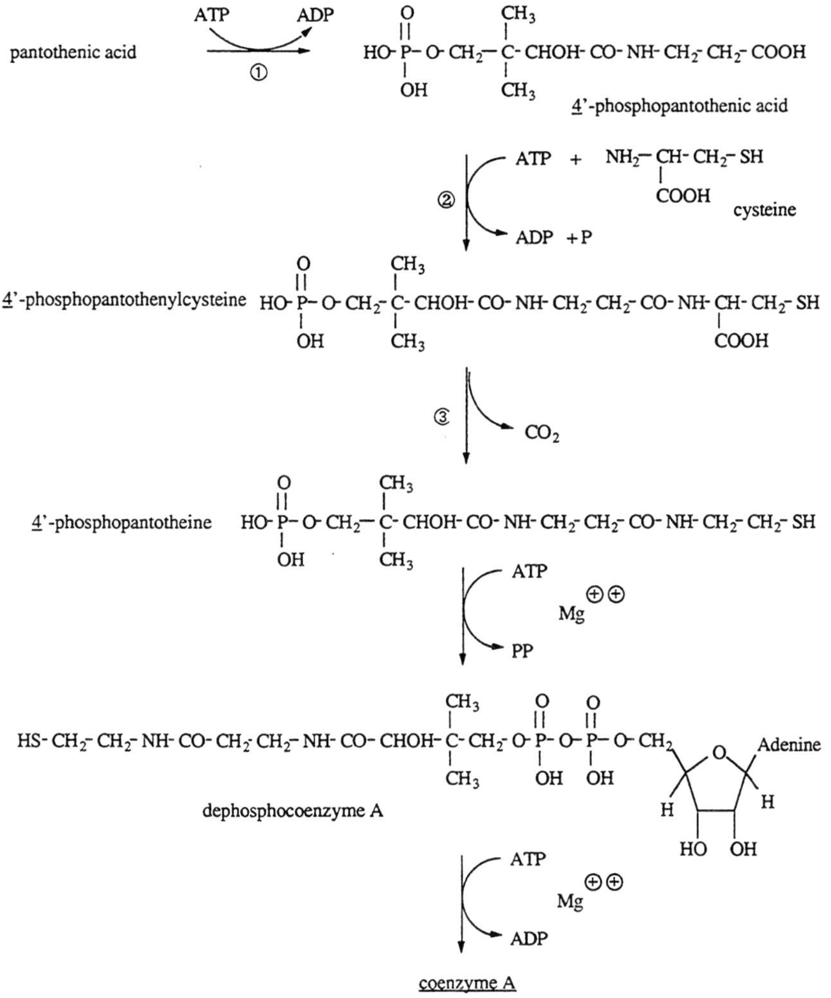

pantothenic acid —(① ATP → ADP)→ $HO-P(=O)(OH)-O-CH_2-C(CH_3)_2-CHOH-CO-NH-CH_2-CH_2-COOH$

4'-phosphopantothenic acid

② ATP + $NH_2-CH(COOH)-CH_2-SH$ (cysteine) → ADP + P

4'-phosphopantothenylcysteine $HO-P(=O)(OH)-O-CH_2-C(CH_3)_2-CHOH-CO-NH-CH_2-CH_2-CO-NH-CH(COOH)-CH_2-SH$

③ → CO_2

4'-phosphopantotheine $HO-P(=O)(OH)-O-CH_2-C(CH_3)_2-CHOH-CO-NH-CH_2-CH_2-CO-NH-CH_2-CH_2-SH$

ATP → PP (Mg^{2+})

$HS-CH_2-CH_2-NH-CO-CH_2-CH_2-NH-CO-CHOH-C(CH_3)_2-CH_2-O-P(=O)(OH)-O-P(=O)(OH)-O-CH_2$-(ribose: H, H, HO, OH)-Adenine

dephosphocoenzyme A

ATP → ADP (Mg^{2+})

coenzyme A

in redox reactions. More than 150 of these dehydrogenases are known and many have been obtained in crystalline form. They catalyze reactions of this type [5.26].

[5.26]

$$\text{reduced substrate} + NAD^{\oplus} \rightleftharpoons \text{oxidized substrate} + NADH + H^{\oplus}$$

The pyridine nucleotide dehydrogenases reversibly transfers two reducing equivalents from a substrate to the oxidized form of the pyridine nucleotide. One of these appears in the reduced pyridine nucleotide in the form of a hydrogen atom; the other hydrogen atom torn from the substrate is liberated into the medium in the form of a hydrogen ion.

NAD and NADP were isolated for the first time in 1935 by O. Warburg. The structure of NAD, or coenzyme I, has been determined thanks to the collaboration of several groups: Warburg (1936), von Euler (1936), von Euler and Karrer (1936) and Schlenk (1942).

The structure of NADP, or coenzyme II, was established by Kornberg between 1948 and 1950 [5.27].

[5.27]

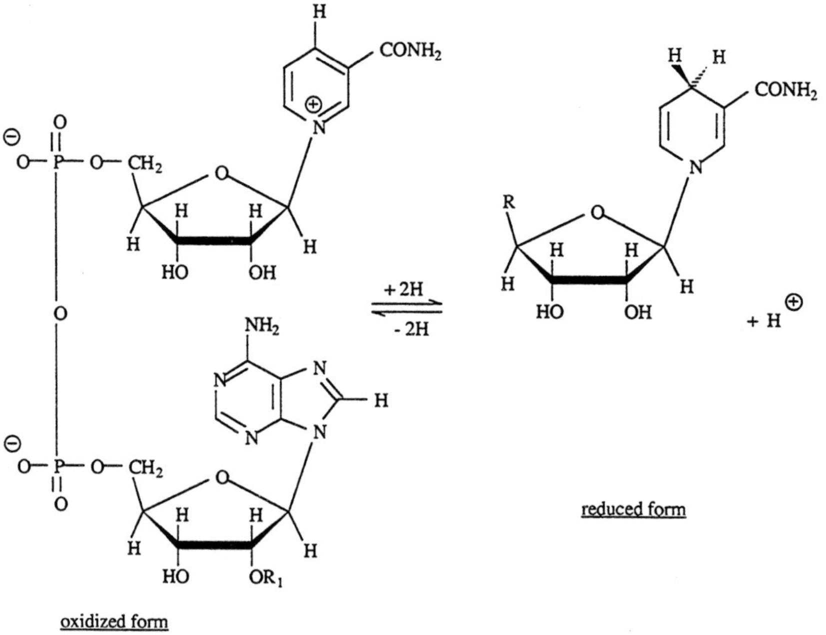

oxidized form

+2H / -2H

reduced form

$+ H^{\oplus}$

R_1 = -H nicotinamide-adeninedinucleotide = NAD

R_1 = $-PO_3H$ nicotinamide-adeninedinucleotide phosphate = NADP

The first part of these coenzymes to be isolated was nicotinic acid, which exists in these coenzymes in the form of nicotinamide (still called vitamin PP or antipellagic vitamin due to its function as an indispensible factor in mammalian nutrition) [5.28].

[5.28]

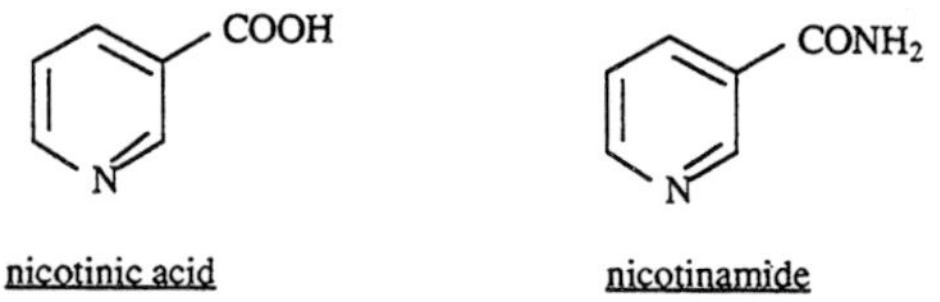

nicotinic acid

nicotinamide

The synthesis of NAD and of NADP was achieved by Todd in 1957. Lord Todd and his co-workers obtained NAD in its two anomeric forms α and β in the ratio of 4 to 1, whereas only the non-natural α form of NADP was obtained.

NAD and NADP are found in all types of cells with NAD generally being at a higher concentration than that of NADP. Dehydrogenation of NAD intervenes especially in respiration and it participates in the transfer of electrons from substrates to oxygen. The NADP dehydrogenases are principally associated with the transfer of electrons in biosynthetic reductions.

The characteristic physicochemical modifications that appear during enzymatic reduction of NAD^+ and $NADP^+$ can be exploited for their dosage. In their oxidized forms, these coenzymes only absorb intensely near to 250 nm (adenine nucleus). In their reduced forms, a new band appears at 340 nm with the other diminishing slightly—the band at 340 nm corresponds to the reduction of the pyridinic nucleus of the nicotinamide [5.29].

[5.29]

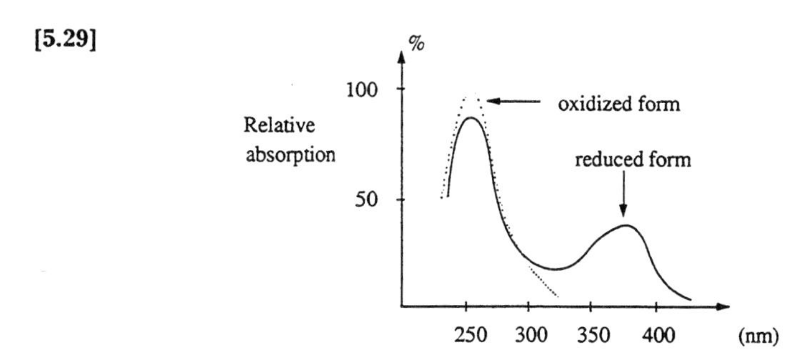

A rather simple example of the intervention of NAD is found in the oxidation of glyceraldehyde 3-phosphate to give 1,3-diphosphoglycerate. This is one of the most important steps in the sequence of glycolysis, because it permits the oxidation energy of the aldehyde group of glyceraldehyde 3-phosphate to be conserved in the form of an energy-rich phosphate bond in the product, 1,3-diphosphoglycerate. The mechanism of this reaction was demonstrated between 1937 and 1938 by Warburg and co-workers and is considered one of the most important discoveries of biochemistry because it revealed for the first time, a chemical and enzymatic mechanism through which energy liberated in the oxidation of an organic molecule could be conserved in the form of ATP.

The enzyme which catalyzes this reaction, glyceraldehyde–3-phosphate dehydrogenase, is readily obtained in its crystalline form (NW = 140,000). The overall reaction catalyzed by this enzyme is the following [5.30]:

[5.30]

$$\text{(D)-glyceraldehyde-3-phosphate} + NAD^{\oplus} + Pi \rightleftharpoons \text{1,3-diphosphoglycerate} + NADH + H^{\oplus} \quad ①$$

The variation in standard free enthalpy in this reaction amounts to $\Delta G'_0 = +1.56$ $Kcal.mole^{-1}$ (Pi represents a molecule of inorganic phosphate) [5.31].

It is worthwhile to give some attention to 1,3-diphosphoglycerate, formed in the course of the enzymatic degradation of cumbustible cellular molecules (anaerobic fermentation of glucose, for example). In intact cells, 1,3-diphosphoglycer-

[5.31]

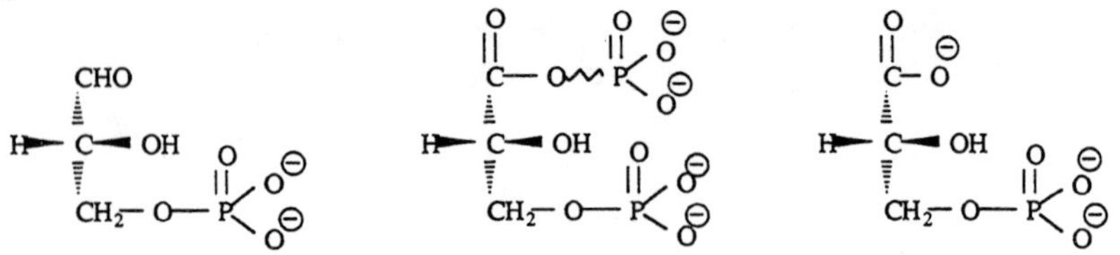

(D)-glyceraldehyde-3-phosphate

1,3-diphosphoglycerate
(mixed anhydride)

3-phosphoglycerate

ate is not hydrolyzed but its phosphate group in position 1 is transferred to ADP, thereby furnishing ATP and 3-phosphoglycerate. This reaction is catalyzed by 3-phosphoglycerate kinase ("ATP: 3-phosphoglycerate phosphotransferase" in systematic nomenclature).

1,3-diphosphoglycerate + ADP ↔ 3-phosphoglycerate + ATP ②

At pH 7.0, the equilibrium constant of reaction 2 is K'eq = 2070. From this, a standard free enthalpy of

$$\Delta G'_o = -2.303\ R.T \log K'_{eq} = -2.303 \times 1.987 \times (273° + 27°) \lg 2070$$
$$= \Delta G'_o = -2.303 \times 1.987 \times 300 \times 3.32 = -4557 \text{ cal.mole}^{-1}$$
$$= -4.56\text{-Kcal.mole}^{-1}.$$

Furthermore, the difference in the standard free enthalpy of the hydrolysis reaction of ATP to ADP and inorganic phosphate has been the subject of numerous experimental studies. It is [5.32]:

[5.32]

ATP + H_2O ⟶ ADP + Pi $\Delta G_0' = -7.3 \text{ Kcal.mole}^{-1}$ ③

the sum of 2 and 3 gives 4

1,3-diphosphoglycerate + H_2O ⟶ 3-phosphoglycerate + Pi + H + 4

$\Delta G_0' = -11.86 \text{ Kcal.mole}^{-1}$

Hydrolysis of the 1-phosphate group of 1,3-diphosphoglycerate is therefore a strongly exothermic reaction, clearly superior to the hydrolysis of ATP.

Return to reaction 1 for which $\Delta G'_o = + 1.56 \text{ Kcal.mole}^{-1}$. Such a value for the difference in standard free enthalpy indicates that this reaction can easily evolve in both senses, according to the relative concentrations of the starting materials and the products.

In order to analyze the energetic differences in this reaction, two distinct processes should be separated [5.33, 34].

By taking the sum of the two reactions 5 and retro–4, one obtains [5.35].

[5.33]

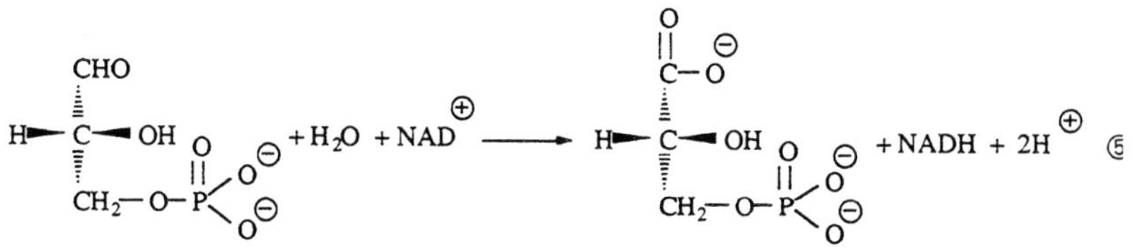

CHO; H–C–OH; CH_2–O–PO_3^{2-} + H_2O + NAD^+ ⟶ COO^-; H–C–OH; CH_2–O–PO_3^{2-} + NADH + $2H^+$ ⑤

$\Delta G_1^{o'} = -10.3$ Kcal.mole^{-1}

[5.34]

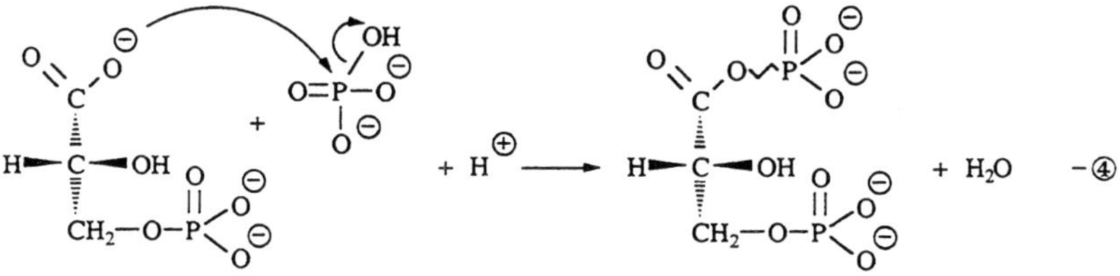

COO^-; H–C–OH; CH_2–O–PO_3^{2-} + HO–PO_3^{2-} + H^+ ⟶ C(=O)–O~PO_3^{2-}; H–C–OH; CH_2–O–PO_3^{2-} + H_2O –④

[5.35]

R–CHO + Pi + NAD^+ ⇌ R–C(=O)–O~P(=O)(O$^-$)$_2$ + NADH + H^+ ①

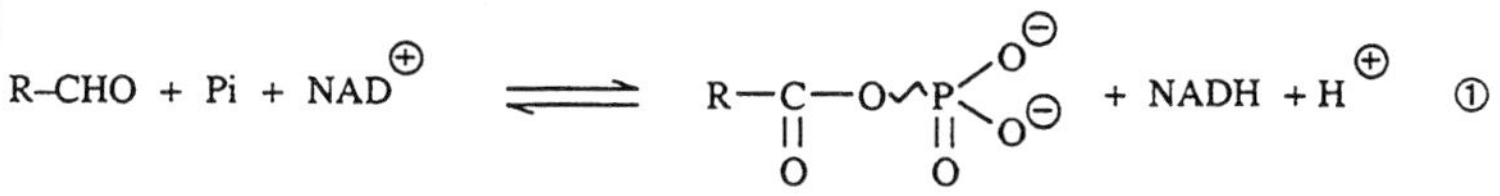

for which the free enthalpy difference corresponds to:

$$\Delta G_S^{0'} = \Delta G_1^{0'} + \Delta G_2^{0'} = -10.3 + 11.86 = +1.56 \text{ Kcal.mole}^{-1}.$$

Oxidation of R-CHO into R-COO^- by NAD^+ is a highly exothermic process ($\Delta G_1^{o'}$ = 10.3 Kcal.mole^{-1}) and this should normally evolve in the sense shown. The formation of 1,3-diphosphoglycerate (reverse of reaction 4) is strongly endothermic and can normally evolve in the senses shown. In the overall reaction, the endothermic process is coupled to the exothermic one and the energy liberated by the oxidation of the aldehyde is thereby transformed into an energy-rich phosphate bond (~) in 1,3-diphosphoglycerate.

Stereospecificity of the reduction of NAD^+ and $NADP^+$

Before embarking on this section, it is recommended, for a better understanding, to read Appendix 1 concerning the concept of prochirality.

In the course of the last 10 years, an increasing number of enzyme-catalyzed reactions have demonstrated the stereochemistry of processes in which identical atoms or groups attached to the same carbon atom behave differently. The simplest example is that of the hydrogen atoms of the methylene group of ethanol [5.36].

[5.36]

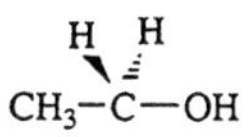

CH_3–C(H)(H)–OH

This concept signifies that enzymatic reactions are capable of distinguishing between these two hydrogen atoms.

Taking D (+)-glyceraldehyde as a more complex example, the configuration is R, as shown [5.37].

[5.37]

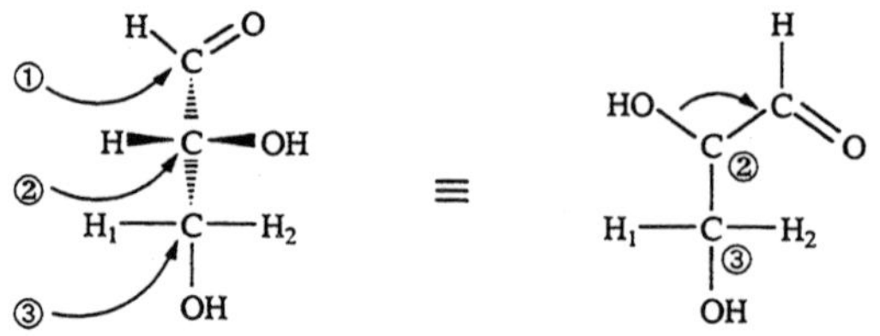

Carbon 3 is prochiral. If H_1 is replaced, for example, by deuterium (see the projection in which H_2 is above the plane), it is said that H_1 is pro R [5.38].

[5.38]

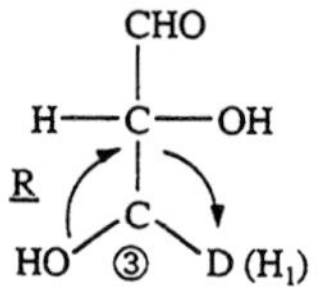

If the same is done with atom H_2, it can be said that this atom is pro S [5.39].

[5.39]

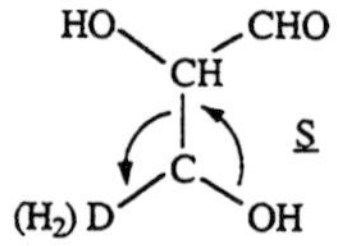

It has just been seen that under the action of an NAD dehydrogenase, a reduced substrate furnishes a proton and, at the same time, a hydride ion is transferred to the NAD^+ (or $NADP^+$). This is a direct transfer and isotopic labeling (D or T) has shown that it does not occur through the mediation of a free hydrogen ion. This was demonstrated in the following manner: when ethyl alcohol is oxidized by NAD^+ in the presence of D_2O, the NADH thus formed contains no deuterium [5.40].

[5.40]

$$CH_3\text{–}CH_2OH + NAD^{\oplus} \underset{}{\overset{D_2O}{\rightleftharpoons}} CH_3\text{–}CHO + NADH + H^{\oplus}$$

It is completely different, however, when the ethyl alcohol substrate is labeled with deuterium or when the methylene group is replaced specifically with CD_2; in this latter case, the following is observed [5.41].

[5.41]

$$CH_3\text{–}C\underline{D}_2OH + NAD^{\oplus} \rightleftharpoons CH_3C\underline{D}O + NAD\underline{D} + H^{\oplus}$$

When the pyridine nucleus of NAD^+ is reduced by chemical means (non-enzymatically), a hydrogen atom of the substrate attaches itself to carbon 4 of this nucleus, indiscriminately from one side or the other [5.42].

[5.42]

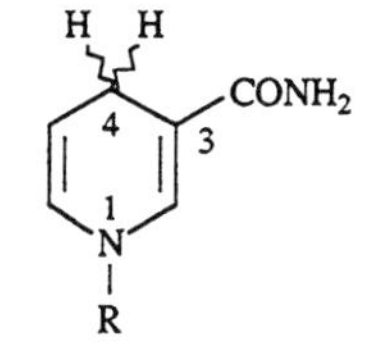

H H
CONH$_2$
4 3
1
N
R

This is not the case when the reduction is effected enzymatically. It can be noted that carbon 4 is prochiral and there are two possible outcomes depending on whether transfer occurs from *face A* or *face B* [5.43].

[5.43]

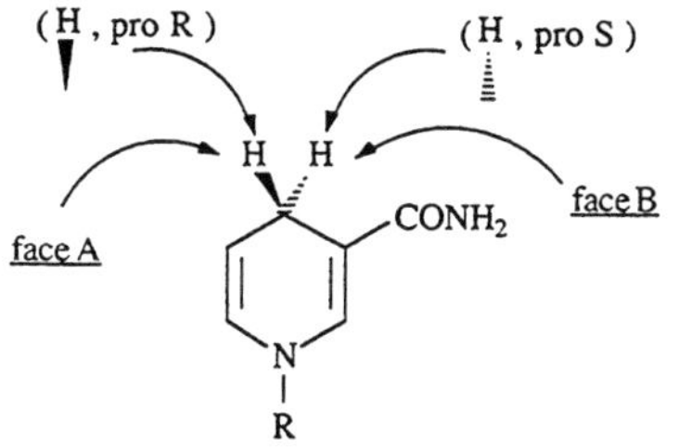

(H, pro R) (H, pro S)
H H
CONH$_2$
face A face B
N
R

Reduction of 2-(R) [2–^{2}H] ethanol by NAD^+ in the presence of the enzyme alcohol:NAD oxidoreductase (alcohol dehydrogenase) gives acetaldehyde and 4-(R) [4–^{2}H] NADH [5.44].

[5.44]

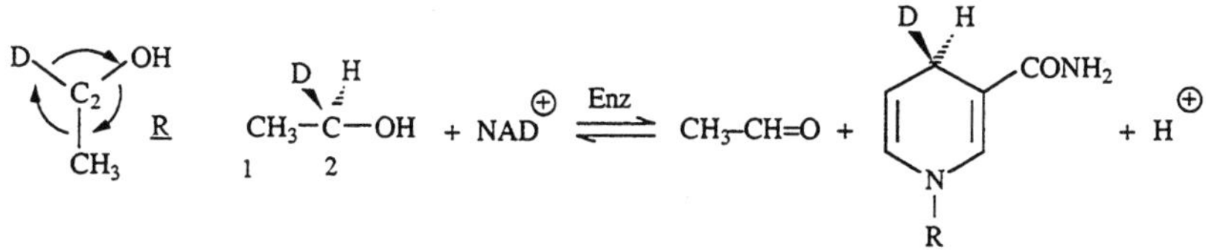

D OH C$_2$ R CH$_3$ — D H, CH_3–C–OH (1 2) + $NAD^\oplus$ $\overset{Enz}{\rightleftharpoons}$ CH_3–CH=O + (D H, CONH$_2$, N, R) + $H^\oplus$

The two faces, A and B, correspond, respectively, to two classes of enzymes: those which are capable of transferring pro R hydride, and those which are capable of transferring pro S hydride.

With regard to the terpeneoids, it seems that the first observation of acetic acid was observed by Schopper and Grob in 1950. They fed Phycomyces with labeled acetic acid and observed that it was incorporated into the carotene formed by this species.

A similar fact had been noted some years previously with regard to cholesterol but it was Bloch et al. who had already forseen in 1942 that acetic acid must be the precursor of this substance. At that time, it was not at all clear that the biosyntheses of terpenoids and steroids passed through the same initial steps. It was also

Bloch who suggested, in 1952, that the isoprenoid intermediates were implicated in the biosynthesis of steroids.

One of these intermediates, formed from three acetate molecules, was, according to Bloch, β-methyl-hydroxyglutaric acid which after decarboxylation formed the necessary 5-carbon intermediate [5.45].

[5.45]

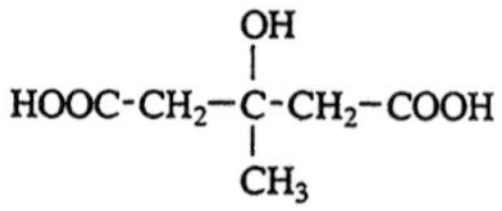

$$HOOC\text{-}CH_2\text{-}C(OH)(CH_3)\text{-}CH_2\text{-}COOH$$

In fact, the veritable and most active precursor (which results from the reduction of one of the carboxyl functions of this diacid) was discovered some years later (1956) by researchers in the Merck group under the direction of Folkers. This was mevalonic acid as well as its lactone, mevalolactone [5.46].

[5.46]

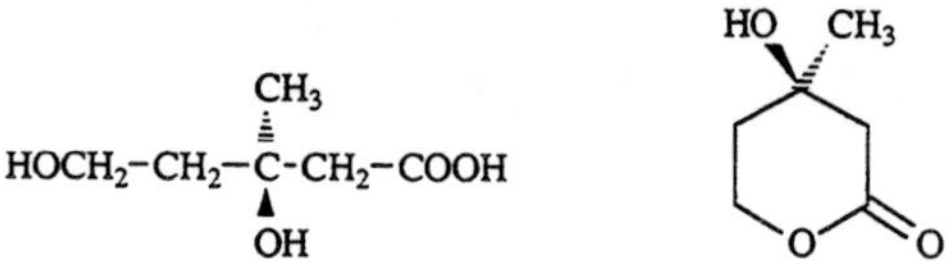

$$HOCH_2\text{-}CH_2\text{-}C(CH_3)(OH)\text{-}CH_2\text{-}COOH$$

(mevalolactone: six-membered lactone ring bearing HO and CH_3)

Before going further, it should be discussed how mevalonic acid is formed and by what mechanisms it is transformed into terpenoids.

The formation of mevalonic acid can be represented by the following scheme [5.47]:

[5.47]

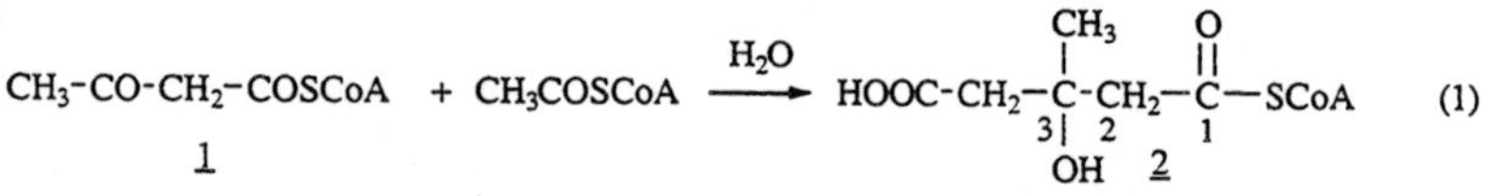

$$\underset{1}{CH_3\text{-}CO\text{-}CH_2\text{-}COSCoA} + CH_3COSCoA \xrightarrow{H_2O} \underset{2}{HOOC\text{-}CH_2\text{-}\overset{3}{C}(CH_3)(OH)\text{-}\overset{2}{CH_2}\text{-}\overset{1}{C}(=O)\text{-}SCoA} \quad (1)$$

This first reaction is irreversible and catalyzed by an enzyme, 3-hydroxy–3-methylglutaryl-CoA synthase (H–3, M–3 glutaryl CoA:acetoacetyl CoA lyase) [5.48].

[5.48]

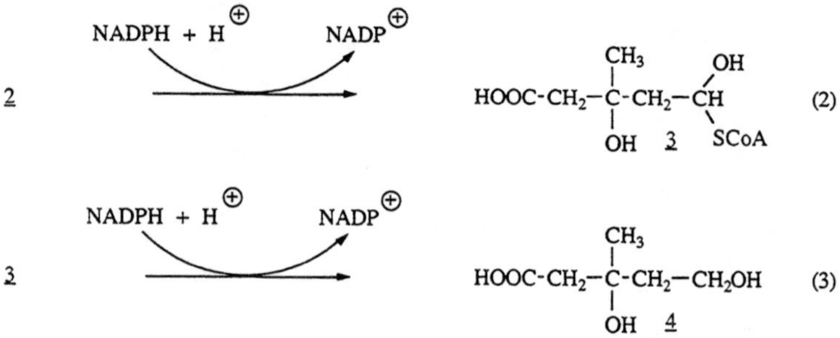

$$2 \xrightarrow{NADPH + H^{\oplus} \;\rightarrow\; NADP^{\oplus}} \underset{3}{HOOC\text{-}CH_2\text{-}C(CH_3)(OH)\text{-}CH_2\text{-}CH(OH)(SCoA)} \quad (2)$$

$$3 \xrightarrow{NADPH + H^{\oplus} \;\rightarrow\; NADP^{\oplus}} \underset{4}{HOOC\text{-}CH_2\text{-}C(CH_3)(OH)\text{-}CH_2\text{-}CH_2OH} \quad (3)$$

Reactions 2 and 3 are catalyzed by another enzyme, 3-hydroxy–3-methylglutaryl-CoA reductase (mevalonate: NADP oxidoreductase) which is not yet designated as an enzyme of class A or B. The synthesis of 3-hydroxy–3-methylglutaryl-CoA 2, the precursor of mevalonic acid 4, must be a stereospecific process because its reduction leads only to 3(R)-mevalonic acid 4'.

It is therefore possible to predict that the absolute configuration of the enzymatic synthesis of hydroxymethylglutaryl-CoA at carbon 3 is S and results from the addition of acetyl-CoA to the si face of C_3 of acetoacetyl-CoA [5.49].

[5.49]

si face; $H^{\oplus}$; O; CH_3; CH_2; C=O; CoA–S; $CH_2^{\ominus}$; =O; CoA–S → ① HO, CH_3, C, ② CH_2, ③ CH_2, CO, CoA–S, C, O, OH; priority given to S; S; + CoASH

$2H^{\oplus}$ | 2 NADH ↓

① HO, CH_3, C, ③ CH_2, ② CH_2, HO-CH_2, COOH; R; + CoASH + 2 $NADP^{\oplus}$

This stereospecific synthesis has been confirmed in the following way. If indeed hydroxymethylglutaryl-CoA is synthesized by chemical means, a mixture of two diastereoisomers is obtained of which only one reacts under the action of a cleavage enzyme to give acetoacetic acid and acetyl-CoA. On the other hand, the product synthesized enzymatically is entirely cleaved.

One very important point has not yet been discussed. In the equilibrium (1) we have admitted the existance of acetoacetyl-CoA 1. But, under what conditions is it formed?

A first demonstration was provided by the group of Lynen as the result of a long series of experiments involving a detailed study of the biosynthesis of fatty acids. These experiments permitted from 1960 onwards the conclusion that the fundamental element in the elaboration of the carbon chains of fatty acids was malonyl-CoA. After formation of acetyl-CoA, the next phase in the biosynthesis is a carboxylation process in which biotin intervenes as a coenzyme and as a transporter of CO_2.

It is not known for certain if the carboxylation of biotin by bicarbonate ions is the result of a concerted mechanism or one which occurs in two steps. In the following process, the carboxybiotin-enzyme complex thus formed, transfers the carboxyl group to acetyl-CoA to give malonyl-CoA 7.

Concerted mechanism [5.50]

[5.50]

biotin

enzyme biotin complex

(ATP)

+ ADP

(4)

Two-step mechanism

[5.51]

1) + ATP + ADP

2)

(4A)

(5)

Acetoacetyl-CoA is finally obtained by acylation of malonyl-CoA in the presence of acetyltransferase [5.52].

The equilibrium constant K_1 of this reaction has been determined:

- in acid media and at 0°C: K_1 is approximately 2×10^5
- in neutral media (pH 7) and at 0°C: K_1 is approximately 2×10^{-2} ($\Delta G°'$ approx. 2.1 Kcal.mole^{-1}).

It is also possible to see how acetoacetyl-CoA is formed by the condensation of

[5.52]

$$^{\ominus}O-\overset{O}{\overset{\|}{C}}-CH_2-COSCoA \;+\; CH_3-\underset{O}{\underset{\|}{C}}-SCoA \;\;(H^{\oplus})\;\; \underset{}{\overset{K_1}{\rightleftharpoons}} \;\; CO_2 + CH_3-CO-CH_2-COSCoA + HSCoA \quad (6)$$

two molecules of acetyl-CoA. In such a process, two aspects of the reactivity of thioesters are involved [5.53].

[5.53]

$$CH_3-\underset{SCoA}{\underset{|}{\overset{O}{\overset{\|}{C}}}} + \underset{H}{\underset{|}{C}}H_2-COSCoA \;\overset{K_2}{\rightleftharpoons}\; CH_3-CO-CH_2-COSCoA + HSCoA \quad (7)$$

In fact, under biological conditions, the equilibrium of this reversible reaction will be displaced far to the left, which means that it will correspond to the degradation of acetoacetyl-CoA (which can be achieved by the action of β-oxathiolase).

At pH 7 and at 0°C, K_2 is approximately 1.6×10^{-5} ($\Delta G°'$ approx. 6)

Detailed knowledge of the thermodynamics of this equilibrium has demonstrated the impossibility of constructing polyketomethylene chains and fatty acids by this mechanism. This is contrary to what was initially believed.

If we compare values for K_1 and K_2 (at pH 7 and 0°C), we see, in fact that K_1 is about $1{,}000\,K_2$. This does not deter from the fact that reaction (6) is nonetheless endothermic. In such circumstances, how can it proceed? In order to understand this, it is indispensible to give some details about coupled reactions.

Coupled reactions

If the following reaction is considered [5.54]

[5.54]

$$A + B \rightleftharpoons C + \underline{D} \text{ endothermic } (\Delta G° > 0)$$

under standard conditions, it could not evolve towards the right hand side. To do this, it would be necessary to provide the required free enthalpy by means of another exothermic reaction ($\Delta G°'<0$) in such a way that, overall, the two reactions would be exothermic. This is the principle of coupled reactions. However, this is not sufficient in itself. Both reactions must share a common intermediate. If there existed a second reaction [5.55].

[5.55]

$$\underline{D} + E \rightleftharpoons E + F$$

which is exothermic to the extent that the sum of the $\Delta G°'$ values for both is greater than 0, the equilibrium of the intitial reaction will be displaced to the right.

These are fundamental biochemical principles.

Let us return to the case of reactions (4) or (4a) and (5), which can be summarized in the following manner [5.56].

[5.56]

$$CH_3{-}COSCoA + CO_3H^{\ominus} \xrightleftharpoons[\text{enzyme-biotin}]{ATP \to ADP} Pi + {}^{\ominus}OCO{-}CH_2{-}COSCoA$$

and can be broken down as follows [5.57].

[5.57]

$$CH_3{-}COSCoA + CO_3H^{\ominus} \rightleftharpoons {}^{\ominus}OCO{-}CH_2{-}COSCoA + H_2O$$

$$\text{and} \quad ATP + H_2O \longrightarrow ADP + Pi \qquad \Delta G^{o\prime} = -7.30 \text{ Kcal.mole}^{-1}$$

Reaction (6) has a ΔG°' of the order of +2.1 Kcal.mole^{-1} so the balance of the two reactions is –5 Kcal.mole^{-1}. In this decarboxylation process, it is therefore the hydrolysis of ATP to ADP which constitutes the energy reservoir.

The formation of acetoacetyl-CoA in the case of the biosynthesis of fatty acids and phenols has been demonstrated by Lynen and co-workers. Rudney and collaborators have, however, proposed that the very unfavorable equilibrium of reaction (7) (ΔG°' around 6) could be surmounted in the biosynthesis of isoprenoids thanks to intimate coupling of reaction (6) with the irreversible reactions (1), (2) and (3). Nevertheless, there are no precisely known thermodynamic data on these three reactions. They are as follows [5.58].

[5.58]

$$2\ CH_3{-}COSCoA \rightleftharpoons \underset{\underline{1}}{\underline{CH_3{-}CO{-}CH_2{-}COSCoA}} + HSCoA \qquad (6)$$

$$\underset{\underline{1}}{\underline{CH_3{-}CO{-}CH_2{-}CO{-}SCoA}} + CH_3{-}COSCoA \xrightarrow{H_2O} HOCO{-}CH_2{-}C(CH_3)(OH){-}CH_2{-}C(=O){-}SCoA\ \ \underline{2} \qquad (1)$$

$$\underline{2} \longrightarrow HOCO{-}CH_2{-}C(CH_3)(OH){-}CH_2{-}CH(OH)(SCoA)\ \ \underline{3} \qquad (2)$$

$$\underline{3} \longrightarrow HOCO{-}CH_2{-}C(CH_3)(OH){-}CH_2{-}CH_2OH\ \ \underline{4} \qquad (3)$$

The sum of the ΔG°' values for reactions (1), (2) and (3) will be well below 6 Kcal.mole^{-1} and the common intermediate shared with reaction (6) is acetoacetyl-CoA 1 [5.59].

[5.59]

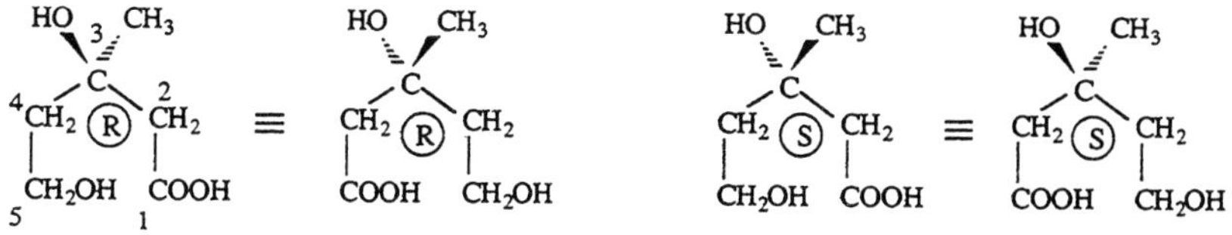

3(R)-mevalonic acid

3(S)-mevalonic acid

Biosynthetic Mechanisms Leading to Terpenoids

Elaboration of Synthetic Schemes for the Sesquiterpenoids

What then are the biosynthetic reactions that permit access to the terpenoids from (R)-mevalonic acid? Some of these mechanisms which will be discussed here have been the fruit of important studies by numerous researchers, notably Cornforth, Popjak, Eggerer, Arigoni, and others.

It is obviously impossible to analyze all of these studies in detail here, however, they are rapidly summarized as follows [5.60, 61, 62].

Among the acyclic sesquiterpenes that can be cited as examples, trans β-farnesene can be isomerized by rhodium chloride trihydrate to give trans,trans-farnesene [5.63].

Oxygenated aliphatic sesquiterpenes are also found. Examples are davanone (found in Davana oil) and a toxic sesquiterpenoid, (–) ngaione [5.64]. Other oxygenated sesquiterpenoids such as torreyal and dendrolasine can be cited. Their syntheses are described here [5.65].

Synthesis of torreyal

The condensation of β-furfuryl alcohol with ethoxyisoprene, an ethoxy diene (see Appendix 3, on the chemistry of enol ethers) in the presence of $Hg(OAc)_2$, gives an allylic enol ether which on pyrolysis undergoes a Claisen rearrangement followed by a Cope rearrangement [5.66].

A monoterpene furan aldehyde is obtained which in turn gives its corresponding alcohol on reduction with $LiAlH_4$. When this alcohol is subjected to the same treatment with ethoxyisoprene, torreyol is formed. This can then be transformed into dendrolasine (thioacetal of aldehyde function followed by treatment with Raney nickel). The aldehyde can also be reduced to alcohol by treating its tosylate with $LiAlH_4$ [5.67].

This sequence does not, however, result in a single product. In fact, a mixture of the E and Z isomers in the 6, 7 double bond is obtained.

Synthesis of dendrolasine

A stereospecific synthesis of dendrolasine can be performed using a method described by Julia. In the case of dendrolasine [5.68].

[5.60]

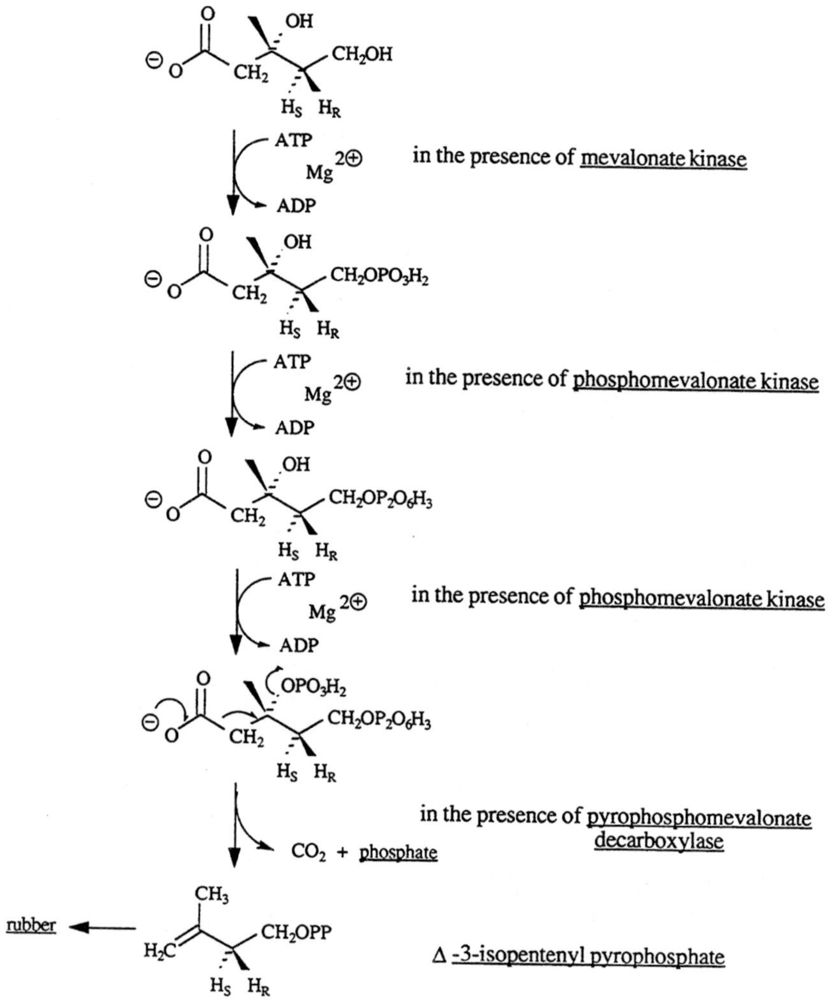

$^{\ominus}O-C(=O)-CH_2-C(OH)-CH_2OH$ (H_S H_R)

ATP, $Mg^{2\oplus}$ → ADP — in the presence of mevalonate kinase

$^{\ominus}O-C(=O)-CH_2-C(OH)-CH_2OPO_3H_2$ (H_S H_R)

ATP, $Mg^{2\oplus}$ → ADP — in the presence of phosphomevalonate kinase

$^{\ominus}O-C(=O)-CH_2-C(OH)-CH_2OP_2O_6H_3$ (H_S H_R)

ATP, $Mg^{2\oplus}$ → ADP — in the presence of phosphomevalonate kinase

$^{\ominus}O-C(=O)-CH_2-C(OPO_3H_2)-CH_2OP_2O_6H_3$ (H_S H_R)

→ CO_2 + phosphate — in the presence of pyrophosphomevalonate decarboxylase

rubber ← $H_2C=C(CH_3)-CH_2OPP$ (H_S H_R) — Δ -3-isopentenyl pyrophosphate

[5.61]

CH_3 H_2C C C CH_2OPP $H^{\oplus}$ H_S H_R * E

nucleophilic site of the enzyme
isopentenyl pyrophosphoisomrase

CH_3 H_3C C C CH_2- OPP H *

dimethyl allyl pyrophosphate

CH_3 H_2C C C CH_2OPP H_S H_R E

nucleophilic site of the enzyme
dimethyl allyl transferase
(DMAT)

$Mg^{2\oplus}$ Pi

monoterpenes

CH_3 CH_3 CH_3 C C CH_2 CH_2 C C CH_2- OPP H * H *

geranyl pyrophosphate

CH_3 H_2C C C CH_2OPP H_S H_R E

DMAT + $Mg^{2\oplus}$

Pi

sesquiterpenes

CH_3 CH_3 CH_3 CH_3 C C CH_2 CH_2 C C CH_2 CH_2 C C CH_2OPP H * H * H *

farnesyl pyrophosphate

[5.62]

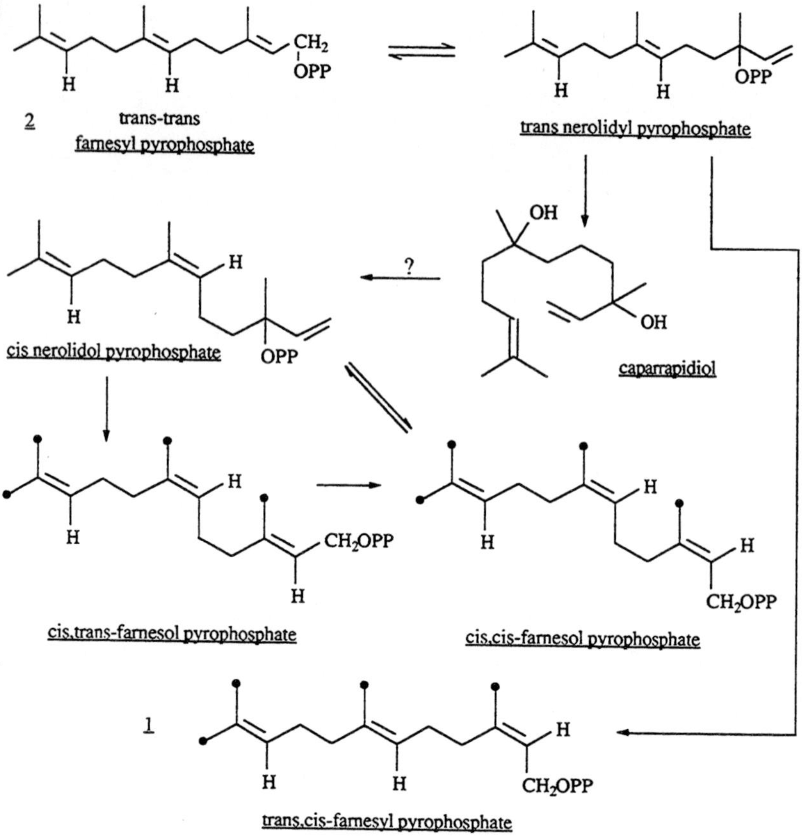

2 trans-trans farnesyl pyrophosphate ⇌ trans nerolidyl pyrophosphate

caparrapidiol

? cis nerolidol pyrophosphate

cis,trans-farnesol pyrophosphate

cis,cis-farnesol pyrophosphate

1 trans,cis-farnesyl pyrophosphate

[5.63]

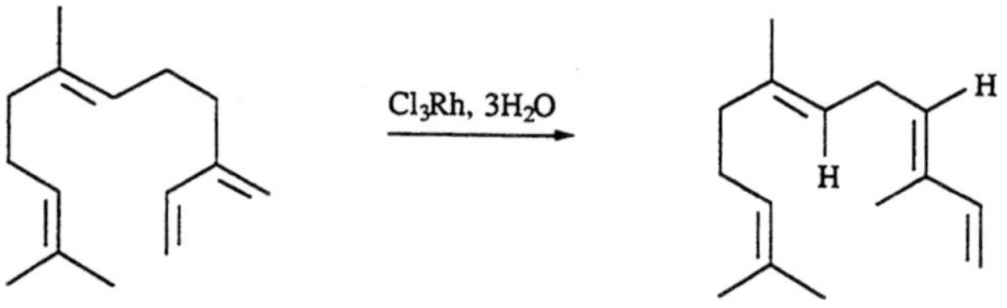

β-farnesene → (Cl_3Rh, $3H_2O$) → α-farnesene

[5.64]

davanone

ngaione

[5.65]

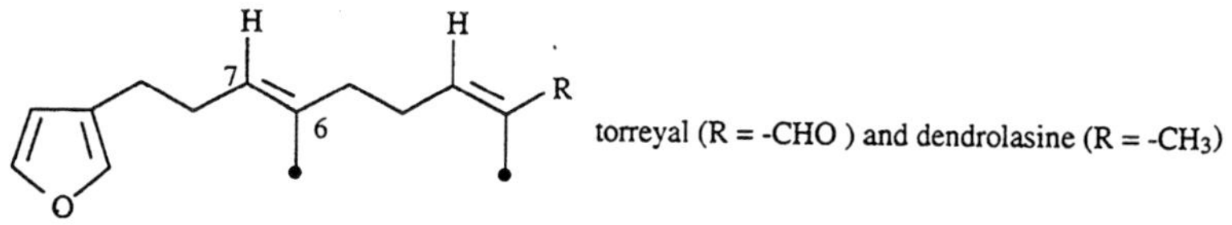

H H 7 6 R

torreyal (R = -CHO) and dendrolasine (R = $-CH_3$)

[5.66]

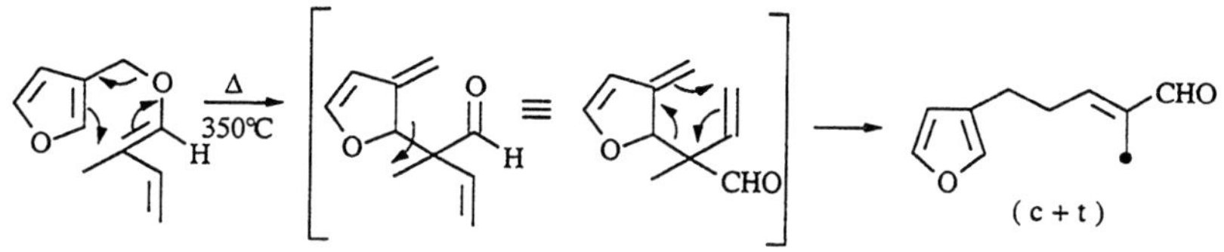

O H Δ 350°C

O H ≡ CHO

CHO

(c + t)

[5.67]

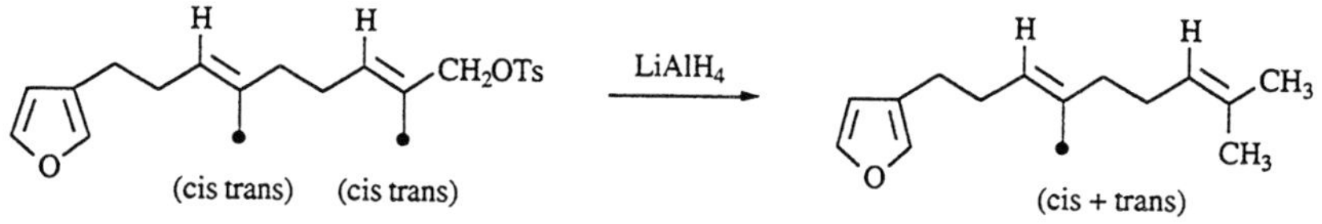

H H CH_2OTs

(cis trans) (cis trans)

$LiAlH_4$

H H CH_3 CH_3

(cis + trans)

[5.68]

CH_2– Br

O COOEt

1) NaH
2) $Ba(OH)_2$
3) $H_3O^{\oplus}$

O

OH

$H^{\oplus}$ OH $^{\ominus}$Br

H Br

*

H CHO

$\phi_3P{=}C(CH_3)_2$

* *

H H

trans-dendrolasine

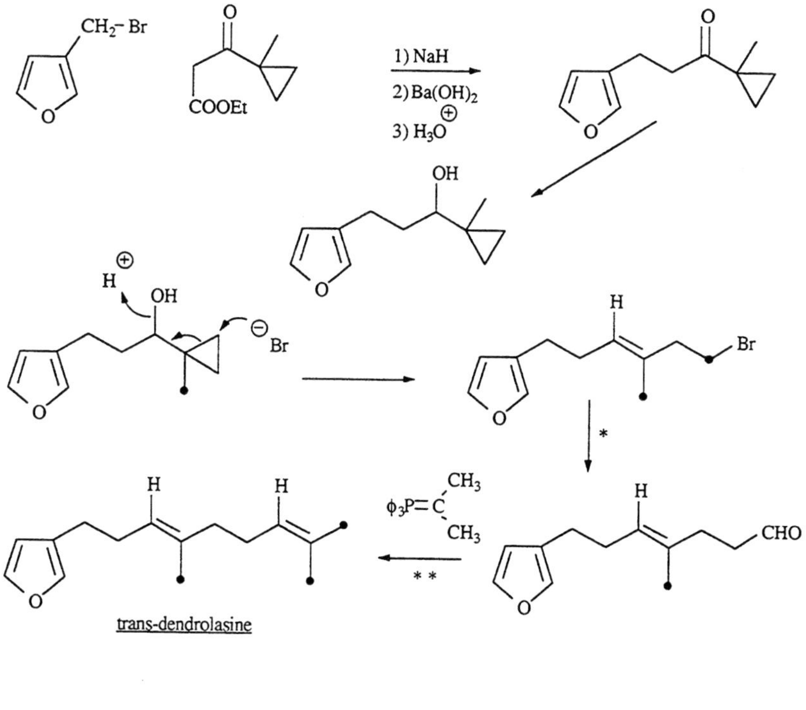

- This transformation can be achieved, for example, by the action of sodium cyanide followed by reduction of the resulting nitrile with diisobutylaluminum (Dibal).
- Wittig reaction takes place (see explanation in Appendix 2 for a better understanding).

The industrial syntheses of nerolidol and farnesol will be dealt with in a later section as well as the α- and β-sinesals, which are important constituents of orange oil (see Appendix 3) [5.69].

[5.69]

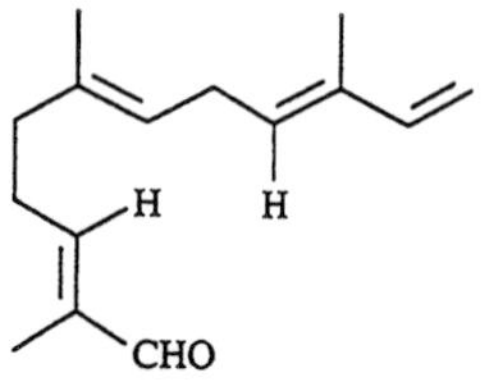

α-sinensal 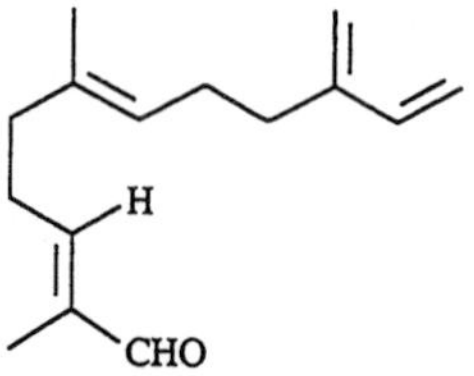 β-sinensal

Starting with the pyrophosphates of farnesol, notably those of trans,cis-farnesyl 1 and trans,trans-farnesyle 2, it is possible to derive the quasi-totality of the sesquiterpenic compounds. The following table indicates the principal carbocations implicated in these transformations [5.70].

As an example, the stabilization of the bridged cation 4 leads to sesquicarene [5.71].

Sesquicarene has been synthesized by nine different routes. All of these syntheses are based on the same principle which consists of elaborating the bicyclo [4.1.0] heptane skeleton by intramolecular insertion of a carbene. This is a stereospecific reaction. These syntheses can be schematized as follows [5.72]:

The simplest synthesis is that of Corey. In this synthesis, the hydrazone of cis,trans-farnesal (hydrazine + triethylamine in ethanol) is oxidized by an excess of activated MnO_2 to give an intermediate diazo compound which is transformed into a carbene in the presence of cuprous iodide in THF. Intramolecular addition onto the double bond gives sesquicarene [5.73].

The carbocation 6 is also the precursor of a large number of sesquiterpenes. First, direct intramolecular cyclization followed by stabilization involving loss of a proton leads to the bergamotenes or the santalenes, depending on whether the cyclization is of the Markownikoff or the anti-Markownikoff type [5.74].

β-*santalese:* the first sesquiterpene for which the exact structure was proposed by Semmler in 1910.

This same carbocation 6 can stabilize itself by losing a proton, thereby yielding γ-bisabolenes which can then reprotonate in two different ways to give 6a and 6b [5.75, 6].

Cation 6 is, above all, the precursor of the monocyclic sesquiterpenes of the bisabolane type. Several of these are represented here [5.77].

[5.70]

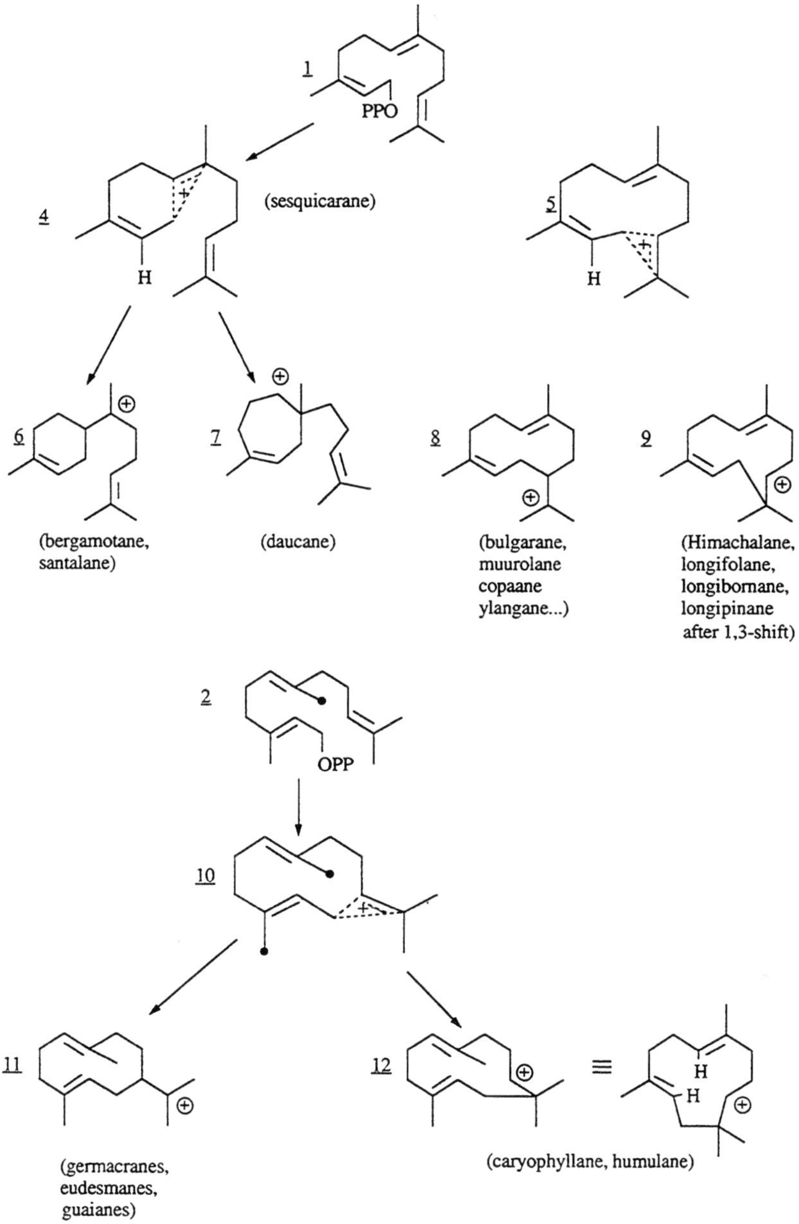

1

PPO

4

(sesquicarane)

H

5

H

6

(bergamotane, santalane)

7

(daucane)

8

(bulgarane, muurolane copaane ylangane...)

9

(Himachalane, longifolane, longibornane, longipinane after 1,3-shift)

2

OPP

10

11

(germacranes, eudesmanes, guaianes)

12

H

H

(caryophyllane, humulane)

[5.71]

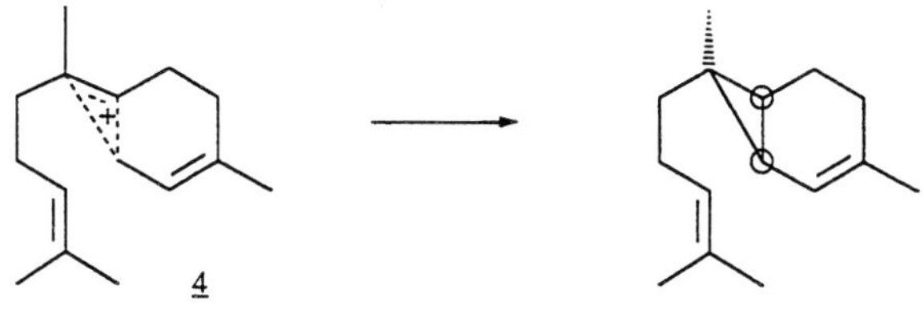

4

[5.72]

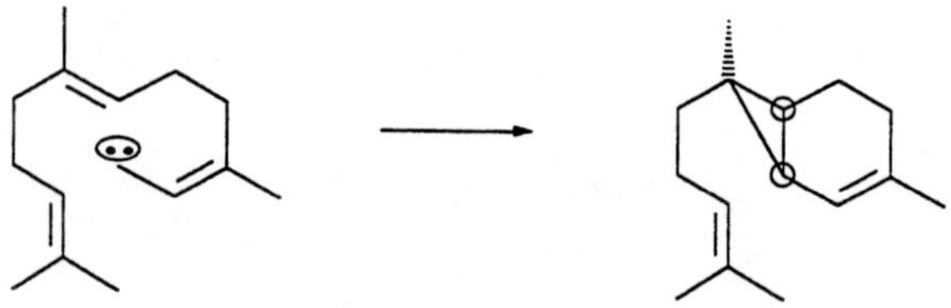

[5.74]

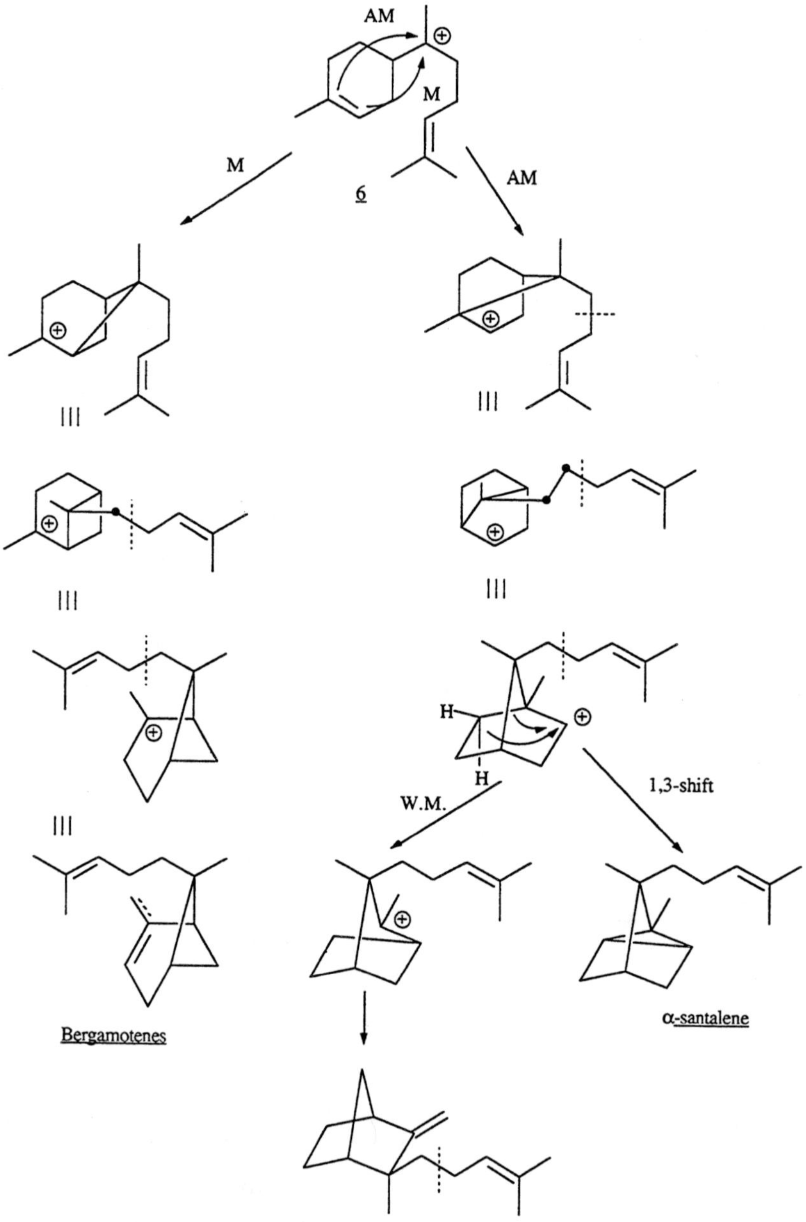

AM

M

M

AM

6

III

III

III

III

III

W.M.

1,3-shift

Bergamotenes

α-santalene

[5.73]

$H_2N-N=CH$ $\xrightarrow[CH_2Cl_2]{MnO_2}$ $N_2=CH$ ≡ N_2 $\xrightarrow[THF]{CuI}$

[5.75]

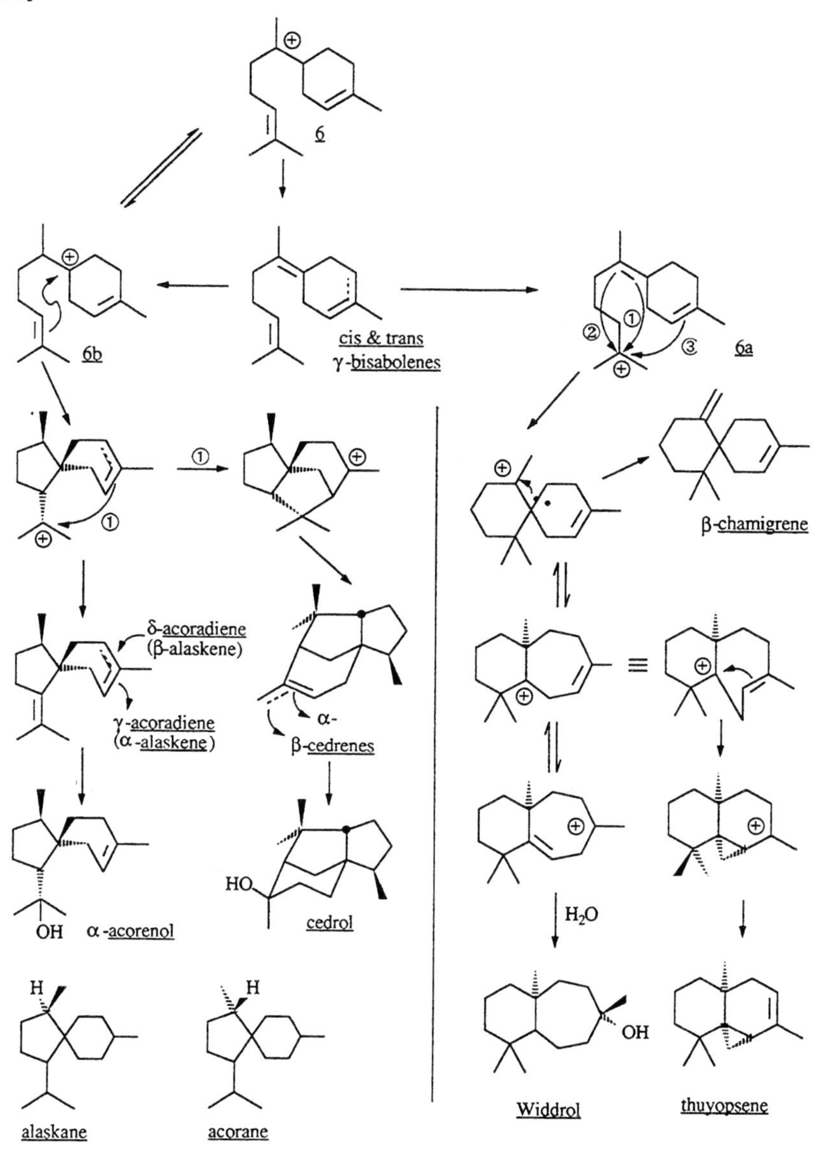

[5.76]

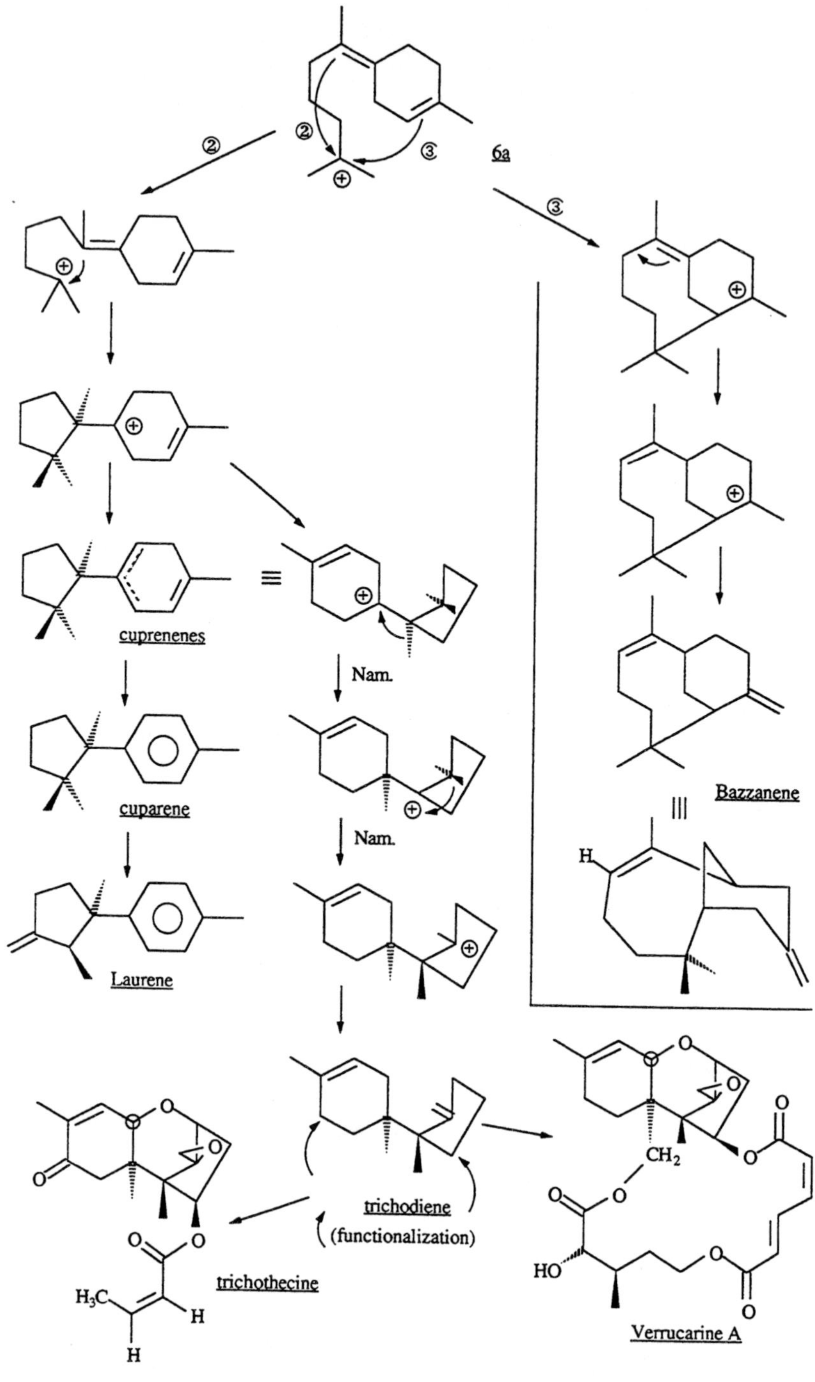

6a

cuprenenes

cuparene

Laurene

Nam.

Nam.

Bazzanene

trichodiene

(functionalization)

trichothecine

Verrucarine A

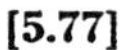

[5.77]

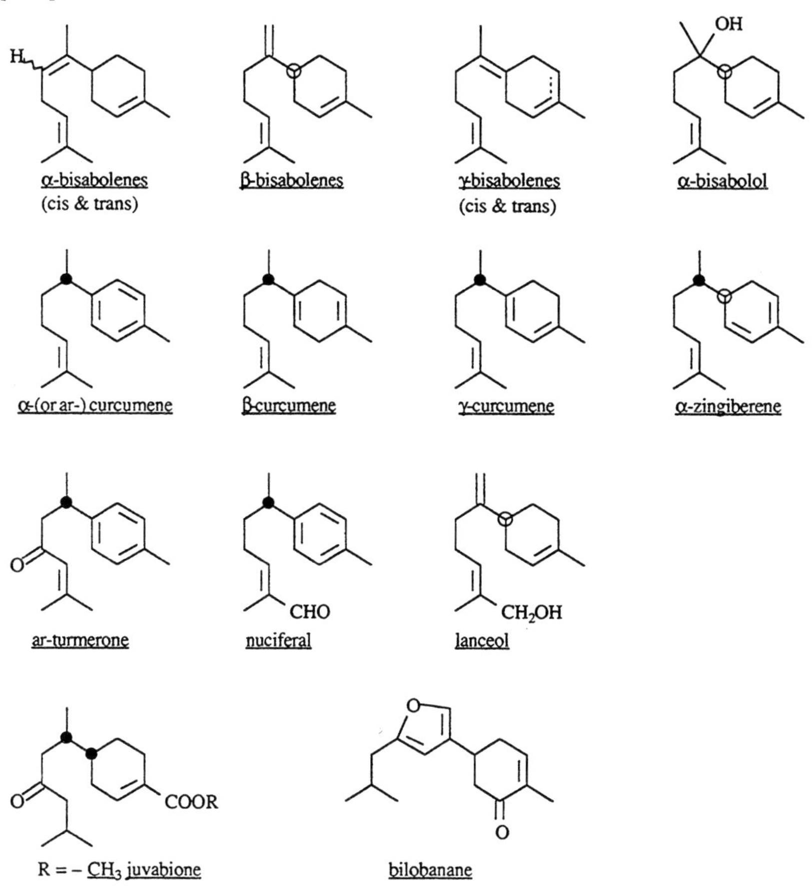

The syntheses of some of these sesquiterpenes will now be briefly described; α-zingiberene will be dealt with first, followed by those of the curcumenes.

The first synthesis makes use of the special reactions of enamines, while the others make use of Birch's method for the reduction of aromatic compounds [5.78].

[5.78]

H H H

(+)-β-bisabolene (-)-α-zingiberene (-)-β-curcumene

The reactions of electrophilic alkenes (alkenes conjugated to electron-withdrawing groups) with enamines can lead to different products. Take the case of methylvinylketone, which can undergo three different types of reaction:

- simple acylation a—this is the case with relatively weak bases such as morpholine;
- 1,2-cycloaddition b;
- 1,4-cycloaddition (c or d), the latter predominating when the base is a strong one such as pyrrolidine or piperidine [5.79].

[5.79]

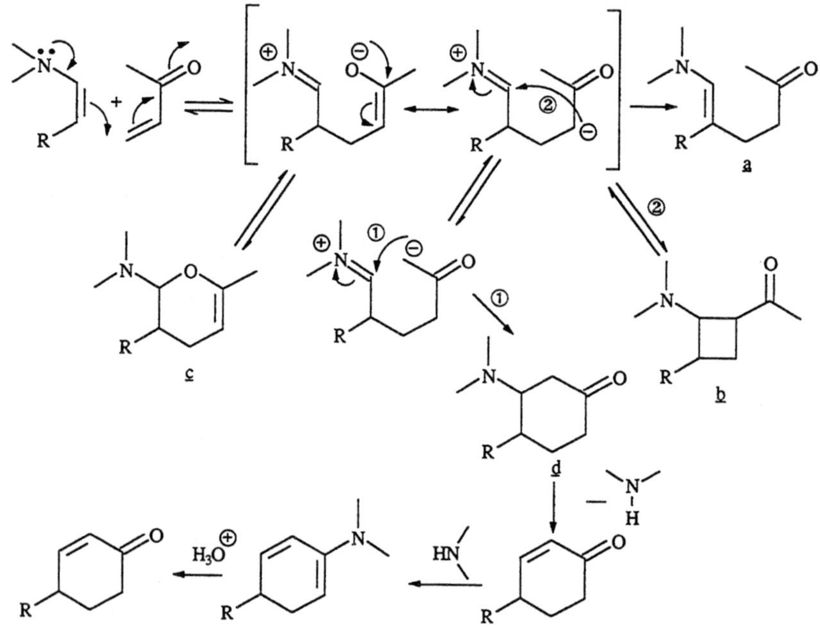

If the enamine of (+)-citronellal reacts with methylvinylketone, the product is (–)α-zingiberene, although the stereochemistry at carbon 4 is a mixture (Joshi-Kulkarni) [5.80].

[5.80]

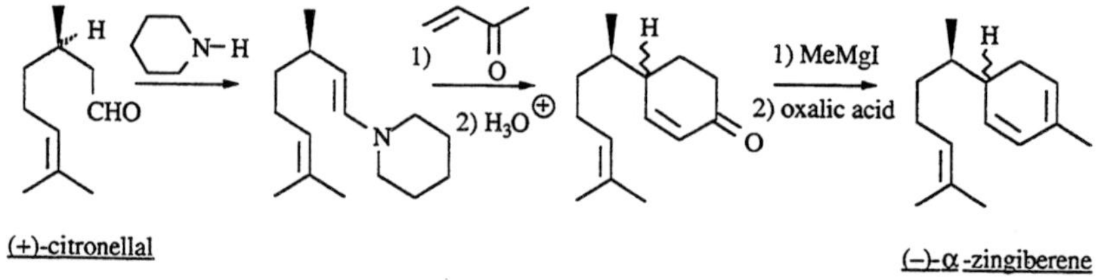

(+)-citronellal

(–)-α-zingiberene

Application of the Birch reaction to the synthesis of the curcumenes

Note on the Birch reaction: The most probable mechanism for the reduction of aromatic compounds under these conditions (reaction with an alkali metal [M] in liquid ammonia either in the presence or the absence of an alcohol) can be depicted as follows [5.81].

[5.81]

1) M + NH_3 ⇌ $M^{\oplus}$ ······ $^{\ominus}$ (NH_3) (NH_3)

2) $M^{\oplus}$ ······ $^{\ominus}$ (NH_3) (NH_3) + ⇌ ······] $^{\oplus}$ (NH_3) (NH_3)

3) ······] $^{\oplus}$ (NH_3) (NH_3) + ROH ⟶ H H (NH_3) + ROM (NH_3)

4) H H (NH_3) + $M^{\oplus}$ ······ $^{\ominus}$ (NH_3) (NH_3) ⟶ H H ······] $^{\oplus}$ (NH_3) (NH_3)

5) H H ······] $^{\oplus}$ (NH_3) (NH_3) + ROH ⟶ H H H H + ROM (NH_3)

isolated aromatic nuclei

6) H H H H $NH_2^{\ominus}$ ⇌ $H^{\oplus}$; $H^{\oplus}$ ⇌ $NH_2^{\ominus}$ H HH H M , ROH ⟶ in the absence of alcohols

7) ······] $^{\oplus}$ + $M^{\oplus}$ ······ $^{\ominus}$ (NH_3) (NH_3) (NH_3) (NH_3) ⇌ ······ 2 [M $^{\oplus}$ (NH_3) (NH_3)

8) ······ 2 [M $^{\oplus}$ (NH_3) (NH_3) + ROH ⟶ H H H H + 2 ROM (NH_3)

When an alkali metal (M) is dissolved in liquid ammonia (step 1), the solvated cation and its separated electron remain weakly attached to each other, forming an "ion pair." This reacts with aromatic derivatives by a process of electron transfer to give a radical anion which takes the place of the electron in the ion pair (2). This radical anion is an extremely unstable entity due to the weak electron affinity of the aromatic nucleus. As a consequence, the equilibrium in step 2 can be displaced towards the right only if the radical anion is constantly eliminated from the equilibrium mixture by irreversible protonation (hence, the role of the alcohol [step 3]. Ammonia can not donate a proton because of its weak acidity (pK_a about 34), whereas the acidity of alcohols (pK_a around 16–18) is sufficient.

The radical obtained in the course of step 3 will accept immediately a second electron and thus be converted into an anion (step 4) which on protonation results in the 1,4-dihydro derivative (step 5).

If any NH_2- ions are present in the reaction medium, the 1,4-dihydro derivative, could be anionized and the 1,2-dihydro derivative, which is more stable thermodynamically due to its conjugation, will be formed. This derivative will, however, like all conjugated dienes, be subsequently reduced under these reaction conditions to give the tetrahydro derivative (step 6). In the majority of cases, however, an alcohol is present in the mixture and it will act as an acid buffer (compared with ammonia)—thereby preventing formation of NH_2.

The radical anion formed in step 2 can also accept a second electron and become a dianion (step 7), which can then undergo double protonation (step 8) leading also to a 1,4-dihydro derivative. However, as we have seen, the weak electron affinity of the aromatic nucleus makes this eventuality quite improbable in the case of simple aromatic compounds. On the other hand, in the case of naphthalene and its derivatives, this is the actual reaction path (i.e., steps 1 → 2 → 7 → 8).

The influence of substituents

The Birch reaction is used for the reduction of a range of various aromatic compounds differing in their degree of substitution and in the nature of their substituents.

Reduction of benzene and its derivatives generally gives the corresponding cyclohexadienes with yields which can reach 90%. In this type of process, the orientation of the reduction is extremely important.

From the mechanisms proposed above, it follows that the structure of the products will be determined by the direction of addition of protons to the radical anion in step 3 and to the anion in step 5. It is evident that in these two cases, the ring carbons with the highest electron density will be preferentially protonated (the electron density of each ring carbon can be calculated by molecular orbital methods). Predicting the products of a Birch reaction can then be reduced to determining which substituents are electron-donating and which are electron-withdrawing.

Electron-donating substituents:

R = CH_3, Et, t-Bu, OMe, NH_2, NHe_2. . . [5.82].

[5.82]

R R

Electron-withdrawing substituents
R = COOH, COOEt, $CONH_2$, $CO-CH_3$. . . [5.83].

[5.83]

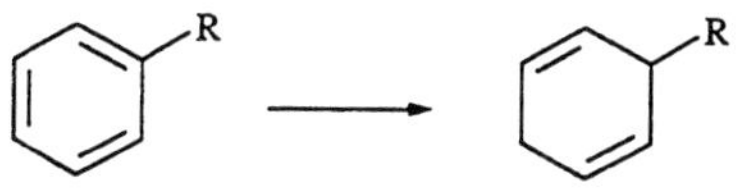

Secondary reactions: The following can be observed:

- reduction of double bonds which are conjugated with the ring
- hydrogenolysis of some substituents such as hydroxy and acetoxy groups.

Example [5.84]:

[5.84]

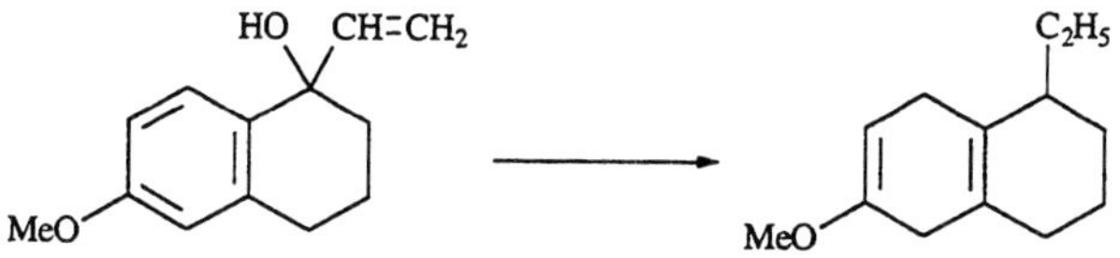

HO CH=CH_2 ... MeO → C_2H_5 ... MeO

Synthesis of α-, β- and γ-curcumenes (Birch) is shown in scheme [5.85].

[5.85]

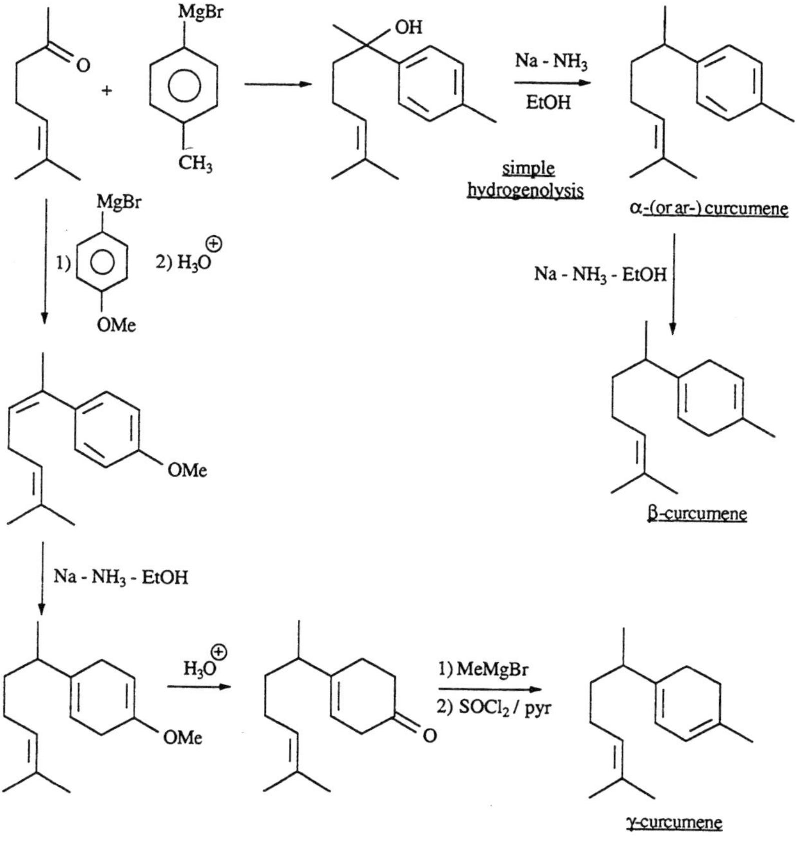

MgBr; CH_3; OH; Na - NH_3 / EtOH; simple hydrogenolysis; α-(or ar-) curcumene

1) MgBr / OMe 2) $H_3O^{\oplus}$; OMe

Na - NH_3 - EtOH; β-curcumene

Na - NH_3 - EtOH

OMe; $H_3O^{\oplus}$; O; 1) MeMgBr 2) $SOCl_2$ / pyr; γ-curcumene

Synthesis of carotol and daucol from the cation 7

Cation 7 is the precursor of sesquiterpenes of the series of carotane or daucane [5.86]. The passage from carotol to daucol can be rationalized in the following way. The epoxidation of carotol occurs from the least hindered face (α = face). The resulting product is an γ-epoxy alcohol which has known instability due to the participation of the hydoxy to give a tetrahydrofuran or tetrahydropyran derivative [5.87].

[5.86]

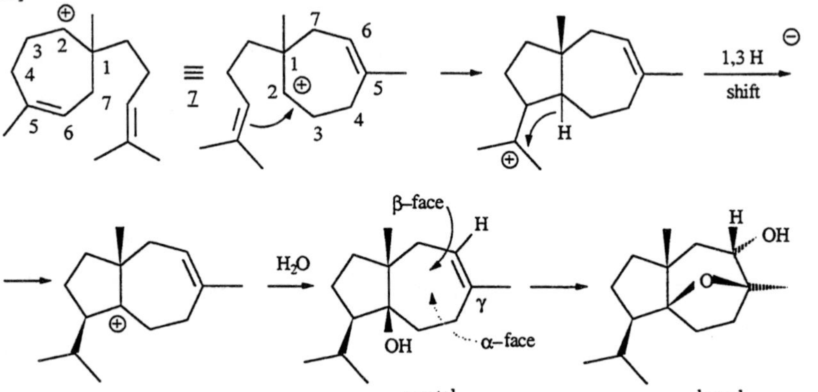

1,3 H shift

H_2O

β–face

α–face

carotol

daucol

[5.87]

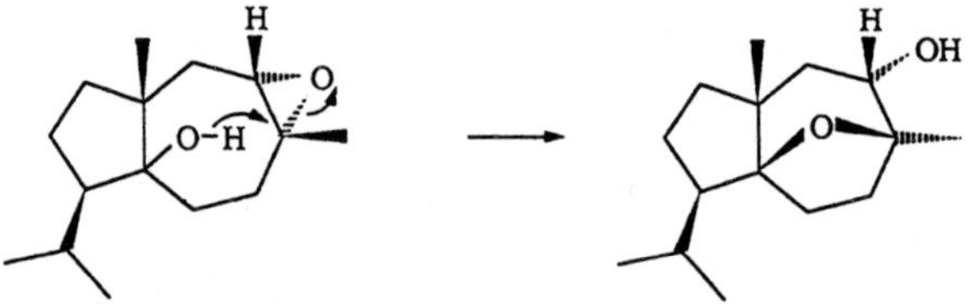

daucol

Laserpitine is an example of a substance belonging to this group of derivatives [5.88].

[5.88]

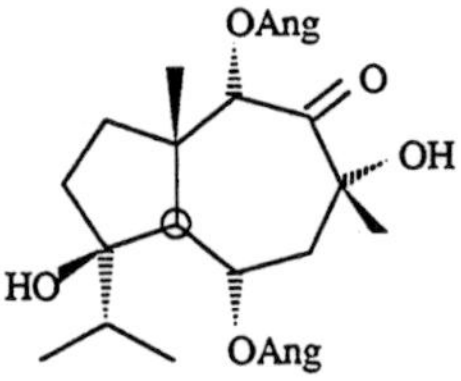

OAng

OH

HO

OAng

(Ang = angelyl)

Carotol and daucol are among the major oxygenated constituents of the essential oil obtained from carrot seeds.

(–)-Cyclocopacamphene is the enantiomer of the (+)-form which is derived from cyclocopacamphenic acid found in vetyver oil. It can be seen that these four sesquiterpenes are derivatives of camphene (shown in bold-face in the above structures) containing a third isoprene unit (saturated in this case). It will be recalled that camphene is obtained by Wagner-Meerwein rearrangement of borneol [5.89].

[5.89]

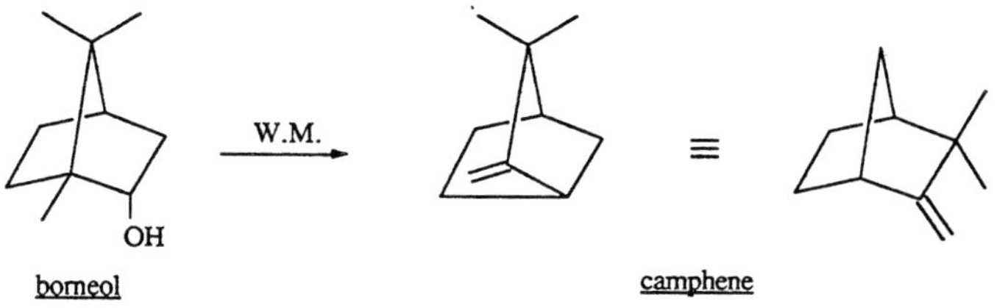

W.M.

≡

OH

borneol

camphene

(–)-Copacamphene is obtained in the same way from (+)-copaborneol, the main sesquiterpenic alcohol in the oil of the common pine (*Pinus sylvestris*) and in turpentine oil obtained as a by-product in the manufacture of paper pulp in Sweden [5.90].

[5.90]

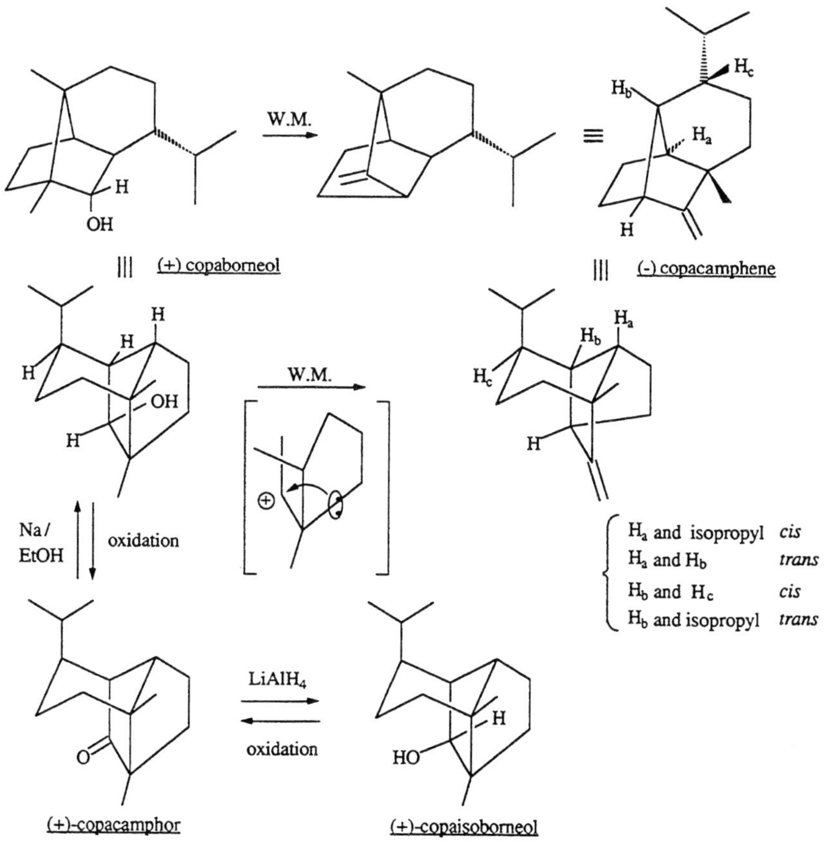

H

OH

W.M.

≡

H_c

H_b

H_a

H

||| (+) copaborneol

||| (-) copacamphene

H

H

H

OH

H

W.M.

H_a

H_b

H_c

H

Na / EtOH

oxidation

H_a and isopropyl *cis*
H_a and H_b *trans*
H_b and H_c *cis*
H_b and isopropyl *trans*

LiAlH$_4$

oxidation

O

H

HO

(+)-copacamphor

(+)-copaisoborneol

Bicyclic, tricyclic and tetracyclic derivatives which have cation 8 as precursor

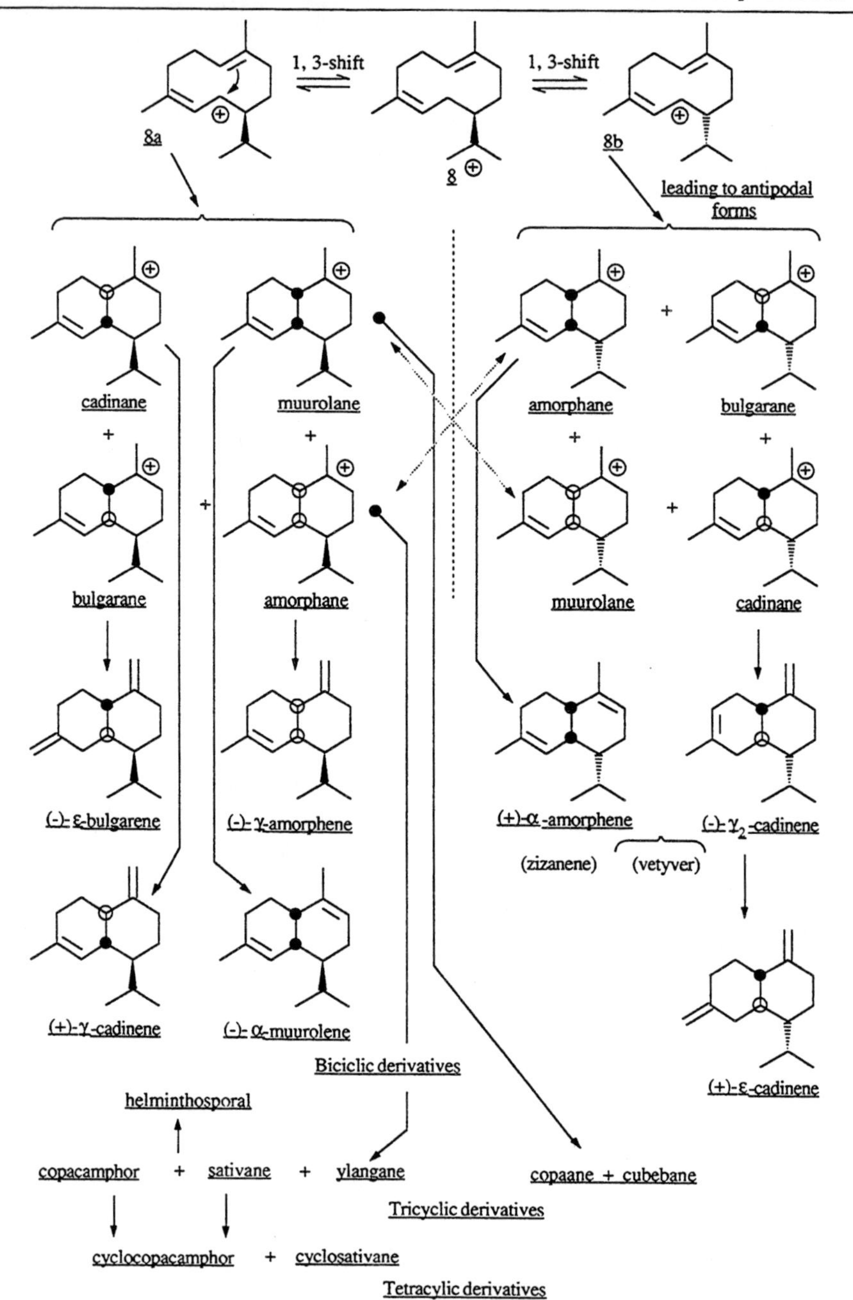

Tricyclic derivatives

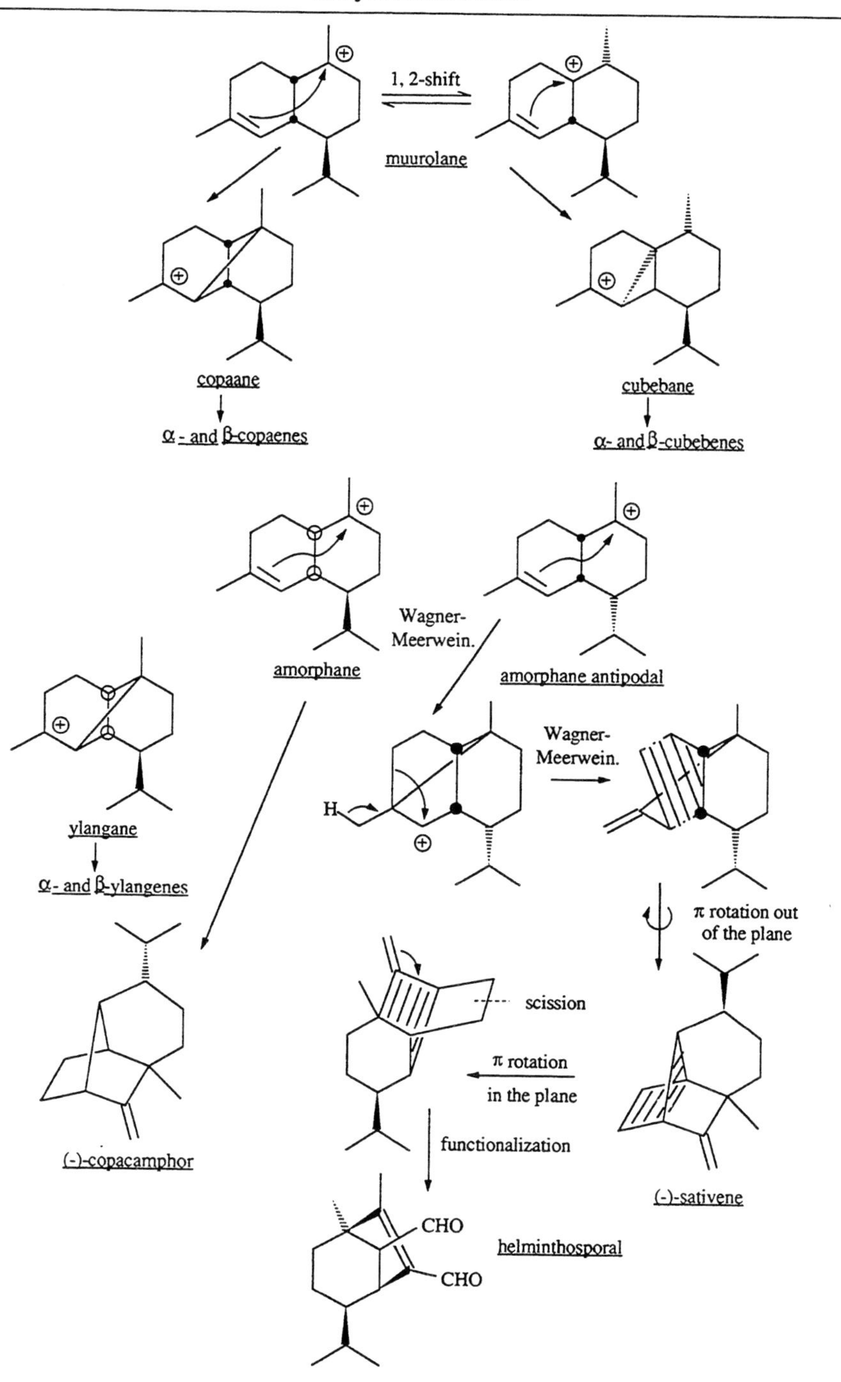

Tetracyclic derivatives

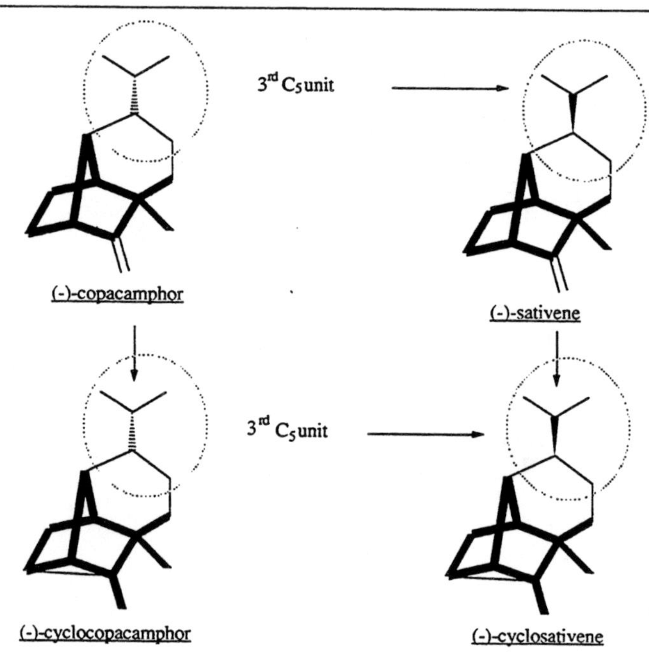

CHAPTER

6

The Chemistry of Sesquiterpenes and of Some Macrocyclic Diterpenes

Introduction

The synthesis of sesquiterpenoids has already been discussed and schematized in the previous chapter—including a study of their biosynthesis. Chapter 7 will deal with macrocyclic ketones and lactones. It should be noted now, however, that the terpenoids themselves provide numerous examples of rings containing an elevated number of elements. The sesquiterpenes can attain 11-membered rings, but 7-, 8-, 9- and 10-membered rings are more numerous and 8- to 11-membered rings are considered "medium sized."

The diterpenes can attain a maximum of 15 ring elements. This chapter will only consider the known diterpenes with 14-membered rings. This is the basic skeleton of cembrane which is found in extracts of tobacco and olibanum.

Cyclizations and interconversions in the sesquiterpene series

Before going into detail in the area of sesquiterpenes, it is perhaps useful to make some general remarks on the special properties of medium-sized rings.

The particular properties of these rings originate from steric effects. Prelog and Dünitz used X-ray studies to demonstrate that these substances possess structures in which the normal tetrahedral valence angle is exceeded—with the result that vicinal groups (1,2-positioned) are partially eclipsed.

An important consequence of this geometry is the existence of steric interactions between non-adjacent atoms. These energetically unfavorable properties of medium-sized rings have been termed "I strain" by H. C. Brown. They provide an explanation of the numerous chemical properties of these compounds.

X-ray analysis does not furnish direct proof of the existance of these transannular interactions, because the exact position of the hydrogen atoms cannot be determined by this method. However, calculations based on the molecular geometry determined from the crystal show that certain hydrogen atoms attached to carbons in positions 1,3; 1,4 and 1,5 must be very close, especially in 9- and 10-membered rings. These calculations have been confirmed by NMR spectroscopy, particularly from studies of the Nuclear Overhauser Effect (NOE).

These effects of proximity are responsible for the trans-annular reactions that characterize these compounds. For eight-membered rings, the solvolyses of two bromo derivatives 1 and 2 with the longifolene 3 skeleton can be cited as examples [6.01].

[6.01]

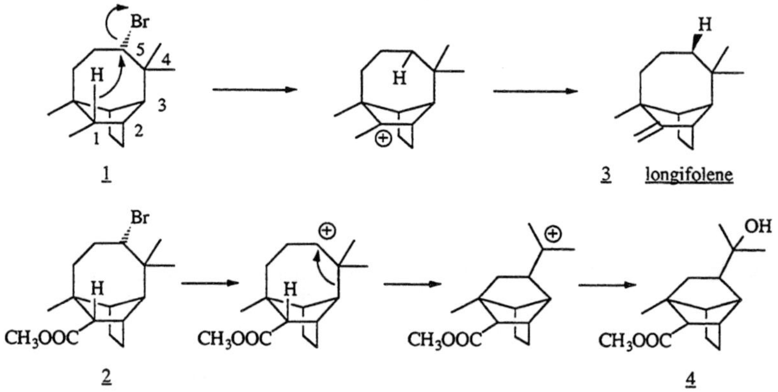

Br 5 4 H 3 1 2 H H
1 3 longifolene
Br H H OH
CH_3OOC CH_3OOC CH_3OOC CH_3OOC
2 4

Solvolysis of bromo derivitive 1 leads to longifolene with an almost quantitative yield via trans-annular 1,5-hydride transfer. Solvolysis of bromo derivative 2 in which a CH_3 has been substituted by a carbomethoxy group leads to compound 4 by ring contraction. In the first case, the CH_3 stabilizes the intermediate tertiary carbenium ion. In the second case, the electron-withdrawing $COOCH_3$ group has a destabilizing effect; thus, the reaction evolves differently.

With regard to nine-membered rings, an example of a trans-annular reaction is the hydrolysis of epoxycaryophyllene 5a in acidic media and of the epoxy-ketone 6 (obtained from the oxidation of caryophyllene[5]) in alkaline media [6.02].

[6.02]

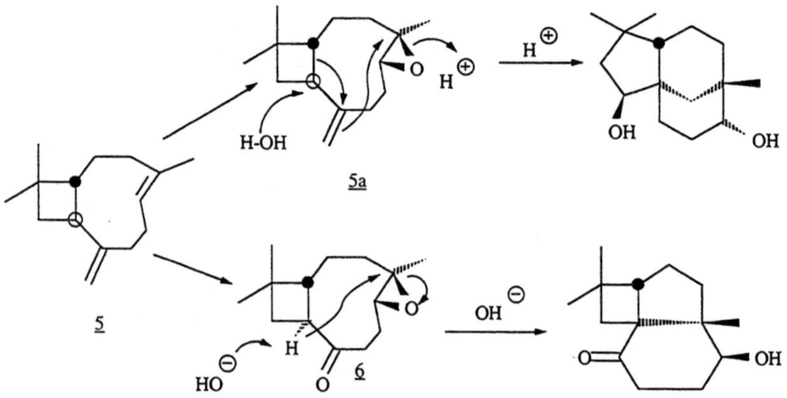

O H H H-OH 5a OH OH
5 H HO O 6 OH O OH

In 1950, Sòrm's group demonstrated that caryophyllene contains a 9-membered ring. In 1951, Sòrm and Clémo independently discovered that humulene contained an 11-membered ring and that its structure was closely related to that of caryophyllene. These cases will be evoked during later discussion on rings of this size. Sesquiterpenes containing 10-membered rings nevertheless offer the most interest in the field of monocyclic sesquiterpene compounds.

A. Sesquiterpenes with himachalane, longifolane, longipinane and longibornane skeletons

Of the sesquiterpene compounds whose biogeneses derives from cis,trans-farnesyl pyrophosphate, we will describe those for which carbocation 1a is the precursor. This arises from a 1,3-shift in the carbocation 1.

The four principal types of sesquiterpenic compound thus formed belong to the skeletal families of himachalane, longifolane, longipinane and longibornane.

These transformations are represented in the following scheme [6.03].

Longifolene (Simonsen), α-longipinene (Erdtman and Westfeld), longicyclene (Nayak and Sukh-Dev) and longiborneol (Naffa and Ourisson) have been isolated from different *Pinus* species.

Sukh-Dev isolated α- and β-himachalenes from the essential oil of the Himalayan cedar (*Cedrus deodora*) while Bredenberg and Erdtmann found them in the oil of Atlas cedar (*Cedrus atlantica*). From the same oil, the Roure Bertrand Dupont group in Grasse has isolated γ-himachalene, 7,8-dehydroarylhimachalene, 6,7α-epoxy–2,3-himachalene as well as the ketone c [6.04].

This ketone could result from a rearrangement of 6,7α-epoxy–2,3-himachalene according to the above mechanism.

The oil from the Himalayan cedar also contains the sesquiterpene alcohols, (+)-himachalol and (+)-allohimachalol, which could result from cation 2 either directly or after rearrangement [6.05].

Indian turpentine oil is obtained from *Pinus longifolia* and contains about 30% longifolene. This is the principal natural source in which Simonsen was able in the 1920s to report the presence of this sesquiterpene. The structure of this compound was determined 30 years later thanks to the work of the group Dupont, Dulou, Naffa and Ourisson at the Ecole Normale Superiéure de Paris and also to Moffet and Rogers' radiocrystallographic study of its hydrochloride. Finally, E. J. Corey realized its total synthesis in 1961 (summarized later in this chapter).

The analogy between natural longifolene and (+)-camphene is one of its essential characteristics. It is manifested in the optical properties (IR spectra, optical rotatory dispersion, optical rotation) of many of its derivatives. It is also found in numerous aspects of its reactivity. Yet, the structure of camphene is rigid and presents no conformational ambiguity while the large bridge of longifolene is able to adopt several conformations. Detailed analysis shows that, in fact, one of these conformations presents sufficiently few interactions between non-adjacent atoms, to be by far the most preferred [6.06, 7].

[6.03]

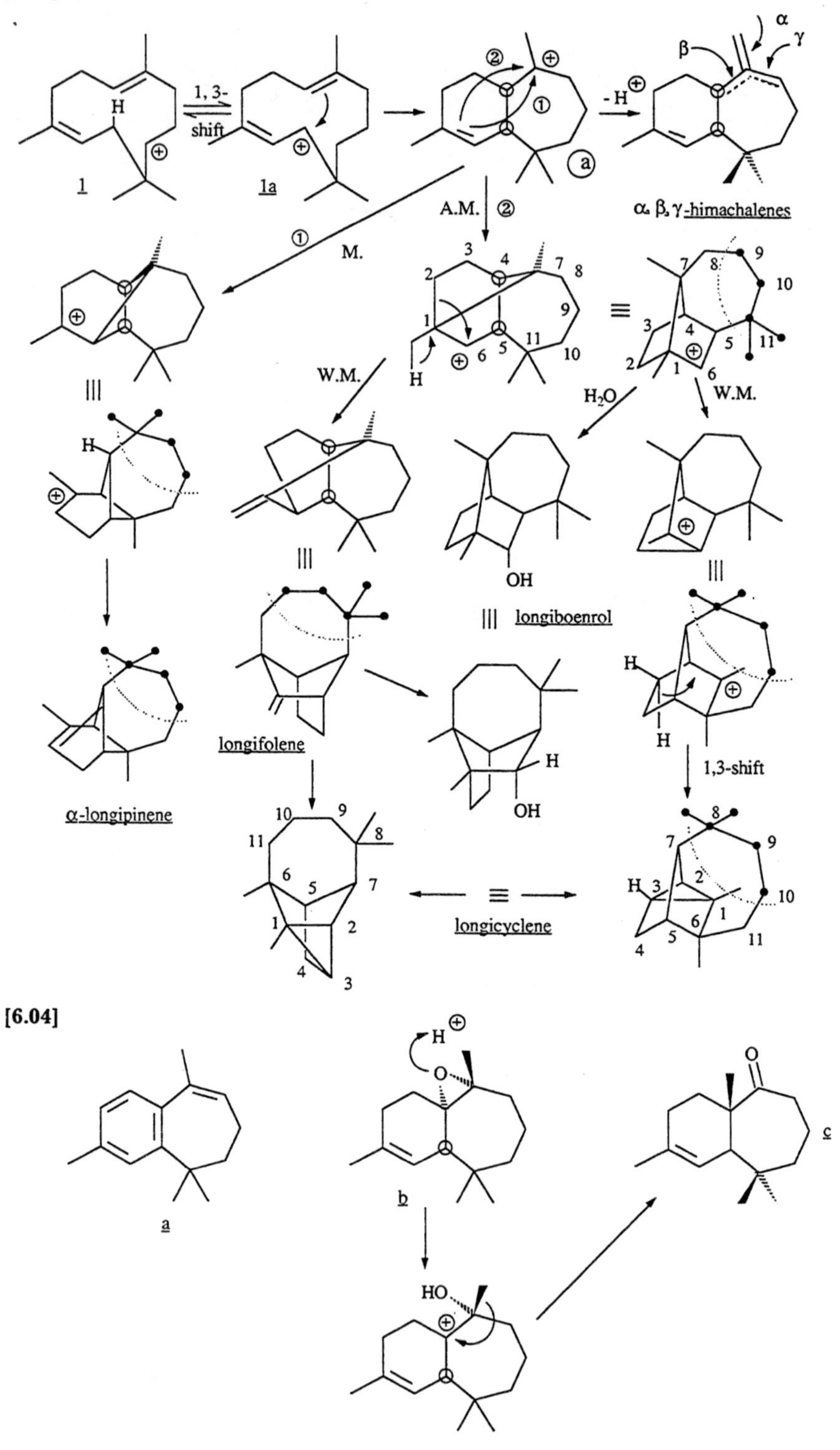

1, 3-
shift
1
1a
a
- H
α
β
γ
α, β, γ-himachalenes
A.M.
M.
W.M.
H
H_2O
W.M.
OH
longiboenrol
α-longipinene
longifolene
H
OH
1,3-shift
longicyclene

[6.04]

H
O
HO
a
b
c

[6.05]

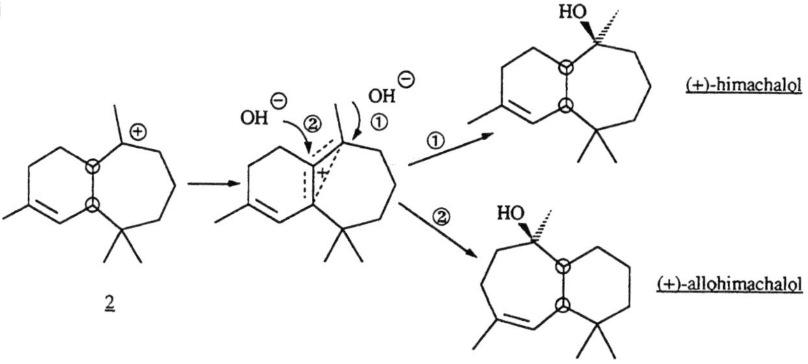

(+)-himachalol

(+)-allohimachalol

[6.06]

(+)-camphene

(-)-isoborneol (+)-camphor (+)-borneol

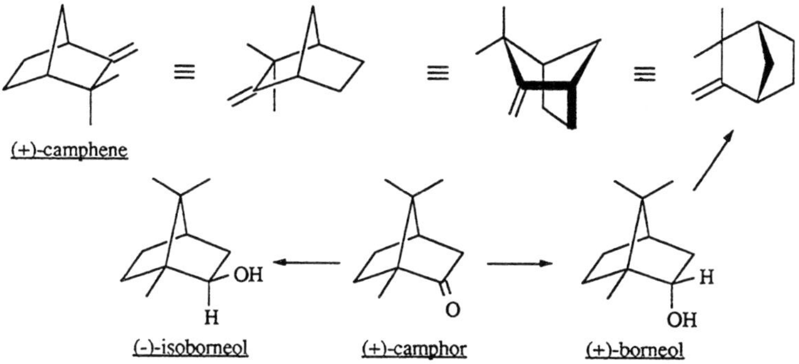

[6.07]

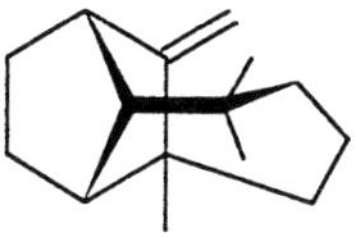

The big eight-membered bridge permits the classification of longifolene among the medium-sized ring compounds which show quite special properties. The most spectacular of these are based on the possiblity of trans-annular reactions. It may be useful before studying these reactions, to provide some general aspects of strain in these medium-sized rings.

B. Some general aspects of medium-sized rings

Strain

This term can designate either the energy due to deformation of the valence angles (Baeyer strain), or the energy which corresponds to the barrier of rotation of the angles of torsion (Pitzer strain). In some cases, both types of strain intervene simultaneously without it being possible to determine the relative contribution of each.

Baeyer strain

This intervenes each time the valence angle corresponding to a given hybridization state varies as a result of the molecular structure. The valence angles should, in principle, have the following values [6.08].

[6.08]

Nature of the orbitales		angle
p^3		90°
sp^3		109° 28'
sp^2		120°
sp^3d	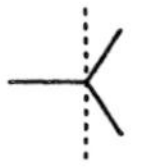	90° and 120°

In the course of a reaction, any change in the hybridization, no matter how temporary, will involve a change in the valence angle. When this is not compatible with the strain in the overall structure, the reaction proceeds with difficulty. Conversely, when it is favored by the structure, the reaction becomes facile [6.09].

[6.09]

Example : (cyclopropanone) ⇌ (hydrate, OH, OH)

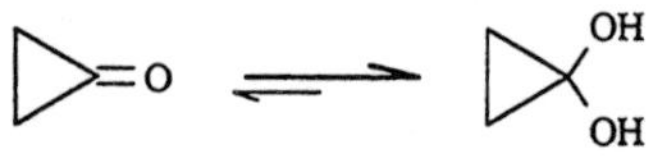

The factor responsible for the strong tendency for cyclopropanone to form its ketone hydrate (usually unstable relative to the corresponding ketone) is the passage from a trigonal to a tetragonal carbon. It is also the Baeyer strain that opposes the transition from a tetrahedral hybridization state to a trigonal state in bridgehead carbons in rigid cyclic compounds (Bredt's rule) [6.10].

[6.10]

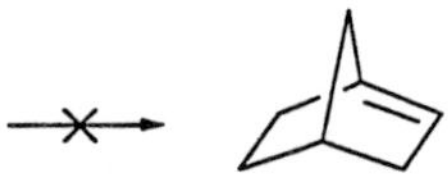

prohibited structure

Pitzer strain

This results from the eclipsed (face to face) position of bonds coming from contiguous carbons. Any reaction that tends to minimize this type of tension in a molecule will be favored. Hence, in cyclopentane, which has a fairly flat form,

Pitzer interactions are important. Transformations requiring rehybridization of a ring carbon from the tetrahedral to the trigonal state will be favored. The cyanohydrin of cyclopentanone is less stable than the cyanohydrin of cyclohexanone because the Pitzer strain in the latter is nil.

The ring effect

As a result of Baeyer and Pitzer strains, the reactivity of functional groups attached to a ring will be a function of ring size. The tetrahedral hybridization state is favorized for rings made up of 3, 4 and 6 carbon atoms whereas the trigonal hybridization state is favored for rings of 5, 7, 8, 9, 11, 12, etc. atoms. This effect is maximized in the so-called medium-sized rings (8–11). This can be represented schematically as follows [6.11].

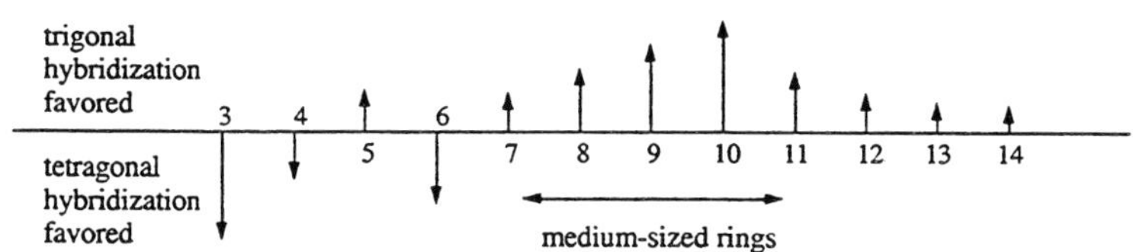

The heat of combustion of cycloalkanes gives a precise image of internal strain (total strain) [6.12].

[6.12]

Size : n	3	4	5	6	7	8	9	10	11	13	15
Heat of combustion by CH_2: $\frac{Hc}{n}$	166.6	164	158.7	157.4	158.3	158.6	158.8	158.6	158.4	157.8	157.5
Total strain $n\left(\frac{Hc}{n} - 157.4\right)$	27.6	26.2	6.5	0	6.3	9.6	12.6	12.0	11.0	5.2	1.6

Examples: Abnormal opening of 1,2-epoxycyclooctane under acidic conditions to give 1,4-dihydroxycyclooctane [6.13].

The solvolysis of the bromo derivative A leads to longifolene in an almost quantitative yield via a trans-annular 1,5-hydride transfer.

On the other hand, solvolysis of the bromo derivative B (with silver nitrate), in which a CH_3 has been substituted by a carbomethoxy group, leads to the compound D by ring contraction. In the first case, the CH_3 of A stabilizes the intermediate tertiary carbenium ion. In the second case, the electron-withdrawing $COOCH_3$ group of B has a destabilizing effect and thus the reaction evolves differently without any trans-annular reaction.

These trans-annular reactions also play an important role in the chemistry of sesquiterpenes having 9, 10 and 11 carbon units, which correspond to derivatives of caryophyllane, germacrane and humulane. It will be recalled that these three

[6.13]

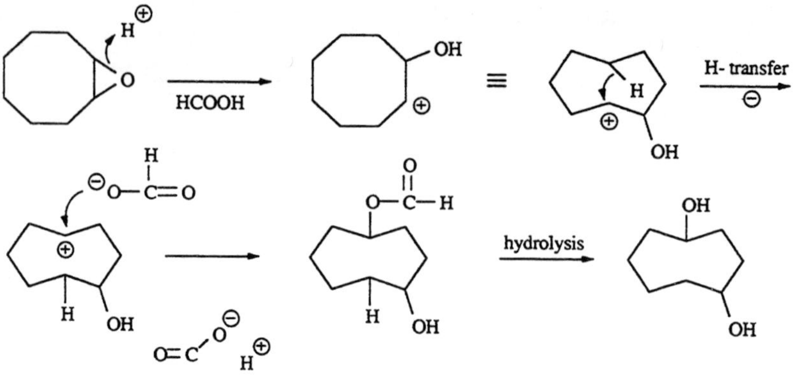

HCOOH

H- transfer

hydrolysis

groups of sesquiterpenes arise biosynthetically from trans,trans-farnesyl pyrophosphate [6.14].

[6.14]

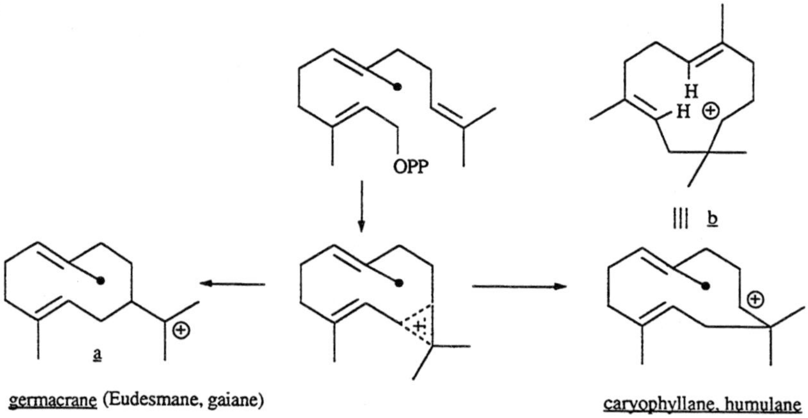

OPP

germacrane (Eudesmane, gaiane)

caryophyllane, humulane

C. Sesquiterpenes with the germacrane skeleton (10-membered ring)

In 1953, when no 10-membered derivative of the germacrane type was known, Ruzika and co-workers suggested the the cation A should play an important role in the biogenesis of derivatives of elemane, eudesmane and gaiane [6.15].

[6.15]

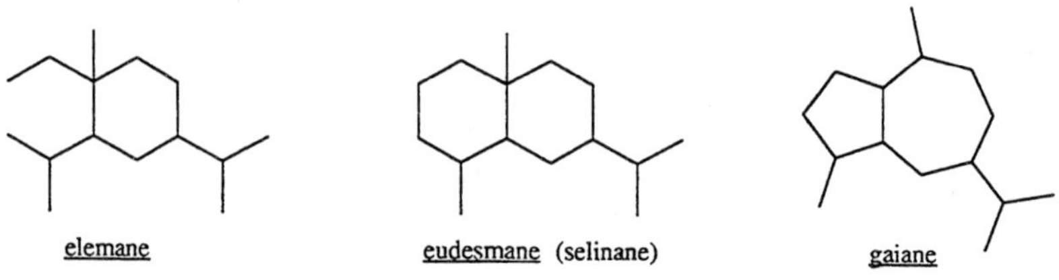

elemane

eudesmane (selinane)

gaiane

The first sesquiterpene compound with a skeleton belonging to the germacrane family, pyrethrosine, was isolated by Barton and de Mayo in 1958–1960 [6.16].

[6.16]

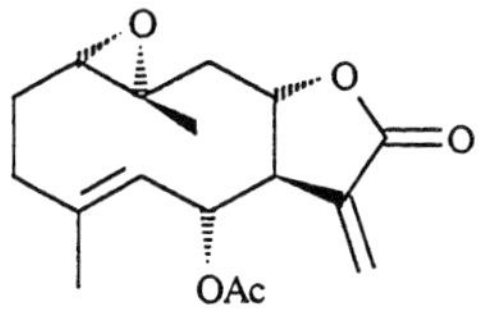

OAc

Simultaneously, the Czech group of Sòrm isolated from the essential oil of *Geranium macrorrhizum* (or "Zdravetz") a ketone containing a 10-membered ring which they called germacrone (1957) from which originates the name "germacrane" given to all compounds with this skeleton [6.17].

[6.17]

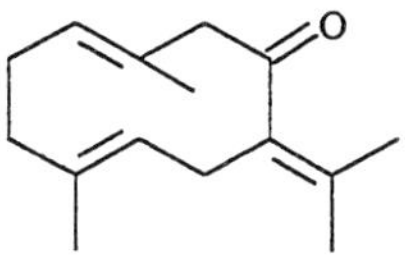

As in the case of pyrethrosine, these compounds often carry many oxygenated functional groups. They can be classed into three distinct series: the 6α–12 olides, the 8α–12 olides and the furanogermacranes. The first two classes have been designated by Sòrm as "germacranolides." They are widely distributed in the Compositæ family [6.18].

[6.18]

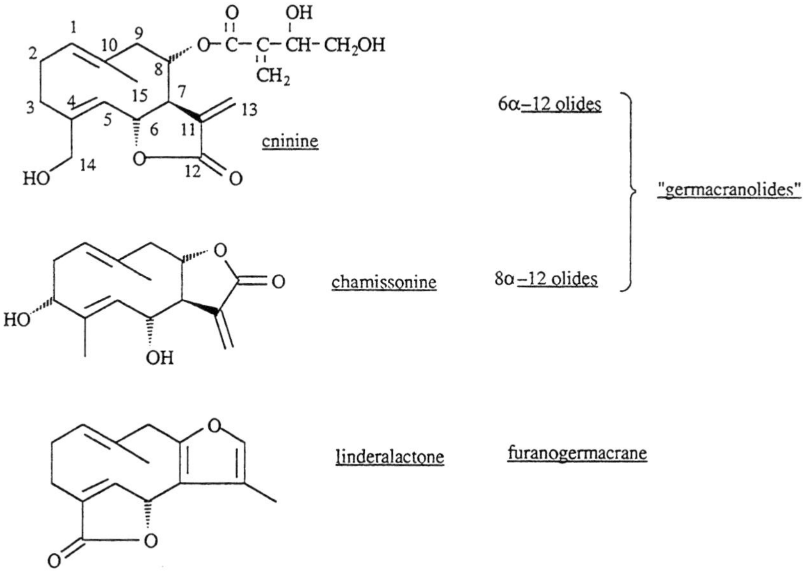

O-C(=O)-C(=CH₂)-CH(OH)-CH₂OH

cninine — 6α–12 olides

chamissonine — 8α–12 olides

"germacranolides"

linderalactone — furanogermacrane

The simplest of the 6α–12 germacranolides is costunolide, isolated from oil of costus [6.19].

[6.19]

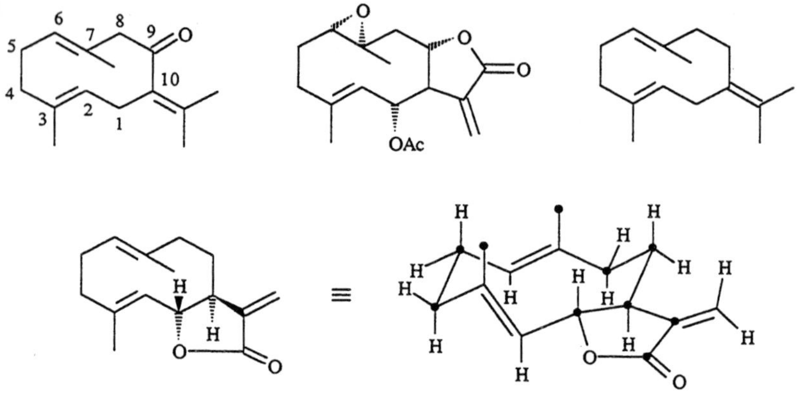

The 1,5-cyclodecadienic system present in many of these compounds is characteristic of these products. The exact geometry of this system has recently been unambiguously demonstrated by X-ray analysis of the π-complexes that some of these derivatives form with silver nitrate. A notable case is that of costunolide in which it can be noted that the two methyl groups are oriented almost vertically to the median plane of the cyclic skeleton. For this reason they are said to be in the syn position. Moreover, the two double bonds are virtually orthogonal to each other. Hydrocarbons with a germacrane skeleton and their oxygenated derivatives are frequently and simultaneously found in nature; their double bonds are frequently epoxidized.

"Germacrenes," however, have only recently been isolated. These hydrocarbons are in fact, extremely unstable and frequently give polymers in their pure state. Their isolation therefore requires particularly mild conditions. They also show a susceptibility to transform into many other bicyclic sesquiterpenes for which they are probably also the biological precursors. The principal germacrenes known at this time are designated germacrene A, B, C and D and cyclogermacrene [6.20].

[6.20]

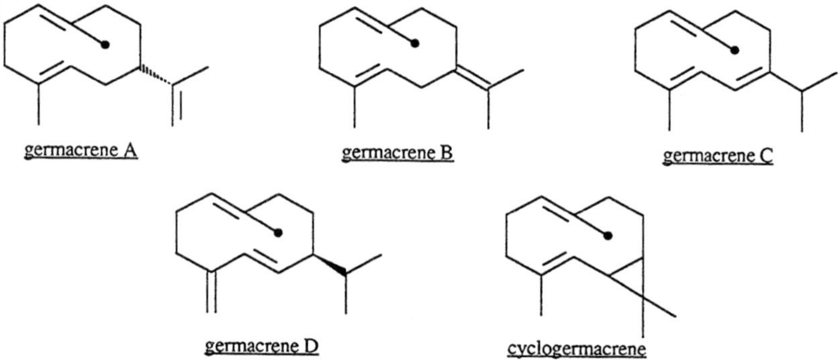

germacrene A

germacrene B

germacrene C

germacrene D

cyclogermacrene

A direct correlation between cyclogermacrene and the gaianic and cadinanic skeletons was made in 1969 when this substance was first discovered. When this hydrocarbon is treated in weak acidic media, it is transformed into a mixture of δ-cadinene and ledene (which has an "aromadendrane" skeleton closely related to the skeleton of "gaiane") [6.21].

[6.21]

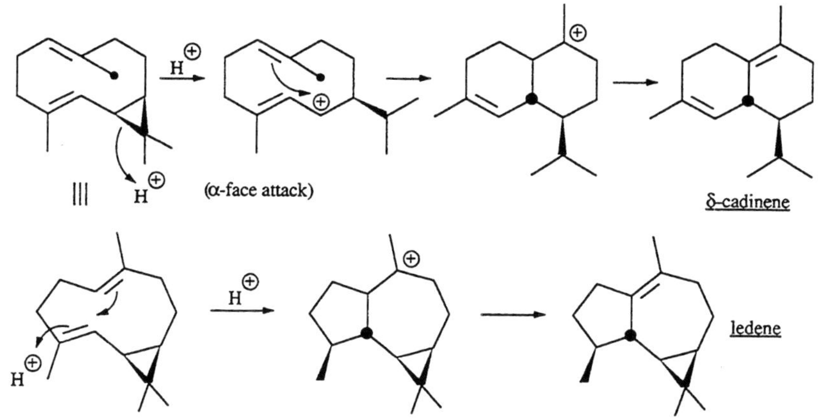

(α-face attack)

δ-cadinene

ledene

We have seen that the different germacrenes have not been known until recently. Germacrene A was first isolated in 1970 by Weinheimer et al. from *Eunicea mammosa* Lamoureux, a type of gorgonaire (marine animals). This germacrene, which had been sought in vain for so long, could only be isolated by extremely delicate chromatographic techniques. It was obtained in its levorotatory form along with (–)β-selinene. On heating, (–)-germacrene A gives (+)β-elemene [6.22].

[6.22]

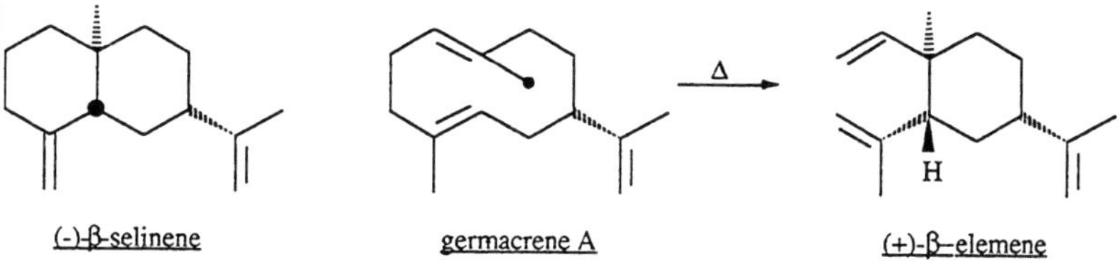

(-)-β-selinene

germacrene A

(+)-β-elemene

(+)β-Elemene and (–)β-selinene are antipodal to the products isolated from terrestial sources; it can therefore be anticipated that terrestial germacrene A, which has not yet been discovered, will be dextrorotatory.

Germacrene B was first isolated in 1969 from hop oil by Hartley and Fawcett, in addition to the two selinadienes shown below [6.23].

Germacrene C was obtained by Morikawa and Hirose in 1969 from the dried fruit of *Kadsura japonica.* This substance is converted by heat into optically inactive δ-elemene. It forms two selinadienes on contact with silica gel [6.24].

[6.23]

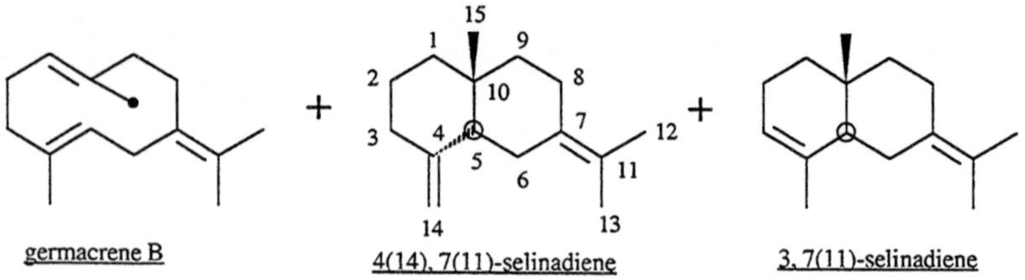

germacrene B + 4(14), 7(11)-selinadiene + 3, 7(11)-selinadiene

[6.24]

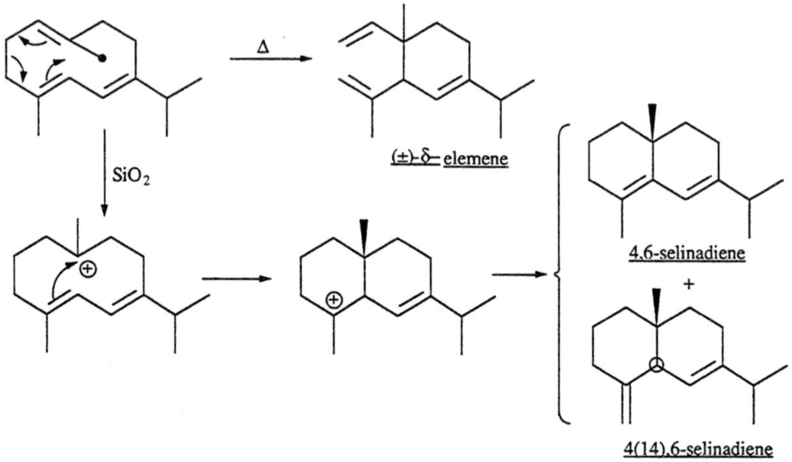

Δ

(±)-δ- elemene

SiO_2

4,6-selinadiene

+

4(14),6-selinadiene

Germacrene D was also isolated from the same source in 1969 by Yoshihara and Hirose. At the Roure Bertrand Dupont Research Center, we extracted it from ylang-ylang oil (third distillate) and from clary sage oil. Germacrene D polymerizes rapidly (or dimerizes) and it rearranges thermally in the absence of oxygen in neutral media or on contact with silica gel to give a complex mixture of sesquiterpene hydrocarbons—among which it has been possible to identify δ-cadinene, γ-cadinene, α-amorphene, α-muurolene and γ-muurolene. In addition, photoisomerization of germacrene D affords a mixture containing α- and β-bourbonenes and β-copaene. These three hydrocarbons have also been found in the essential oil of *geranium Bourbon.*

It should be noted that no derivative of bulgarane has been discovered as the result of any reaction of germacrene D [6.25, 26].

[6.25]

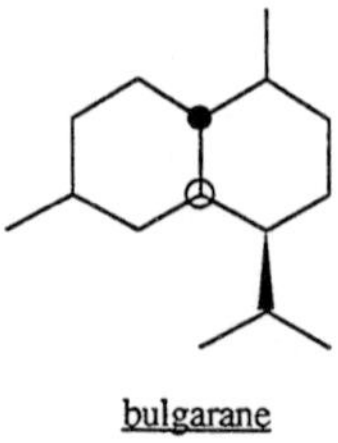

bulgarane

[6.26]

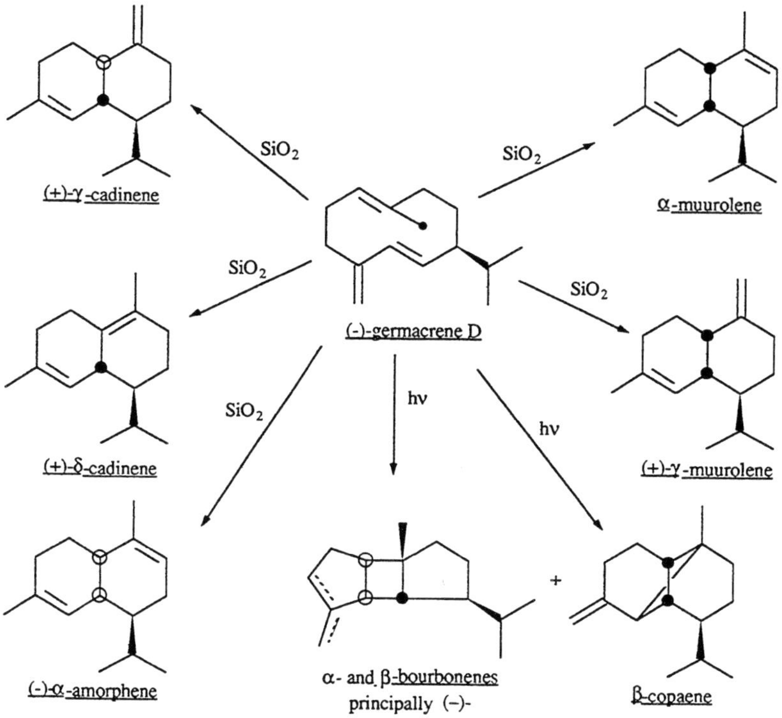

SiO_2 SiO_2 SiO_2 SiO_2 SiO_2 hν hν

(+)-γ-cadinene

α-muurolene

(-)-germacrene D

(+)-δ-cadinene

(+)-γ-muurolene

(-)-α-amorphene

α- and β-bourbonenes principally (-)-

+

β-copaene

Probable mechanisms:

a) ionic

[6.27]

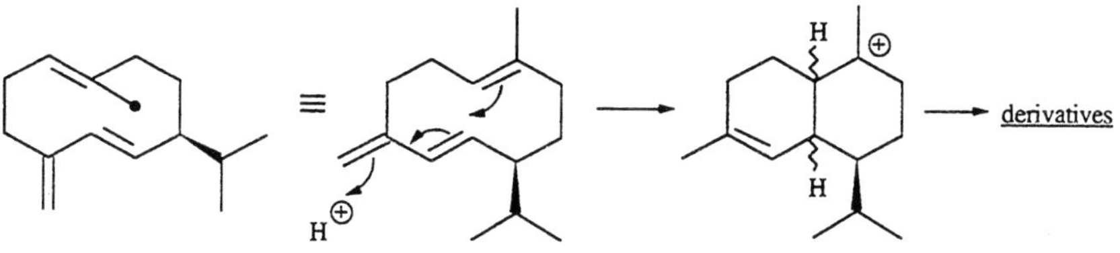

≡ H⊕ H H ⊕ derivatives

b) photochemical

In addition to the above trans-annular ionic mechanism, another trans-annular effect becomes evident from examination of the ultraviolet spectrum of germacrene D [6.28].

[6.28]

$\nu_{max}^{n\text{-hexane}}$ ~ 260 nm   ε ~ 4,500

This spectrum indicates the existence of an important trans-annular interaction between the two double bonds situated in the cyclodecadiene ring and this suggests that the preferred conformation is the following [6.29]:

[6.29]

hν

isomerization of the (E)-double bond into its (Z)-isomer

β-bourbonene (major)

α-bourbonene (minor)

hν

β-copaene

In addition to germacrene D, it seems that germacrene B plays an important role in the biosynthesis of many sesquiterpenes derived from trans,trans-farnesyl pyrophosphate for which the biological hydroxylation gives a tertiary cyclodecadienyl alcohol (H). Bates and collaborators have suggested that the a-, β- and γ-eudesmols as well as bulnesol and gaiol all derive from this alcohol (H). This hypothesis is supported by the fact that these five alcohols are simultaneously present in different proportions in numerous essential oils—for example, for gaiacwood oil in which gaiol and bulnesol predominate. These five alcohols are also found in the sesquiterpene fraction of galbanum oil [6.30].

At the experimental level, it is interesting to note certain reactions of germacrene B which indicate that the 1,10-double bond is the most susceptible of the three to electrophilic attack. This is the case for its reaction with N-bromosuccinimide in aqueous acetone in which germacrene B gives a bicyclic bromohydrin with a eudesmane skeleton having trans geometry at the ring junction, as in the majority of the sesquiterpenes of this group. Debromination of this bromohydrin furnishes an alcohol corresponding to the product of mono-dehydration of cryptomeridiol [6.31].

The results that will now be discussed have definitively demonstrated the role of the cyclodecadienyl alcohol (H). In 1968, Jones and Sutherland showed that when the leaves of *Hedycarya angustifolia* A. Cum. are treated with a volatile solvent at normal temperatures, the principal product is a novel alcohol named hedycaryol which has the same structure as the alcohol (H).

When this alcohol is heated for two hours in refluxing diethyl ether containing 1% p-toluenesulfonic acid, a 90% yield of a mixture of α-, β- and γ-eudesmols plus three unidentified minor products is obtained.

[6.30]

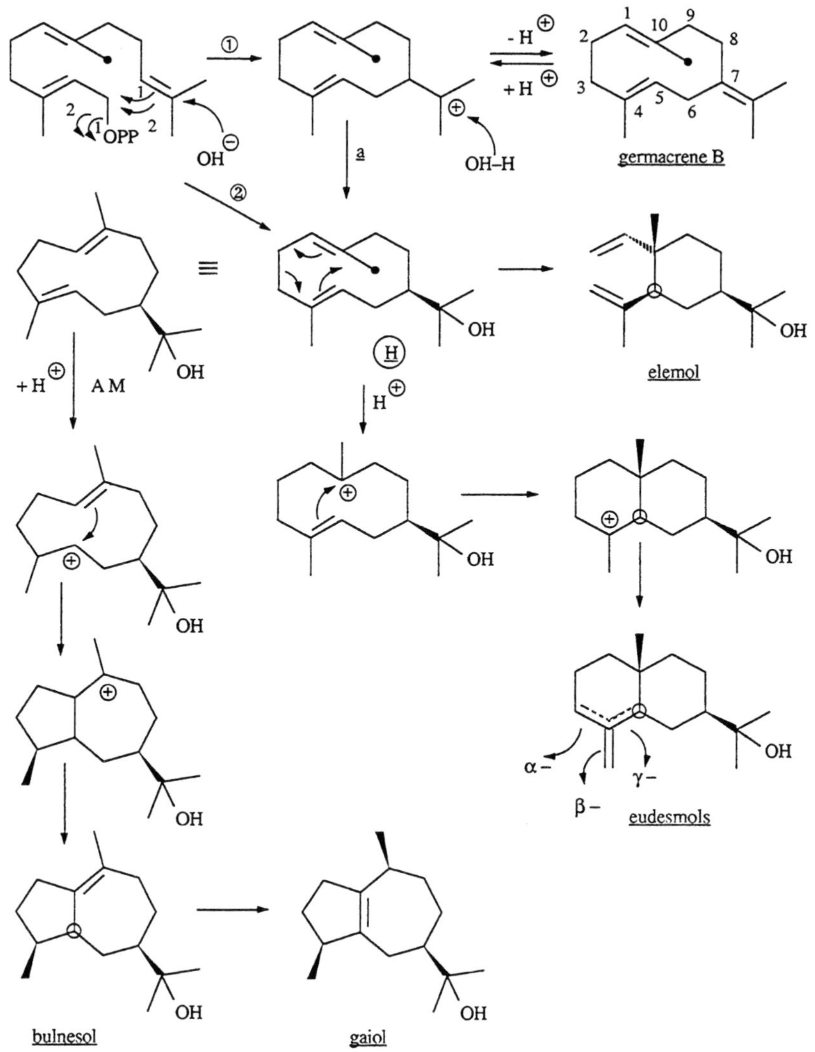

In addition, it is well known that the essential oil obtained by steam distillation of *Hedycarya angustifolia* is a very rich source of elemol. In this precise case, elemol must therefore be an artifact resulting from a thermal Cope rearrangement of hedycaryol [6.32].

It is evident, at least in the present case, that the derivatives of the germacrenes are the direct precursors of the monocyclic sesquiterpenes of the elemane type. There is therefore a possiblitity of the direct generation of bicyclic sesquiterpenes of the eudesmane type. It should be emphasized that already in 1953, Ruzicka had postulated that elemol had to be derived from the alcohol (H) which was itself derived from germacrene B.

[6.31]

NBS

germacrene B

H–OH

cryptomeridiol

bromhydrine (trans)

[6.32]

Δ
~100°C

H

elemol

A case identical to this has also been observed by Jones and Sutherland with the leaves of *Geijera parviflora* Lind., which were extracted with the aid of volatile solvents permitting the isolation of pregeijerene, a C–12 hydrocarbon derived from hedycaryol by loss of an isopropylhydroxy group. Steam distillation of these leaves gave geijerene. The exact stereochemistry of these two hydrocarbons has been demonstrated by Sim thanks to X-ray crystallography of their complexes with silver nitrate [6.33].

[6.33]

pregeijerene

(±) geijerene

Consequently, prudence should be shown in attributing the presence of elemane derivatives in essential oils as being other than artifacts. While on this subject, it should be noted that despite the presence of two asymmetric centers, geijerene is nonetheless optically inactive. This case is comparable to that of δ-elemene (obtained by heating germacrene C) found in several essential oils, in particular that of gurjun balsam.

We have indicated that germacrene B shows susceptibility to electrophilic attack at its 1, 10-double bond. Recently, Brown and Sutherland have shown that germacrene B gives two mono-epoxides at 1,10 and at 4,5. In acidic media, these two epoxides provide either eudesmane derivatives or gaiane derivatives [6.34].

[6.34]

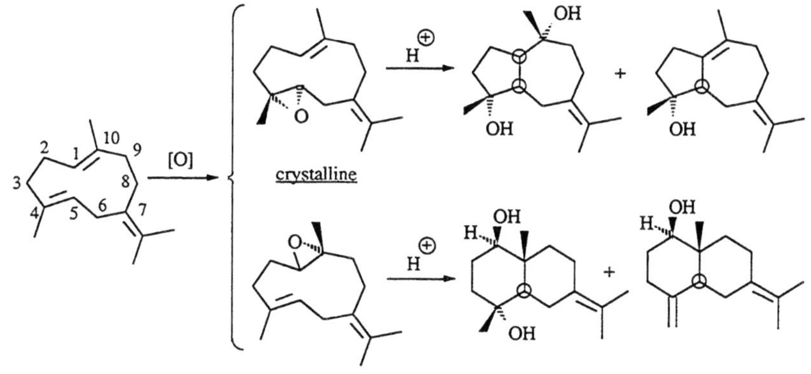

crystalline

liquid

These experimental results strongly suggest that the epoxides of the germacrenes in their preferred conformations are involved in the biosynthesis of these compounds [6.35].

The thermal isomerization reaction is one of the most remarkable properties of the trans,trans–1,5-cyclodecadiene systems. This is a [3,3] suprafacial sigmatropic Cope rearrangement, first observed in the Prague laboratories of Sòrm, who showed that germacrone gives β-elemenone on pyrolysis [6.36].

The specific conformation of the cyclodecadienic ring is extremely important in determining the steric course of the Cope reaction. If the conformation cannot be achieved, enantiomeric compounds are obtained; for instance, with linderalactone and litsealactone. This latter isomerizes to isolitsealactone on heating [6.37].

Some germacranolides may, however, assume several conformations. As an example of this, Yoshika and Mabry demonstrated that by using NMR at room temperature, isabeline exists in two conformations [6.38].

It is interesting to note in this context that when isabeline is irradiated at 253.7 nm, photoisabeline is obtained as the result of a concerted ($\pi^2_s + \pi^2_s$) cycloaddition of conformer b. In addition, the partial hydrogenation of isabeline gives dihydroisabeline which assumes only one conformation at room temperature. When this is then irradiated, a mixture of two products with a ratio 1 to 3 is obtained. These are, respectively, dihydrophotoisabeline and lumidihydroisabeline. Pyrolysis of dihydrophotoisabeline forms pyrodihydroisabeline (opening permitted by symmetry) along with a small amount of dihydroisabeline (forbidden by symmetry) [6.39].

Recent studies on the Cope rearrangement of some germacranes have provided evidence on one hand for its reversibility and on the other hand for the conformational control of the stereochemistry of the reaction. In 1970, McCloskey and coworkers demonstrated this reversibility in three systems of the costunolide type [6.40].

[6.35]

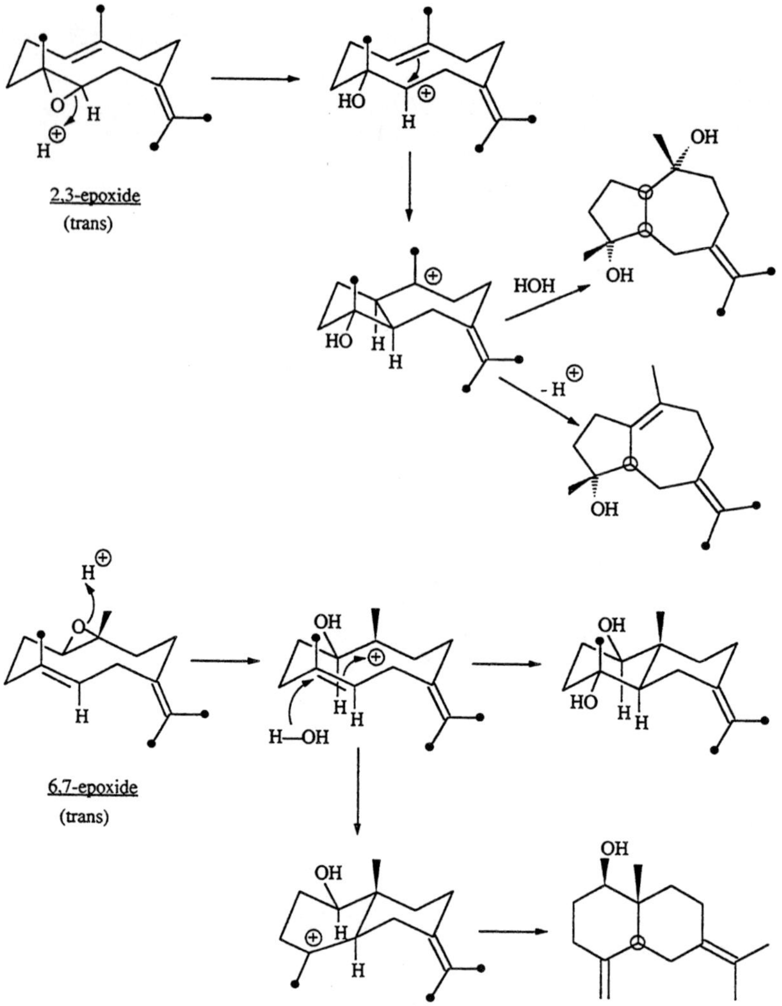

2,3-epoxide
(trans)

HOH

- H⊕

6,7-epoxide
(trans)

H—OH

[6.36]

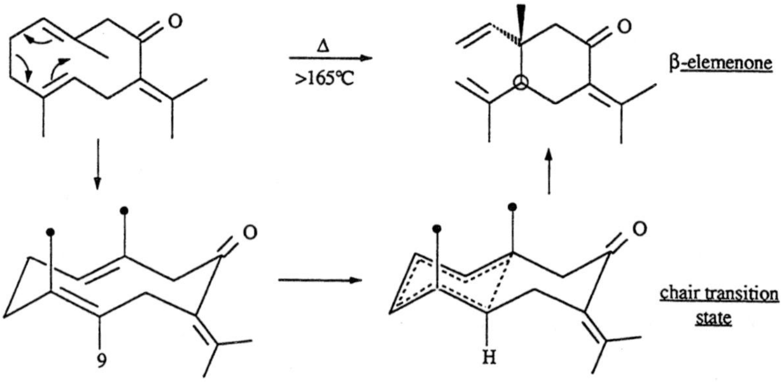

Δ
>165°C

β-elemenone

chair transition state

[6.37]

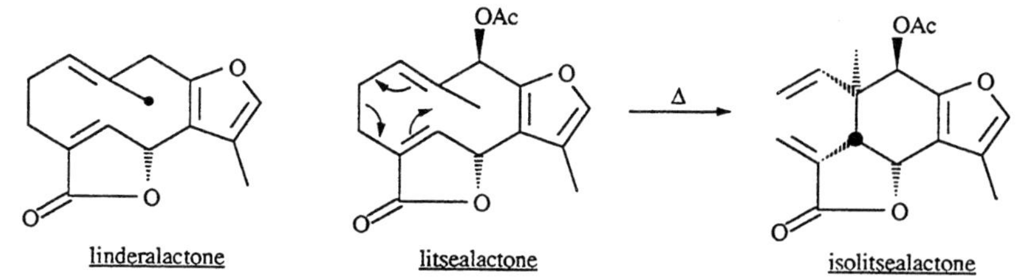

linderalactone

litsealactone

Δ

isolitsealactone

[6.38]

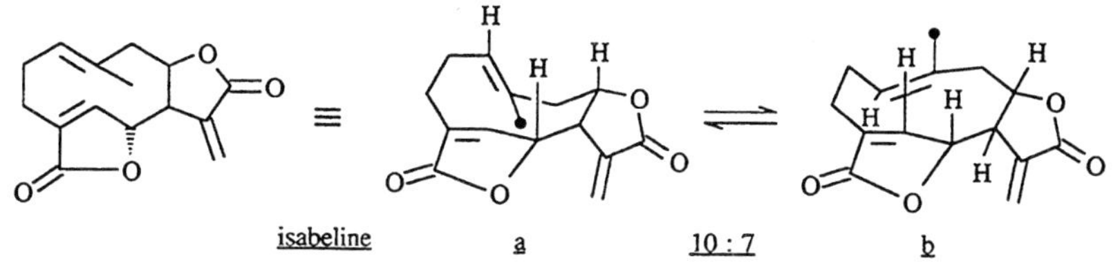

isabeline ≡ a 10 : 7 b

[6.39]

b hν → photoisabeline

isabeline

H_2

H_2

dihydrophotoisabeline Δ (1) → dihydroisabeline (minor) + pyrodihydroisabeline (preponderant)

thermodynamically forbidden photochemically allowed

hν (2)

dihydrophotoisabeline 1 : 3 lumidihydroisabeline

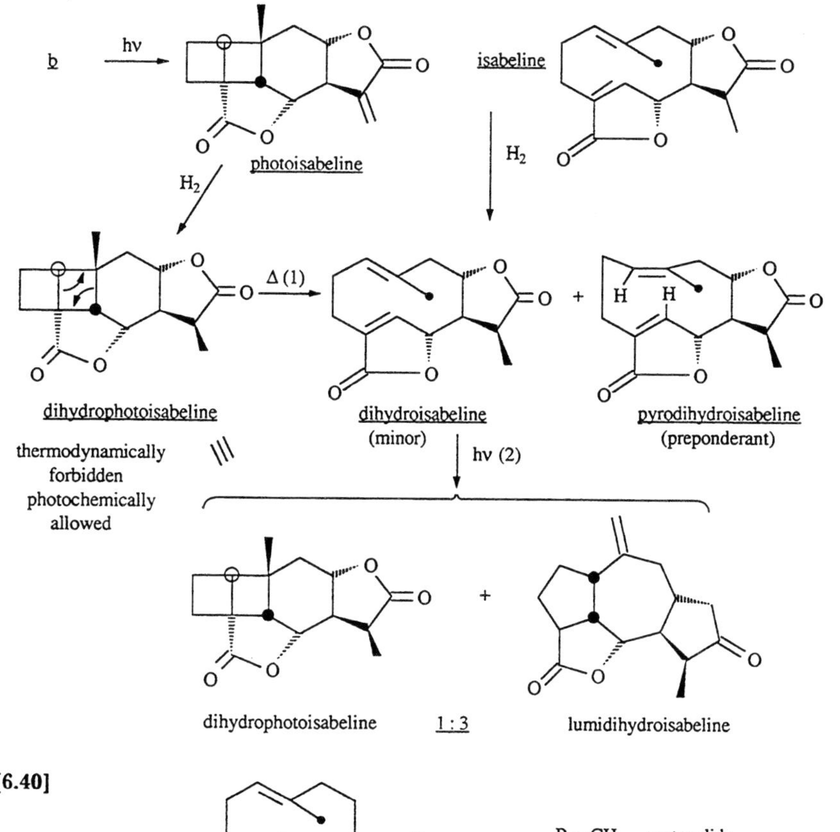

[6.40]

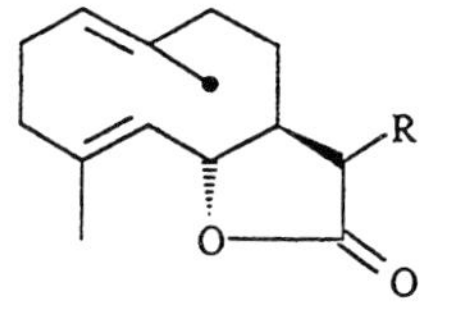

$R = CH_2$ costunolide

$R = -\alpha Me$

$R = -CH_2N(Me)_2$

Takeda and collaborators (1970) not only observed the reversibility of the Cope reaction, but they also discovered that the relative stereochemistry of certain furanoelemadienes obtained in the course of this rearrangement varied from case to case. For instance, linderalactone and the related ether E gave, respectively, isolinderalactone and the isomeric ether Ea. Yet, the diol D gave the isomeric diol Da [6.41].

[6.41]

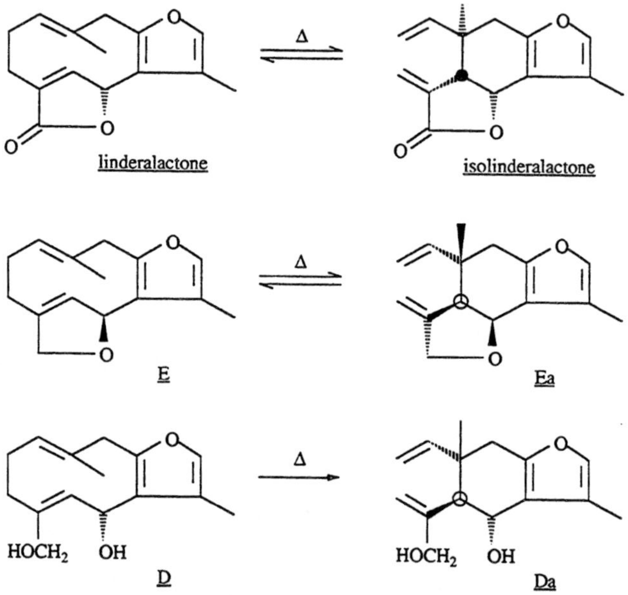

linderalactone ⇌ (Δ) isolinderalactone

E ⇌ (Δ) Ea

D → (Δ) Da

$HOCH_2$ OH

On the basis of the hypothesis of a chair type transition state for the Cope rearrangement, it is possible to justify these results by examining molecular models of these three furanodienes. The precise conformation of linderalactone has been deduced from examination of its Nuclear Overhauser Effect (NOE) [6.42].

[6.42]

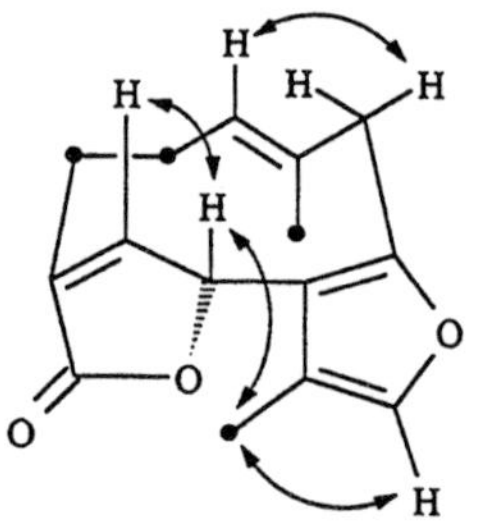

It should be noted that the reversibility of the Cope rearrangement also depends on the favorable conformation of the resulting elemane. A similar reasoning can be invoked to explain the isomerization of (–)δ-elemenol into (+)-epi-δ-elemenol through the intermediation of germacratrienol [6.43].

[6.43]

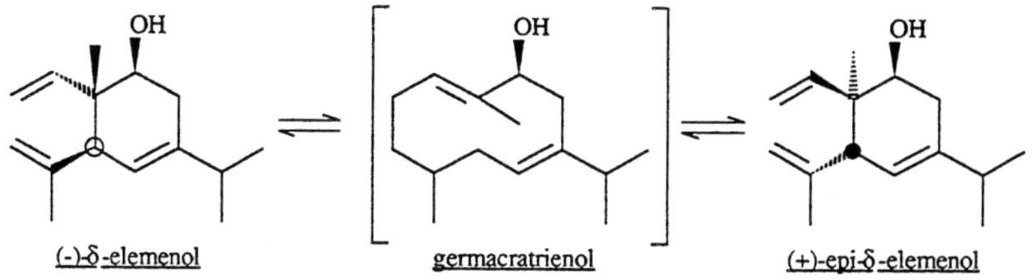

(-)-δ-elemenol ⇌ [germacratrienol] ⇌ (+)-epi-δ-elemenol

Two particularly interesting examples of thermolysis involving double electrocyclic processes have recently been discovered. The materials undergoing thermolysis are shiromodiol acetate and 4,5-epoxygermacrene B [6.44].

[6.44]

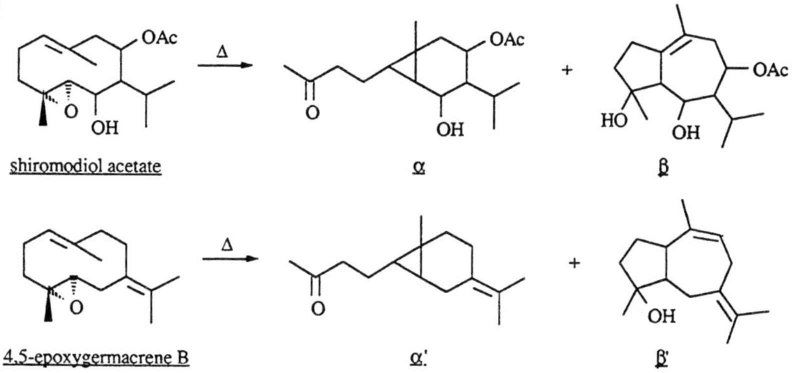

shiromodiol acetate → (Δ) α + β

4,5-epoxygermacrene B → (Δ) α' + β'

In the latter case, for example, the mechanism is the following: A product called carabrone resembling α and α' is found in nature [6.45].

[6.45]

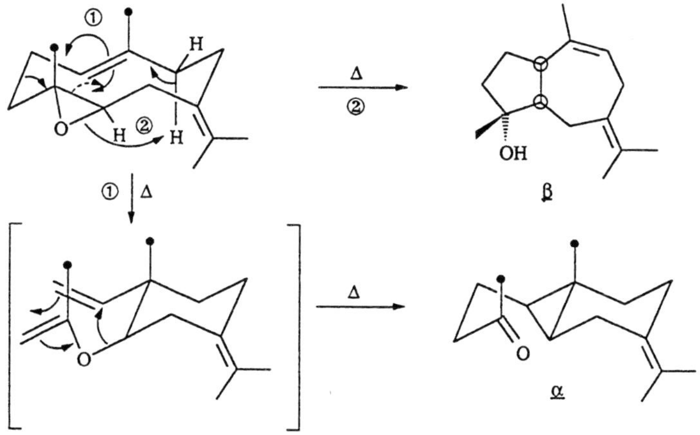

The above-mentioned reactions raise the question of the biogenesis of this ketone, particularly as to whether dehydroxy–2-baileyine could act as its precursor [6.46].

[6.46]

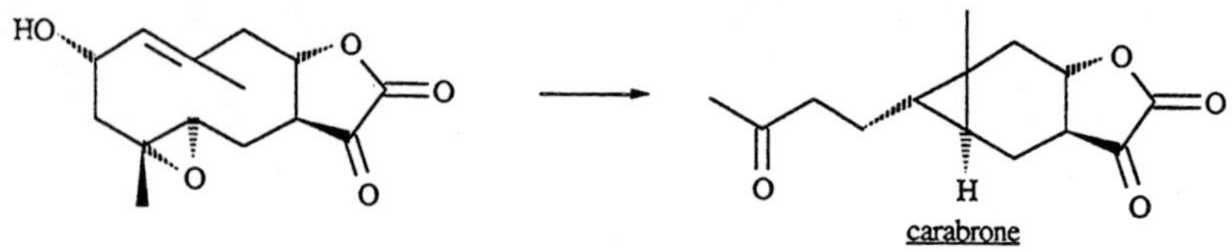

carabrone

In addition to these thermal and photochemical reactions, there is another reaction characteristic of the medium-sized rings which can also manifest itself due to another possibility discovered by Prelog for trans-annular reactions. This is the reaction under acidic conditions which results in cyclization for which several examples have already been shown. This type of reaction was first noted by Barton and Mayo in 1957 in the case of pyrethrosine which cyclizes to give isopyrethrosine [6.47].

[6.47]

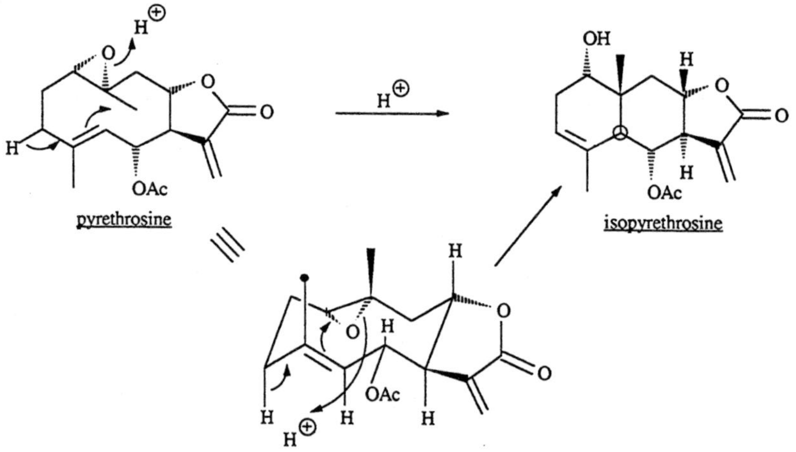

pyrethrosine

isopyrethrosine

The epoxide ring is extremely sensitive to trans-annular nucleophilic attack. The same occurs when costunolide is hydrogenated in acidic media to give a mixture of santanolides A and C [6.48].

[6.48]

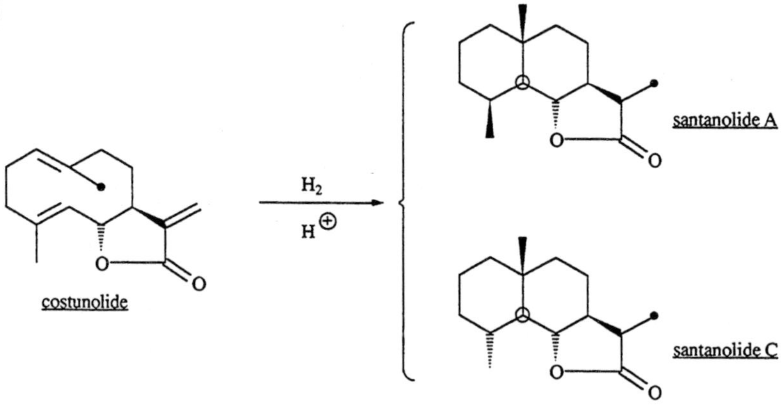

costunolide

santanolide A

santanolide C

It is highly probable that both thermal and ionic processes intervene in the following case. In 1969, Iguchi and co-investigators separated preisocalamendiol and four related substances designated here as α, β, γ and δ from the rhizomes of Acorus calamus. It would seem that the ketones C_1 and C_2 are the precursors of these products [6.49].

[6.49]

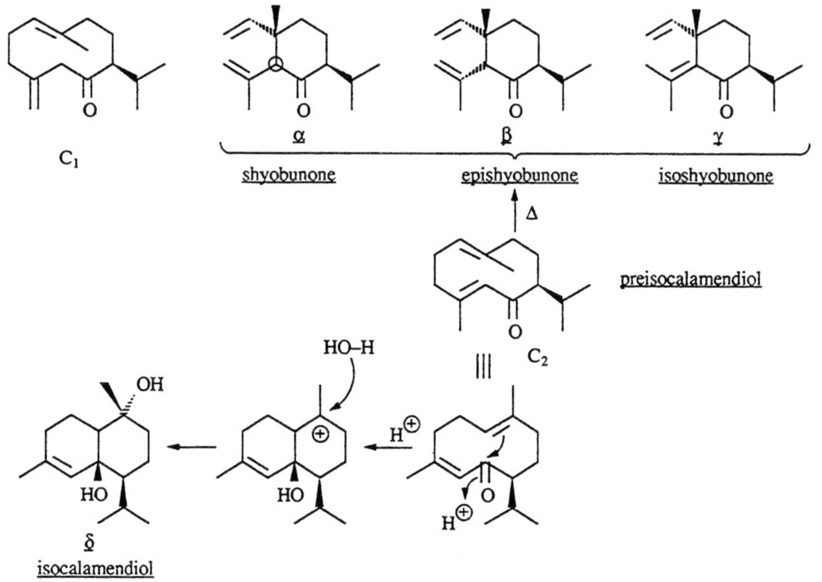

C_1

α shyobunone

β epishyobunone

γ isoshyobunone

Δ

preisocalamendiol

C_2

HO–H

OH

HO

HO

$H^{\oplus}$

$H^{\oplus}$

δ isocalamendiol

We can finish this discussion of the germacrane derivatives by mentioning an extremely important biological property of certain sesquiterpene compounds. Kupchan and co-workers at the University of Wisconsin have isolated two new sesquiterpenoids (elephantine and elephantopine) from the plant *Elephantus elatus* [6.50].

[6.50]

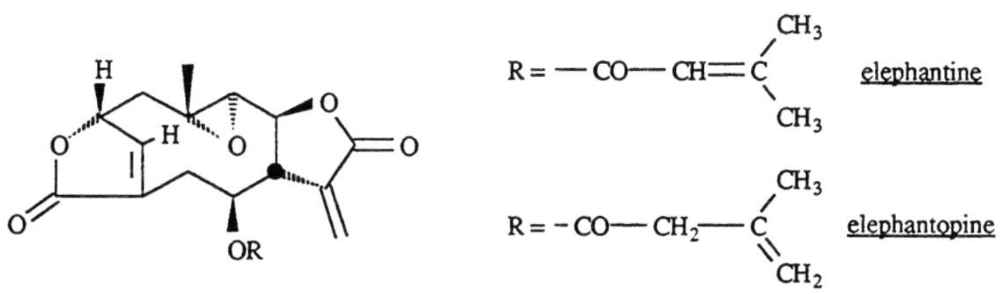

H H O O O O OR

R = —CO—CH=C(CH_3)$_2$ elephantine

R = —CO—CH_2—C(CH_3)=CH_2 elephantopine

These two substances show the remarkable property of inhibiting tumors in rats, in particular Walker's carcinoma 256. The precise stereochemistry of these substances has been determined by McPhail and Sim using X-rays. In the course of their work on anti-tumor agents, Kupchan and collaborators have also isolated three dilactones with modified elemane-type structures. They are called vernolepine, vernomenine and vernodaline [6.51].

[6.51]

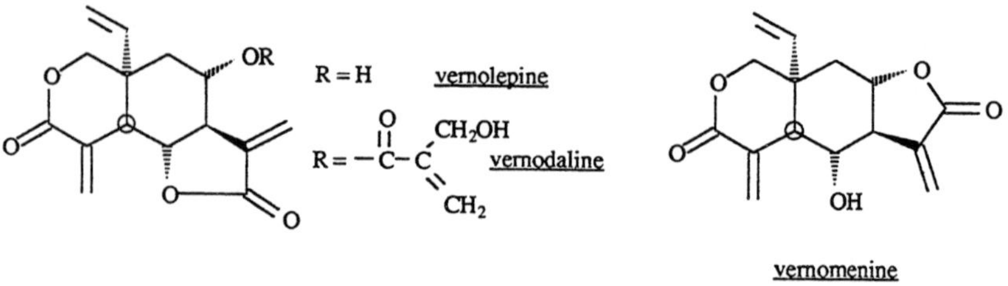

R = H vernolepine

R = $-C(=O)-C(CH_2OH)=CH_2$ vernodaline

vernomenine

It is possible that these compounds are formed in the course of their isolation from precursors of the 14,15-germacradienolide type [6.52]. It can be readily seen that such precursors realize all of the stereochemical details of the three compounds shown above.

[6.52]

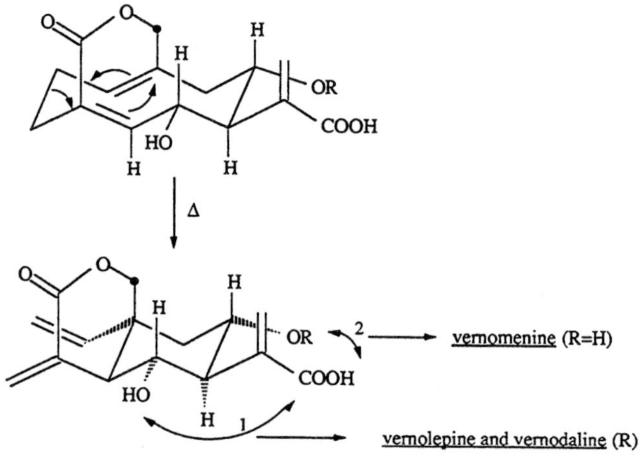

Δ

vernomenine (R=H)

vernolepine and vernodaline (R)

A very important series of sesquiterpenes with irregular isoprenic skeletons can also be considered derivatives of the germacradiene skeleton. They are eremophilane, valencane and vetispirane. The precursor in this case is a 1,6-cyclodecadienyl tertiary alcohol which can exist in two conformations: C_1 in which the methyl groups are syn and syn relative to the isopropylol group and C_2 in which the methyl groups are syn and trans relative to the isopropylol group [6.53].

[6.53]

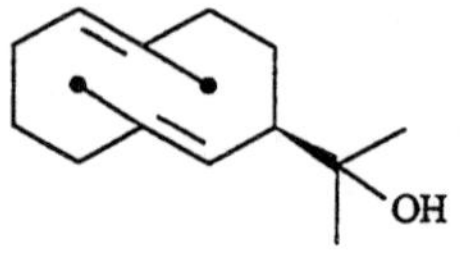

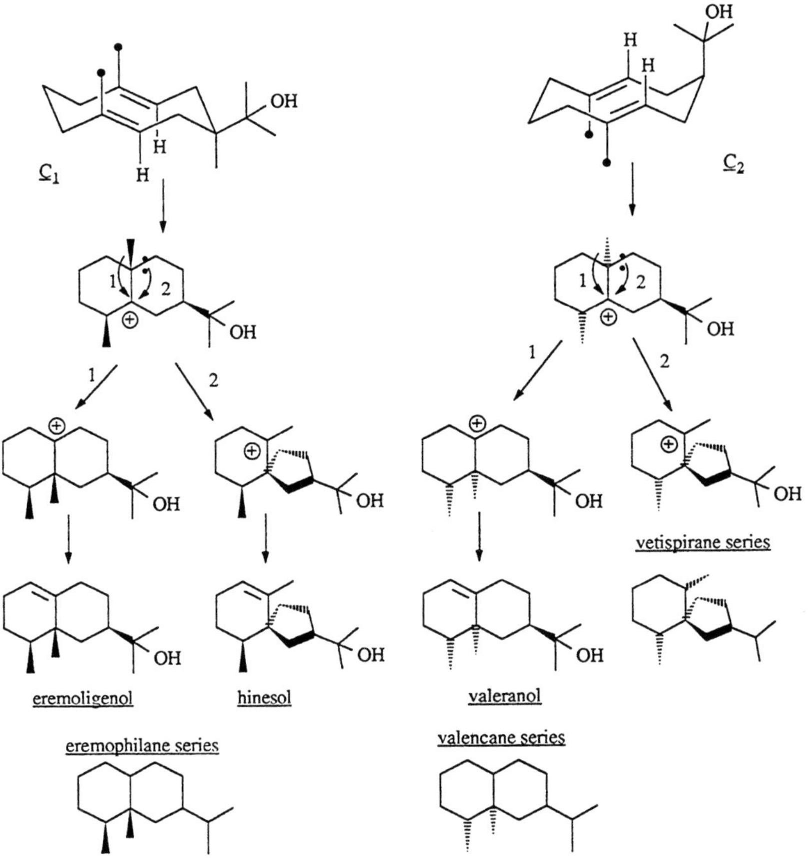

In the eremophilane series there is, above all, eremophilone, which was isolated by Simonsen in 1932 from the oil of *Eremophila mitchelli* [6.54]. Its structure was determined by Penfold and Simonsen in 1937 and it was the first example of a sesquiterpene with an irregular isoprenic skeleton.

[6.54]

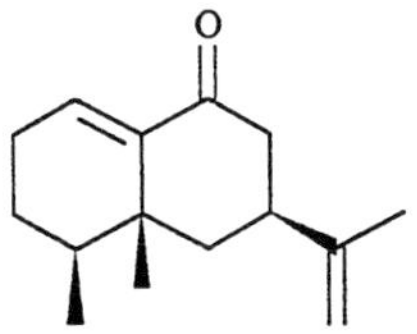

In the valencane series, there are valencene (orange oil), nootkatone (grapefruit), α-vetivone (vetyver) as well as nootkatene, aristolochene and α- and γ-vetivene [6.55]. The principal member of the vetispirane series is β-vetivone [6.56].

In support of this hypothesis of biogenesis there is one particularly significant experimental observation—dehydration of α- and β-rotunol gives a dienone with a vetispirane structure [6.57].

[6.55]

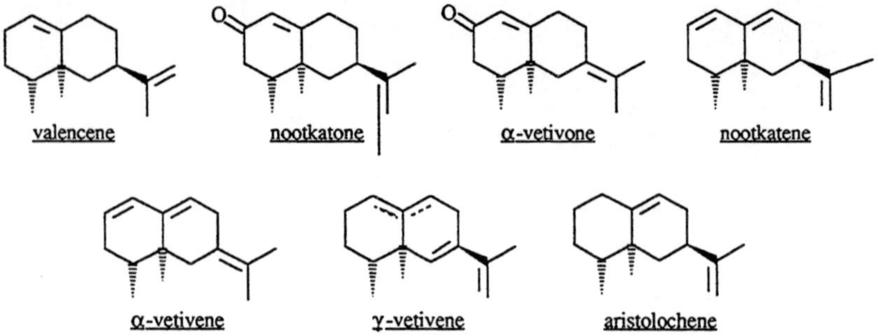

valencene nootkatone α-vetivone nootkatene

α-vetivene γ-vetivene aristolochene

[6.56]

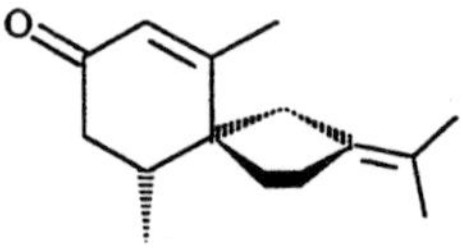

[6.57]

- H_2O

β-rotunol α-rotunol dienone

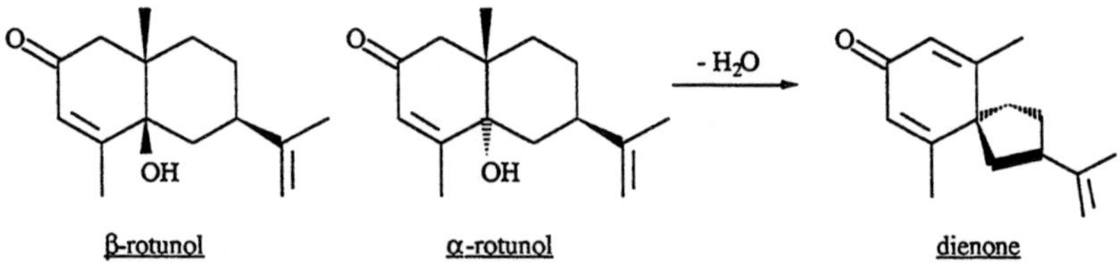

D. Sesquiterpenes with the humulane (11-membered ring) and caryophyllane skeletons

Given its regular isoprene coupling, this class of monocyclic sesquiterpenes must by necessity possess an 11-membered ring [6.58].

[6.58]

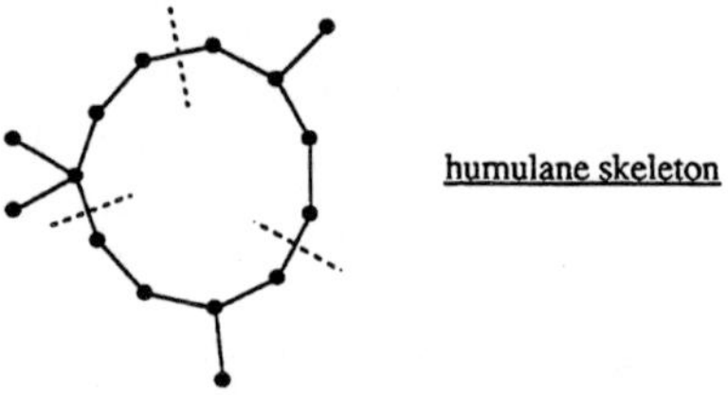

humulane skeleton

Its biogenesis from trans,trans-farnesyl pyrophosphate should be remembered [6.59].

Today, this group contains only a very small number of representatives: α-humulene and its epoxides (principally 6,7-epoxyhumulene), β-humulene, zerum-

[6.59]

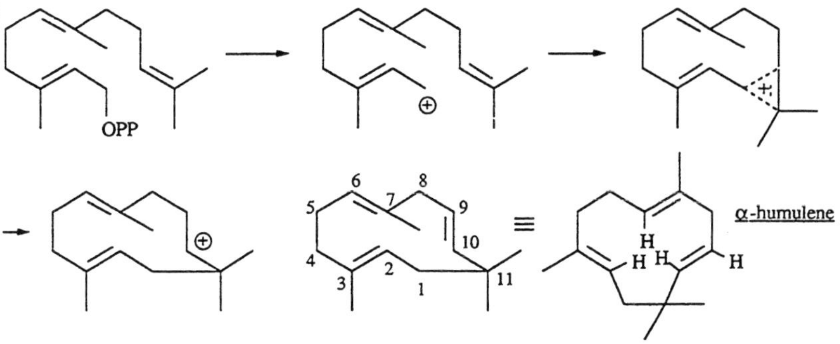

OPP

α-humulene

bone and humulenol. In addition to these compounds which have been known for some time, Naya and Kotake isolated three new derivatives from the essential oil of Japanese hops in 1969—humulenone, humuladienone and humulol [6.60].

[6.60]

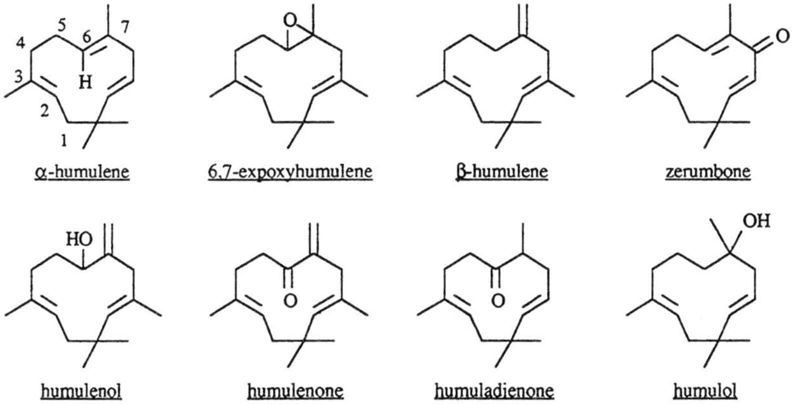

α-humulene | 6,7-expoxyhumulene | β-humulene | zerumbone

humulenol | humulenone | humuladienone | humulol

The stereochemistry of α-humulene had been the source of numerous controversies which were finally resolved by X-ray crystallographic studies on its adduct with silver nitrate (made from 1 mole of α-humulene and 2 moles of $AgNO_3$ with each Ag^+ forming a π-complex with either the 6,7 or the 9,10-double bonds) [6.61].

It will be recalled that for germacrene B, the composition of the adduct was equimolar, but in this case the crystal matrix contains two (D) molecules and two (L) molecules which are joined to each other by the Ag^+ and NO_{3-} ions. Each Ag^+ ion is associated with the 6,7-double bond of one molecule and the 2,3-double bond of a second molecule. This is similar to the binding found in the 1/1 adduct with cyclooctatetraene [6.62].

The average distance Ag^+-C is 2.53 Å (2.65 Å in the case of cyclooctatetraene). Compared to the distance of 2.38 Å observed for α-humulene, this latter value indicates a clearly higher degree of coordination.

[6.61]

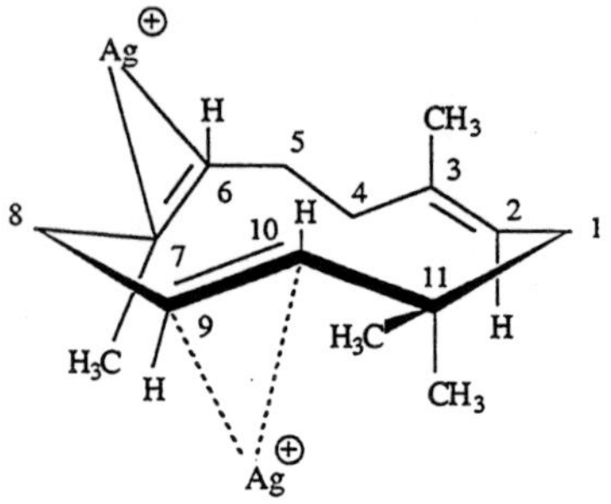

the average distance $Ag^{\oplus}$—C is 2.38 Å

[6.62]

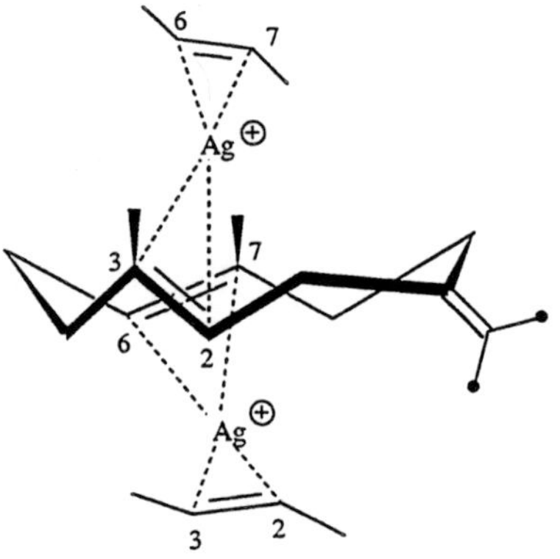

It may appear surprising to speak of the (D) and (L) forms of germacrene B when this substance contains no asymmetric carbon atom. In fact, this molecule possesses no plane of symmetry. This new type of asymmetry can be easily analyzed in the case of cyclooctene [6.63].

The two enantiomers are optically stable and can be separated (Cope). They can isomerize by two possible processes: torsion around the double bond, and torsion around the single bonds (rotation of the hexamethylenic chain of 180° around the double bond starting at one end and finishing at the other).

The latter case is more favorable (smaller barrier) for cyclic alkenes with larger rings such as in the case of cyclodecene which rapidly racemizes (this is also true for the cyclodecadienes. This explains why it has never been possible to isolate their optically active forms which exist in equal proportions in both the liquid and the crystalline states).

The stereochemistry of α-humulene has important consequences with regard to the biogenesis of this group of sesquiterpenes. In fact, if the 2,3-double bond is Z, as shown in the structure A attributed by Hendrickson, then α-humulene derives from cis–2,3-trans–6,7-farnesol. However, if this double bond is E as indicated earlier (proposition of Sutherland and Waters), then α-humulene will originate from trans,trans-farnesol. This study has confirmed this second structure and, thus, this latter biogenetic origin [6.64].

[6.63]

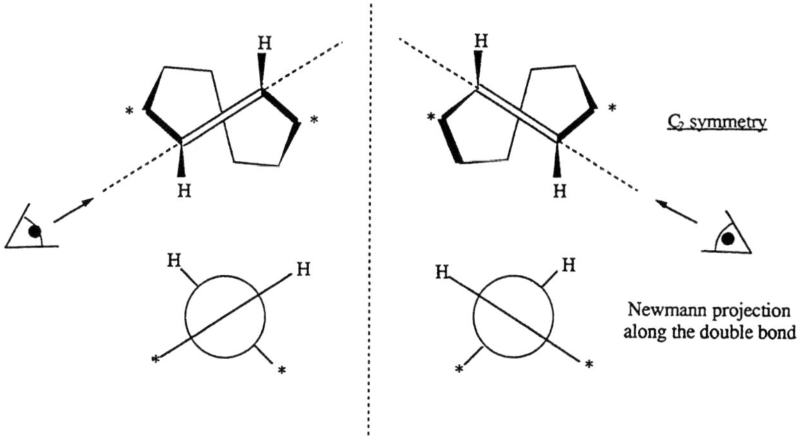

C_2 symmetry

Newmann projection along the double bond

[6.64]

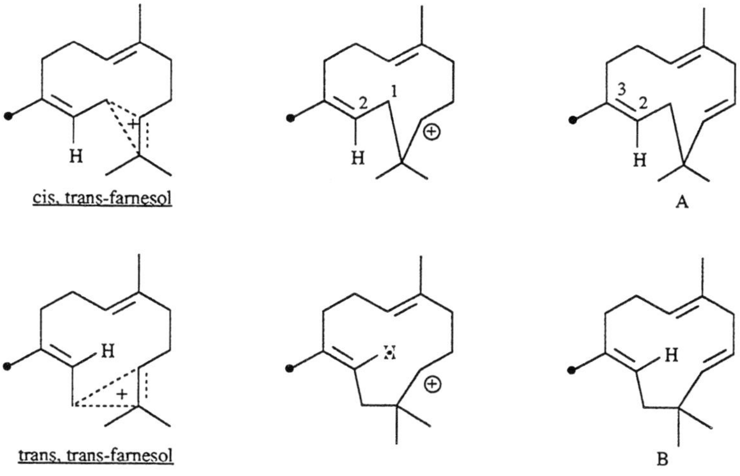

cis, trans-farnesol

A

trans, trans-farnesol

B

The conformations of α-humulene may of course be different in the liquid state (or in solution) than they are in crystalline complexes. The conformational mobilities of α-humulene and zerumbone as a function of temperature have been studied by NMR. The free enthalpies of activation for inversion of these two compounds are calculated as $\Delta G^{\ddagger}$ = 10.6 and 15.9 Kcal.mole^{-1}, respectively. The isomer of humulene is 11.9 Kcal.mole^{-1} [6.65].

While on the subject of biogenesis, it can be observed that all-natural products containing humulene also contain isocaryophyllene and caryophyllene. In fact, caryophyllene is nearly always the predominant one [6.66].

This observation has led some authors to postulate that α-humulene might be the precursor of caryophyllene. It is simpler to state that the ion A can lead just as well to α-humulene as to caryophyllene.

[6.65]

$\Delta G^{\ddagger} = 11.9$ Kcal.mole^{-1}

[6.66]

α-humulene

caryophyllenes

For this reason, Sutherland et al. embarked upon a study of the action of aqueous acetone solutions of N-bromosuccinimide on α-humulene (as had been done for germacrene B). They obtained a bromohydrin whose chemical reactivity was studied in detail (summarized here). In addition to this bromohydrin, another bromo derivative is formed [6.67].

[6.67]

bromohydrin

bromo derivative

On treatment with $LiAlH_4$, the bromo derivative regenerates α-humulene along with β-humulene. Its oxidation by DMSO (allylic bromide) gives humulenone which is identical to the natural ketone described above. An aldehyde which originates from an allylic migration of the bromide is also obtained [6.68].

Examination of the kinetics of the hydrolysis of the bromohydrin demonstrates that this reaction is of the SN_1 type. This rapid hydrolysis with retention of configuration is due to the effect of proximity which is well known in cyclopropanes. This means that the reaction must pass through a bicyclobutonium ion intermediate as demonstrated with other substrates in the fundamental works of Winstein

[6.68]

Br O CHO +

on one hand and Roberts on the other [6.69].

[6.69]

Br OH Cl

H_2O / Na_2CO_3 / CH_3COCH_3

3 hours; yield= 100%

HO HO A

$POCl_3$

$POBr_3$

Br

$LiAlH_4$/THF

reflux 2 days

yield ~ 80%

B

103

α β γ

$H^{\ominus}$ $H^{\ominus}$ $H^{\ominus}$

SiO_2

yield = 100%

H

$H^{\oplus}$

C

50% 30% 15%

caryophyllene

α-humulene

The reduction of the bromo derivative B by $LiAlH_4$ in refluxing THF over two days gives an 80% yield of a mixture containing 50% hydrocarbon C, 30% caryophyllene and 15% α-humulene. All of these reactions can be interpreted by invoking the involvement of a bicyclobutonium cation ($LiAlH_4$ - LiH + AlH_3 is also a Lewis acid) which is itself a possible precursor of the cationic species designated α, β and γ. Each of these is then attacked by a hydride ion provided by $LiAlH_4$.

Separation of the three hydrocarbons obtained in this reaction is a delicate operation. Indeed, if no special precautions are taken during chromatography of this mixture, the silica quantitatively isomerizes hydrocarbon C into α-humulene (recalling the example of germacrene D). This sequence of reactions permits the conversion of α-humulene into caryophyllene.

We have seen that the hydrolysis of the bromohydrin gives the diol A with retention of configuration. It should be noted that this diol, called tricyclohumuladiol, was discovered by Naya and Kotake in the essential oil of the Japanese hop along with humulenone, humuladienone and humulol. We have also noted that 6,7-epoxyhumulene is also a natural product (essential oil of *Zingiber zerumbet*). This observation led McKervey and Wright to study the action of acids on this epoxide. Sulfuric acid at 20% in acetone solution and at room temperature, gives a mixture of three products with the following yields: tricyclohumuladiol (28%), humulenol (15%) and another glycol, D (7%).

These results are displayed as follows [6.70]:

[6.70]

O H ⊕ OH OH H—OH H_2SO_4 0,4 N OH OH OH OH OH (MP = 201°–203°C) D

Nigam and Levi observed that the mono-epoxide of humulene is converted into humulenol during its chromatography on activated alumina. This result led McKervey and Wright to propose that 6,7-epoxyhumulene could also be involved in the biosynthesis of tricyclohumuladiol, caryophyllene and certain other oxygenated derivatives of humulene.

It should be noted, however, that α-humulene does not give caryophyllene when treated with dilute acid. Instead, it leads to α-caryophylenic alcohol, which is presently called apollan–11-ol. The exact structure of this alcohol has recently been unambiguously and independently demonstrated by the group of Nickon and by Gemmel and Sim.

For a long time, this alcohol had been considered a result of the acid-catalyzed rearrangement of carophyllene (hence, its name). Thanks to the work of Nickon, it is now known that it originates from α-humulene. He gave it the name apollan–11-ol because of the analogy between the symmetry of this substance and that of Apollo 11, which was the first spacecraft to land on the Moon in July 1969, during the course of this work [6.71, 72].

[6.71]

OH ← $H_3O^{\oplus}$ — H, H

A final substance possessing a humulane skeleton is caucalol diacetate, which has been isolated from the benzene extract of the seeds of *Caucalis scabra* Makino, however, its structure does not seem to be known with the same assurance as the previously mentioned substances. Controlled basic hydrolysis of this diacetate gives the monoacetate which can be reacetylated by Ac_2O to obtain the intial diacetate. On the other hand, complete hydrolysis gives an alcohol which is different from caucalol, called isocaucalol. This substance, when acetylated, gives a diacetate that is not identical to the initial diacetate. This can be summarized in the following way [6.73].

Structural studies were undertaken first on isocaucalol using mass spectrometry. Its structure, as proposed in 1966 by the Japanese group of Sasaki, Nakanishi and co-workers, is the following [6.74].

Its mass spectrum shows the following fragments:

			High resolution
molecular ion at m/e:	254	(weak)	
basic peaks at m/e:	43	(100%)	
	84	(38%)	
important peaks at m/e:	109	(30%)	109.102 C_8H_{13}
	110	(18%)	110.110 C_8H_{14}
	127	(13%)	123.076 $C_7H_{11}O_2$
			127.111 $C_8H_{15}O$

These results suggest the following fragmentations [6.75].

This peak is derived from the preceding fragment as demonstrated by the presence of a metastable peak at

$$m/e : 93.5 = \frac{109^2}{127}$$

[6.72]

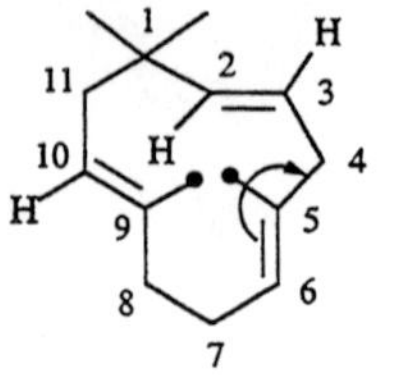

1) isomerization of the double bond from position 5,6 to position 4,5

2) attack on the 2,3-double bond

3) trans-annular raction of the 9,10-double bond

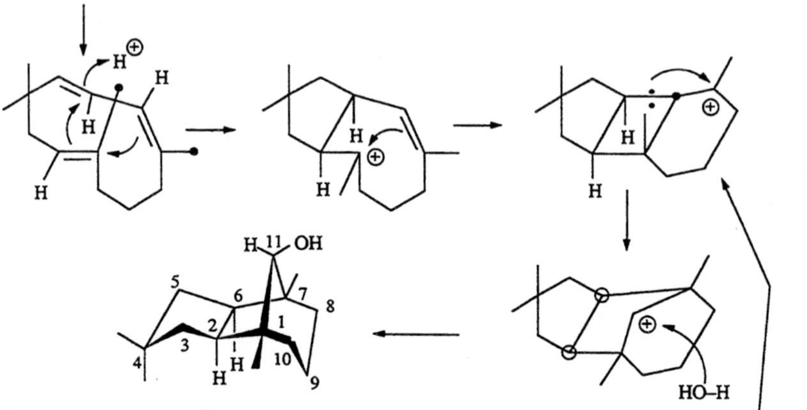

H⊕

HO–H

H 11 OH

apollan-11-ol

HO

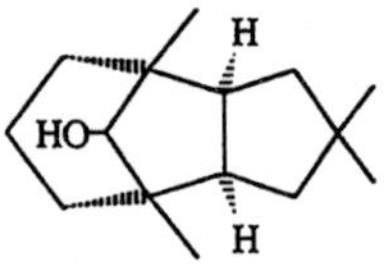

The total synthetis was carried out by E. J. Corey in the following manner:

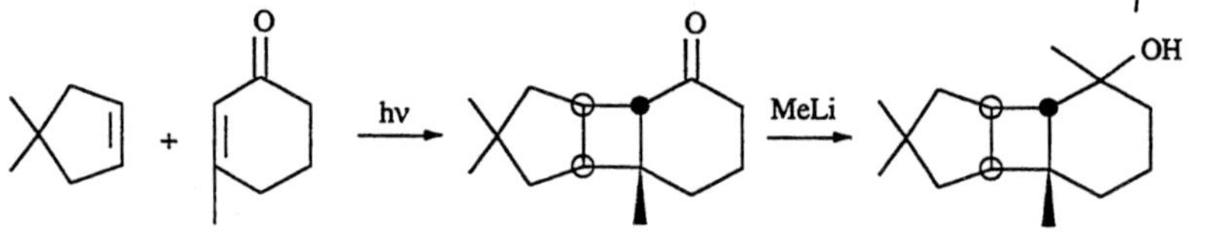

hν

MeLi

OH

[6.73]

caucacol diacetate	$\xrightarrow{OH^{\ominus}}$ / $\not\leftarrow$ Ac_2O	isocaucalol
$C_{19}H_{30}O_5$		$C_{15}H_{26}O_3$
MP = 121°–122°C		MP = 120°–121°C
$OH^{\ominus}$ ↓ ↑ Ac_2O		Ac_2O ↓ ↑ $OH^{\ominus}$
monoacetate caucalol		isocaucalol diacetate
$C_{17}H_{28}O_4$		$C_{19}H_{30}O_5$
MP = 188.5°–189°C		MP = 85.5°–86°C

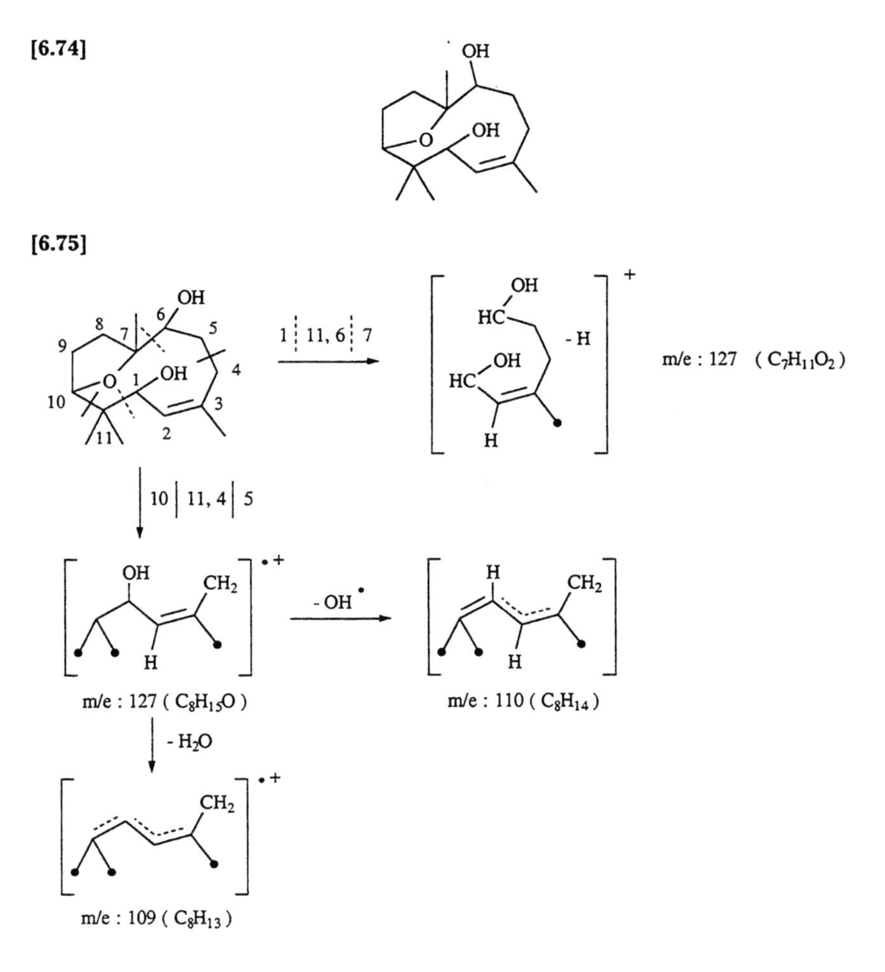

This structure for isocaucalol led the Japanese authors to propose the following structure for caucalol diacetate (R = Ac) [6.76], and for the opening of the epoxide to give the diol [6.77].

[6.76]

The epoxy glycol is not stable under alkaline conditions and a trans-annular reaction between the hydroxyl at position 10 and the epoxide at 6,7 affords isocaucalol. On the basis of these results, the structure of the monoacetate of caucalol (controlled hydrolysis) would therefore be as follows (hydrolysis of the allylic acetate) [6.78].

The structure of the diacetate of caucalol has of course been confirmed by the

[6.77]

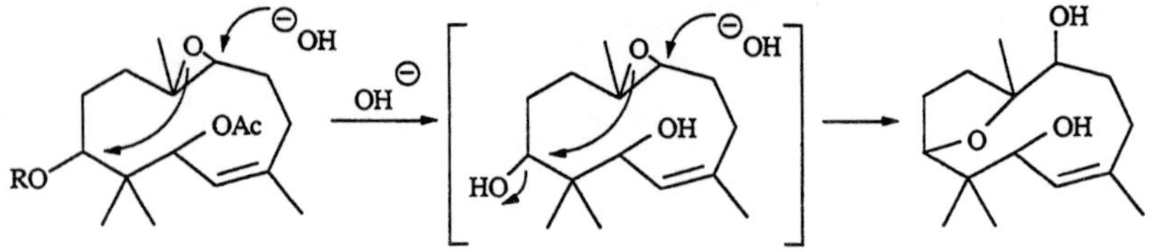

(opening of epoxides in basic media)

[6.78]

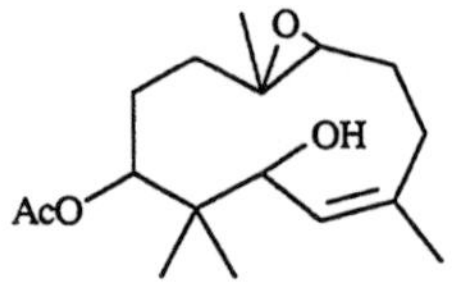

study of its NMR spectrum and by its high resolution mass spectrum. In this representation, it can be seen that the 2,3-double bond is cis. According to this hypothesis, this substance would therefore originate from trans,cis-farnesol (in other words, it would correspond to the humulene of Hendrickson). In fact, these authors provide no serious proof concerning the cis geometry of the 2,3-double bond. It would therefore be more reasonable to envisage this substance as derived from trans,trans-farnesol and thus, having the following structure [6.79]:

[6.79]

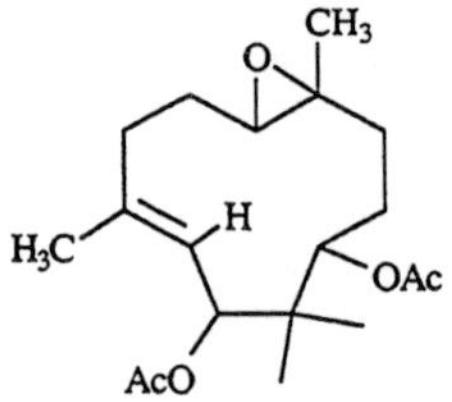

This hypothesis would render the biogenesis of all of the oxygenated derivatives of α-humulene more homogeneous.

Study of some rearrangements of caryophyllene

1) Acid-catalyzed rearrangements in diluted acids [6.80]

When caryophyllene is in its preferred conformation C, the hydrogen on carbon 1 is very close to carbon 7 (destined to be protonated). This can be stabilized by a 1,5-migration of hydride, thereby leading to the formation of a bond between carbons 1 and 7 [6.81].

2) Rearrangements of 1-caryolanol (or of caryophyllene) under the action of polyphosphoric acid

a) Rearrangement of caryolanol [6.82]

[6.80]

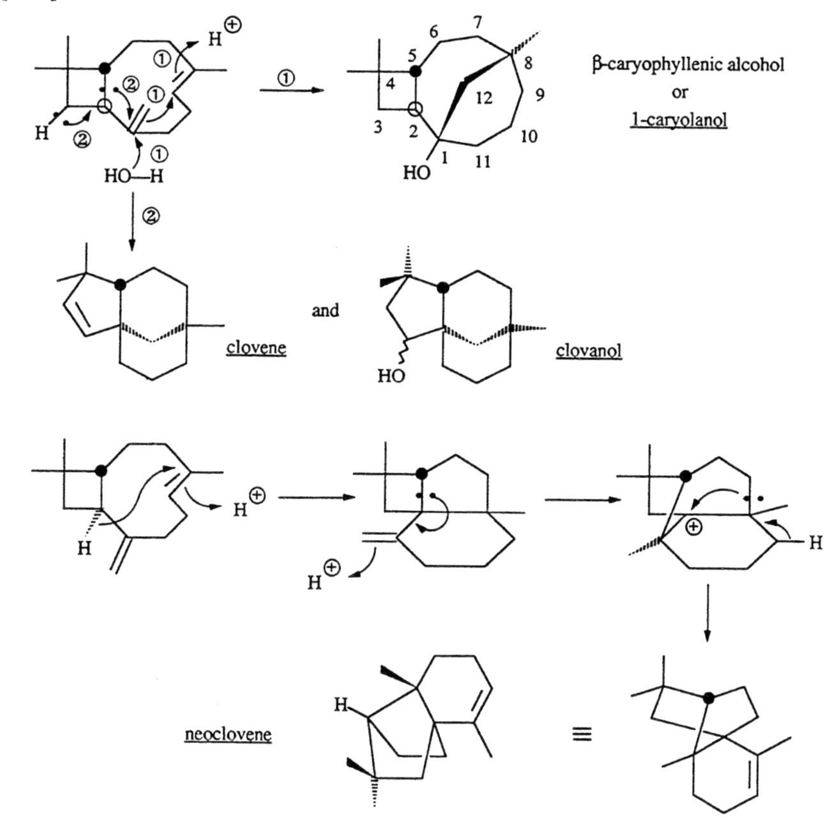

β-caryophyllenic alcohol
or
1-caryolanol
clovene
and
clovanol
neoclovene

[6.81]

C
α-attack
by H⊕
caryolanol

[6.82]

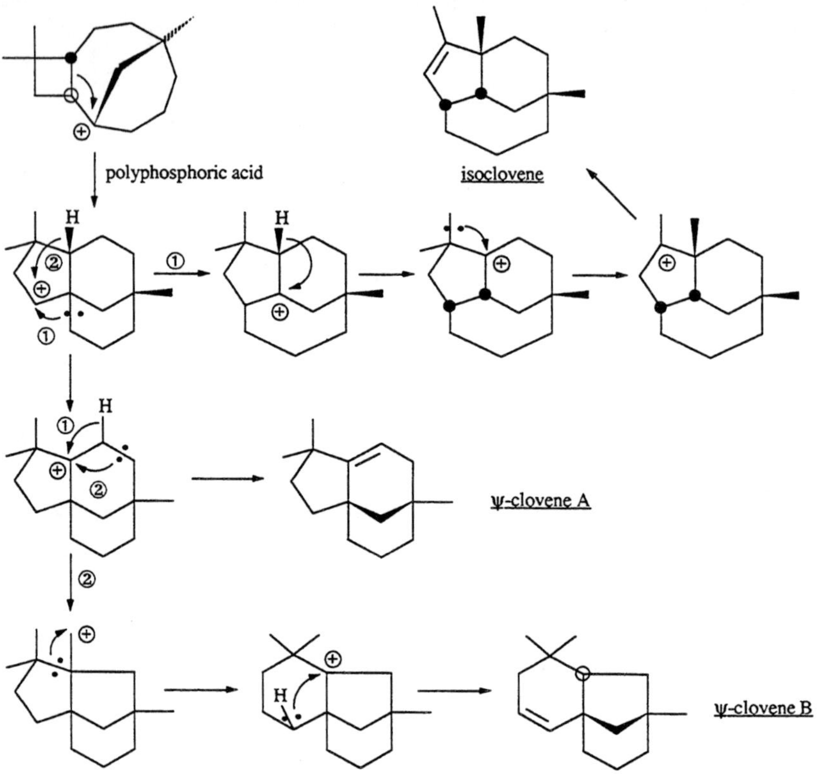

polyphosphoric acid

isoclovene

ψ-clovene A

ψ-clovene B

b) Rearrangement of isocaryophyllene [6.83]

[6.83]

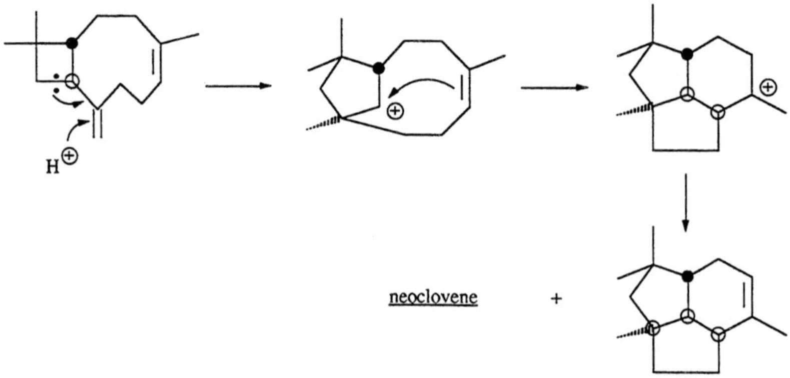

neoclovene +

3) Rearrangement of epoxycaryophyllene and of epoxyisocaryophyllene by diluted mineral acids [6.84]

[6.84]

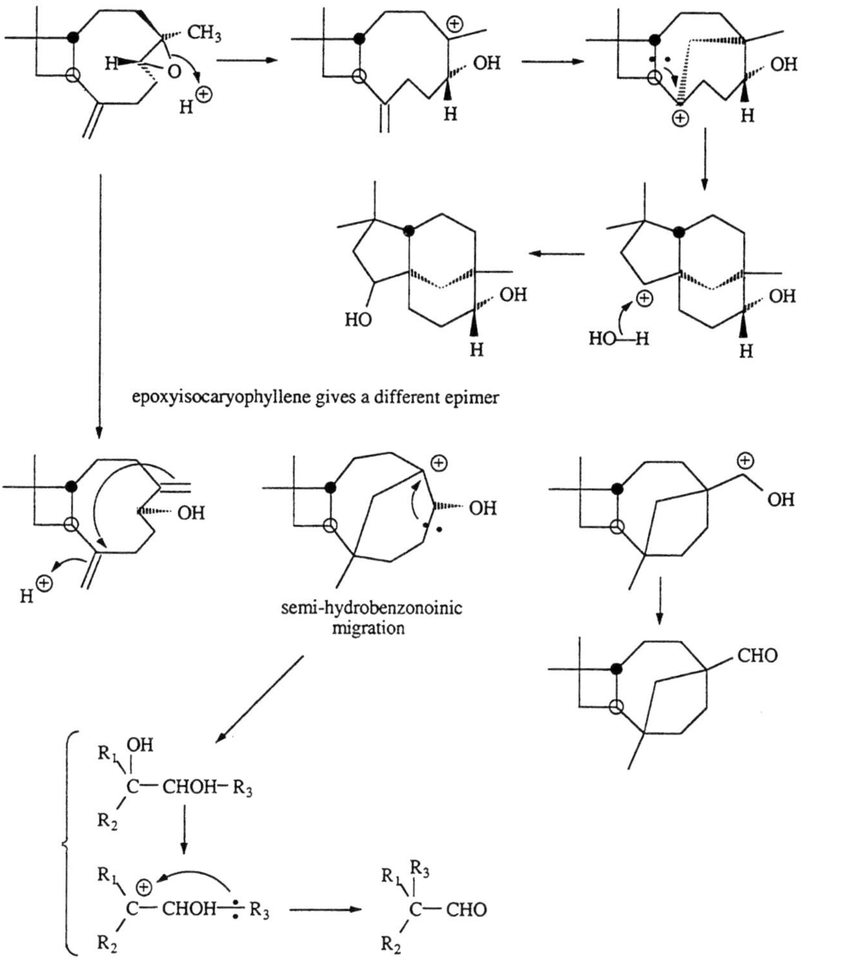

epoxyisocaryophyllene gives a different epimer

semi-hydrobenzonoinic migration

E. Sesquiterpenes with the gaiane skeleton

We have already had occasion to mention derivatives of "gaiane" [6.85]. In this series are found gaiol, bulnesol as well as the corresponding hydrocarbons α-gaiene and α-bulnesene [6.86]. Other oxygenated products such as kessane and, in particular, the gaianolides and the pseudogaianolides are also found in this series.

[6.85]

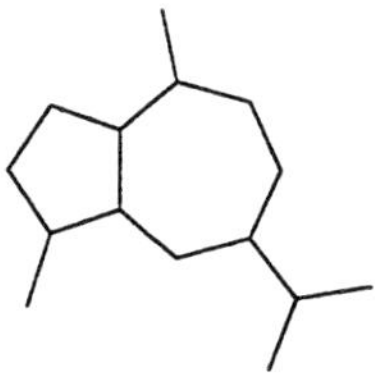

[6.86]

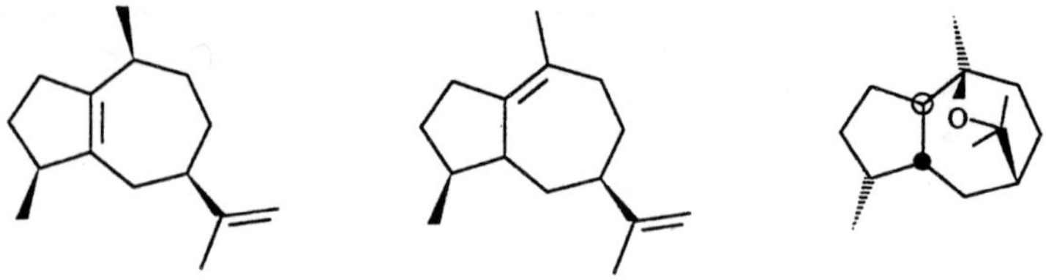

Kessane

For example, gaianolides (work of the Czech group of Sòrm and Herout and of the Japanese group of Takeda) [6.87]:

[6.87]

2 groups

A

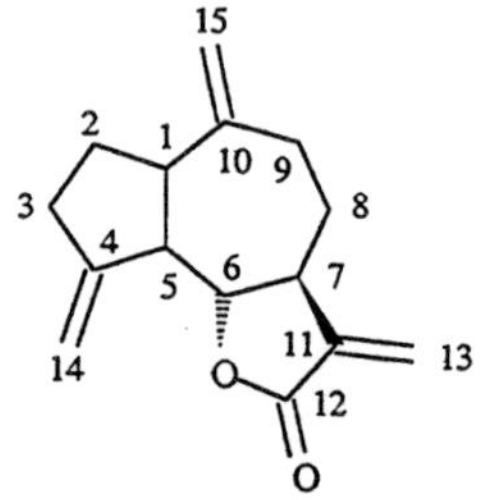

dehydrocostuslactone found in the roots of costus (*Saussurea lappa* Clarke, compositæ)

B

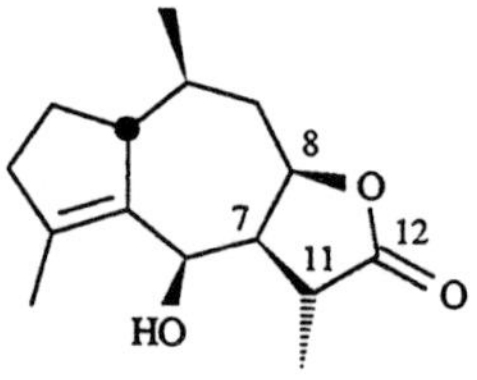

geimerine

Dehydrogenation obtains the following azulenes [6.88]:

[6.88]

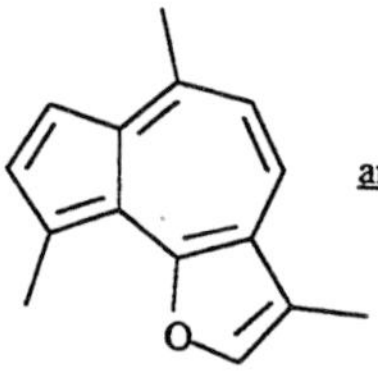

artemazulene

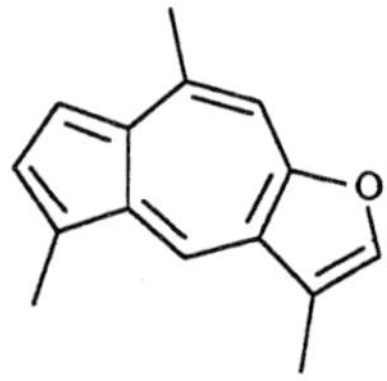

linderazulene

It should be noted that, for a long time, an azulene has been known to be responsible for the dark blue coloration of the oil of *Chamomile matricaria.* It is called chamazulene and it manifests anti-inflammatory properties [6.89].

[6.89]

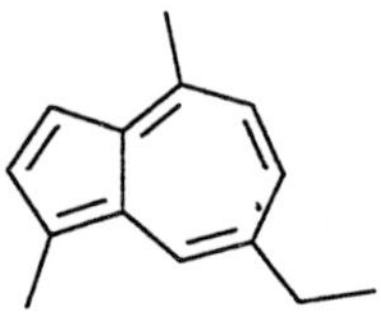

Yet, chamazulene (a C^{14} and not a C^{15}) does not exist in the plant, but forms in the course of the steam distillation process from unstable precursors identified as artabsine (*Artemisia absinthum* L.) and matricine (*Matricaria chamomilla* L.) [6.90, 91].

[6.90]

OH

- CO_2

COOH

oxidation

$OCOCH_3$

$H^{\oplus}$

- CO_2

OH

COOH

chamazulene

gaiazulenic acid
(isolated from *Achillea millefolium*)

[6.91]

H

HO

COOH

isolated and characterized in blue chamomile (*Matricaria chamomilla* L.)

m/e 29

COOH

Pseudogaianolides (work of the Mexican group of Romo de Vivar) are also found in two types [6.92]. In the course of the biogenesis there is a migration of a methyl group from position 4 to position 5.

Among the compounds with gaianic structures, the constituents of patchouli oil represent a particularly interesting group which merits more detailed discussion. Patchouli oil holds an important place in the perfumery industry. It is obtained by the steam distillation of the dried leaves of a labiate, *Pogostemon cablin* Benth.

[6.92]

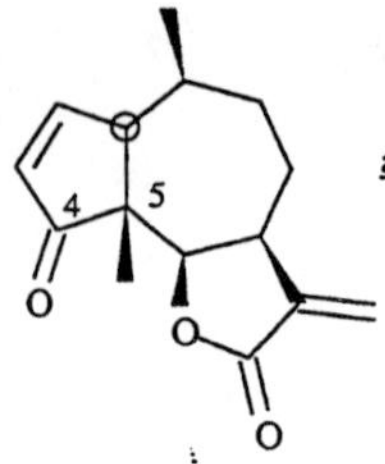

Gal was the first to isolate the main constituent of this oil, in 1869. For a long time this substance was called "camphor of patchouli" because of its "crystalline" aspect which was similar to that of camphor. It is now known as "patchoulol." In 1877, Montgolfier established its exact molecular formula as $C_{15}H_{26}O$ and, in 1894, Wallach proposed the name "patchouli alcohol."

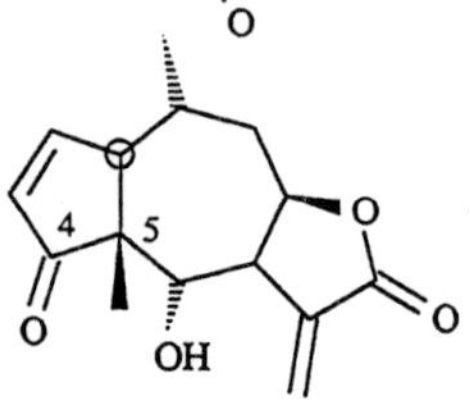

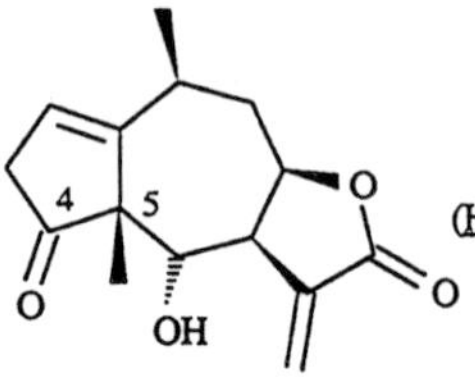

In 1912, Semmler suggested a tricyclic structure for this alcohol and in 1949, Treibs proposed structure A. Büchi and Erickson, after having proposed structure B in 1956, succeeded in making its first synthesis in 1962 using homocamphor as the starting material.

Actually, neither structure A nor structure B is correct. What is remarkable, however, is that during this synthesis, a rearrangement occurred without the authors' knowledge, to give the exact structure C [6.93].

[6.93]

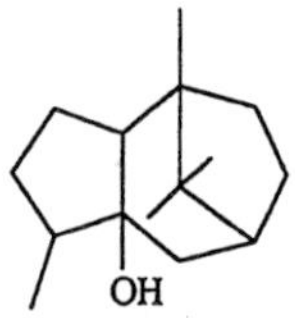

Büchi's attribution of structure B is consistent with the following experimental facts:

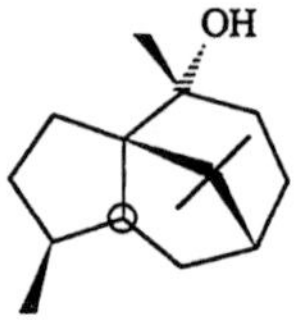

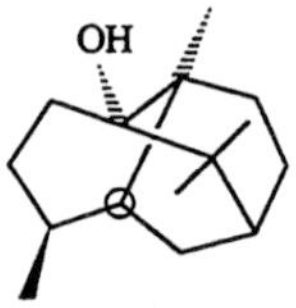

1) Treatment under acidic conditions

Under the action of these acids, particularly hot boric acid, patchoulol gives β-patchoulene in an excellent yield (Wagner-Meerwein rearrangement) [6.94].

2) Pyrolysis of patchoulol acetate (obtained from reaction with ketene)

The major product of this reaction is α-patchoulene. The acetate of B could give this in the following manner [6.95].

The structures of α- and β-patchoulene were established with certitude by

[6.94]

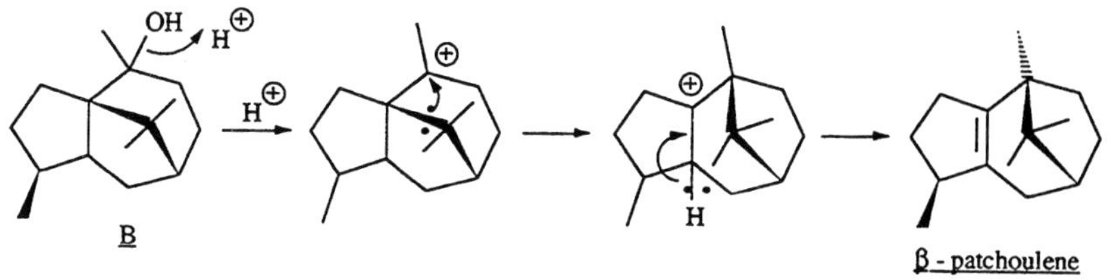

OH H⊕ H⊕ ⊕ ⊕ H

B

β - patchoulene

[6.95]

CH_3 C O O H

Δ 300°C

α - patchoulene (52%) + γ - patchoulene (46%) + β - patchoulene (4%)

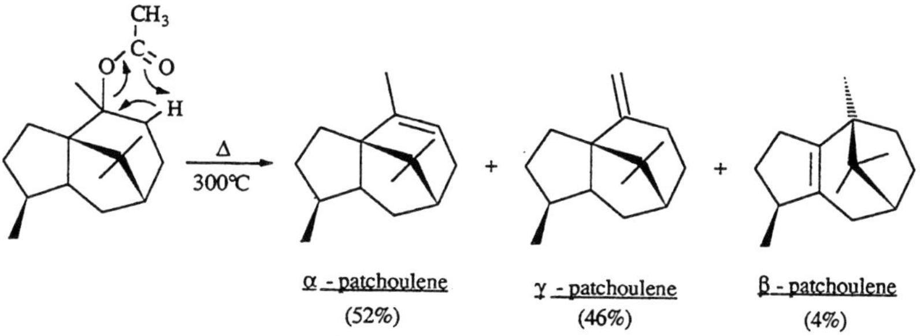

Büchi thanks to considerable work involving chemical degradation associated with spectrometry.

β-patchoulene: by degradation, this hydrocarbon could be correlated with homocamphoric acid, camphoric acid and camphoric anhydride [6.96].

[6.96]

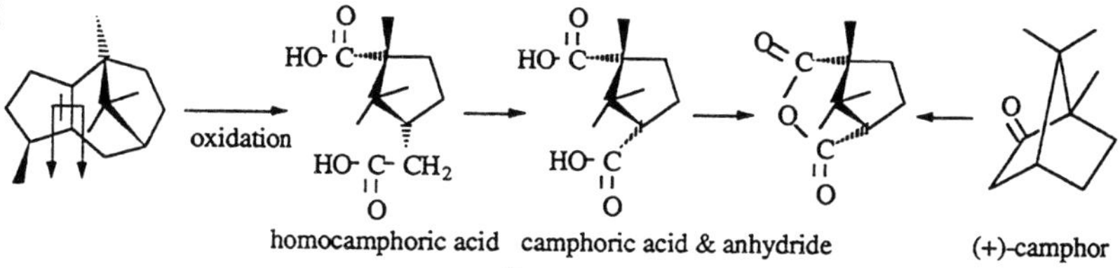

oxidation

HO-C(=O) HO-C-CH_2 HO-C(=O) HO-C(=O) O C O C O

homocamphoric acid camphoric acid & anhydride (+)-camphor

α-(and γ-) patchoulene: These were correlated with α-cedrene which had been synthesized strereospecifically by Stork [6.97]. The syntheses of α- and β-patchoulene were achieved from homocamphor, by Büchi in 1962 [6.98].

The β-patchoulene obtained was identical in all aspects, including optical rotation, to β-patchoulene obtained from the action of acids on patchoulol. Consequently, this synthesis established the absolute configuration of this compound. Under the action of peracids, β-patchoulene furnished α-epoxypatchoulane [6.99].

3) Synthesis of patchoulol (Büchi) [6.100]

The product obtained by Büchi at the end of this reaction sequence is from all points of view, identical to natural patchoulol.

In 1967, the group of Dünitz in Zurich attempted to determine the Cr-O-C

[6.97]

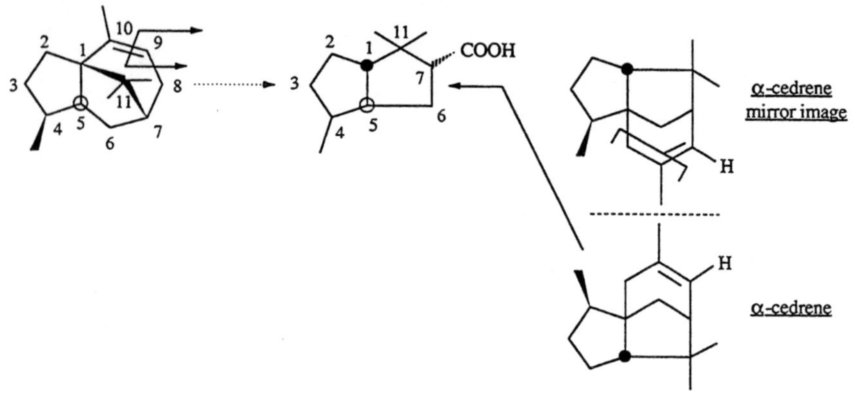

α-cedrene mirror image

α-cedrene

[6.98]

(+)-camphor

homocamphor

β-patchoulene

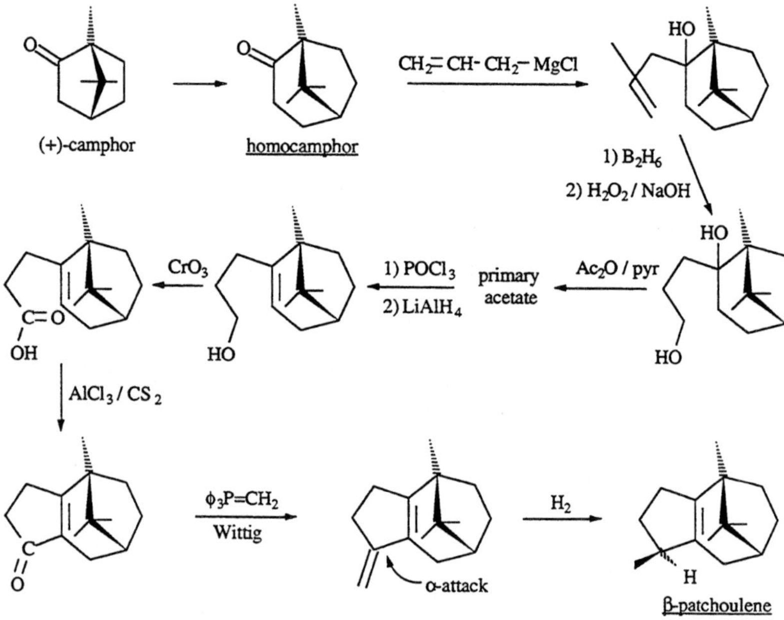

angle in alkyl chromates by performing a radiocrystallographic study with X-rays on the chromate of patchoulol. This chromate had been obtained in a particularly crystalline form (MP = 117°–118°C) by Scholz in Leipzig in 1930. There was considerable astonishment when it was seen that the structure of patchoulol deduced from this study was not compatible with that proposed by Büchi. From the structure of the chromate [6.101], it can be concluded the the exact structure C is the following [6.102].

But then what had happened:

[6.99]

β-patchoulene → ≡

1) BF_3
2) H_2O

H, H, HO, $F_2B—F$

Ac_2O hot pyridine

B_2H_6 H_2O_2 (α-attack)

OAc, OH, OH, OH

H_2 / Pt

H_2 in α

OAc

hydrolysis

OH

Δ

+ one other olefin

α-patchoulene

[6.100]

α-patchoulene

peracide

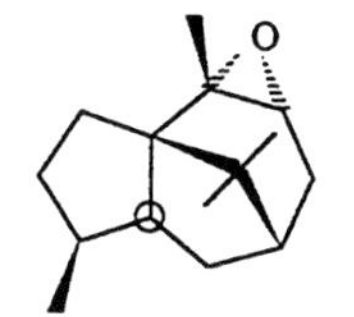

$H_3O^{\oplus}$

OH OH

X

Ac_2O / Pyr

OH OAc

Δ

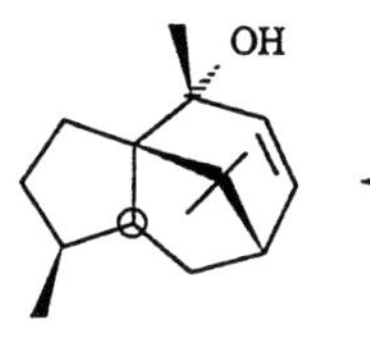

H_2

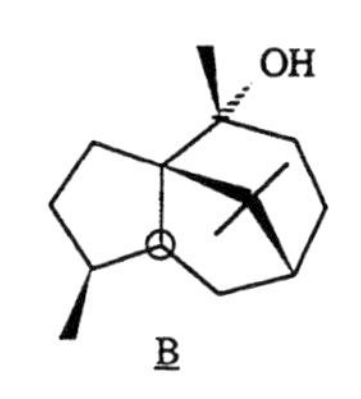

B

[6.101]

O, Cr, O, O, O

[6.102]

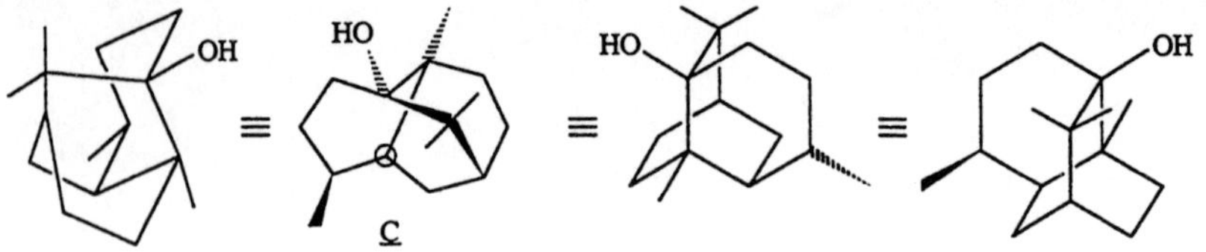

HO OH HO OH

C

a) in the course of determining the structure of patchoulol?
b) in the course of its synthesis?
c) in the course of pyrolysis of patchoulyl acetate?

a) In the course of determining its structure
On the one hand, the dehydration of patchoulol to β-patchoulene is accompanied, in fact, by *two* successive Wagner-Meerwein rearrangements and not a single one [6.103].

[6.103]

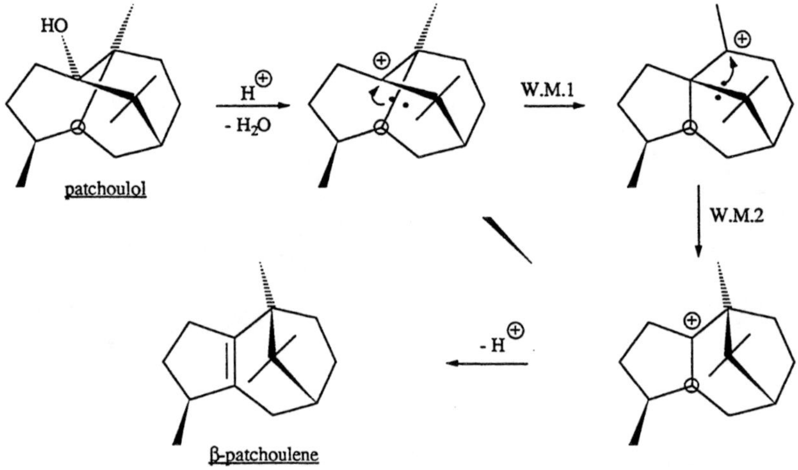

HO

patchoulol

$H^{\oplus}$ $-H_2O$

W.M.1

W.M.2

$-H^{\oplus}$

β-patchoulene

On the other hand, the exact structure of patchoulol implies that the pyrolysis of patchoulol acetate is also accompanied by a rearrangement [6.104].

[6.104]

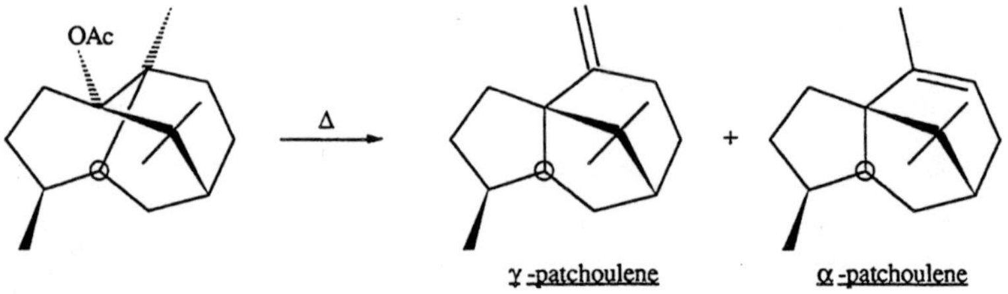

OAc

Δ

+

γ-patchoulene α-patchoulene

b) In the course of the synthesis
During the oxidation of α-patchoulene with peracetic acid, a new rearrangement occurs and this time in the opposite sense, leading not to the 1,2-diol D, but to the 1,3-diol E [6.105].

[6.105]

CH_3CO_3H

D

E

Ac_2O / pyr

Δ, H_2

C

patchoulol

c) In the course of pyrolysis of patchoulyl acetate
According to the terms used by Dünitz himself, it must involve a rearrangement "without precedent" The course of the pyrolysis of alkyl acetates is usually explicable in terms of a six-centered mechanism which is impossible to write down in the case of patchoulyl acetate. Two personal but possibly inexact ideas based on the large yield of γ-patchoulene in this reaction can be evoked [6.106].

In addition to patchoulol, a norsesquiterpenic alcohol is found in patchouli oil. Although present in small quantities (0.3–0.4% of the oil; therefore only 1% relative to patchoulol), it is the principal vector of the odor of patchouli. This alcohol has been called "norpatchoulenol," and both its structure F and its absolute configuration have been determined from correlation with cyperene and by X-ray analysis of its derivative, the bromoketone G [6.107].

Correlation of norpatchoulenol and cyperene through the intermediate norcyperene
In order to facilitate our discussion of these results, they will be described in terms of this structure. It should be recalled, however, that all of the chemical reactions in which norpatchoulenol is engaged, are intended to bring a maximum of information regarding its structure.

The diol 1 can be obtained either by hydroboration-oxidation or by epoxida-

[6.106]

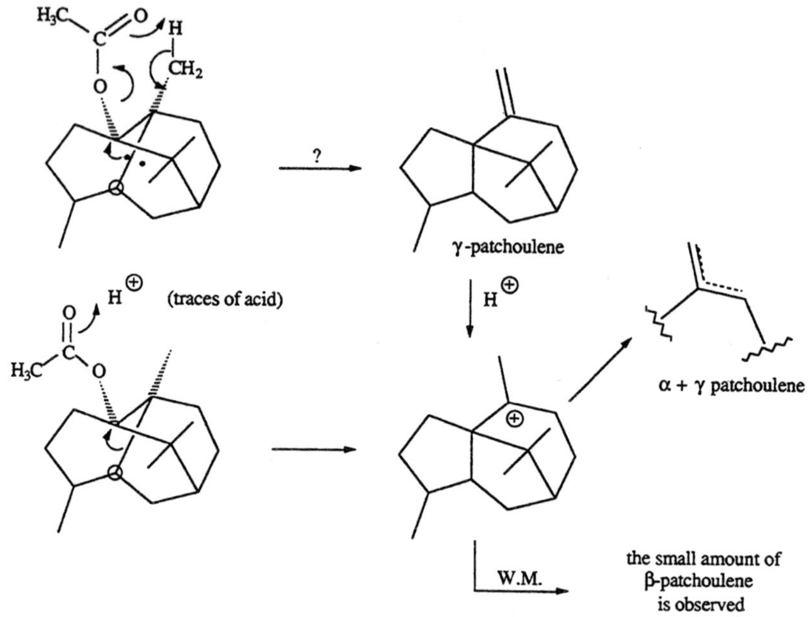

?

γ-patchoulene

$H^{\oplus}$ (traces of acid)

$H^{\oplus}$

α + γ patchoulene

W.M.

the small amount of β-patchoulene is observed

[6.107]

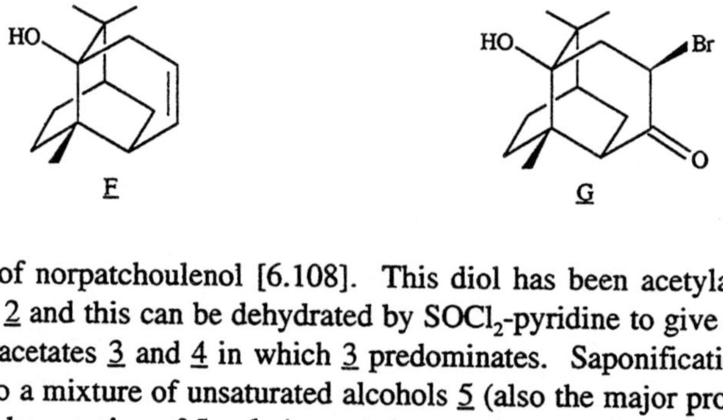

E

G

tion-reduction of norpatchoulenol [6.108]. This diol has been acetylated to an acetate-alcohol 2 and this can be dehydrated by $SOCl_2$-pyridine to give a mixture of unsaturated acetates 3 and 4 in which 3 predominates. Saponification of this mixture leads to a mixture of unsaturated alcohols 5 (also the major product) and 6. Catalytic hydrogenation of 5 + 6 gives a mixture of saturated alcohols 7 and 8. These are epimeric at position 10 with 7 being the major of the two [6.109].

[6.108]

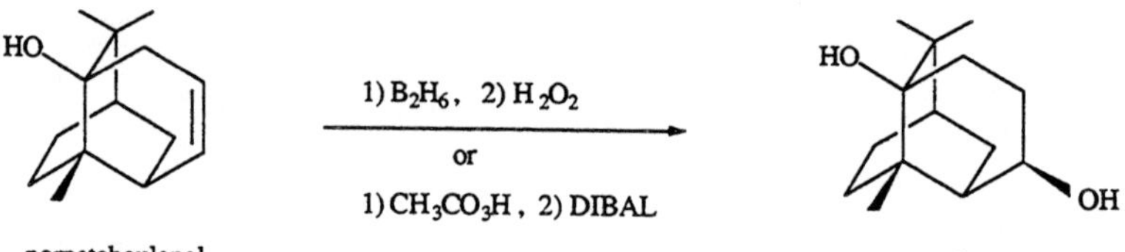

1) B_2H_6, 2) H_2O_2

or

1) CH_3CO_3H, 2) DIBAL

norpatchoulenol

1

The stereochemistry of the methyl group at position 10 is in agreement with the result of the hydrogenation of α-patchoulene under the same conditions. In this latter case, a mixture of patchoulane and isopatchoulane is obtained in the respective proportions of 25:75 [6.110].

[6.109]

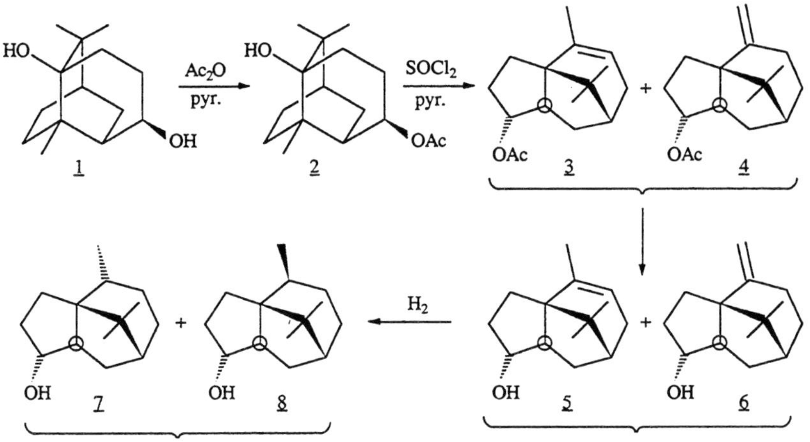

[6.110]

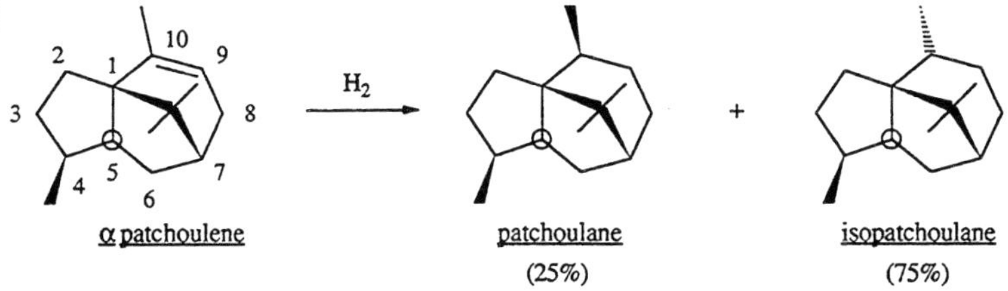

It has not been possible to separate the two saturated alcohols 7 and 8. After oxidation, however, a mixture of ketones 9 and 10, in which the former is strongly predominant, is obtained and these can be separated by chromatography.

The ketone 9 has, by its spectroscopic properties and optical rotation, been shown to be identical to the ketone obtained from natural cyperene 12. This hydrocarbon, when treated with selenium oxide in acetic anhydride, affords cyperenyl acetate 14. Saponification of this, followed by oxidation, gives cyperenal 15. Treatment of this aldehyde with tris-triphenylphosphine rhodium leads to norcyperene 16.

Epoxidation of norcyperene furnishes the epoxide 17 which gives the ketone 9 by isomerization. In addition, the saturated alcohol 11 is obtained from this ketone following addition of methyllithium. Dehydration of 11 leads to a mixture of three hydrocarbons of which there is cyperene 12 and isocyperene 13 (the major product and the one obtained by isomerization of cyperene using perchloric acid) [6.111].

This correlation establishes the structure of "norpatchoulenol" and shows that this alcohol is found in the same antipodal series as patchoulol.

X-ray analysis of the bromoketone G

These studies were undertaken by Oberhänsli and Schönholzer of Hoffmann-La Rôche in Basel and were based on the work of Bijvoet. Because of the abnormal dispersion of bromine in the radiocrystallographic study of a monocrystal of the bromoketone G, they deduced that this norsesquiterpenic alcohol (which is the

[6.111]

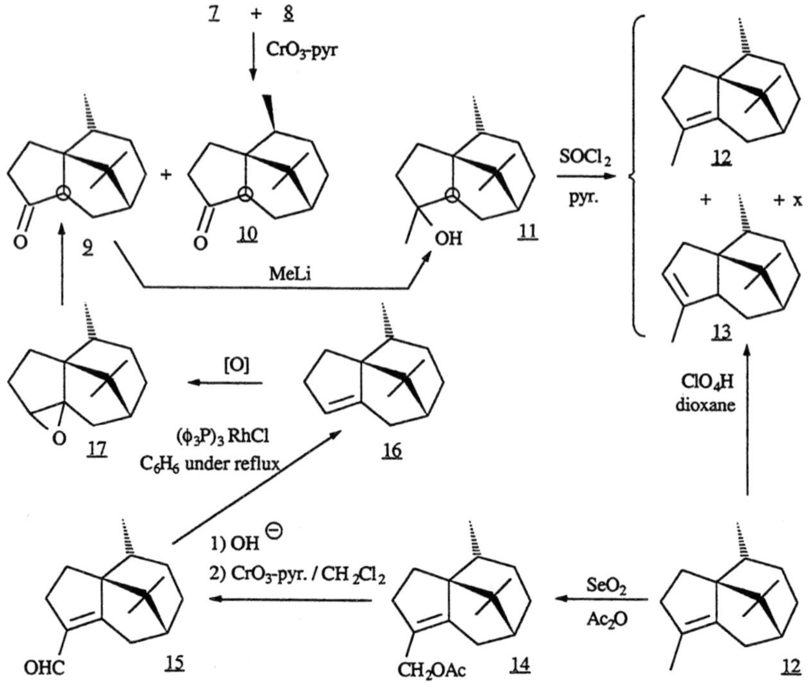

vector of the odor of patchouly) does indeed have the structure F which had been assigned.

Detailed Study of some Syntheses of Sesquiterpenes of Major Interest

A. Total syntheses by Piers: sesquiterpenes of the copabornane and copacamphane series

The total syntheses of (–)-copacamphene, (–)-cyclocopacamphene, (+)-copacamphor, (+)-copaborneol and (+)-copaisoborneol have all been realized by the group of Piers in Vancouver. The structural determination of copaborneol had previously been established by Westfelt and collaborators in Stockholm—they also synthesized this substance using santalol as a starting material.

The syntheses of these five sesquiterpenic compounds will be briefly described here, starting with the oxygenated products (synthesis of Piers).

Comments on this synthesis

It is possible to pass with ease from copacamphor 14 to copaisoborneol and copaborneol and visa versa.

Furthermore, it can be noted that the described synthesis contains 16 steps and that despite relatively good yields in each step, the final yield of (+)-copacamphor 14 relative to (+)-carvomenthone 1 is only 8% at the end. This is unfortunately a characteristic of the majority of syntheses of complex sesquiterpenes (multistage syntheses).

Protection of the methylene in position 3 of (+)-carvomenthone 1 is aimed at permitting acylation of carbon α of the carbonyl group. The blocking group is in the form of π-butylthiomethylene and is generated using the method of Ireland and Marshall (perfected for the synthesis of natural products). This strategy consists of first preparing the hydroxymethylene using the classical method (Claisen reaction with ethyl formate) then treating it with π-butylmercaptan in the presence of para-toluenesulfonic acid in refluxing benzene to eliminate water. This method is particularly useful for the bridgehead alkylation of octalones [6.112, 113].

Alkylation of position 1 is achieved by treatment with methyl α-iodopropionate. This occurs with high regiospecificity and high stereospecificity to give compound 3. Deprotection of the methylene in position 3 (retroaldolization) is effected by aqueous potassium hydroxide in refluxing ethylene glycol and gives the ketoacid 4 which must be remethylated to accede to the ketoester 5. The sodium derivative of hexamethyldisilazane [sodium bis-(trimethylsilyl) amide] liberates the anion at position 3 and this then reacts with the ester function by an internal Claisen condensation to give bicyclo [3.2.1] octadione 6 (MP = 76°–77°C, p = 90%).

Further reaction of this reagent on the dione 6 generates the enolate of the carbonyl at position 6 (the carbonyl at position 8 is incapable of enolization). It is immediately captured in the form of the enol acetate 7 by reaction with acetyl chloride [6.114].

Hydrogenation of this compound (cis-addition of H_2 from the least hindered face) gives the keto acetate 8. When the phosphorane obtained by the action of chloromethyl ether with triphenylphosphine (Wittig modified by Levine) reacts with this keto-acetate, the enol ether 9 is formed. Reduction of this compound with $LiAlH_4$ regenerates the secondary alcohol function at carbon 6 and treatment with perchloric acid in ether regenerates the aldehyde function. Finally, the action of sodium carbonate in hydro-methanolic solution permits the aldehyde function to equilibrate to the thermodynamically more stable equatorial position in 10. The reaction of methylenephosphorane with 10 leads to the derivative 11 for which Brown's hydroboration-oxidation furnishes a primary diol (equivalent to anti-Markownikoff hydration). When this diol is treated with a slight excess of para-toluenesulfonyl chloride (for the primary alcohol function), the tosylate 12 is obtained selectively and its secondary alcohol function is then oxidized by Collin's reagent (CrO_3^{-2} × pyridine) to give the keto-tosylate 13. The anion of DMSO in this solvent reacts with 13 to give (+)-copacamphor via a new internal Claisen reaction.

The syntheses of (–)-copacamphene and of (–)-cyclocopacamphene from the dione 6 can be summarized in the following scheme [6.115, 116].

Reduction of the diketone 6 by an equivalent of $NaBH^4$ in methanol only occurs on the carbonyl at position 8 (less hindered) and leads to an axial alcohol 15. This keto alcohol furnishes a tosylhydrazone 16 when it is treated with para-

[6.112]

(+) carvomenthone 1 → 2 → 3 (CH_3-CH(I)-$COOCH_3$, t-BuOK / t-BuOH)

3 → 4: KOH at 25% / H_2O, diethylene glycol, under reflux

1 → 4: yield = 60%

4 → 5: $Me_3Si-N(Na)-SiMe_3$, $MeOCH_2-CH_2OMe$, under reflux

5 → 6: yield = 90%

6 → 7: $Me_3Si-N(Na)-SiMe_3$ (1éq/C_6H_6), excess of CH_3COCl, yield = 90%

7 → 8: H_2, PtO_2 / EtOH

8 → 9: ϕ_3P=CHOMe, Et_2O, Wittig

9 → 10: 1) $LiAlH_4$ 2) ClO_4H / Et_2O 3) CO_3Na_2 / H_2O / MeOH

7 → 10: yield = 50%

10 → 11: $\phi_3P=CH_2$ ↔ $\phi_3\overset{\oplus}{P}-\overset{\ominus}{C}H_2$, DMSO, yield = 92%

11 → 12: 1) hydroboration-oxydation 2) p.C$_6H_4SO_2$Cl

12 → 13: CrO_3, 2 Pyr (Collins)

11 → 13: yield= 45%

13 → 14: $CH_3-S-\overset{\ominus}{C}H_2$, DMSO, yield = 82%

14 (+)-copacamphor

[6.113]

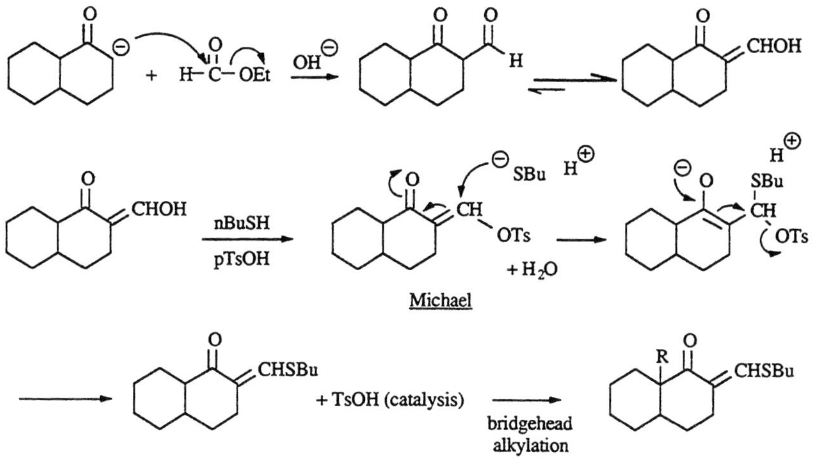

+ H–C(=O)–OEt $\xrightarrow{OH^{\ominus}}$ CHO ⇌ =CHOH

=CHOH $\xrightarrow[\text{pTsOH}]{\text{nBuSH}}$ SBu, $H^{\oplus}$, CH–OTs, $+ H_2O$ (Michael) ⟶ SBu, CH, OTs

⟶ =CHSBu + TsOH (catalysis) $\xrightarrow{\text{bridgehead alkylation}}$ R, =CHSBu

[6.114]

$Na^{\oplus}$ ······ $O^{\ominus}$, $CH_3CO^{\oplus}$ $Cl^{\ominus}$

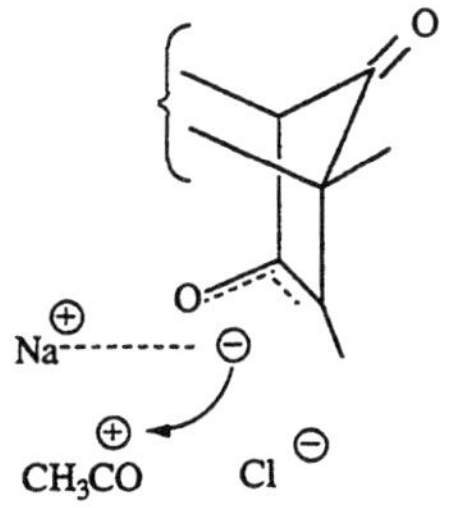

tosylhyrazine in benzene. This compound reacts with methyllithium in THF to give an olefinic alcohol 17 in which the double bond is in position 6,7. This reaction, which is referred to as the Shapiro modification of the Bamford-Stevens reaction, involves a series of consecutive processes [6.117].

The elimination process requires two equivalents of alkyllithium: the first to form a salt of the tosylhydrazone and the second to effect the actual elimination. Passage from the olefinic alcohol 18 to the tosylate 21 requires the intervention of a sequence already seen for the conversion of 10 into 12. Because of the proximity of the 6,7-double bond in 21, there is a tendency for spontaneous elimination with participation of the π electron pair. If the crude tosylate is filtered on a column of silica gel, total elimination accompanied by cyclization occurs to give (–)-copacamphene.

The olefinic alcohol 20 can be oxidized to the aldehyde 22 for which the tosylhydrazone 23 can be prepared. At temperatures between 120° and 140°C and under reduced pressure (P = 0.25 mm Hg), pyrolysis of the lithium salt of this hydrazone permits distillation of the pyrazoline 24. As a general rule, pyrolysis of salts of para-tosylhydrazones permit the preparation of diazoalkanes in the following way [6.118].

Here again, the proximity of the 6,7-double bond allows facile 1,3-dipolar

[6.115]

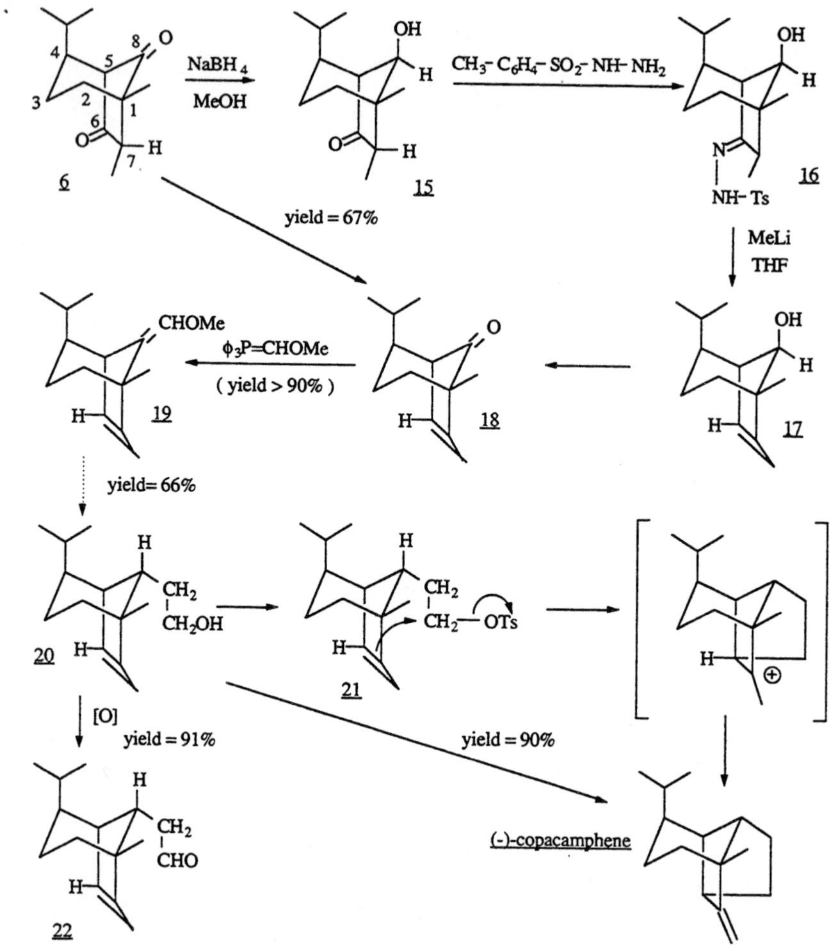

6 — NaBH$_4$ / MeOH → 15 (yield = 67%) — CH_3–C_6H_4–SO_2–NH–NH_2 → 16 — MeLi / THF → 17 → 18

6 → 18

18 — ϕ_3P=CHOMe (yield > 90%) → 19 (=CHOMe)

19 — yield = 66% → 20 (CH_2–CH_2OH) → 21 (CH_2–CH_2–OTs) → [cation intermediate] → (-)-copacamphene

20 — yield = 90% → (-)-copacamphene

20 — [O], yield = 91% → 22 (CH_2–CHO)

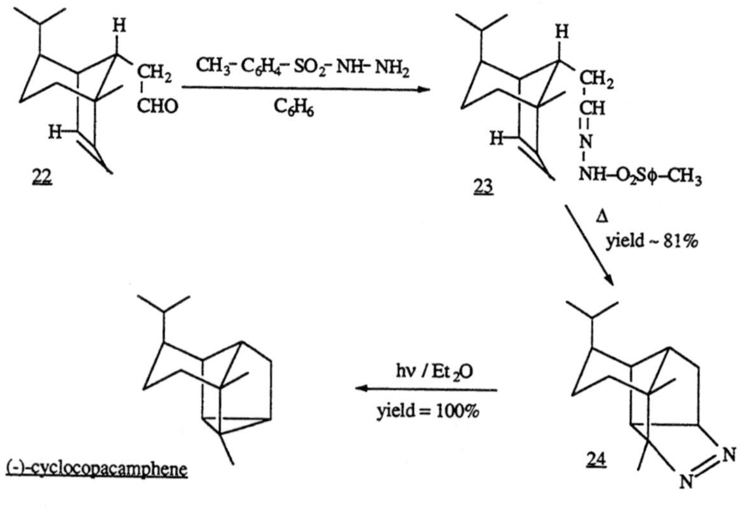

[6.116]

22 — CH_3–C_6H_4–SO_2–NH–NH_2 / C_6H_6 → 23 (CH_2–CH=N–NH–$O_2S\phi$–CH_3)

23 — Δ, yield ~ 81% → 24

24 — hν / Et_2O, yield = 100% → (-)-cyclocopacamphene

[6.117]

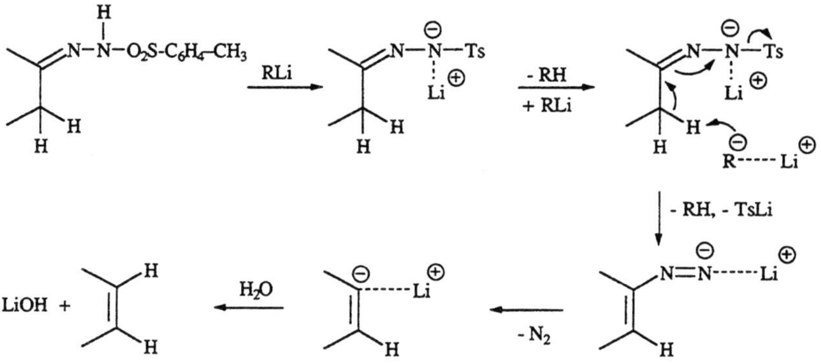

[6.118]

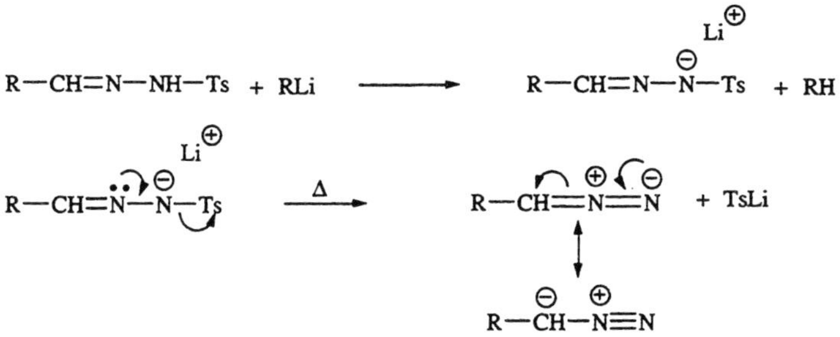

addition of the diazo compound to obtain the pyrazoline 24. It is possible to submit pyrazolines like this to thermal decomposition at temperatures around 200°C, but the yields are, in general, rather small. However, their photolysis is generally quantitative and probably passes through an intermediate 1,3-biradical (a) [6.119].

[6.119]

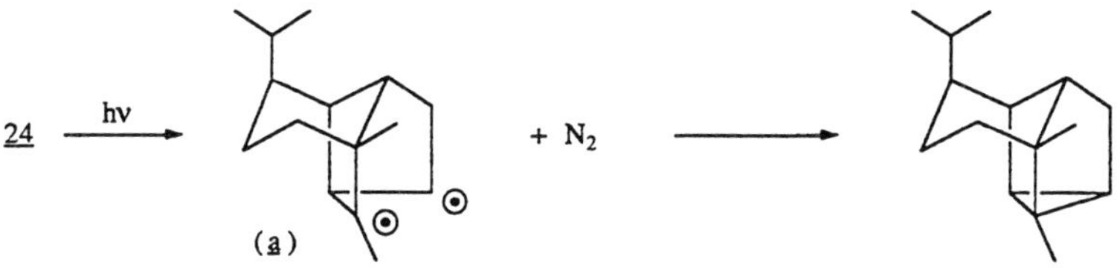

It should also be noted that the direct photolysis of the sodium salt of a tosylhydrazone will also give a diazo compound which loses a molecule of nitrogen to give an alkylcarbene. In this case, the end product is (–)-cyclocopacamphene by $\omega^2 + \pi^2$ addition to the double bond [6.120].

B. Synthesis of longifolene (Corey)

The key step in this synthesis is an internal Michael addition on ketone A to form the dione B [6.121].

[6.120]

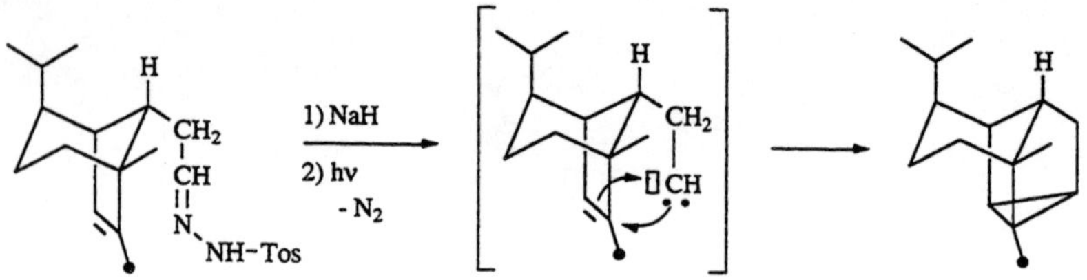

1) NaH

2) hv

- N_2

[6.121]

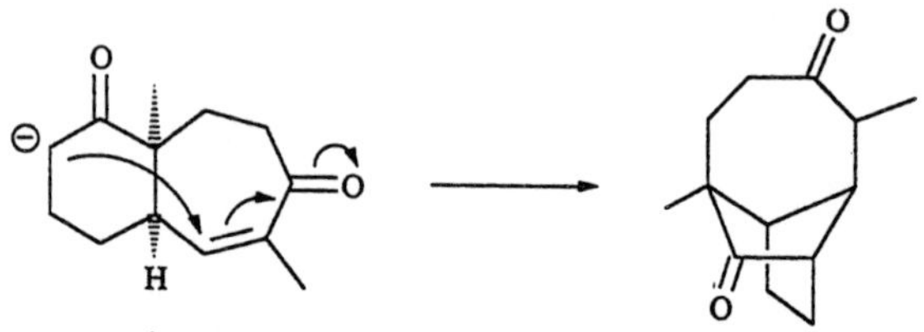

A

B

There are other examples of internal Michael additions; notably that in which santonine is transformed into santonic acid [6.122].

[6.122]

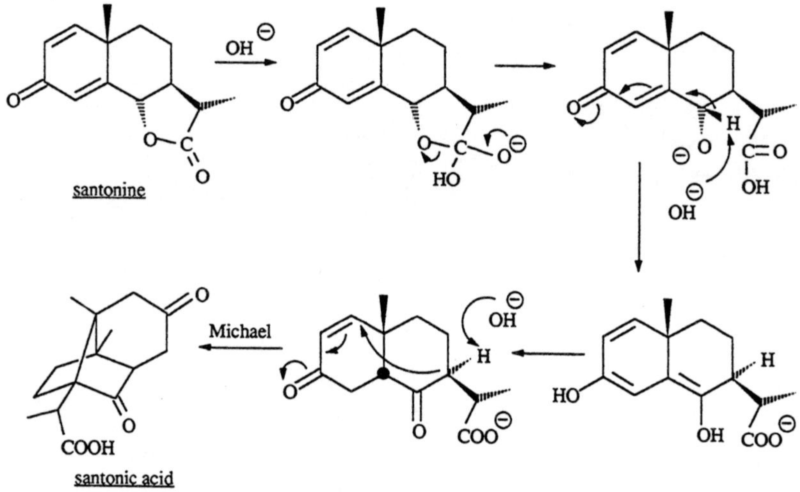

santonine

santonic acid

It is therefore necessary to synthesize the dione A [6.123].

[6.123]

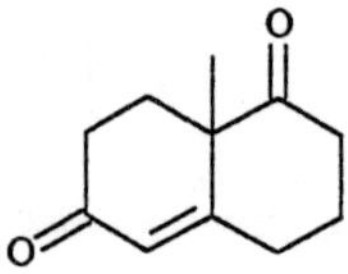

C

The Wieland-Mischler ketone C is chosen as starting material. This ketone has been frequently used in the synthesis of sesquiterpenes such as those of atractylone and lindestrene (Minato and Nagasaki), sativene and cyclosativene (McMurray), copaene and ylangene (Heathcock) and seychellene (Piers) [6.124].

[6.124]

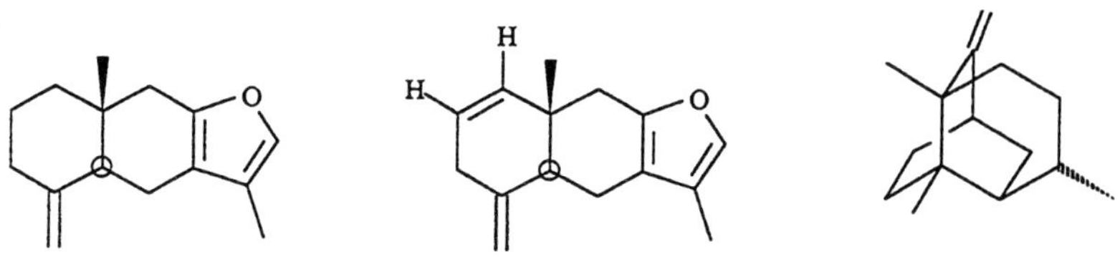

atractylone lindestrene seychellene

The Wieland-Mischler ketone was in fact originally prepared for use in the total synthesis of steroids (rings C and D), which requires a contraction of ring D [6.125].

[6.125]

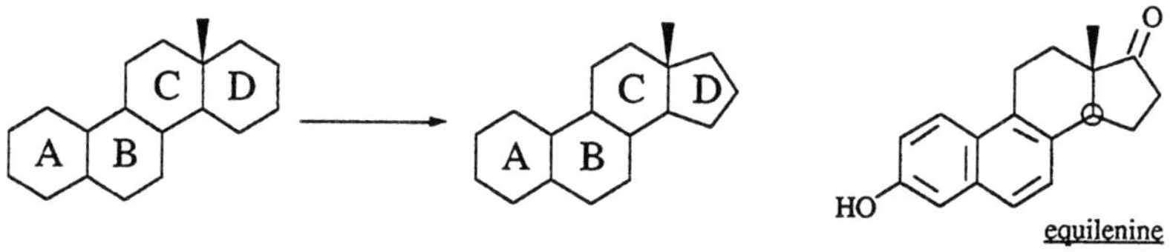

equilenine

This ketone C is readily obtained from resorcinol followed by transformation into methylcyclohexanedione which is then condensed with methylvinylketone [6.126].

[6.126]

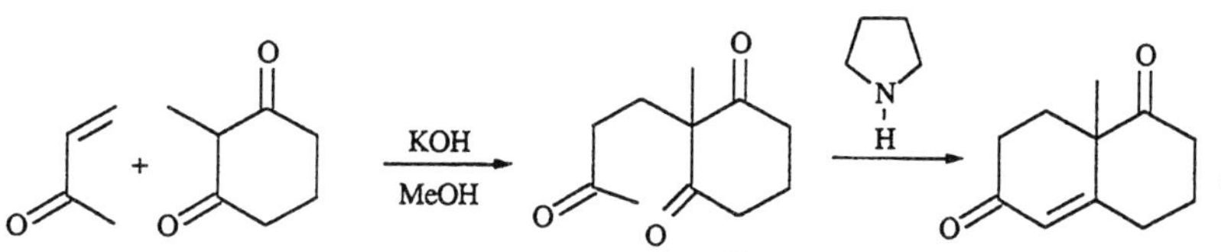

KOH / MeOH D C

It is also possible to prepare optically active forms of the Wieland-Mischler ketone. The unconjugated carbonyl is reduced with $LiAlH(t\text{-}BuO)_3$ and the ketol thus obtained is separated using the brucine salt of its acid phthalate [6.127].

It is also possible to prepare the optically active ene-dione (C) by asymmetric induction. If the Robinson annelation of the trione D is carried out in the presence of an optically active amino acid, dione C is obtained in optically active form. It is sufficient only to have catalytic quantities of the amino acid and if its absolute configuration is (S) or (R), it will induce formation of the (S) or (R) forms of the ketone C. In this way, L-proline (S) induces the formation of the (S) dione C [6.128].

Nevertheless, Corey's synthesis used the racemic ketone C according to the

[6.127]

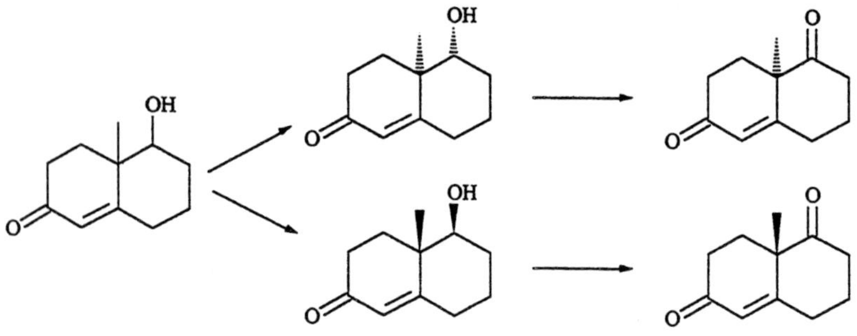

[6.128]

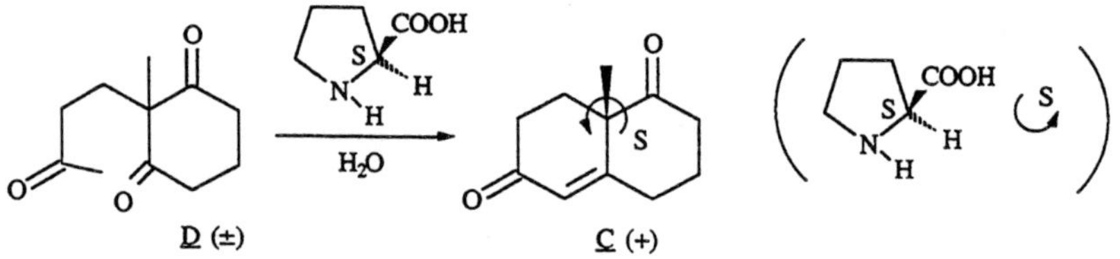

H_2O

D (±)

C (+)

following sequence [6.129].

The ethylenic dione A is thus obtained with a cis ring junction. This is of fundamental importance because even though this is not necessarily the most stable ring junction, it can be obtained as the result of delicate equilibrium. This is the weak point of this synthesis.

Cyclization of the dione A is also critical. Indeed, strong bases produce intense resinification and it is necessary to use triethylamine in ethyleneglycol in a sealed tube at 225°C over 24 hours; however, the yield is low (about 20%) [6.130].

This structure shows that the absolute configuration of natural longifolene is the (–) form. It is thus related to (+)-camphene.

C. Synthesis of sesquiterpenes with the santalane skeleton

Another series of very important compounds originating from trans,cis-farnesyl pyrophosphate and in particular from the cation shown below, are constituents of sandalwood oil. They are α- and β-santalol and the corresponding hydrocarbons α- and β-santalene [6.131].

Before discussing several syntheses of these compounds, a very interesting general method should be pointed out. Developed by Corey, it permits the selective coupling of different carbonyl groups. It involves complexes of nickel metal for which the electronic structure in its fundamental state is . . .$3s^2\ 3p^6\ 3d^8 4s^2\ 4p^0$ and which forms π-allylic complexes in three different ways.

[6.129]

C

Wittig

$\phi_3P{=}CH{-}CH_3$

$CH_3{-}HC$

OsO_4 / Et_2O-pyridine

at 0°C

double bond more accessible sterically than the other

ClTs

OH

OH

① OH

OTs

①

②

Ring extension by pinacol rearrangement is carried out using ClO_4Li / THF / CO_3Ca

OH

HCl 2N, 24 hours

minor product (~ 5%)

A

a) Square planar: bis complex [6.132]

b) Square planar: halide of the Niπ-allylic dimer [6.133]

c) Penta-coordinated π-allylic monomeric complex [6.134]

Corey's method uses the tetra-coordinated complexes of the second type. Allyl bromides react with an excess of metalcarbonyl in anhydrous benzene solution (50°C, 2–3 hours). The solvent is eliminated and the crude product is recrystallized in diethyl ether at –70°C to obtain the π-allyl nickel bromide with a yield of 80–95%.

Example [6.135]:

These complexes are relatively inert with alkyl halides in hydrocarbon or etherial solvents. On the other hand, in polar and coordinating solvents such as DMF or N-methylpyrrolidone, a facile reaction occurs between these complexes and a wide range of halides, with iodides being the most reactive.

[6.130]

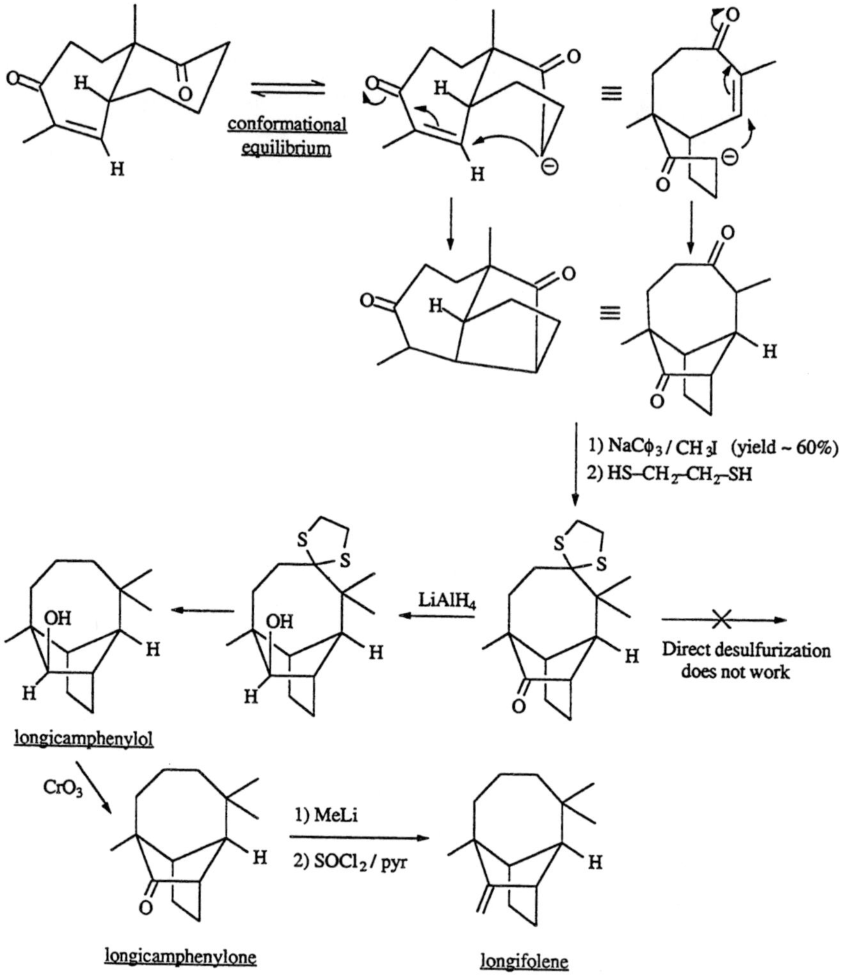

conformational equilibrium

1) $NaC\phi_3$ / CH_3I (yield ~ 60%)
2) $HS–CH_2–CH_2–SH$

$LiAlH_4$

Direct desulfurization does not work

longicamphenylol

CrO_3

longicamphenylone

1) MeLi
2) $SOCl_2$ / pyr

longifolene

[6.131]

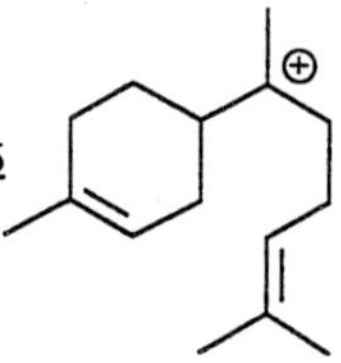

6

[6.132]

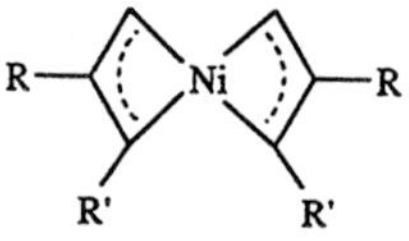

[6.133]

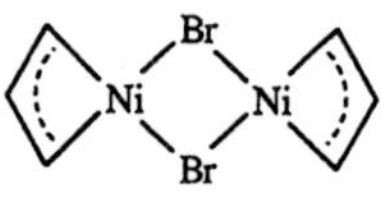

[6.134]

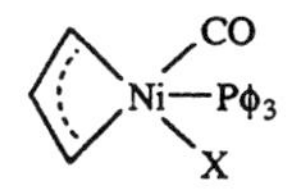

$$\text{(allyl)Ni(CO)(P}\phi_3\text{)X}$$

[6.135]

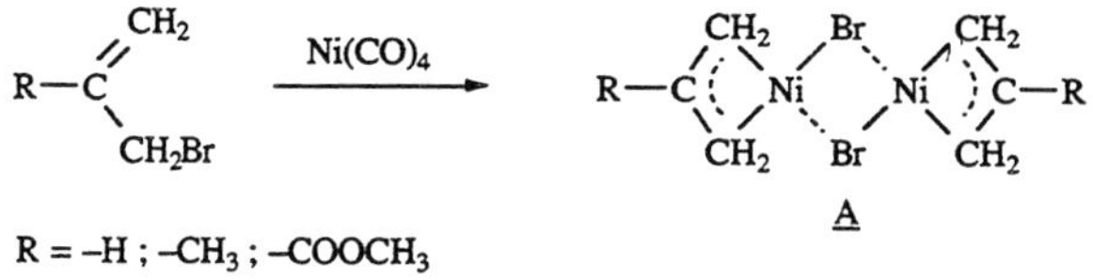

$$R{-}C(=CH_2)CH_2Br \xrightarrow{Ni(CO)_4} [(R{-}C(CH_2)_2)Ni(\mu{-}Br)_2Ni(C(CH_2)_2{-}R)] \quad \underline{A}$$

$R = -H ; -CH_3 ; -COOCH_3$

Example [6.136]:

[6.136]

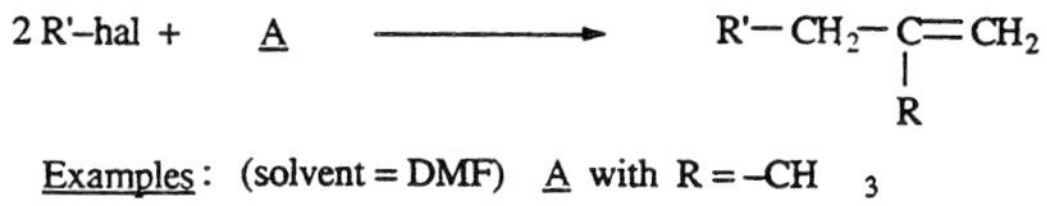

$$2\,R'{-}hal + \underline{A} \longrightarrow R'{-}CH_2{-}C(R){=}CH_2$$

Examples: (solvent = DMF) A with $R = -CH_3$

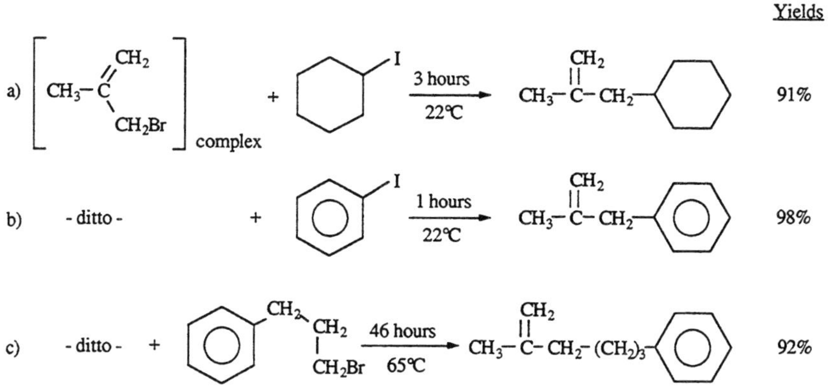

	Reaction	Yields
a)	$[CH_3{-}C(=CH_2)CH_2Br]$ complex + cyclohexyl iodide $\xrightarrow[22°C]{3\ hours}$ $CH_3{-}C(=CH_2){-}CH_2{-}$cyclohexyl	91%
b)	- ditto - + iodobenzene $\xrightarrow[22°C]{1\ hours}$ $CH_3{-}C(=CH_2){-}CH_2{-}C_6H_5$	98%
c)	- ditto - + $C_6H_5{-}CH_2{-}CH_2{-}CH_2Br$ $\xrightarrow[65°C]{46\ hours}$ $CH_3{-}C(=CH_2){-}CH_2{-}(CH_2)_3{-}C_6H_5$	92%

This reaction was used by Corey in his syntheses of α-santalene and epi-β-santalene [6.137].

Comments on these syntheses

- Preparation of the π-iodotricyclene [6.138].
- Preparation of iodocamphene.

When π-bromo camphor, whose preparation has just been described, is reduced with $NaBH^4$, it gives π-bromo borneol which can then be dehydrated (accompanied by a Wagner-Meerwein rearrangement) using a mixture of $POCl_3$-pyridine. The resulting bromocamphene can then be transformed into the iodocamphene [6.139].

Another method described by Corey permits the synthesis of β-santalene and epi-santalene. It is based on the following reaction sequence starting with norbornanone A [6.140].

[6.137]

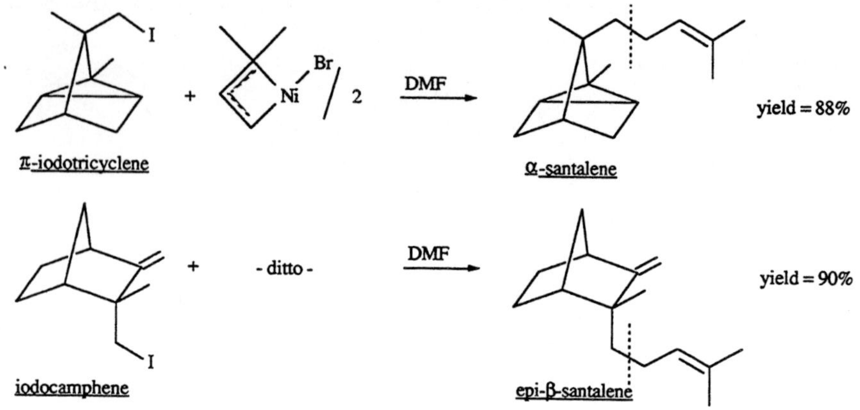

I

Br

Ni

2

DMF

\+

yield = 88%

π-iodotricyclene

α-santalene

\+

\- ditto -

DMF

yield = 90%

I

iodocamphene

epi-β-santalene

[6.138]

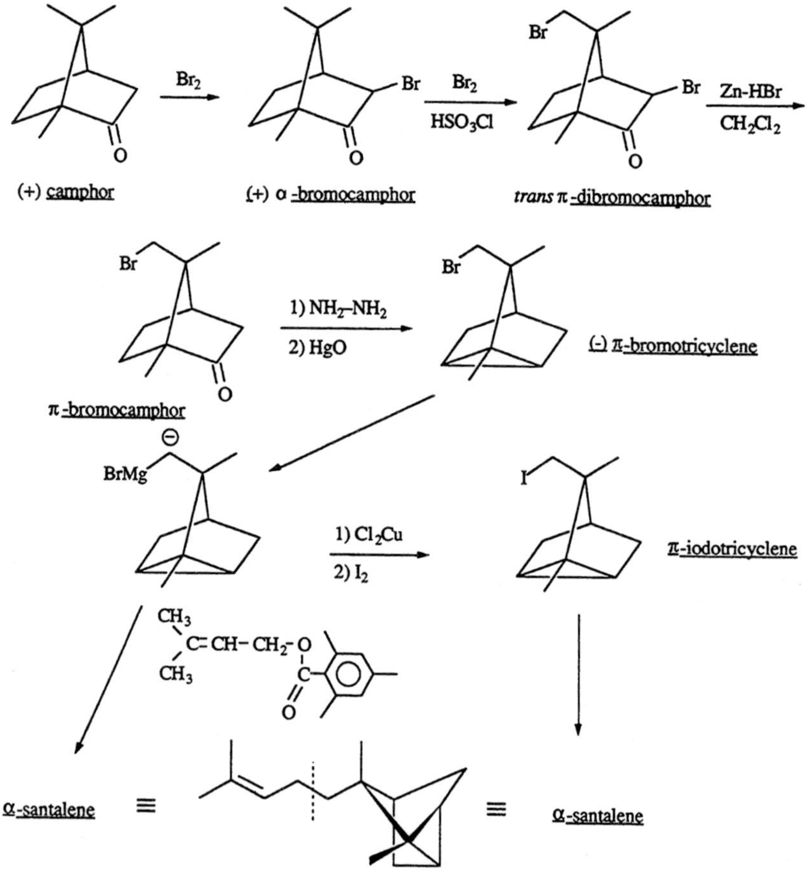

Br_2

Br

Br_2

HSO_3Cl

Br

Br

Zn-HBr

CH_2Cl_2

O

O

O

(+) camphor

(+) α -bromocamphor

trans π -dibromocamphor

Br

Br

1) NH_2–NH_2

2) HgO

(-) π-bromotricyclene

O

π -bromocamphor

⊖

BrMg

I

1) Cl_2Cu

2) I_2

π-iodotricyclene

CH_3

C= CH– CH_2– O

CH_3

C

O

α-santalene

≡

≡

α-santalene

[6.139]

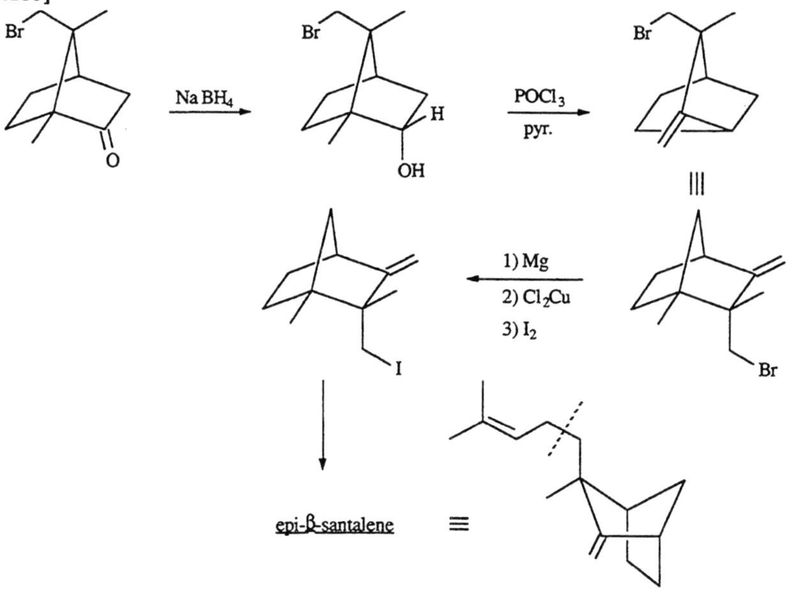

NaBH$_4$ — POCl$_3$ / pyr. — 1) Mg 2) Cl$_2$Cu 3) I$_2$

epi-ß-santalene ≡

[6.140]

NaNH$_2$ / CH$_3$I — B — NaNH$_2$ / RX — B$_1$

A

NaNH$_2$ / RX — C — NaNH$_2$ / CH$_3$I — C$_1$

1) CH$_3$Li 2) SOCl$_2$

ß-santalene

epi-ß-santalene

with R = $(CH_3)_2C{=}CH{-}CH_2{-}CH_2{-}$

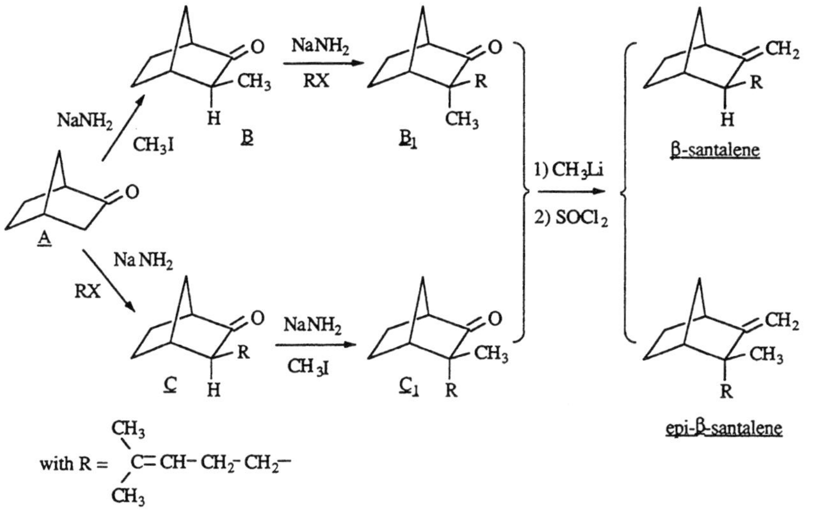

For the preparation of the bromide $(CH_3)_2{=}CH{-}CH_2{-}CH_2{-}Br$, see the homoallylic migration described by Julia. The stereochemistry of the alkylation of the enolate is the crucial aspect of these syntheses [6.141].

In fact, even with any alkylating agent as small as CH_3, the "exo" attack is still kinetically favored by a factor over 30. For this reason, when norbornanone is treated with CH_3I in the presence of $NaNH_2$, the derivative B is obtained. In the same way, alkylation with the 4-methyl–3-pentenyl bromide furnishes derivative C. However, if either of these derivatives, B or C, is treated in the same way with

[6.141]

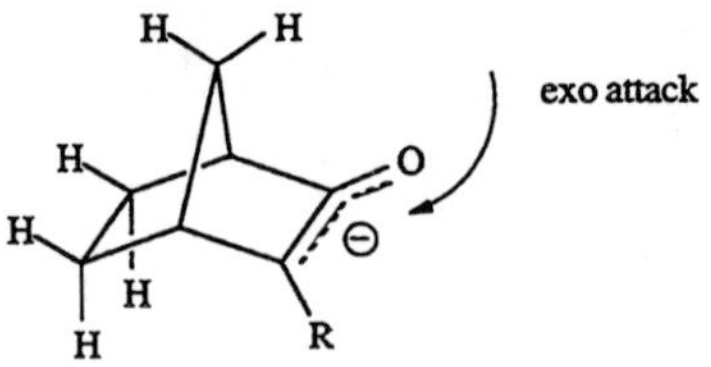

exo attack

the above mentioned bromide or with CH_3I, the respective derivatives $\underline{B}_1$ and $\underline{C}_1$ are obtained for the same reason (exo attack on the enolate).

Note: The exo derivatives are "kinetic" products, but are not the more thermodynamically stable ones. When indeed they are treated with $NaOCH_3$ in refluxing methanol, an equilibrium is established [6.142].

[6.142]

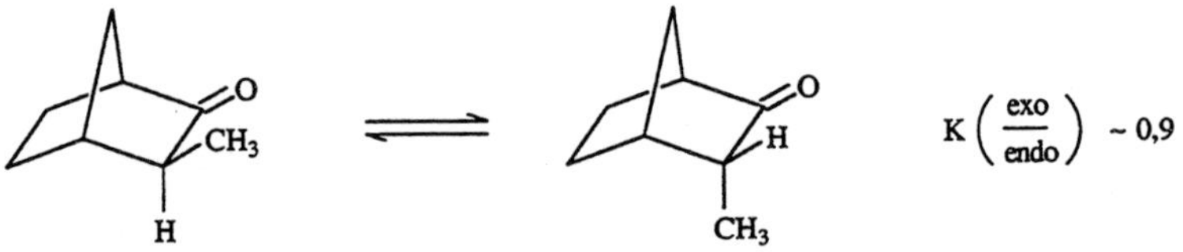

$$K\left(\frac{exo}{endo}\right) \sim 0,9$$

This strategy has been employed by Erman to synthesize (+)β-santalol in the form of a cis + trans mixture. The derivative B is first treated with Büchi's bromide which is prepared by the action of HBr acid on acrolene in the presence of ethylene glycol [6.143, 144].

[6.143]

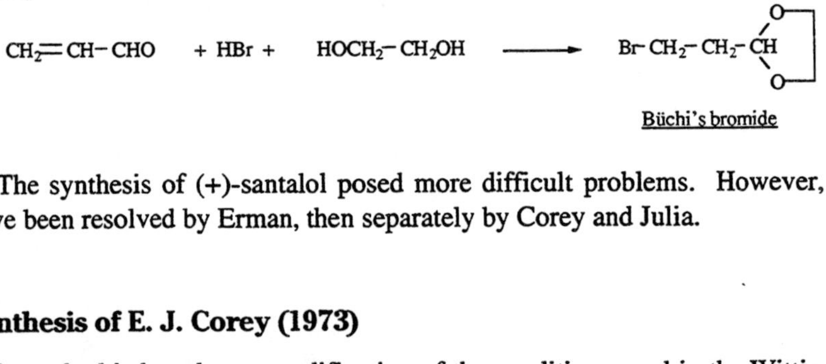

$$CH_2{=}CH{-}CHO + HBr + HOCH_2{-}CH_2OH \longrightarrow Br{-}CH_2{-}CH_2{-}CH(OCH_2CH_2O)$$

Büchi's bromide

The synthesis of (+)-santalol posed more difficult problems. However, they have been resolved by Erman, then separately by Corey and Julia.

Synthesis of E. J. Corey (1973)

This method is based on a modification of the conditions used in the Wittig reaction and will be briefly summarized here [6.145, 146]. It is possible to obtain (with stereospecificity) the colorless betaine which is the lithium salt of β,β'-dioxidophosphonium shown here. This erythro betaine is the precursor of olefins with (Z) geometry [6.147].

When this sequence is applied to tricycloekasantalal, it gives (±)-α-santalol. Tricycloekasantalal can be prepared in the following way [6.148]. The relative proportions of the two acetylenes, one terminal and the other bisubstituted, depend on the operating conditions [6.149].

[6.144]

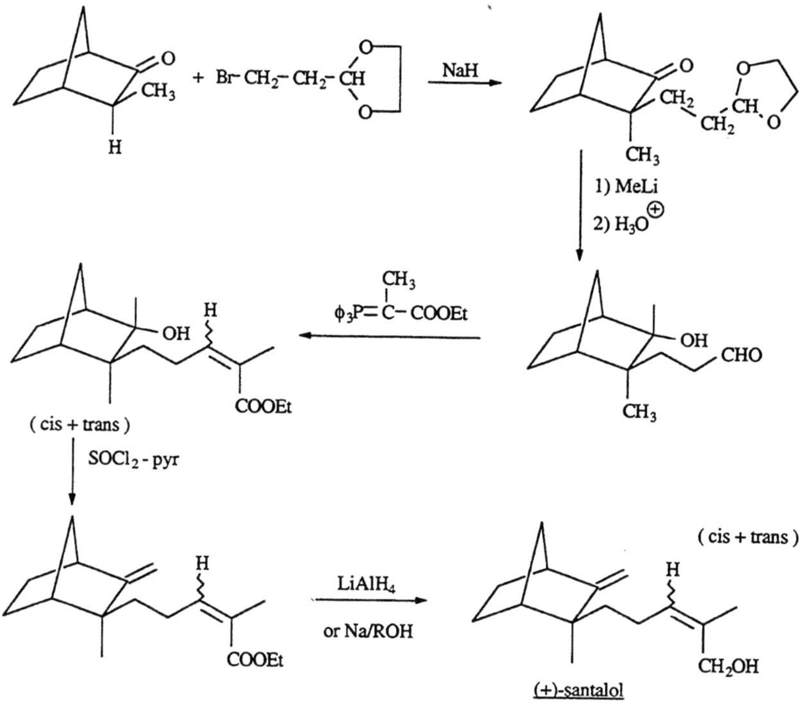

+ Br−CH₂−CH₂−CH (dioxolane) —NaH→

1) MeLi
2) $H_3O^{\oplus}$

$\phi_3P=C(CH_3)-COOEt$

(cis + trans)

$SOCl_2$ - pyr

LiAlH₄ or Na/ROH

(cis + trans)

(+)-santalol

[6.145]

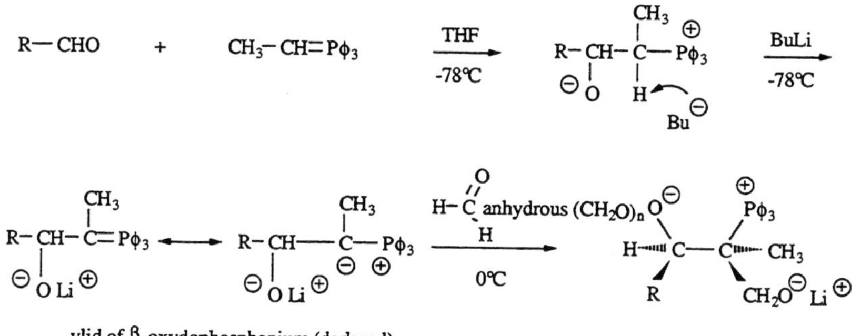

$$R-CHO + CH_3-CH=P\phi_3 \xrightarrow[-78°C]{THF} R-CH(O^{\ominus})-C(CH_3)(H)-P^{\oplus}\phi_3 \quad Bu^{\ominus} \xrightarrow[-78°C]{BuLi}$$

$$R-CH(O^{\ominus}Li^{\oplus})-C(CH_3)=P\phi_3 \longleftrightarrow R-CH(O^{\ominus}Li^{\oplus})-C^{\ominus}(CH_3)-P^{\oplus}\phi_3 \xrightarrow[0°C]{H-CHO \text{ anhydrous } (CH_2O)_n}$$

$$H-C(O^{\ominus})(R)-C(P^{\oplus}\phi_3)(CH_3)-CH_2O^{\ominus}\ Li^{\oplus}$$

ylid of β-oxydophosphonium (dark red)

[6.146]

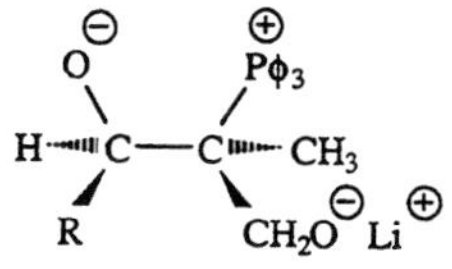

$$H-C(O^{\ominus})(R)-C(P^{\oplus}\phi_3)(CH_3)-CH_2O^{\ominus}\ Li^{\oplus}$$

[6.147]

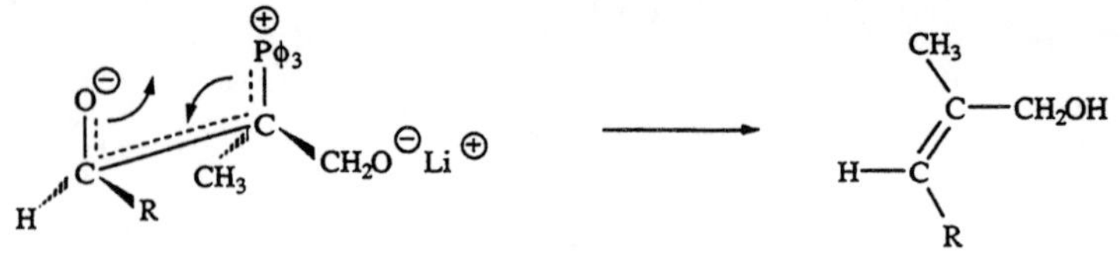

[6.148]

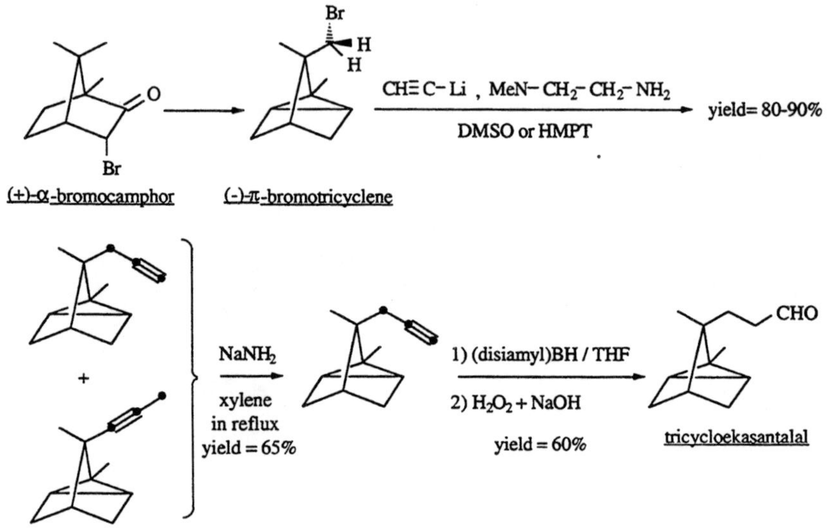

(+)-α-bromocamphor

(-)-π-bromotricyclene

CH≡C–Li , MeN–CH_2–CH_2–NH_2

DMSO or HMPT

yield= 80-90%

$NaNH_2$

xylene in reflux

yield = 65%

1) (disiamyl)BH / THF

2) H_2O_2 + NaOH

yield = 60%

tricycloekasantalal

[6.149]

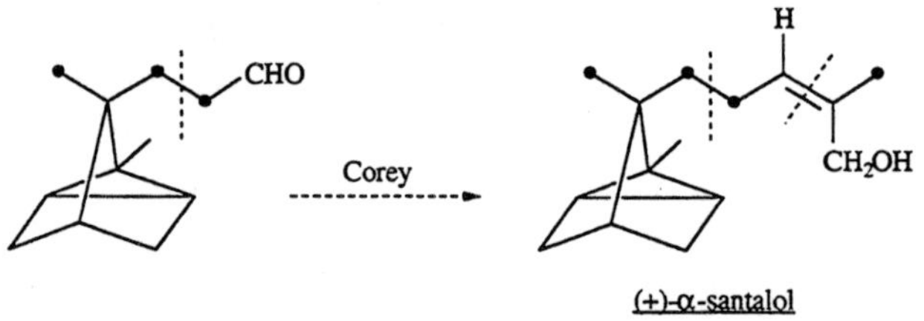

Corey

(+)-α-santalol

Synthesis of Marc Julia (1973)

This synthesis uses the phenylsulfone group to permit anionization of the neighboring carbon atom. It is thus possible to establish bonds between two prefabricated elements ("Synthons") to establish a reaction related to a disymmetric Wurtz condensation [6.150]. The last step consists of hydrogenolyzing the sulfonyl group by using a sodium amalgam in alcohol at low temperatures [6.151].

With (–)-π-bromotricyclene as one of the prefabricated elements, this strategy leads either to α-santalene or to α-santalol. The use of methylbutenyl chloride leads to α-santalene with a yield of around 60%. Substitution of the neopentyl

[6.150]

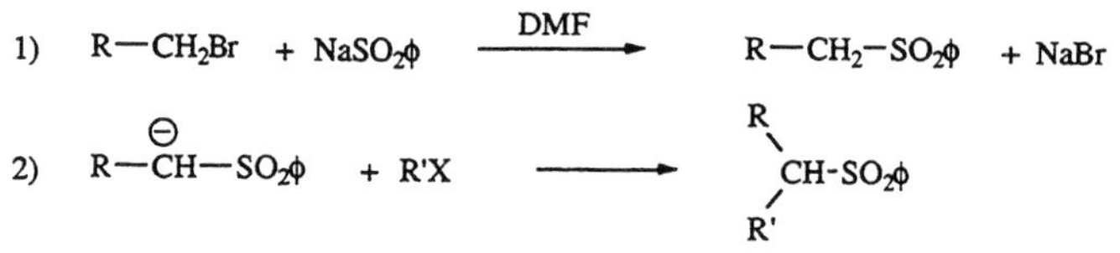

1) $R{-}CH_2Br + NaSO_2\phi \xrightarrow{DMF} R{-}CH_2{-}SO_2\phi + NaBr$

2) $R{-}\overset{\ominus}{C}H{-}SO_2\phi + R'X \longrightarrow RR'CH{-}SO_2\phi$

[6.151]

3) $RR'CH{-}SO_2\phi \longrightarrow R{-}CH_2{-}R' + NaSO_2\phi$ which is thus regenerated

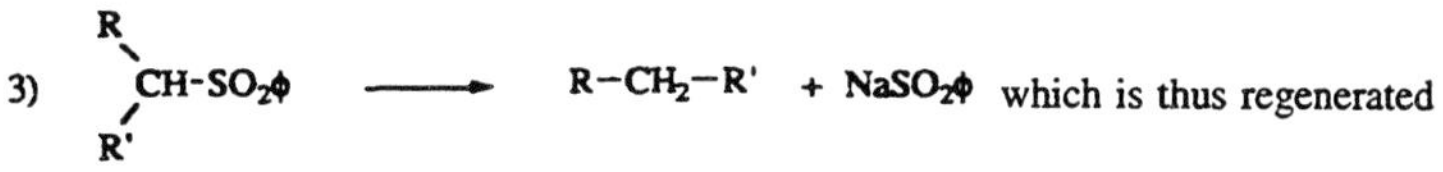

bromine is possible in DMF. The anion is then obtained by the action of BuLi in THF "doped" with 20% HMPT [6.152].

[6.152]

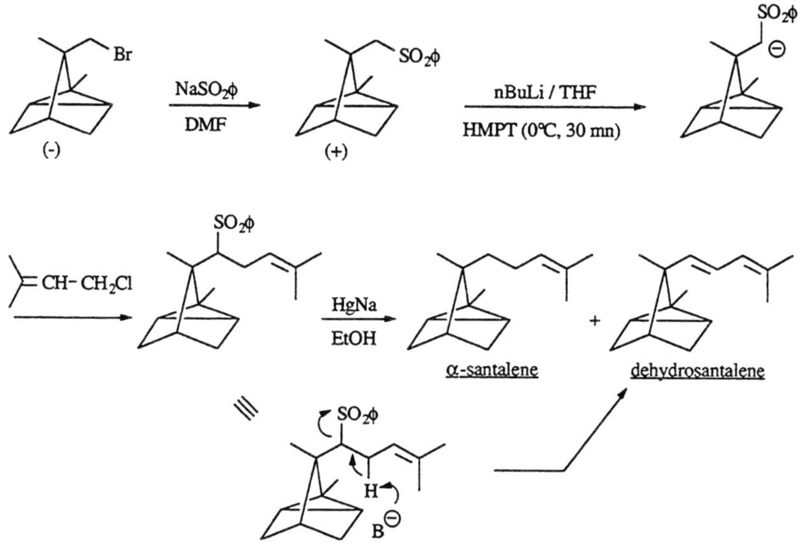

Br; (-); $NaSO_2\phi$ / DMF; $SO_2\phi$; (+); nBuLi / THF, HMPT (0°C, 30 mn); $SO_2\phi$

$(CH_3)_2C{=}CH{-}CH_2Cl$; $SO_2\phi$; HgNa / EtOH; α-santalene + dehydrosantalene

$SO_2\phi$; H; $B^{\ominus}$

In this case, about 30% 10,11-dehydrosantalene is formed. The following unit is used to synthesize α-santalol [6.153], which can be obtained by way of the following schemes [6.154].

[6.153]

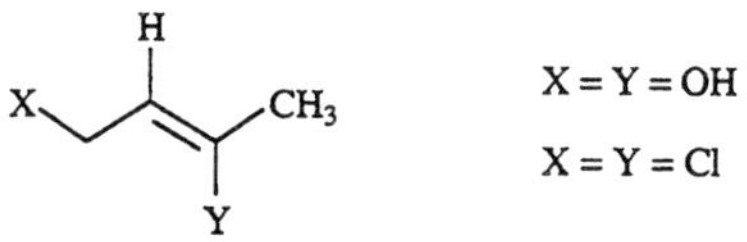

X, H, CH_3, Y

X = Y = OH

X = Y = Cl

This dichloride reacts in a completely regioselective way with the (+)-phenylsulfone derived from (–)-π-bromotricyclene. The factors which determine this regioselectivity are not yet understood [6.155].

Passage from the santalols to the copaborneols

We have already mentioned that L. Westfelt and colleagues used the natural san-

[6.154]

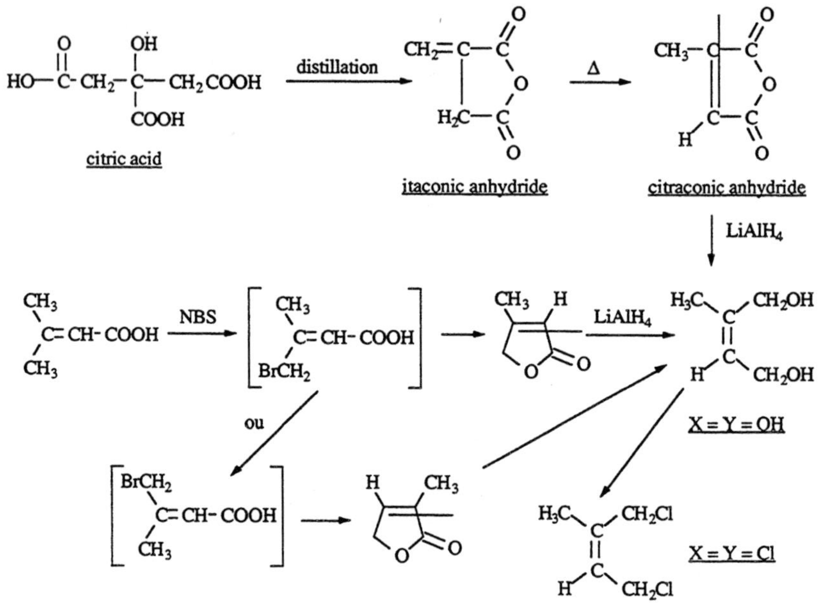

citric acid — distillation → itaconic anhydride — Δ → citraconic anhydride — $LiAlH_4$ →

$(CH_3)_2C=CH-COOH$ — NBS → [$BrCH_2(CH_3)C=CH-COOH$] → lactone — $LiAlH_4$ → $H_3C(H)C=C(CH_2OH)_2$ X = Y = OH

ou [$BrCH_2(CH_3)C=CH-COOH$] → lactone

$H_3C(H)C=C(CH_2Cl)_2$ X = Y = Cl

[6.155]

$SO_2\phi$ (anion) + $CH_3(H)C=C(CH_2Cl)_2$ → 1) AcONa 2) HgNa 3) KOH → α-santalol

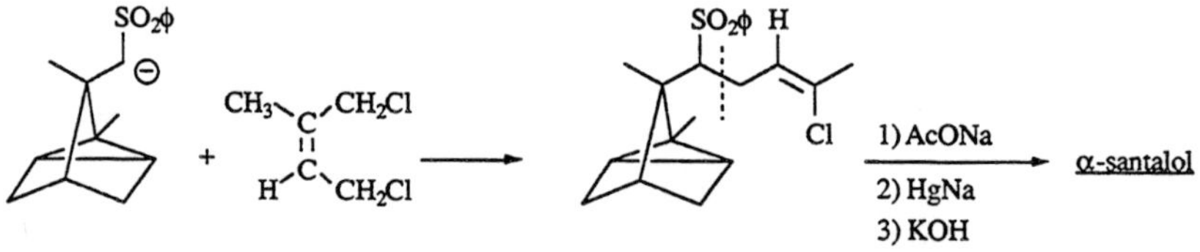

talols as starting materials in the synthesis of (+)-copaborneol. This correlation has permitted the absolute stereochemistry of (+)-copaborneol to be determined from that of the santalols. It will be recalled that the mixture of natural santalols is made up of about 70% (+)-α-santalol and 30% (–)-β-santalol; the latter being made up of about 70% (–)-β-santalol (exo) and 30% (–)-epi-β-santalol (endo) [6.156].

The above mixture can be oxidized by SeO_2 or, better still, MnO_2 to give the corresponding santalals. This mixture of aldehydes can then be converted into the corresponding santalic acids by the action of silver oxide. Treatment of these crude acids with anhydrous formic acid, first at room temperature and then for eight hours at 45°C, affords formates. In the case of (–)-santalol, for example, the following acid-ester is obtained [6.157].

Alkaline hydrolysis leads to acid-alcohols for which the former function can be esterified with methanol and the latter oxidized to the ketone. The end result is a mixture of three oxo-esters a, b and c, summarized as follows [6.158].

[6.156]

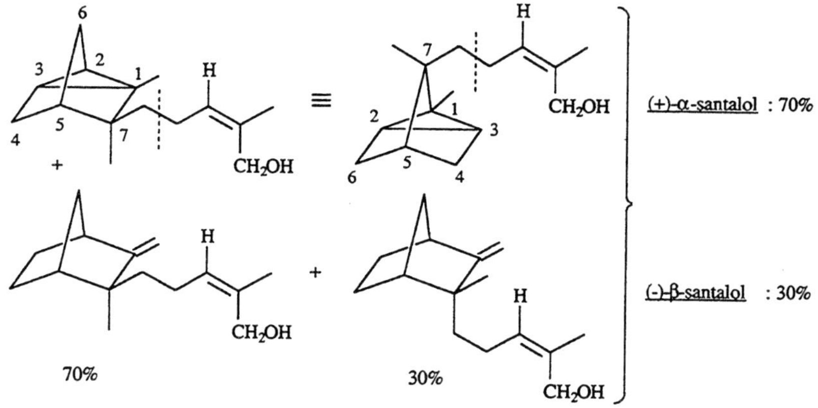

6 3 2 1 H 5 4 7 + CH_2OH ≡ H 7 2 1 3 CH_2OH 6 5 4

(+)-α-santalol : 70%

H CH_2OH + H CH_2OH

70% 30%

(-)-β-santalol : 30%

[6.157]

O ⊖ O–C–H ⊕ HCOOH R COOH COOH OCOH

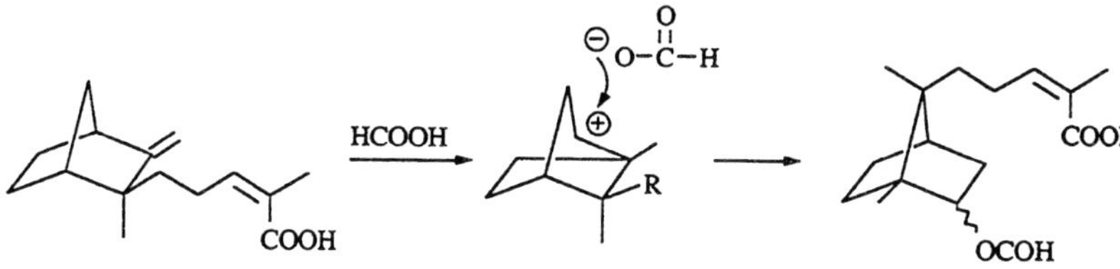

[6.158]

② ①

opening of side ②

COOMe

~ 35%

a (anti)

(+)-α CH_2OH

opening of side ①

COOMe

~ 55%

(-)-β-exo CH_2OH O b (syn)

MeOOC

~ 10%

(-)-β-endo O c (anti)

CH_2OH

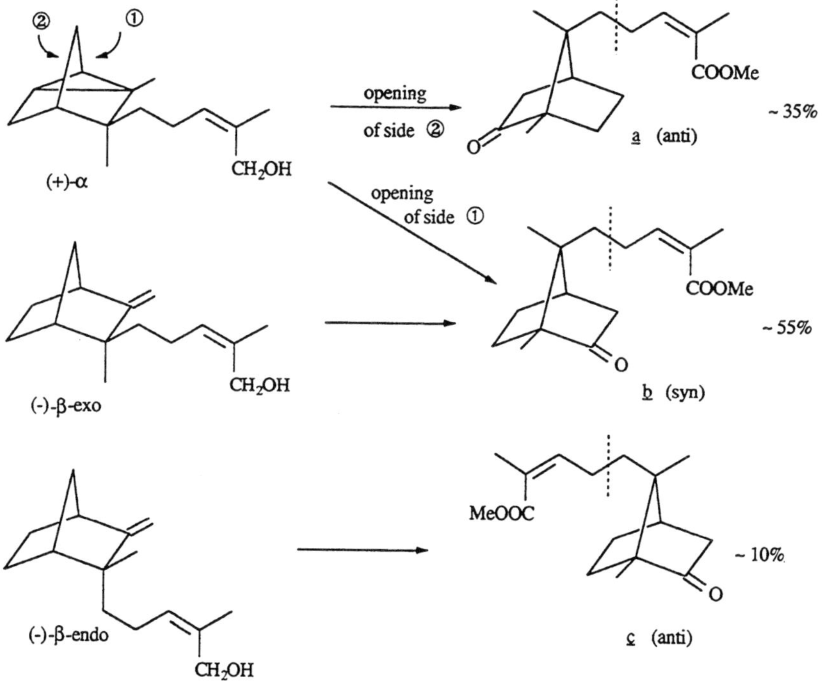

When this mixture of syn and anti oxo-esters is treated for 45 minutes with potassium tertiary butylate in dioxan at room temperature under a nitrogen atmosphere, only the syn oxo-ester b undergoes cyclization, while its isomers a and c remain practically unchanged [6.159].

[6.159]

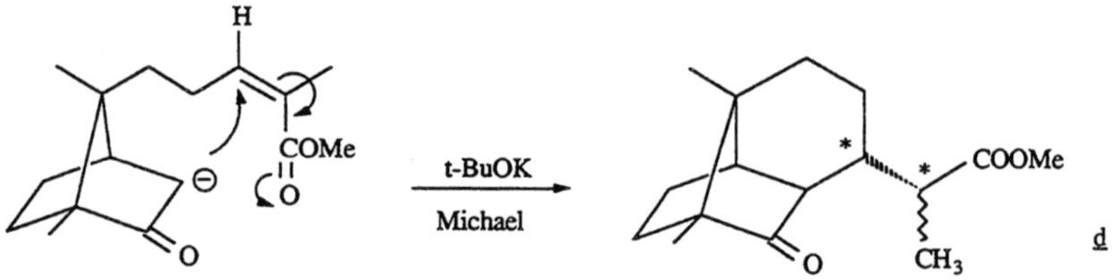

The cyclized product can be separated from the oxo-esters a and c by chromatography on silica gel. It constitutes a diastereomeric mixture of two oxo-esters d.

The carbonyl function of this compound is not very reactive and can be protected by ketalization. This weak reactivity allows the transformation of the ester function into the free acid by alkaline hydrolysis. Subsequent treatment with thionyl chloride in benzene gives the corresponding acid chloride. This can then be reduced to the alcohol by sodium borohydride without affecting the free carbonyl function of the ketone. The mesylate of this keto-alcohol can be prepared using methanesulfonyl chloride in pyridine. Subsequent treatment with $LiAlH_4$ finally gives the desired product by reducing the mesylate group to a CH_3 and the carbonyl function to its alcohol [6.160].

D. Syntheses of elemol and hedycaryol

1. *Total synthesis of elemol*

The critical part of this synthesis is in the initial preparation of compound C [6.162]. This requires the generation of two subunits A and B [6.163]. Unit A can be obtained in the following way [6.164]. Unit B is obtained according to the following scheme [6.165].

Treatment of compound C with 1.3 equivalents of methanolic KOH for 10 hours at 25°C hydrolyzes only the ortho toluyl group. Disubstituted malonic esters are, in fact, difficult to hydrolyze [6.166].

When a solution of the derivative D in N-methylpyrrolidone is added to an excess of Ni(CO) 4 under a CO atmosphere at a temperature between 47° and 50°C, an 85% yield of a mixture of three products, E, F and G is obtained. These can be separated by chromatography on SiO_2 impregnated with $AgNO_3$ [6.167].

The derivative G is either cis,cis or cis,trans. Indeed, if it was all trans (structure H), its pyrolysis at relatively low temperatures would give rise to derivative E. Product G can, however, only be pyrolyzed at 285°C; thus, the product is F.

[6.160]

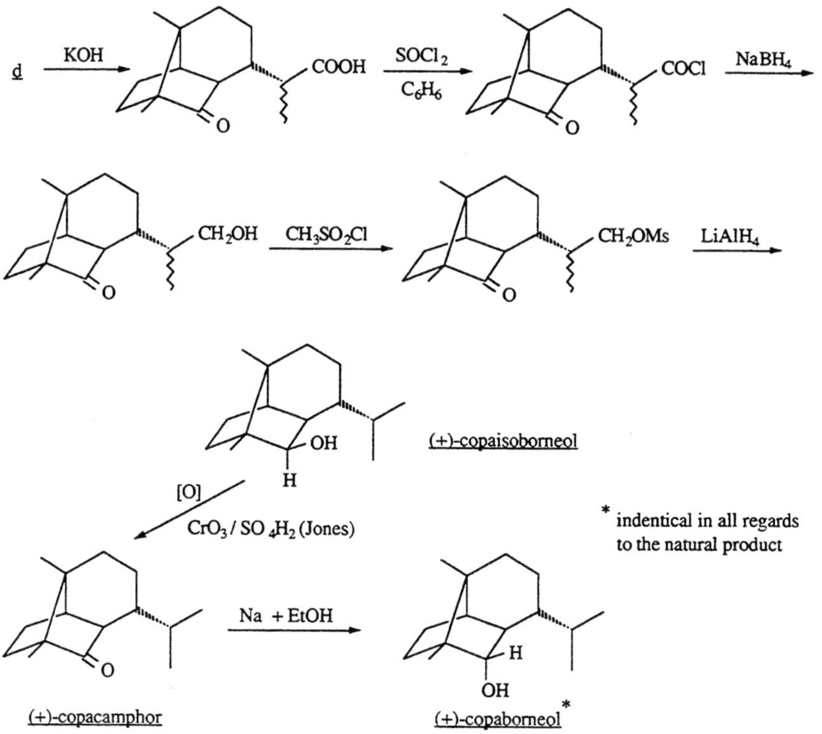

d —KOH→ [keto acid, COOH] —$SOCl_2$ / C_6H_6→ [COCl] —$NaBH_4$→

[CH_2OH] —CH_3SO_2Cl→ [CH_2OMs] —$LiAlH_4$→

[OH, H] (+)-copaisoborneol

[O]
CrO_3 / SO_4H_2 (Jones)

[O] (+)-copacamphor —Na + EtOH→ [H, OH] (+)-copaborneol*

* indentical in all regards to the natural product

[6.161]

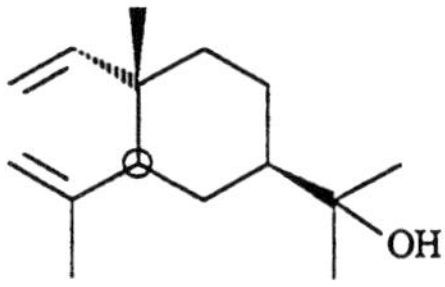

OH

[6.162]

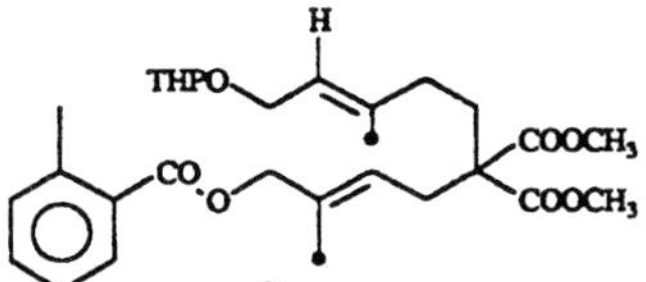

H
THPO
$COOCH_3$
$COOCH_3$
CO
O
C

[6.163]

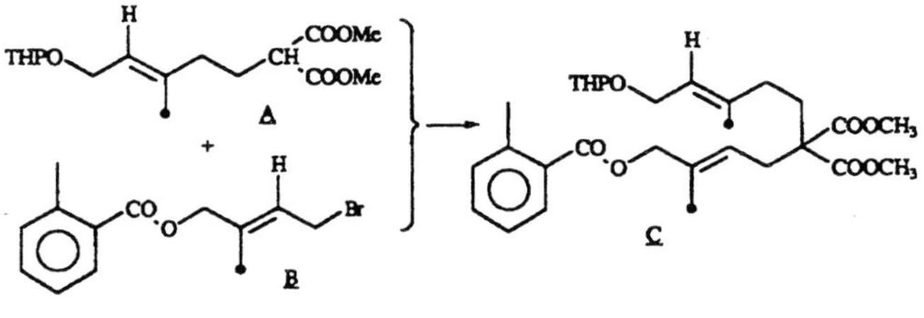

H
THPO
CH
COOMe
COOMe
A
+
H
CO
O
Br
B

H
THPO
$COOCH_3$
$COOCH_3$
CO
O
C

[6.164]

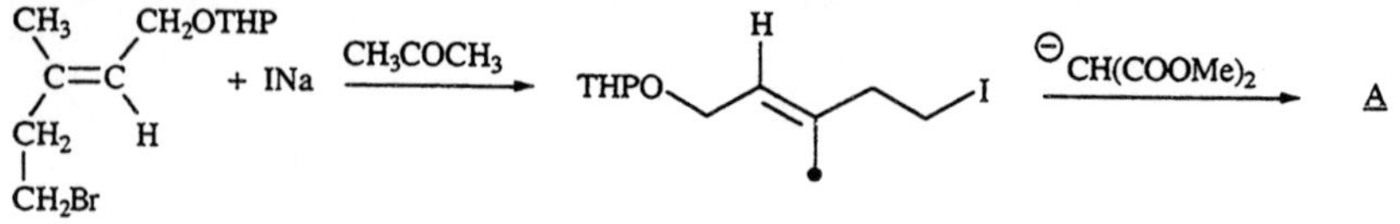

[6.165]

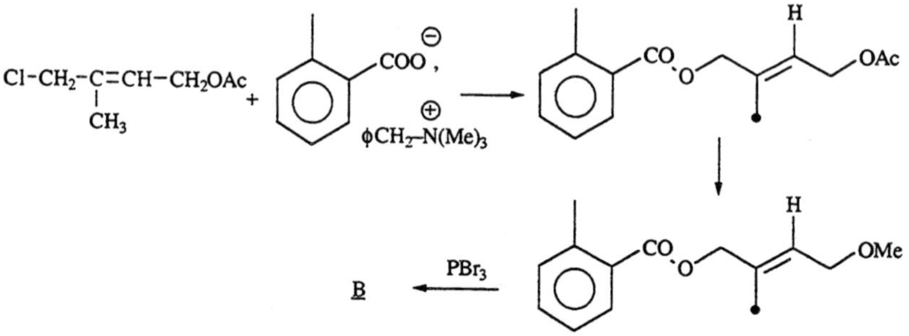

[6.166]

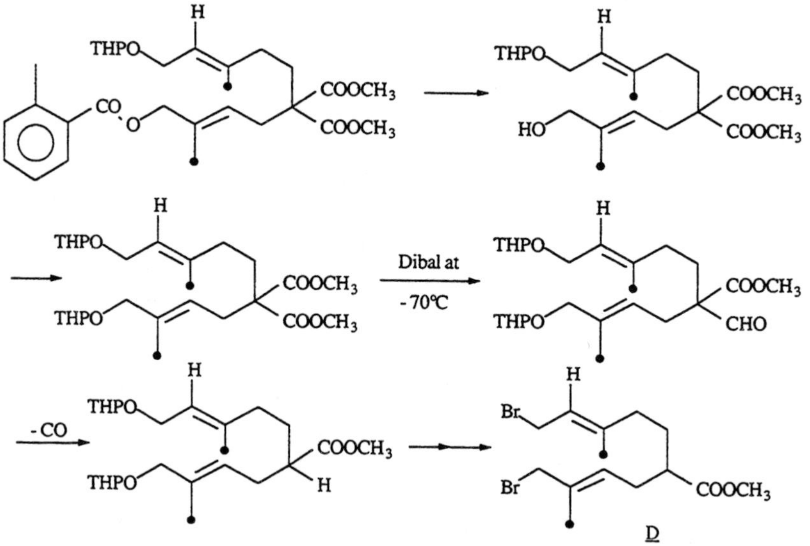

D

The action of an excess of methylmagnesium bromide on compound E gives elemol [6.168].

How can the formation of the three products E, F and G be explained? In fact, from the results already described, the cyclization of the dibromo allylic derivative D should have formed exclusively the compounds G and/or H. In the case at hand, it could be supposed that the product H is indeed formed but then is transformed into product E under the prevailing experimental conditions. This, how-

[6.167]

rf = 0.08 ; yield ~ 32%

COOCH$_3$

E

Δ

rf = 0.23 ; yield ~ 23%

COOCH$_3$

F

Δ = 285°C

COOCH$_3$

H

H

H

rf = 0.45 ; yield ~ 11%

COOCH$_3$

G

[6.168]

BrMgCH$_3$

COOCH$_3$

CH$_3$

OH

CH$_3$

E

elemol

ever, is very improbable because the temperature never exceeds 50°C. Furthermore, this does not take into account the formation of compound F. The explanation must therefore be found elsewhere. For this, we can return to our discussion on the action of halogenated derivatives with π-allylic complexes [6.169].

[6.169]

R$_1$ Br R$_2$

R$_1$ 2 3 1 R$_2$ Ni Br 2

RX attack at 1

R$_1$ R R$_2$ a

RX attack at 3

R R$_1$ R$_2$ b

This π-allylic derivative can, in fact, be attacked either in position 1 to give a "normal" derivative, or in position 3 to give the "iso" derivative b. Formation of the products E and F can therefore be explained by admitting, for instance, the formation of a π-complex at one of the extremities of the molecule with an allylic migration of bromide occurring at the other extremity. The unmigrated bromide gives the "normal" derivative G [6.170].

[6.170]

2. *Synthesis of hedycaryol*

The biogenetic importance of hedycaryol has understandably encouraged research into ways of making its total synthesis. This has been achieved by Wharton and co-investigators in 1971 and is briefly schematized here. The starting material is 4-hydroxyisophthalic acid, a by-product in the manufacture of salicylic acid [6.171].

The triol obtained from this sequence is in fact a mixture of diastereoisomers from which it is possible to isolate a cystallized triol with a yield of about 50% (MP = 170°C). Treatment with sodium meta-periodate in aqueous methanolic solution obtains a keto-alcohol which is transformed into a mixture of epimeric methyl ketones [6.172].

Reduction of this mixture of ketones using the Wolff-Kischner method permits a correlation to be made with γ-eudesmol and 10-epi-γ-eudesmol (isolated from the essential oil of African geranium) [6.173].

The product, (dl)-hedycaryol, which is thermodynamically unstable, has been purified through the mediation of its π-complex with the Ag^+ ion. It can be regenerated at room temperature with the aid of a concentrated ammonia solution. The product obtained in this way is, apart from its optical rotation, identical to natural hedycaryol. On heating it gives elemol.

E. Total syntheses of α-humulene and of caryophyllene

1. *Total synthesis of α-humulene (Corey)*

The principle of this synthesis is the method developed by Corey involving Ni-π allylic halides (see formulas 11, 12 and 13). As a general rule, it is not necessary to prepare separately this complex. For this reason, this strategy presents advantages for the synthesis of medium-sized rings and macrocycles. In fact, if an

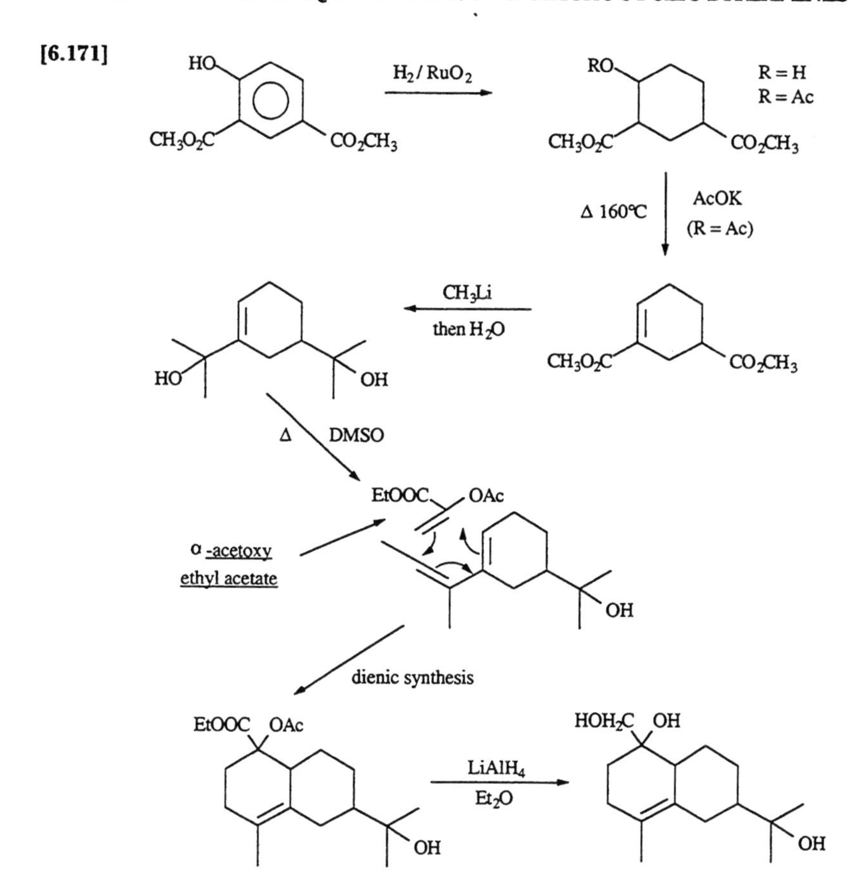

[6.172]

HOH_2C OH

OH

O

OH

NaH + CH_3I

$(CH_2OH)_2$

O

OH

allylic α,ω -dibromide [such as $Br\text{-}CH_2\text{-}CH{=}CH\text{-}(CH_2)_n\text{-}CH{=}CH\text{-}CH_2Br$ which is easily obtainable from the α,ω-dipropargylic glycol $HO\text{-}CH_2\text{-}C{\equiv}C\text{-}(CH_2)_n\text{-}C{\equiv}C\text{-}CH_2OH$] is added to a solution of $Ni(CO)_4$ at 50°C, cyclization occurs with the formation of a 1,5-diene. This method has also been applied by Corey to the synthesis of macrolides.

By slowly adding (over a period of 3.5 hours) the allylic dibromide (Z,Z isomer) to a stirred solution of six equivalents of $Ni(CO)_4$ in N-methylpyrrolidone under an argon atmosphere, the major product is the diene-olide (E,E isomer) with a yield of approximately 70–75%. The dibromide is prepared by the action

[6.173]

γ-eudesmol

10 epi-γ-eudesmol

LiAlH₄ $(-CH_2OMe)_2$ -20°C

pTsOCl / pyr 0°C

B_2H_6 THF

(recrystallized in C_6H_{12})
MP = 124°–125°C

(MP = 62°–68°C)

25°C 7 h

dl-hedycaryol

of PBr_3 in ether at 0°C for 12 hours on the corresponding ethylenic glycol (yield ~ 83%).

The scheme of the synthesis of α-humulene consists first of preparing the unit U (X = Y = Br) cyclized into iso-humulene using $Ni(CO)_4$. This can then be photochemically isomerized into α-humulene [6.174].

[6.174]

$Ni(CO)_4$

hν > 350 nm
ϕS–Sϕ
C_6H_{12}, 25°C
2.5 hours

U

2.5 The crucial step in this synthesis therefore consists of the preparation of the unit U. For this, two subunits, A and B, must be made [6.175].

[6.175]

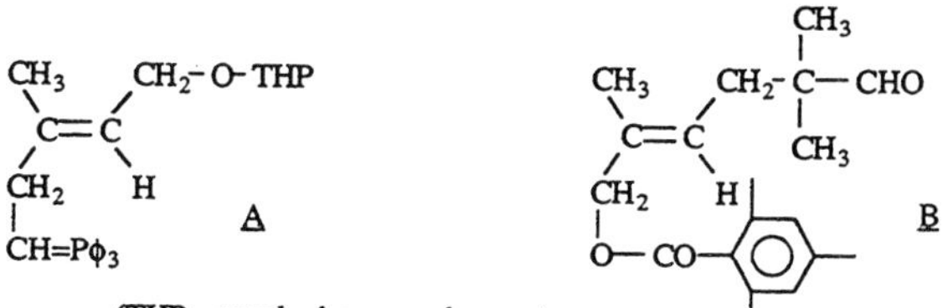

$CH_3(CH=P\phi_3\text{-}CH_2)C=C(H)CH_2\text{-}O\text{-}THP$ **A**

(THP = tetrahydropyranyl group)

$CH_3(CH_2\text{-}O\text{-}CO\text{-}Ar)C=C(H)CH_2\text{-}C(CH_3)_2\text{-}CHO$ **B**

a) Preparation of subunit A

Dimethyl trans-methylglutaconate is the starting material [6.176].

[6.176]

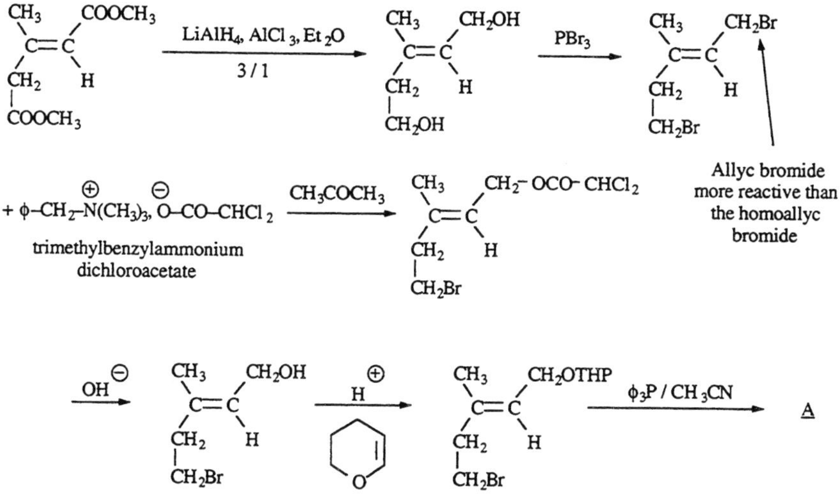

$CH_3(CH_2COOCH_3)C=C(H)COOCH_3 \xrightarrow[3/1]{LiAlH_4,\ AlCl_3,\ Et_2O} CH_3(CH_2CH_2OH)C=C(H)CH_2OH \xrightarrow{PBr_3} CH_3(CH_2CH_2Br)C=C(H)CH_2Br$

Allyc bromide more reactive than the homoallyc bromide

$+ \phi\text{-}CH_2\text{-}\overset{\oplus}{N}(CH_3)_3,\ \overset{\ominus}{O}\text{-}CO\text{-}CHCl_2 \xrightarrow{CH_3COCH_3} CH_3(CH_2CH_2Br)C=C(H)CH_2\text{-}OCO\text{-}CHCl_2$

trimethylbenzylammonium dichloroacetate

$\xrightarrow{OH^\ominus} CH_3(CH_2CH_2Br)C=C(H)CH_2OH \xrightarrow[\text{dihydropyran}]{H^\oplus} CH_3(CH_2CH_2Br)C=C(H)CH_2OTHP \xrightarrow{\phi_3P\ /\ CH_3CN}$ **A**

b) Preparation of subunit B

The starting material for this preparation is methylbutenyl chloroacetate a.

This compound is easily prepared using the method of Oroshnick and Mallory in treating t-butyl-hypochlorite isoprene with tertiobutyl hypochlorite in acetic solution [6.177].

Passage from b to the subunit B uses, at the same time, the work of Stork and of Wittig (action of magnesium compounds on enamines). The enamine of isobutyraldehyde, obtained from cyclohexylamine and isobutyraldehyde, when treated with ethyl magnesium bromide, gives the magnesium derivative of N-(2-methylpropylidene)-cyclohexylamine. When this reagent is then allowed to react with the bromide b, the imine c is formed which corresponds to the aldehyde B. This aldehyde is liberated by the action of 25% oxalic acid (4 hours).

These reactions can be best schematized in the following way [6.178]:

[6.177]

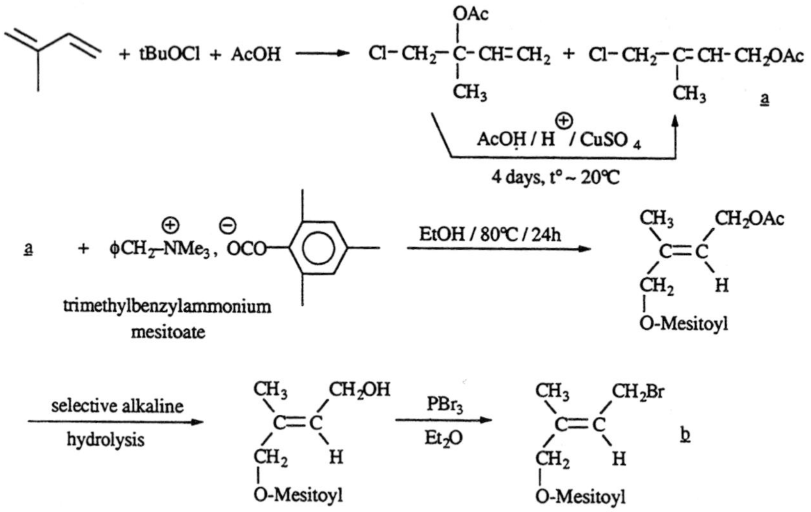

+ tBuOCl + AcOH → Cl−CH2−C(OAc)(CH3)−CH=CH2 + Cl−CH2−C(CH3)=CH−CH2OAc a

AcOH / H⊕ / CuSO4

4 days, t° ~ 20°C

a + ϕCH2−NMe3⊕, ⊖OCO−(mesityl)

trimethylbenzylammonium mesitoate

EtOH / 80°C / 24h

O-Mesitoyl

selective alkaline hydrolysis

PBr3 / Et2O

b

[6.178]

+ H2O

EtMgBr (Stork)

R−Br = b

+ EtH

+ MgBr2

MgBr

Br⊖

H⊕

B

c

O-Mesitoyl

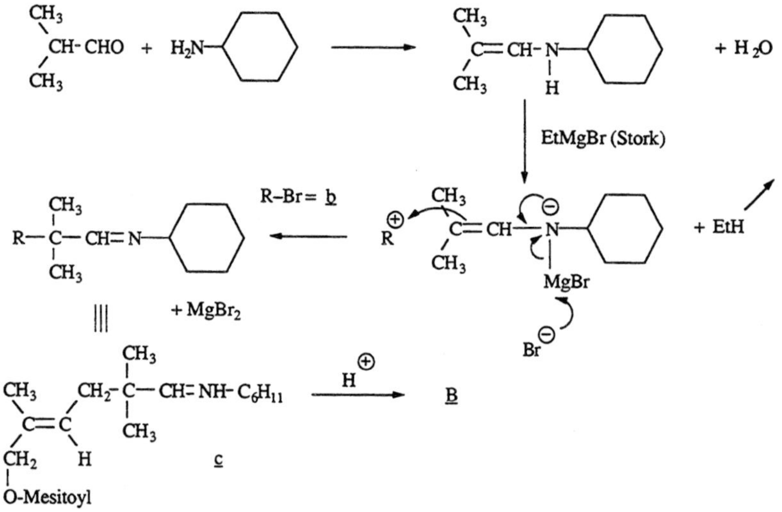

c. Coupling of subunits A and B by a Wittig reaction in DMSO [6.179]

2. *Total synthesis of (+)-caryophyllene (Corey)*

The general scheme for this synthesis follows [6.180].

Concerted fragmentation as the result of the action of bases

This fragmentation which gives rise to the nine-membered ring is the crucial step in this synthesis of caryophyllene [6.181].

[6.179]

A + B —DMSO→ (cis at 98%)

U : Y = THP, X = –O-Mesitoyl

1) $LiAlH_4$, 25°C
2) MeOH, $H^{\oplus}$, 25°C
3) PBr_3

U (X = Y = Br)

isohumulene —hν→ humulene

The cis ring junction, which was the more stable in the case of the four- and six-membered rings, is no longer the more stable for the junction of a four- and a nine-membered ring. In this case, the trans junction is far more stable. The two ketones A and B therefore have to be equilibrated by treatment with t-BuOK in DMSO [6.182].

F. Syntheses of patchoulol other than that of Büchi

It would be overfastidious to describe here in detail all of the syntheses of patchoulol developed since that of Büchi. For memory's sake, we can cite those of Danishevsky and Dumas in 1968 and that of Mirrington and Schmalzl in 1972. Both were based on the cyclization of ε-haloketones and they served as models for later syntheses of norpatchoulenol.

1. Synthesis of Danishevsky and Dumas (1968) [6.183]

2. Synthesis of Mirrington and Schmalzl (1972) [6.184]

The last step of both of these syntheses makes use of nucleophilic attack on a carbon bound to a halogen (I or Br) by a ketyl radical anion, a species which is generated by the reducing action of an alkali metal on a carbonyl [6.185].

In addition to the synthesis of Yamada in 1979 for which the long reaction sequence is described below, we shall spend a bit more time discussing the synthesis of Näf and Ohloff in 1974, which is particularly short and which involves an intramolecular Diels-Alder reaction. This process, which had previous little use, was judiciously applied here to the synthesis of this tricyclic compound.

[6.180]

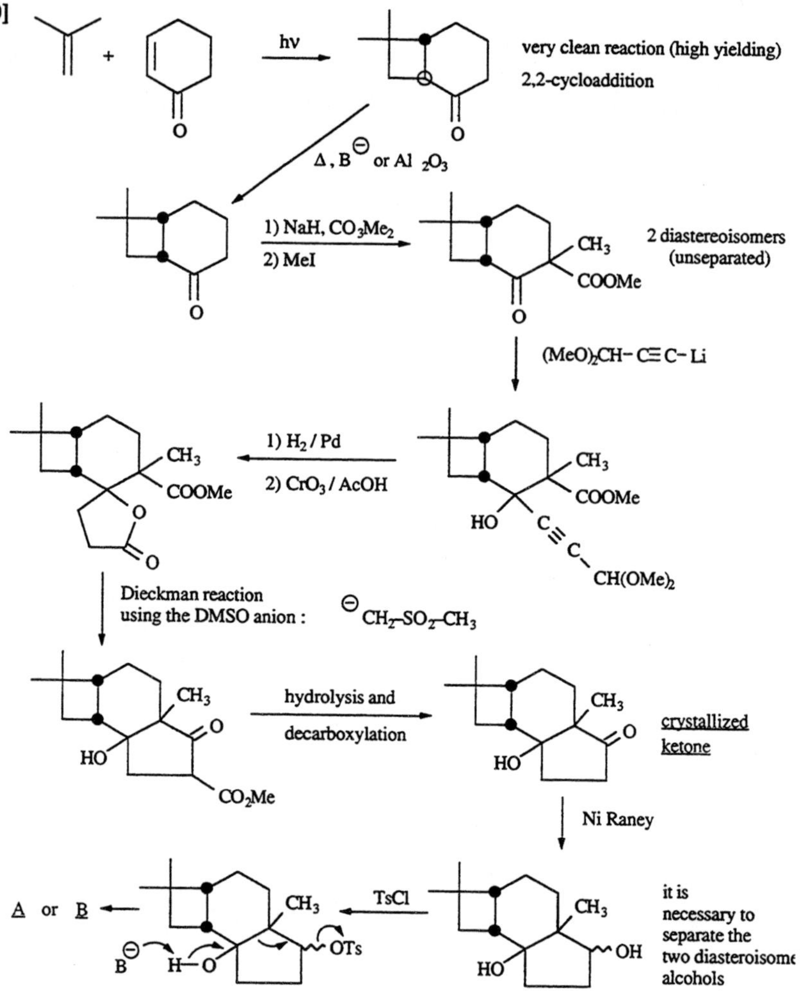
hν
very clean reaction (high yielding)
2,2-cycloaddition
O
O
Δ, B⊖ or Al 2O3
1) NaH, CO3Me2
2) MeI
CH3
COOMe
2 diastereoisomers
(unseparated)
O
O
(MeO)2CH– C≡C– Li
1) H2 / Pd
2) CrO3 / AcOH
CH3
COOMe
O
O
CH3
COOMe
HO
C
C
CH(OMe)2
Dieckman reaction
using the DMSO anion :
⊖ CH2–SO2–CH3
CH3
O
HO
CO2Me
hydrolysis and
decarboxylation
CH3
O
HO
crystallized
ketone
Ni Raney
A or B
CH3
OTs
B⊖
H–O
TsCl
CH3
OH
HO
it is
necessary to
separate the
two diastereoisome
alcohols

[6.181]

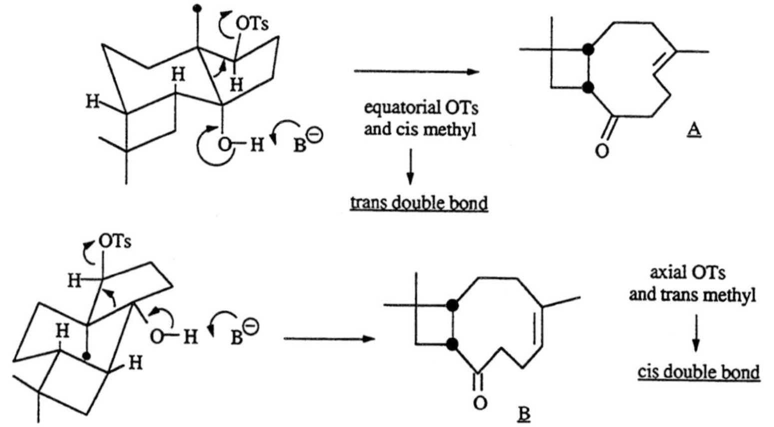
OTs
H
H
H
O–H
B⊖
equatorial OTs
and cis methyl
trans double bond
O
A
OTs
H
H
O–H
B⊖
H
O
B
axial OTs
and trans methyl
cis double bond

[6.182]

A → (ketone) $\xrightarrow[CH_3SO_2CH_2^{\ominus}]{\phi_3P=CH_2}$ caryophyllene

B → (ketone) $\xrightarrow[CH_3SO_2CH_2^{\ominus}]{\phi_3P=CH_2}$ isocaryophyllene $\xrightarrow{h\nu}$ caryophyllene

[6.183]

O, CH_3, D.A., H_2, Pd / C, H_3C, C, O

$MeO^{\ominus}$, MeOH, CH_3, O, Li, CH_3, OH, HCl, Cl, CH_3

H_2O, dioxane, OH, CH_3, H_2 - Pt, $NaNO_2$, OH, CH_3, H, PBr_3, Br, CH_3, H

Na, THF, HO

[6.184]

O

CH_3 D.A. CH_3 O REFORM. COOEt OH CH_3

Ac_2O COOEt OAc CH_3 Δ H COOEt CH_3 H_2 COOEt CH_3 H

$LiAlH_4$ OH CH_3 H ϕ_3CCl / Py $OC\phi_3$ CH_3 H 1) B_2H_6 2) CrO_3 O $OC\phi_3$ CH_3 H

CH_3I / ϕ_3CK O $OC\phi_3$ CH_3 H H_2O O OH CH_3 H ClTs / Py O OTs CH_3 H

NaI / acetone O I CH_3 H Na / THF HO

[6.185]

O X M $M^{\oplus}$ $O^{\cdot -}$ X HO

3. Synthesis of Yamada (1979) [6.186]

[6.186]

COOH; $\phi_3P{=}CH_2$, CH_2N_2; $COOCH_3$; H_2, Pd / C; $COOCH_3$; $LiAlH_4$; OCH_3

CH_2OH; C.C.P.; CHO; E.G.; Li, NH_3; OCH_3

$MeO^{\ominus}$, MeOH; AcOH; CHO; ≡

CHO; tBuOK; OH; Zn, HCl; OH; CrO_3, Py

O; L.D.A., MoO_5, Py, HMPT; HO; 1) NaH, 2) MeI, DME; CH_3O; 1) MeLi, 2) $SOCl_2$

CH_3O; CH_2N_2, Cu chelate; CH_3O; H_2, PtO_2, CH_2N_2; CH_3O

CrO_3, AcOH; HO

4. Synthesis of Näf and Ohloff (1974)

The starting products for this synthesis are 2,6,6-trimethyl–2,4-cyclohexadien–1-one, which was obtained at that time by the α-methylation of the lithium salt of 2,6-dimethylphenol according to the method of Curtin and Stein and racemic 3-methyl–4-pentenyl lithium, which was prepared from crotyl bromide by a Grignard reaction according to the following scheme [6.187]:

[6.187]

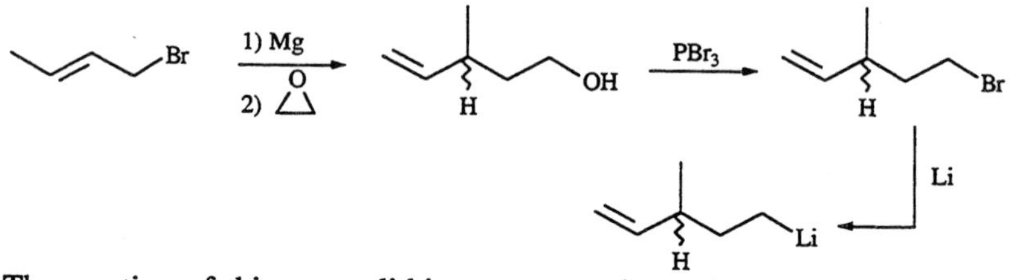

Br
1) Mg
2) O
OH
H
PBr$_3$
Br
H
Li
Li
H

The reaction of this organolithium compound on trimethyl cyclohexadienone leads to a mixture of diastereoisomeric dienols $\underline{D}_1$ and $\underline{D}_2$ [6.188].

[6.188]

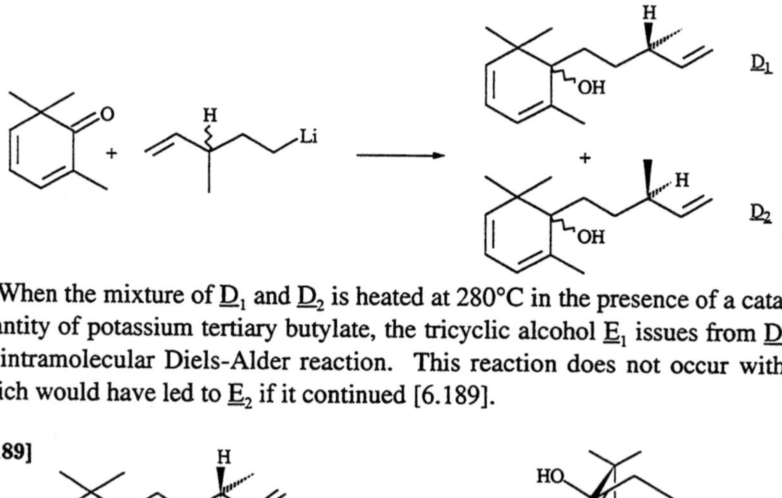

O
H
Li
H
OH
$\underline{D}_1$
H
OH
$\underline{D}_2$

When the mixture of $\underline{D}_1$ and $\underline{D}_2$ is heated at 280°C in the presence of a catalytic quantity of potassium tertiary butylate, the tricyclic alcohol $\underline{E}_1$ issues from $\underline{D}_1$ via an intramolecular Diels-Alder reaction. This reaction does not occur with $\underline{D}_2$, which would have led to $\underline{E}_2$ if it continued [6.189].

[6.189]

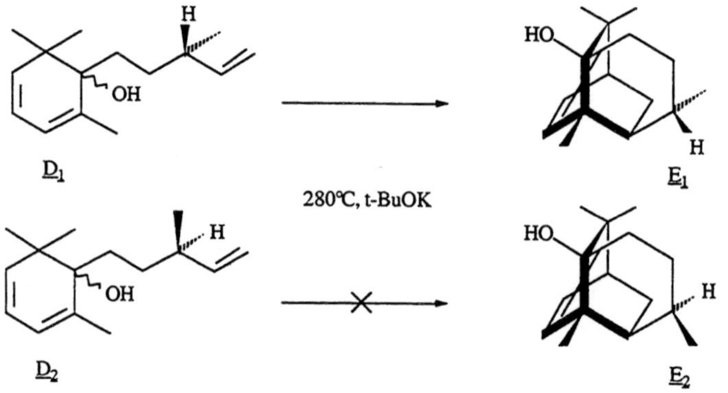

H
OH
$\underline{D}_1$
HO
H
$\underline{E}_1$

280°C, t-BuOK

H
OH
$\underline{D}_2$
HO
H
$\underline{E}_2$

An explanation for this difference of reactivity can be seen when we compare the transition state of the Diels-Alder reactions of each of these dienols [6.190].

[6.190]

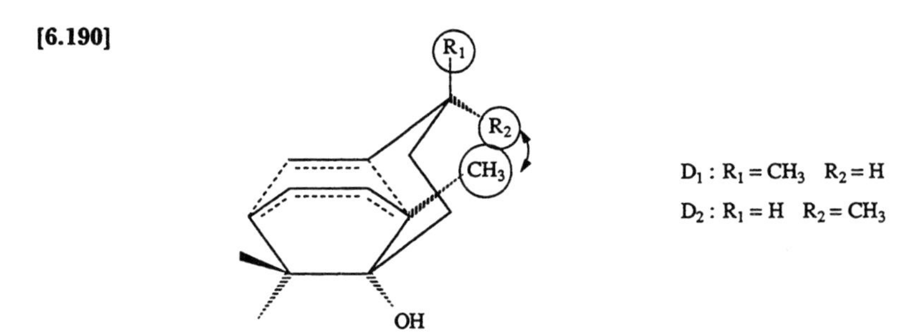

When $R_2 = CH_3$, which is the case in $\underline{D}_2$, a strong steric interaction between the two methyl groups occurs and this is sufficient to impede the formation of the transition state which would have led to $\underline{E}_2$.

Finally, catalytic hydrogenation of $\underline{E}_1$ gives racemic patchoulol [6.191].

[6.191]

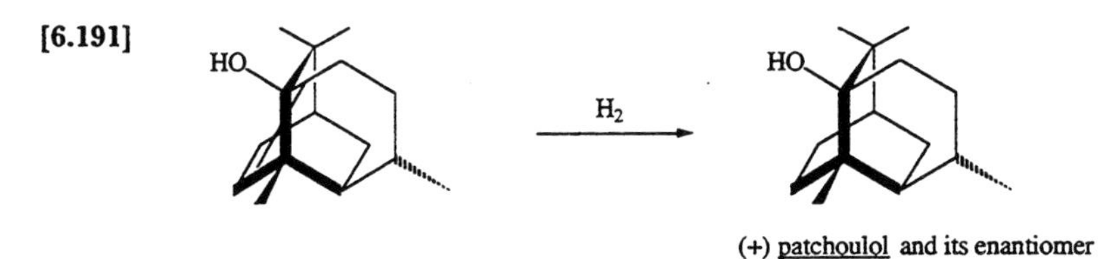

In 1981, the same authors, while reacting trimethylcyclohexadienone with organolithiums prepared from optically active bromides, were able to obtain selectively the two enantiomers (+) and (–) of patchoulol. All of the other reaction parameters were retained unchanged.

G. Syntheses of (±) norpatchoulenol

In 1974, we carried out the first synthesis of norpatchoulenol using a synthetic scheme which was inspired by that used by Mirrington and Schmalzl for the synthesis of patchoulol. It also involved the cyclization of an ε-haloketone in one of its steps [6.192].

Given the difficulty of achieving norpatchoulenol by purely chemical means, it seemed important to obtain it, or rather to obtain a precursor, by biochemical means using patchoulol. This took note of the large difference between the concentrations of these two alcohols in the essential oil of patchouli.

It is well known that steroids and terpenoids can be hydroxylated by microorganisms. Thus, for example, the enzymatic oxidation of camphor by *Pseudomonas putida* gives 5-hydroxy-camphor. In this precise case, the enzyme implicated in this transformation has been isolated and identified as having a hemoprotein as its principal component, cytochrome P_{450}.

[6.192]

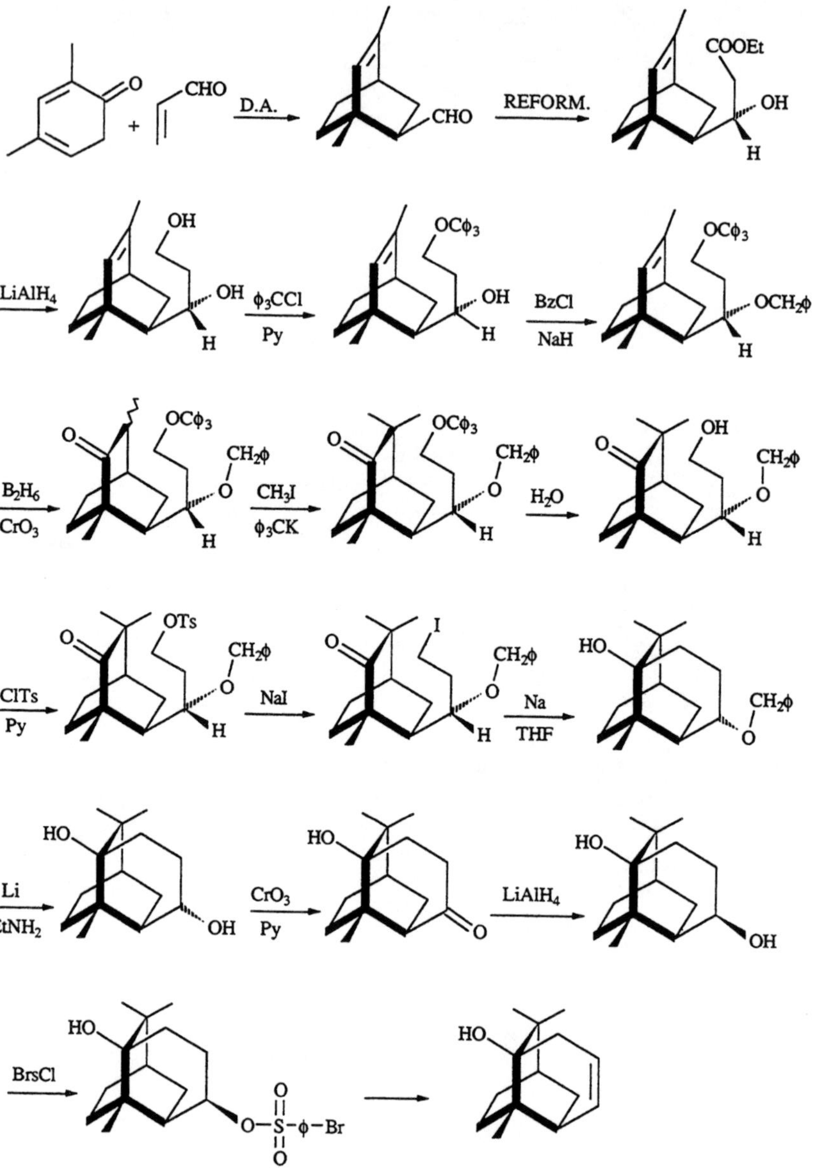

In cases such as these, however, hydoxylation generally occurs at a ring carbon, whereas our strategy necessitated attack on a methyl group.

The presence of two types of methyls in patchoulol led us to believe that they would behave differently with regard to biological oxidation. Thus, if hydroxylation was to occur specifically on the secondary methyl group of patchoulol, the diol H would result and this could be rapidly converted into norpatchoulenol [6.193].

[6.193]

HO

CH_2OH

H

Several microorganisms of vegetable origin, including some taken from the leaves of patchouli, were tested in collaboration with Guillot from Paris, Schuep from Basel and Fujiwara from Kamakura in Japan. The results of these investigations show that the reaction does not, unfortunately, occur with the regiospecificity we had hoped for. While some species gave relatively low yields (17–20%) of the desired diol H, these experiments generally gave a mixture of the four diols I, J, K and L [6.194].

[6.194]

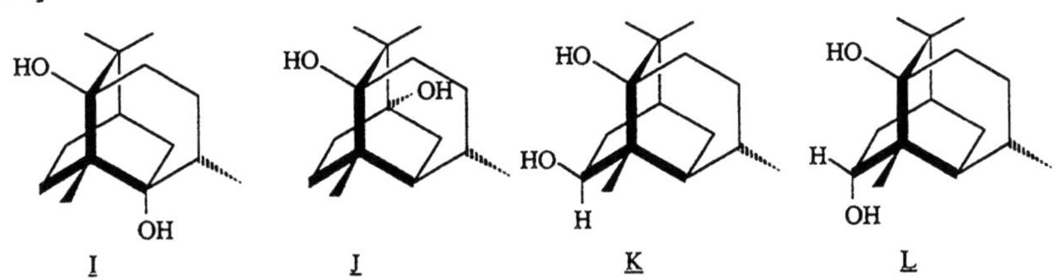

In 1975, studies on animal biological oxidation were undertaken by Bang and Ourisson in Strasburg. After having fed rabbits with patchoulol, they succeeded in isolating two products from the urine of these animals (following acidification to pH 4.5 and hydrolysis of the glucuronates)—on one hand, the hydroxyacid M and on the other, the diol H, which can be converted into M by chemical or catalytic oxidation.

The structure of these two products was confirmed by correlation with patchoulol according to the following scheme [6.195]. Finally, the hydroxyacid M treated with lead tetra-acetate in pyridine in the presence of cupric acetate (method of Bacha and Kochi) gives norpatchoulenol.

Synthesis of norpatchoulenol by intramolecular Diels-Alder reaction

In 1978, Oppolzer synthesized norpatchoulenol by means of an intramolecular Diels-Alder reaction, based on the synthesis of patchoulol by Näf and Ohloff.

Synthesis of Oppolzer (1978) [6.196]

Finally, in 1980, in collaboration with Bertrand, we made a new synthesis of norpatchoulenol starting with trimethylcyclohexadienone: The key step in this syn-

[6.195]

HO
CH_2OH
H
HO
COOH
M
ClMs
LiAlH$_4$
$(AcO)_4Pb$
$(AcO)_2Cu$
HO
CH_2OMs
HO
COOMe
HO
norpatchoulenol
LiAlH$_4$
HO
patchoulol

[6.196]

O
$Li^{\oplus}$
+
$OSiEt_3$
OR
$OSiEt_3$
KF
MeOH
OR
O
Δ
RO
O
AcOH
H_2O
HO
O
1) TsNHNH$_2$
AcOH
2) MeLi
HO

thesis is a cyclization by intramolecular nucleophilic substitution with an SN_2' type allylic rearrangement from a ketyl radical anion. The mixed acetal substituent plays the role of leaving group in this process. This is in fact “an alternative solution to the cyclization of the ε-haloketones” [6.197].

[6.197]

CHO D.A. O CHO 92% H_2 Pd / C O CHO 95%

1) NaH 2) pH 7,0 O CHO 90% MgBr THF O H OH 69%

Cl TEMMA CH_3CN O H O—CH_2 O—CH_3 95% Na THF HO 51%

The Chemistry of Diterpenes with 14-Membered Rings

The last class of large-sized ring terpenoids is that of the diterpenoids. As always, on the basis of the rule of terpene construction, the diterpenes can have theoretically a 15-membered ring as a maximum. Compounds such as this are not known, although several containing 14-membered rings have been known for some years.

The first representative of this series was isolated in 1951 by Haagen-Smit et al. from the oleoresins of *Pinus albicaulis* and other *Pinus* species. This substance is called cembrene and Dauben prepared the compound with the structure A. Unfortunately, cembrene does not give an adduct with silver nitrate and, as a consequence, the stereochemistry of the double bonds could not be rapidly confirmed with certitude. It was only as recently as 1969 that its definitive structure could be determined by a study involving X-ray diffraction by a monocrystal of the hydrocarbon itself. This structure B is slightly different from that proposed by Dauben [6.198].

Structure B is more compatible with the biogenesis of this product starting with all trans geranyl-geraniol [6.199]. The specific reduction of the "conjugated diene" part of cembrene gives two hydrocarbons [6.200].

A second group of important products which have the "cembrane" skeleton has been isolated from tobacco. There are four glycols and two epoxy-alcohols:

[6.198]

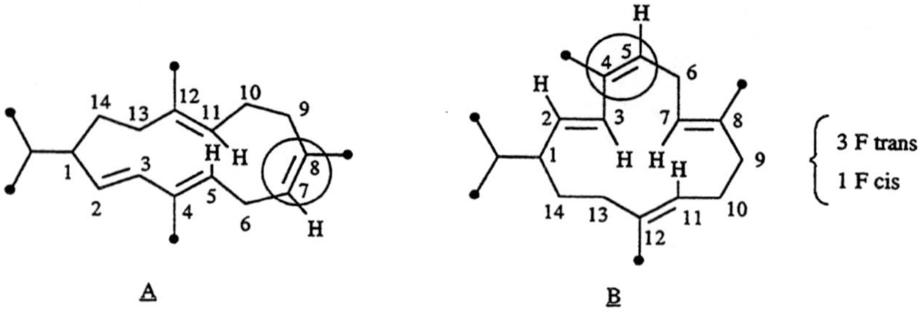

A

B

3 F trans
1 F cis

[6.199]

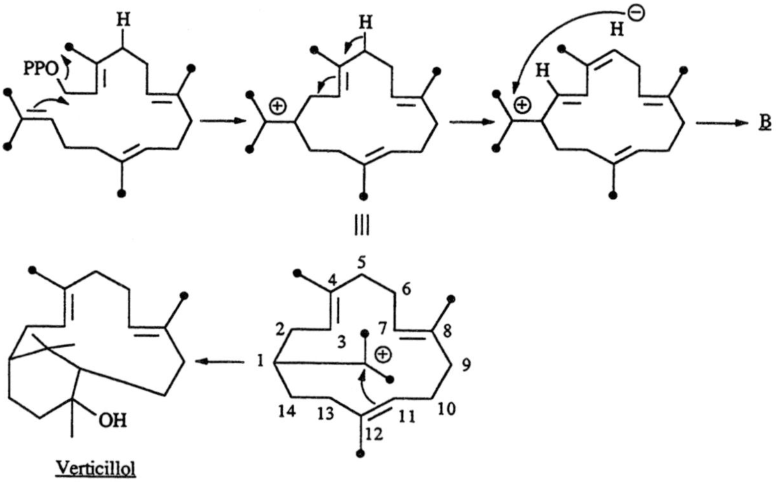

PPO

B

Verticillol

[6.200]

B

Li / NH_3

C

+

D

UV: no longer any diene system

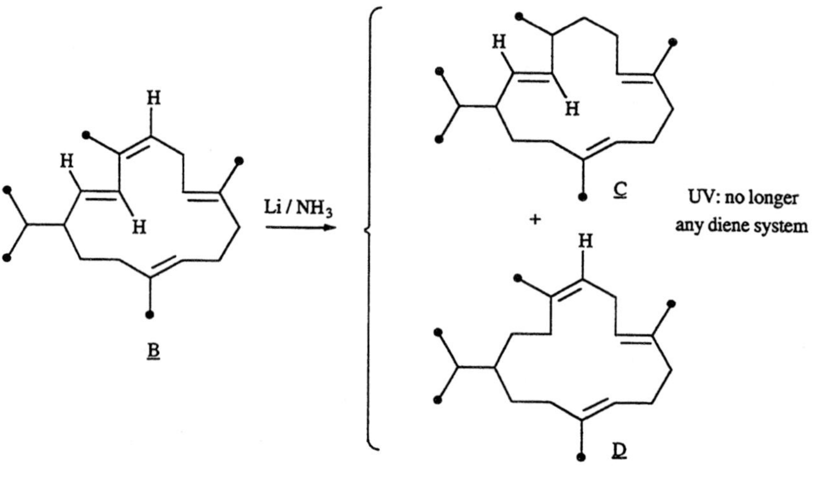

- the α- and β–2,7,11-duvatriene–4,6-diols E
- the α- and β–2,6,11-duvatriene–4,8-diols F
- the two epoxy-alcohols G and H [6.201].

[6.201]

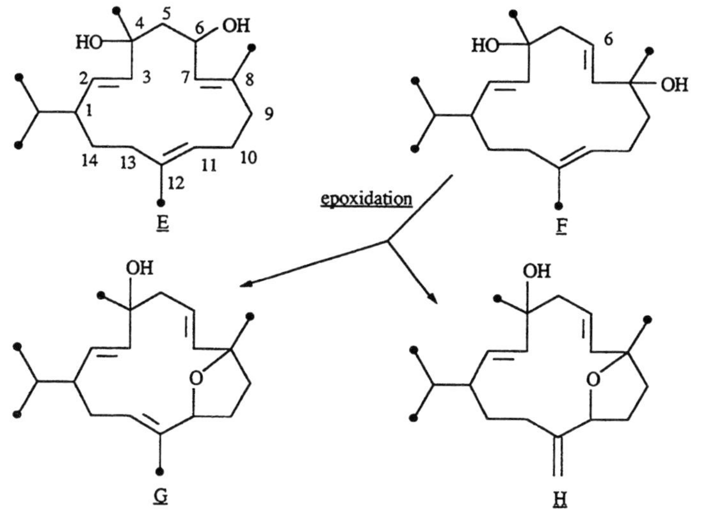

G: l-isopropyl–4,8,12-trimethyl–8,11-oxido–2,6,12-cyclotetradecatriene–4-ol
H: l-isopropyl–4,8-dimethyl–12-methylene–5,8-oxido–2,6-cyclotetradecadi-ene–1-ol.

Recently, it has been shown that these six substances are accompanied in tobacco by a diterpenic aldehyde I with an irregular structure resulting from acid cleavage of the duvatriene diols E [6.202].

[6.202]

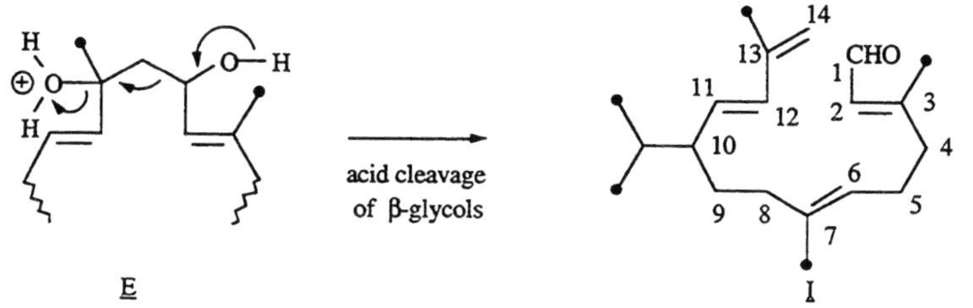

Another diterpenic epoxy-alcohol, incensole J has been isolated from "incense" (olibanum from *Boswellia carteri*) along with other triterpenic substances. This constitutes a rather rare example of the coexistance of di- and triterpenic substances, particularly as it is known that the biosynthesis of these two families of terpenoids is completely different [6.203].

Finally, the last representative of this class of products is a bactereocide of marine origin, eunicine K, which has a cembrane skeleton. Its structure has been confirmed by X-ray analysis of its iodoacetate L [6.204].

[6.203]

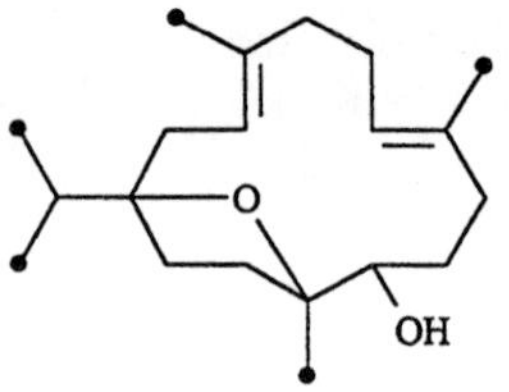

incensole I

[6.204]

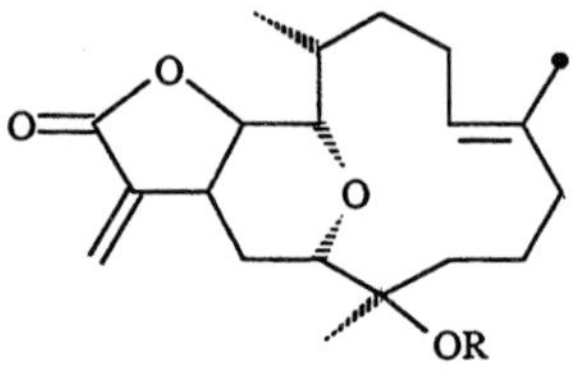

R = H eunicine K

R = —C(=O)—CH_2I L

General principles of the synthesis of cembrene B

The crucial step is the preparation of the macrocyclic ketone 1 [6.205].

[6.205]

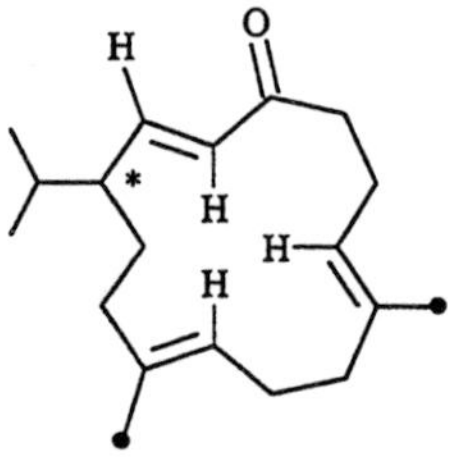

1

This ketone is obtained by cyclization of the allylic dibromide 2 in the presence of $Ni(CO)_4$ [6.206].

[6.206]

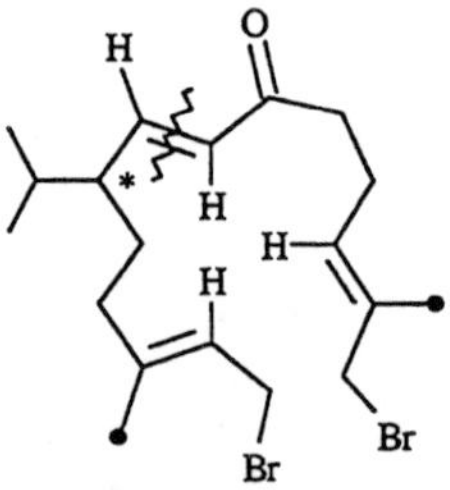

2

The problem here is the preparation of this dibromide and, for this, it is necessary to synthesize the two subunits a and b [6.207].

A Horner-Emmons reaction between these two subunits leads to the formation of the trans-enone c with a yield of 31%. Twenty-four percent of the aldehyde a and 27% of the keto-phosphonate b are recuperated [6.208]. Acid-hydrolysis of the bis-tetrahydropyranyl ether c liberates the corresponding glycol and this, under the action of PBr_3, gives the desired allylic dibromide 2.

Returning to the syntheses of the two subunits a and b:

[6.207]

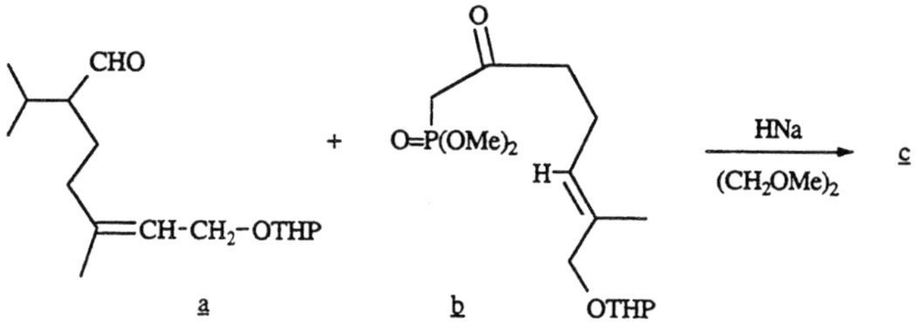

[6.208]

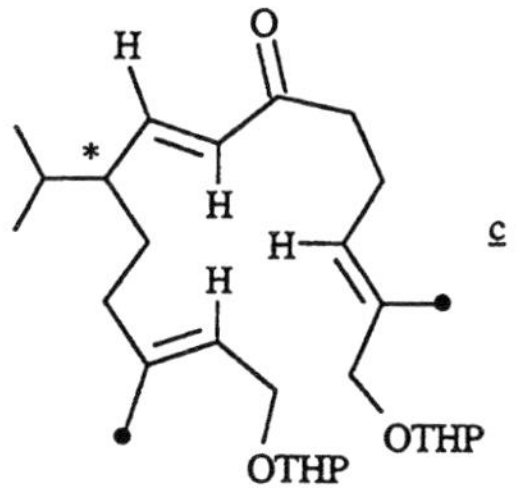

Synthesis of subunit a.

The starting material is optically active piperitone [6.209].

[6.209]

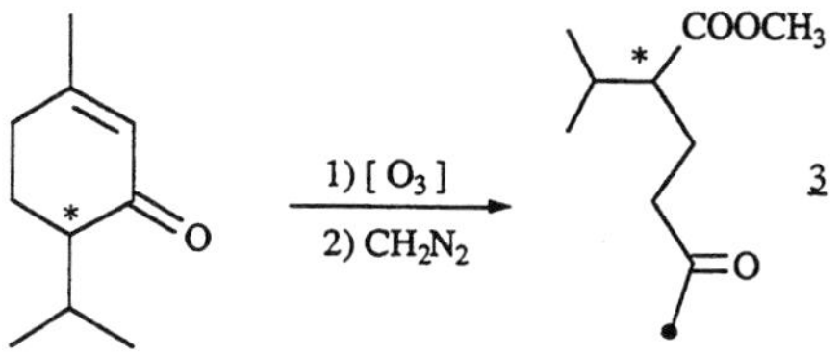

Its ozonolysis furnishes a ketoacid which is converted by methylation into the Δ-ketoester 3. This undergoes an aldol condensation according to the method of Wittig, with the lithium derivative of the imine (Schiff base) of acetaldehyde and cyclohexylamine [6.210].

[6.210]

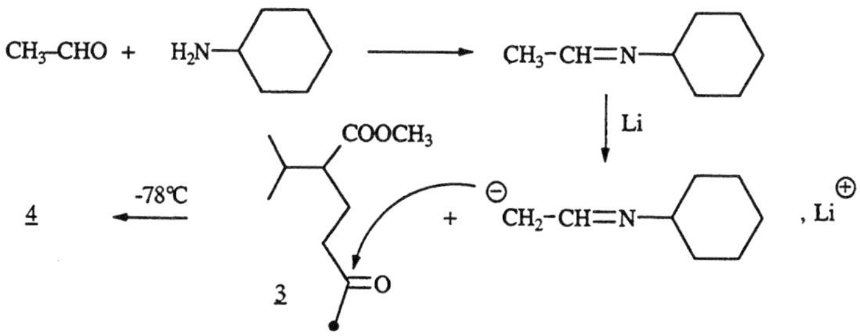

The derivative 4 is obtained in this way and its imine function is hydrolyzed to give the β-hydroxy aldehyde 5 [6.211].

[6.211]

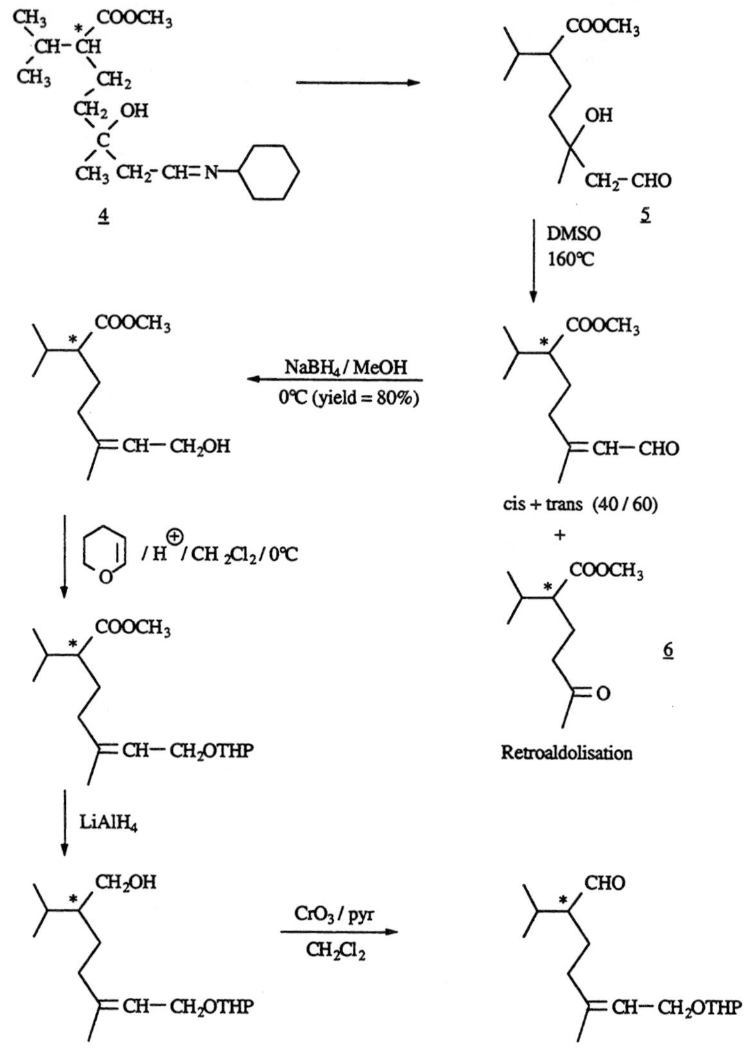

CH_3 $COOCH_3$ CH–CH* CH_3 CH_2 CH_2 OH C CH_3 CH_2–CH=N–(cyclohexyl)

4

$COOCH_3$ OH CH_2–CHO

5

DMSO 160°C

$COOCH_3$ =CH–CHO

cis + trans (40 / 60)

+

$COOCH_3$ =O

6

Retroaldolisation

$NaBH_4$ / MeOH 0°C (yield = 80%)

$COOCH_3$ =CH–CH_2OH

(dihydropyran) / $H^{\oplus}$ / CH_2Cl_2 / 0°C

$COOCH_3$ =CH–CH_2OTHP

$LiAlH_4$

CH_2OH =CH–CH_2OTHP

CrO_3 / pyr CH_2Cl_2

CHO =CH–CH_2OTHP

a

Synthesis of subunit b [6.212]

Once the macrocyclic ketone 1 is available, the last step is to transform it into isocembrol D using, for example, the action of methyllithium in hexane. Dehydration of this alcohol then gives cembrene B along with a small quantity of isocembrene e [6.213].

[6.212]

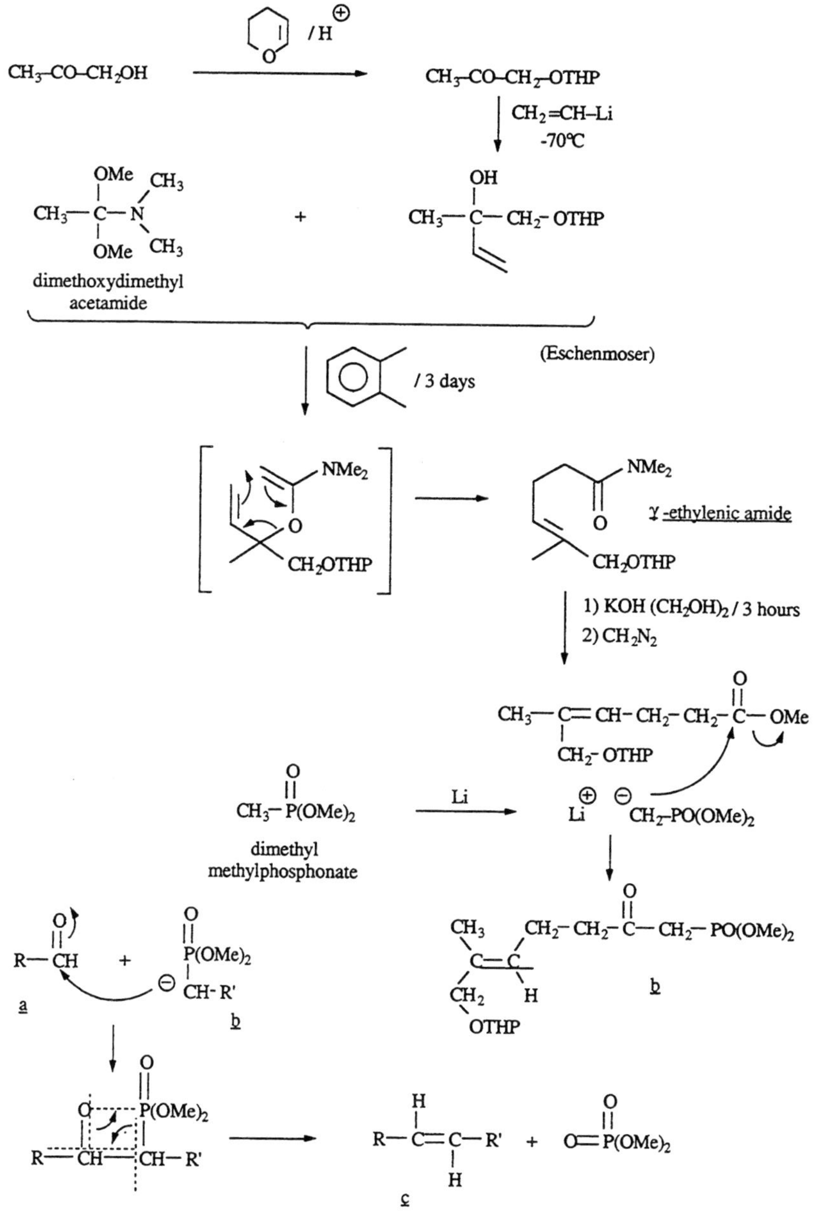

CH_3–CO–CH_2OH → (dihydropyran / $H^{\oplus}$) → CH_3–CO–CH_2–OTHP

CH_2=CH–Li, -70°C

CH_3–C(OH)(CH=CH_2)–CH_2–OTHP

CH_3–C(OMe)$_2$–N(CH_3)$_2$ +

dimethoxydimethyl acetamide

(o-xylene) / 3 days

(Eschenmoser)

NMe$_2$

O

CH_2OTHP

NMe$_2$

O

γ-ethylenic amide

CH_2OTHP

1) KOH $(CH_2OH)_2$ / 3 hours

2) CH_2N_2

CH_3–C(=CH–CH_2–CH_2–C(=O)–OMe)–CH_2–OTHP

CH_3–P(=O)(OMe)$_2$

dimethyl methylphosphonate

Li

$Li^{\oplus}$ $^{\ominus}CH_2$–PO(OMe)$_2$

CH_3(CH$_2$–OTHP)C=C(H)–CH_2–CH_2–C(=O)–CH_2–PO(OMe)$_2$

b

R–CH=O (a) + $^{\ominus}$CH(R')–P(=O)(OMe)$_2$ (b)

O---P(OMe)$_2$

R–CH–CH–R'

R–CH=CH–R' (c) + O=P(=O)(OMe)$_2$

[6.213]

MeLi

1

d

+

cembrene B

isocembrene e

CHAPTER

7

Substances with a Musk Odor

Introduction

The products of animal origin having a musk odor are generally obtained from the sexual glands of certain mammals. These extremely odorant substances play an important role in the sexual life of these animals and are produced in special glands.

In most animals these microscopic glands are disseminated over the entire surface of the body. In some species, however, they are localized in the vicinity of the sexual organs and are considerably developed, becoming veritable "perfumery glands." Their secretion generally augments with increasing sexual activity.

We are referring here essentially to musk and civet. The musk deer or antelope (*Moschus moschiferus*) is found in central Asia, is a ruminant of the Moschidae family, and grows to the size of a goat. In the male, the gland reaches the size of an orange and is visible on the abdomen. Its secretion is a dark brown semiliquid. The pocket in which the secretion is formed and from which the musk material is discharged, only starts to develop in the adult.

When the secretion is too abundant, especially at the time of the rut, the animal rubs himself against trees to rid himself of it, leaving deposits which betray his presence by its penetrating odor. The deer is captured in traps and the gland is removed along with the surrounding skin. It is then dried in the sun and it is in this state that it is found in commerce under the name "musc en poche." The contents of the gland are called "musc en graine" because of its particulate appearance.

Another variety of natural musk is provided by the odorant glands of the American muskrat (*Ondarta zibethicus rivalicius*). This animal is found widely spread over certain areas and after World War II, six million were captured in the state of Louisiana alone. The odorant glands are found in animals of both sexes [7.01].

[7.01]

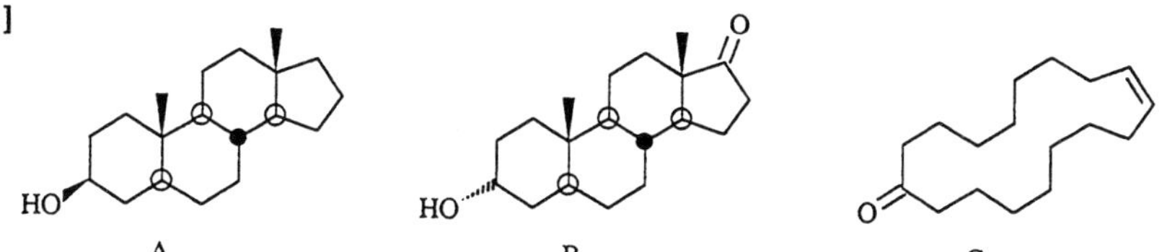

HO A   HO B   O C

In 1942, Stevens discovered a whole series of cyclic polymethylene alcohols (n = 12, 14, 16, 18) in the contents of this gland. They were accompanied by traces of cyclic polymethylene ketones in which the cyclodecaheptanyl derivatives preponderated [7.02].

[7.02]

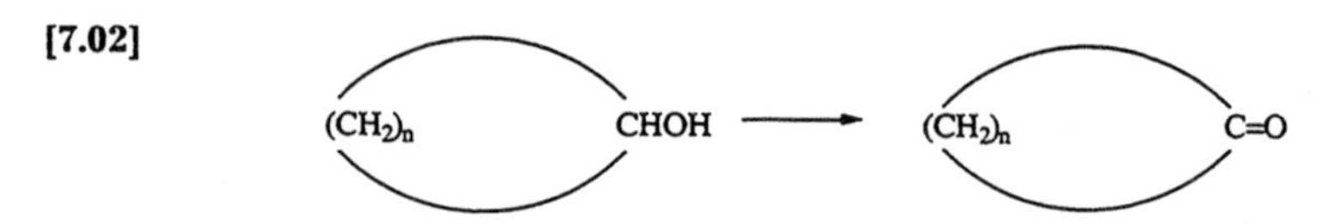

The macrocyclic alcohols were found to be odorless, the product taken from the gland of the muskrat being weakly odorant; however, its oxidation gives a very odorant product rich in cycloheptadecanone.

It is also noteworthy that Prelog and Ruzicka discovered a non-macrocyclic product of animal origin in 1944 called androstenol A, which is contained in pig testicles. It also has a musk-like smell, although according to Stoll, this is not a pure musk odor. The structure of androsterol is related to that of androsterone B which is a urinary metabolite of the male sexual hormone. This relation is interesting because, according to the work of Guillot and Le Magnen, the perception of musk odors by women is linked to the genital cycle. It should be noted that androstenol has a molecular weight of 274 ($C_{19}H_{30}O$) and is at the limit of odorant perception.

A macrocyclic musk which will be discussed later (civettone) has the structure 2 and this can be drawn in the form C which shows a certain degree of analogy with the preceding sterols.

The term "civet" is used at the same time to describe both the animal, *Viverra civetta,* and the secretion of its odorant glands. The animal is a carnivorous mammal of the Viverridae family found in Africa and Asia. It reaches a length of 1 m and a height of 25 to 30 cm. The major part of the production imported into Europe comes from civets raised in Ethiopia. The secretion is formed in a deep pocket near the sexual organs and has the appearance of a yellow-colored grease. It is removed by weekly or monthly curettage according to the season and the age of the animal. Civet arrives on the European market in buffalo horns containing around 1 kg of the odorant material.

In addition to these two principal musk-odor materials of animal origin, there are two others of vegetable origin, the essential oils of angelica and ambrette.

Angelica (*Angelica archangelica* officinalis Hoffm) is a large plant of the Umbelliferae family and is cultivated in Belgium, Holland, Germany, Hungary, France and in the north of India. The dried roots and the fruit (seeds) are treated separately. An absolute and an essential oil are prepared from the roots while the fruit are only used as a source for the production of an essential oil.

Ambrette (*Abelmoschus moschatus* Moench, syn; *Hibiscus abelmoschus*) is a plant from the Malvaceae family which is cultivated in Madagascar (Nossi-Bé), in the Seychelle Islands and principally in the hot regions of Equador and Colombia. As in the case of angelica, the seeds are used. A veritable absolute (redissolution in alcohol of a product which results from extraction with volatile solvents) can be made; however, this is of only minor commercial importance. Actually, the product which is commercially designated by the term "ambrette absolute" in the perfumery industry, has an entirely different origin. When the seeds of ambrette are steam distilled, they yield an essential oil that has the form of a butter called "beurre d'ambrette." This resemblance to a concrete is conferred by the presence of a high proportion of odorless palmitic acid which co-distills with the odorant principles of the seeds. The melting point of the "butter" is a function of its palmitic acid content (MP = 61°C, in similarity to the "butter" of iris). Elimination of this acid by alkaline washing gives an oily substance with a strong musk odor, which constitutes the commercial quality of "ambrette absolute."

Chemistry of musk-smelling substances

In 1896, Ciamician and Silber isolated ω-hydroxypentadecanoic acid by saponification of the oil from angelica seeds.

$$HOCH_2\text{-}(CH_2)_{13}\text{-}COOH$$

This treatment, they observed, made the characteristic musk odor of this oil disappear.

In 1906, Walbaum working with natural musk, isolated a ketone having a strong musk odor. He baptized this substance which has the molecular formula $C_{16}H_{30}O$ and which makes up 0.5 to 2% of natural musk, with the name "muscone." Later, in 1946, Ruzicka and co-workers isolated a nitrogenous base with the molecular formula $C_{16}H_{26}N$ from the same source and named it "muscopyridine." Treatment with $KMnO_4$ afforded pyridine α,α'-dicarboxylic acid, which proves that muscopyridine is an α,α'-disubstituted pyridine. Sack isolated the principle odorant which constitutes 2–3% of civet in 1915—a $C_{17}H_{30}O$ ketone which he named "civettone."

In 1926, Ruzicka demonstrated that the muscone of Walbaum is a macrocyclic ketone, β-methylcyclopentadecanone 1 and that the civettone of Sack is another macrocyclic ketone, 9-cycloheptadecenone 2. This was the first discovery of substances with such a high number of methylene groups in the ring.

The next year, Ruzicka and co-workers reported that civettol, the corresponding alcohol, is also present in civet. This odorless alcohol is transformed into civettone by oxidation. The animal odor of civet is also due in part, to the presence of other odoriferous substances such as skatole and the fatty acids.

Almost simultaneously in 1927, Kerschbaum demonstrated that the characteristic musk odor of "ambrette butter" is linked to the presence of a macrocyclic lactone which he called "ambrettolide." He established its structure as 2-oxa–11-cycloheptadecenone 3. The cis stereochemistry of the double bond in this structure has only been recently confirmed with certainty!

Returning to the studies of Ciamician and Silber, Kerschbaum attributed the musk odor of the essential oil from angelica seeds to the presence of another macrocyclic lactone, cyclopentadecanolide 4 (also called "Exaltolide").

Saponification of this lactone gives ω-hydroxypentadecanoic acid, isolated in 1896 by Ciamician and Silber [7.03].

[7.03]

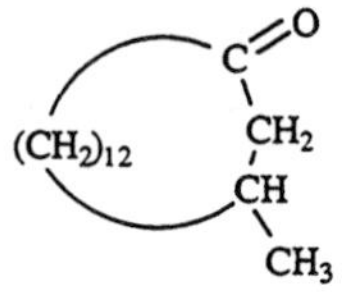

1 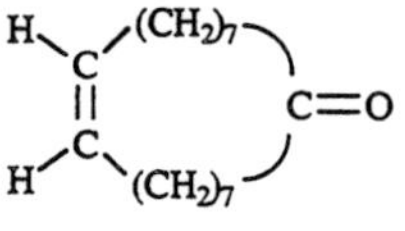 2 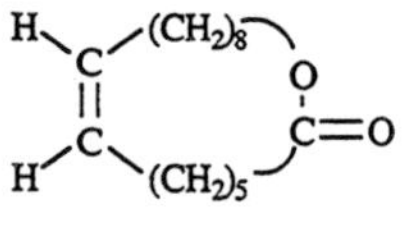 3

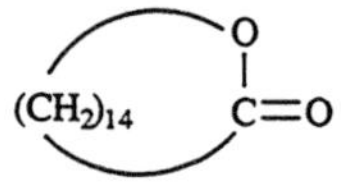

4

The presence of macrocyclic compounds was thus established for the first time nearly simultaneously in both the animal and the vegetable kingdoms. It is also remarkable that the macrocyclic ketones are found only in animal sources, while the macrocyclic lactones are restricted to botanical sources.

After having established the structures of muscone and civettone, Ruzicka and colleagues elaborated with great skill the chemistry of large ring compounds in more than 50 articles between roughly 1926 and 1950. As previously mentioned, this work was compensated when Ruzicka won the 1939 Nobel prize in chemistry for, among other things, "his research into the polymethylenes."

Hundreds of substances in this category are now known where the rings have 14 to 18 members: ketones, lactones, carbonates, diesters, oxides, anhydrides, formals, aldehydes, imines, sulfates, alcohols, etc., with the exception of the alcohols, which also have odors that are more or less musk-like in character.

Of all of these, the ketones and the lactones present by far the greatest importance from the point of view of their olfactive properties. Substances with 11- and 12-membered rings have moldy and camphor odors. When the ring size exceeds 18 units the odor disappears almost completely.

It is therefore understandable that a special effort has been devoted to finding economical methods for manufacturing these two groups of substances: the macrocyclic ketones and lactones. Each of these groups shall be reviewed in turn.

Methods of Synthesis of Macrocyclic Ketones

A. Method of Ruzicka (generalized Piria reaction)

The first attempt to synthesize these ketones was made in 1926 by Ruzicka, Stoll and Schinz by means of the thermal decomposition of rare earth salts of α,ω-dicarboxylic acids (thorium, cerium, ytterbium, etc). The yields obtained are indicated in Table I.

Table I

Ketones	Yields
C_7	45–50%
C_8	20–25%
C_9–C_{13}	0.1 . . . 0.5%
C_{14}–C_{18}	in the order of a few %
C_{18}	decreases again

Examination of Table I shows that yields obtained by this method for ketones containing more than nine ring elements are very low. Furthermore, the cyclization of rare earth salts does not permit a synthesis of muscone because the presence of any methyl groups in the neighborhood of the carboxyl function prevents any chance of cylization.

The difficulties of synthesizing rings with more than nine elements are related to causes which will now be discussed. Before doing so, we should examine Table II, which shows the heats of combustion of gaseous cycloalkanes as well as their total strain energies relative to the corresponding aliphatic hydrocarbons.

This table displays the stability of the different cyclic systems. This stability, however, is in no way synonymous with the facility of forming these cyclic compounds. Among the factors that can intervene, the most important concerns the statistical probability of the extremities of the aliphatic chain approaching each other, which is, of course, imperative for cyclization to occur. This probability decreases as the size of the potential ring increases and therefore can be considered an unfavorable entropy of activation for the formation of medium- and large-sized rings.

Table II

Cycloalkanes $(CH_2)_n$	n	Heat of combustion ($Kcal.mole^{-1}$)	Total strain ($Kcal.mole^{-1}$)
Cyclopropane	3	499.83	27.6
Cyclobutane	4	655.86	26.3
Cyclopentane	5	793.52	6.5
Cyclohexane	6	944.48	0.0
Cycloheptane	7	1,108.20	6.4
Cyclooctane	8	1,269.20	10.0
Cyclononane	9	1,429.50	12.9
Cyclodecane	10	1,586.00	12.0
Cycloundecane	11	1,742.40	11.0
Cyclododecane	12	1,892.40	9.6
Cyclotridecane	13	2,051.40	5.2
Cyclotetradecane	14	2,203.60	0.0
Cyclopentadecane	15	2,362.50	1.5
Cyclohexadecane	16	2,520.00	1.6
n - alcanes		157.4 n + 58.2	0.0

Thus, the ease of formation of cyclic systems depends on two factors. The first is the entropic factor which diminishes monotonally and the second is the enthalpic factor which represents the ring strain. In this way, it can be noted on examining Figure 1, that this second factor increases from three-membered to six-membered rings, then diminishes until we reach nine-membered rings, and finally progressively increases again as the ring size increases. The free energy of cyclization is the resultant of these two factors ($\Delta G = \Delta H - T\Delta S$). Figure 7.04 shows the influence of the enthalpic and entropic effects on the yield of cyclization.

[7.04]

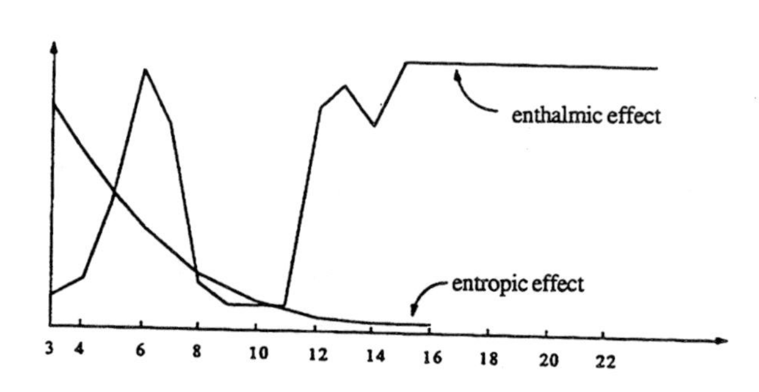

These factors can be illustrated by the following simple examples. The 1,2-halohydrins lead readily to the formation of oxiranes (three-membered rings), while the oxetanes (four-membered rings) form, with much more difficulty, from 1,3-halohydrins. However, oxiranes are much more easily opened than oxetanes due to the much lower thermal stability of three-membered rings compared to four-membered rings. These results are clarified in Figure 7.05.

Another characteristic example concerns the intramolecular condensation of diethyl adipate and pimelate in the presence of sodium (Dieckmann cyclization). Under these identical conditions, diethyl adipate gives 2-ethoxycarbonylcyclopentanone (Dieckmann ester) in a yield over 85%. Diethyl pimelate, on the other hand, gives 2-ethoxycarbonylcyclohexanone in a yield of only 60%. In this case as well, the second β-ketoester is more stable than the first [7.05].

When a cyclic substance with more than nine ring elements is synthesized, a situation arises in which there are two competing reactions. The first is a polycondensation reaction involving a sequence of bimolecular reactions. The second is the cyclization reaction inevitably involving a monomolecular reaction (indeed, this is the case regardless of the size of the ring).

To be more precise, by designating the rates of cyclization and dimerization as v_c and v_d and the molar concentration of the open-chain monomeric precursor as m, it can be shown that:

$$v_c = k_c m \qquad (1)$$
$$v_d = k_d m^2 \qquad (2)$$

[7.05]

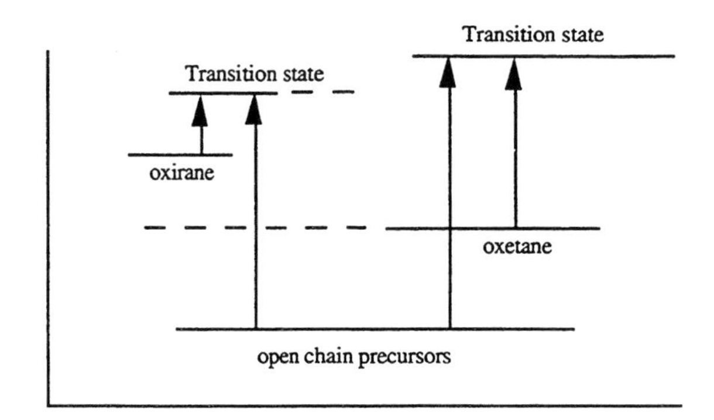

The constant k_c is clearly inferior to k_d because of the very unfavorable entropic factor. As a result, v_c is much smaller than v_d when the concentration is high.

$$\text{If } m = 1, \text{ then } v_c = k_c \text{ and } v_d = k_d.$$

However, if the concentration diminishes; for example,

$$\text{if } m = 1/10, \text{ then } v_c = k_c/10 \text{ and } v_d = k_d/100$$
$$\text{if } m = 1/100, \text{ then } v_c = k_c/100 \text{ and } v_d = k_d/10{,}000.$$

Therefore, when the solutions are sufficiently dilute, v_c can become greater than v_d even though k_c is clearly less than k_d. This "principle of large dilutions" was discovered experimentally in 1912 by Rüggli.

In summary, the yield of the cyclization reaction is equal to the ratio of the reaction rates and is therefore inversely proportional to m:

$$p = v_c/v_d = k_c/k_d \cdot 1/m. \qquad (3)$$

Following the method of Ruzicka, which has conserved only a historical interest, we shall study in detail the principal cyclization reactions which give access to macrocyclic systems. Nearly all of these methods make use of the principle of large dilutions. Figure 3 indicates, for some of these methods, the yields obtained as a function of the number of ring CH_2 elements [7.06].

B. Method of Ziegler (generalized Thorpe reaction)

It was necessary to wait until the work of Ziegler and co-investigators in 1933 before a method was available for synthesizing macrocyclic ketones in satisfactory yields.

These researchers generalized the Thorpe reaction (type of internal Claisen reaction) by cyclizing α,ω-dinitriles in the presence of a lithium amide as base. They applied the principle of large dilutions (equation 3) to this reaction and they studied in detail the problem of the ease of cyclization as a function of chain length. With this method, they were able to synthesize, among other things, racemic muscone with a yield of 55% [7.07].

[7.06]

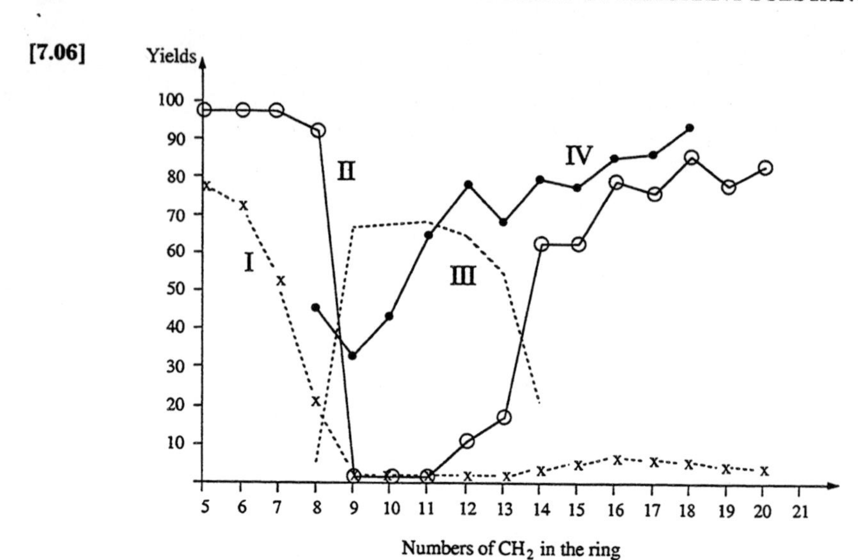

I) Cyclization of dicarboxylic acids according to Ruzicka
II) Cyclization of α,ω-dinitriles according to Ziegler
III) Dimerizing cyclization of dinitriles to give a diketone
IV) Acyloin cyclization according to Stoll

[7.07]

$(CH_2)_{11}$ ring: $CH_2-C\equiv N$ / $\overset{\ominus}{C}H-CN$ / $CH-CH_3$ $\xrightarrow{B^{\ominus}}$ $(CH_2)_{11}$ ring: $CH_2-C{=}NH$ / $CH-CN$ / $CH-CH_3$

$\downarrow$

$(CH_2)_{11}$ ring: $CH_2-C{=}O$ / CH_2 / $CH-CH_3$ (1) $\longleftarrow$ $(CH_2)_{11}$ ring: $CH_2-C{=}O$ / $CH-COOH$ / $CH-CH_3$

The sodium derivative of bis trimethylsilylamide (hexamethyldisilazane) can also be used as base, thereby giving higher yields of cyclization of the dinitriles [7.08].

[7.08]

$$(CH_3)_3-Si-\underset{\displaystyle Na}{\underset{|}{N}}-Si-(CH_3)_3$$

In this manner, suberonitrile, in etherial solution at 25°C, is cyclized to cycloheptanone with a yield of 96%. The base can be prepared by the addition of hexamethyldisilazane to a suspension of sodium amide in benzene.

C. Method of Hunsdiecker

In 1942, Hunsdieker published a new synthesis of macrocyclic ketones. By continuous addition of a ω-halogenated β-ketoester to a large volume of boiling butanone in the presence of a large excess of sodium carbonate, a good yield is obtained of a cyclic β-ketoester. This can then be hydrolized with decarboxylation to give the cyclanone [7.09].

[7.09]

a) $$XCH_2-(CH_2)_n-COCl + \underset{\displaystyle CO-CH_3}{CH_2}-COOC_2H_5 \xrightarrow[CH_3OH]{NaOCH_3} XCH_2-(CH_2)_n-CO-\underset{\displaystyle CO-CH_3}{CH}-COOC_2H_5$$

b) $$\xrightarrow[butanone]{Na_2CO_3} X-CH_2-(CH_2)_n-CO-\overset{\ominus}{C}H-COOC_2H_5 \longrightarrow \overline{CH_2-(CH_2)_n-CO-CH}-COOC_2H_5$$

$$\xrightarrow{H_3O^{\oplus}} \overline{(CH_2)_{n+1}-CO-CH_2}$$

The group X is generally a bromine atom. The reaction of the acid chloride group with the β-ketoester is much faster than that of the terminal bromine which has generally low reactivity (bromo paraffin). Nonetheless, $CH_3O-(CH_2)_n-CO-CH_2-COOC_2H_5$ is a by-product of the ketonic scission reaction of the β-ketoester.

In fact, when X = Br, the yield of the cyclization reaction is low. Also, the bromine is exchanged with iodine (a much better leaving group) after the step involving condensation with ethyl acetoacetate. The yields vary from 40 to 75% for the rings from C_{14} to C_{17}. This method has been applied to the synthesis of racemic muscone using the β-methyl glutaric ester and 11-methoxyundecanoic acid [7.10].

In the Hunsdiecker reaction, Rüggli's principle of large dilutions is equally important. It was later shown that the condensing agent was not, after all, potassium carbonate but rather its contaminant, potassium hydroxide—which explains the need for an enormous excess of carbonate employed by Hunsdiecker.

D. Method of Blomquist

The group of Blomquist et al. also sought ways to obtain macrocyclic ketones—by the cyclization of bis ketenes. Their method also employed the principle of large dilutions, although the yields were generally lower than with the preceding method. Before describing this method, it may be useful to recall that ordinary ketenes dimerize spontaneously to give diketenes, which under the action of alcoholates, give β-ketoesters [7.11].

Taking the synthesis of muscone as an example of the application of Blomquist's method, 3-methylhexadecanoic acid is transformed first into acid

[7.10]

$$CH_3O(CH_2)_{10}-COOH + HOCO-CH_2-CH(CH_3)-CH_2-COOCH_3 \xrightarrow{\text{KOLBE}}$$

$$CH_3O(CH_2)_{11}-CH(CH_3)-CH_2-COOMe \xrightarrow[\text{AcOH}]{\text{HBr}} Br(CH_2)_{11}-CH(CH_3)-CH_2-COO$$

$$\xrightarrow{SOCl_2} Br(CH_2)_{11}-CH(CH_3)-CH_2-COCl \xrightarrow[\text{2) MeONa / MeOH}]{\text{1) } CH_3COCH_2COOC_2H_5}$$

$$Br(CH_2)_{11}-CH(CH_3)-CH_2-CO-CH_2-COOC_2H_5 \xrightarrow[CH_3COC_2H_5]{NaI}$$

$$I(CH_2)_{11}-CH(CH_3)-CH_2-CO-CH_2-COOC_2H_5 \xrightarrow[68\%]{K_2CO_3}$$

$$(CH_2)_{11}-CH(CH_3)-CH_2-CO \text{ (ring closed via) } CH-COOC_2H_5 \xrightarrow[80\%]{H_2SO_4}$$

$$(CH_2)_{12}-CH(CH_3) \text{ / } O{=}C-CH_2 \text{ (cyclic ketone)} + CO_2 + C_2H_5OH$$

[7.11]

$$CH_2{=}C{=}O + CH_2{=}C{=}O \longrightarrow \text{(diketene: } CH_2{=}C-CH_2 \text{ / } O-C{=}O\text{)} \xrightarrow{RO^{\ominus}} CH_3-\overset{O}{\overset{\|}{C}}-CH_2-\overset{O}{\overset{\|}{C}}-OR$$

chloride. This is then introduced into a large volume of diethyl ether in the presence of triethylamine. An α,ω-diketene is obtained as a transient intermediate before undergoing intramolecular dimerization under these conditions [7.12].

[7.12]

$$CH_3-CH(CH_2-COCl)-(CH_2)_{11}-CH_2-COCl \longrightarrow CH_3-CH(CH{=}C{=}O)-(CH_2)_{11}-CH{=}C{=}O \longrightarrow \text{A}$$

A: $CH_3-CH-CH-C{=}O$ / $(CH_2)_{11}$ / $CH{=}C-O$ (four-membered ring, attacked by $RO^{\ominus}$)

$$CH_3-CH(CH{=}C{=}O)-(CH_2)_{11}-CH{=}C{=}O \longrightarrow \text{B}$$

B: $CH_3-CH-CH{=}C-O$ / $(CH_2)_{11}-CH-C{=}O$ (four-membered ring, attacked by $RO^{\ominus}$)

When treated with an alcoholate, the intracyclic dimers A and B give two β-ketoesters C and D which, on hydrolysis, lead to the β-methyl ketone. In conclusion, the result is the same irrespective of the manner in which the α,ω-diketene dimerizes. The formation of the β-methyl ketone is "symmetrized" by the formation of the β-ketoesters [7.13].

[7.13]

$$\underline{A}\ (+\underline{B}) \xrightarrow{RO^{\ominus}} \begin{array}{c} CH_3 \\ | \\ CH-CH-COOR \\ |\quad\quad | \\ (CH_2)_{12}-C{=}O \end{array} + \begin{array}{c} CH_3 \\ | \\ CH-CH_2-CO \\ |\quad\quad\quad | \\ (CH_2)_{11}-CH-COOR \end{array}$$

$$\downarrow H_3O^{\oplus}$$

Muscone 1

Stalbberg-Stenhagen performed a synthesis of the two optical isomers of muscone after separating the two enantiomers of 3-methylhexadecanoic acid. The odor of the two enantiomers of muscone has been judged to show no difference.

E. Method of Prelog and Stoll (generalized acyloin synthesis)

The methods just discussed all present a major disadvantage in that they give mediocre yields in highly diluted media.

This was surmounted in 1943 by the groups of Prelog and Stoll thanks to an adaptation of the acyloin synthesis. This method was originally developed by Bouveault and Locquin for the aliphatic series and already elucidated by Hansley for the alicyclic series. Yields are generally excellent.

The acyloin reaction is about 300 times more rapid than the preceding reactions. It is essentially for this reason that yields are near to 90%, even when the speed of introduction is maintained at around 100 g of the diester per liter and per hour. The lowest yield of around 30% is obtained for 9-membered rings. For 8- and 10-membered rings, it already exceeds 40% (Figure 3).

The reaction sequence shown below describes the application of Stoll's method for the synthesis of (dl)-muscone [7.14].

Before carrying out this synthesis, it was necessary to perfect ways of dehydrating acyloins to α,β-ethylenic ketones. The action of methyl magnesium bromide on this ketone (method of Kharash and Lambert) leads to a mixture of muscone and the tertiary ethylenic alcohol in the ratio 1 to 5. It is very probable that the method of House should lead almost exclusively to (dl)-muscone. This method consists of adding lithium dimethyl cuprate [$Li(CH_3)_2Cu$ or magnesium cuprate] to the α,β-ethylenic ketone with the result that 1,4-addition occurs almost entirely.

It may be useful to return to the mechanism of the "acyloin" condensation. To do this, we should first of all consider an ethyl ester that is subjected to the action of a metal such as sodium. Two separate cases can be anticipated depending on

[7.14]

$$(CH_2)_{13}(COOR)_2 \xrightarrow[Et_2O]{4Na} (CH_2)_{13}[C(ONa){=}C(ONa)] + 2\ NaOR \longrightarrow (CH_2)_{13}[C{=}O,\ CHOH]$$

$$\xrightarrow{Al_2O_3} (CH_2)_{12}[C{=}O,\ CH{=}HC] \xrightarrow[(\text{sels } Cu^+)]{BrMgCH_3} (CH_2)_{12}[C{=}O,\ CH_2,\ CH(CH_3)] + (CH_2)_{12}[C(OH)-CH_3,\ CH{=}HC]$$

whether the reaction is carried out in the presence or the absence of protic solvents.

In the presence of protic solvents

a) reaction of Bouveault and Blanc (C_2H_5OH):

$$R\text{-}COOC_2H_5 \rightarrow R\text{-}CH_2OH$$

b) hydrogenolysis:

$$R\text{-}COOC_2H_5 \rightarrow R\text{-}COOH + C_2H_6$$

In the presence of certain solvents (H_2N-Et) and preferably in the presence of lithium.

In the presence of aprotic solvents

[$(C_2H_5)_2O$, for example], there is the formation of an acyloin

$$2\ R\text{-}COOC_2H_5 \rightarrow R\text{-}CO\text{-}CHOH\text{-}R$$

As a general rule, the action of a metal such as sodium on an ester results in a mesomeric radical anion B ↔ C. The evolution of this radical anion will depend on the nature of the solvent [7.15].

[7.15]

$$\underset{\underline{A}}{R-C(=O)-OC_2H_5} + Na^{\bullet} \longrightarrow \underset{\underline{B}}{R-\overset{\bullet}{C}(OC_2H_5)-O^{\ominus}} \longleftrightarrow \underset{\underline{C}}{R-\overset{\ominus}{C}(OC_2H_5)-O^{\bullet}\cdots\cdots Na^{\oplus}}$$

radical anion

Protic media

a) the radical anion (B ↔ C) can first accept a second electron from the metal to give a dianion D which expulses the EtO⁻ group, with the remaining R-C-=O group then being protonated to give an intermediate aldehyde which is then reduced by the same mechanism to give the alcohol [7.16].

[7.16]

$$[\underline{B} \longleftrightarrow \underline{C}] + e^{\odot} \longrightarrow R{-}\overset{\ominus}{C}(OC_2H_5){-}\overset{\ominus}{O} \cdots\cdots \overset{\oplus}{Na} \longrightarrow C_2H_5\overset{\ominus}{O} + R{-}\overset{\ominus}{C}{=}O$$

$\underline{D}$

$$R{-}\overset{\ominus}{C}{=}O \xrightarrow{H^{\oplus}} R{-}C(H){=}O \longrightarrow R{-}CH_2OH$$

This is the Bouveault and Blanc reaction.

b) the radical anion (B ↔ C) protonates first; this is the case, for example, when NH_2-Et and lithium are present. Under these conditions, the radical E is then formed and this accepts another electron from the metal to give the anion F [7.17, 18].

[7.17]

$$[\underline{B} \longleftrightarrow \underline{C}] + H^{\oplus} \longrightarrow R{-}\overset{\odot}{C}(OC_2H_5){-}OH + \overset{\odot}{Met} \longrightarrow R{-}\overset{\ominus}{C}(O{-}C_2H_5){-}OH \longrightarrow R{-}C(=O){-}OH + \overset{\ominus}{C_2H_5}$$

$\underline{E}$ $\quad$ $\underline{F}$

[7.18]

$$R{-}C(=O){-}OH + \overset{\ominus}{C_2H_5} \xrightarrow{H^{\oplus}} R{-}COOH + C_2H_6$$

This is a hydrogenolysis reaction.

Aprotic media [$(C_2H_5)O$, toluene, etc.]

The radical anion dimerizes through its mesomer B to afford the dianion G which expulses two $C_2H_5O^-$ anions to give the intermediate diketone H [7.19].

[7.19]

$$2\left[R{-}\overset{\odot}{C}(C_2H_5O^-){-}\overset{\ominus}{O}\right] \longrightarrow \begin{matrix} C_2H_5O^- \\ R{-}C{-}O^{\ominus} \\ R{-}C{-}O^{\ominus} \\ C_2H_5O^- \end{matrix} \longrightarrow \begin{matrix} R{-}C{=}O \\ R{-}C{=}O \end{matrix}$$

$\underline{G}$ $\quad$ $\underline{H}$

The action of the metal transforms H into the bis radical anion J [7.20]. Subsequent protonation of this bis radical anion in its mesomeric form J' leads to the acyloin K [7.21]. If the condensation reaction is carried out in the presence of trimethylsilyl chloride $(CH_3)_3SiCl$, the following compound is obtained [7.22]:

[7.20]

$$\begin{matrix} R-C=O \\ | \\ R-C=O \end{matrix} \xrightarrow{2\,Na^{\odot}} \begin{matrix} R-\overset{\odot}{C}-O^{\ominus} \\ | \\ R-\underset{\odot}{C}-O^{\ominus} \end{matrix} \longleftrightarrow \begin{matrix} R-\overset{\ominus}{C}-O^{\odot} \\ | \\ R-\underset{\ominus}{C}-O^{\odot} \end{matrix}$$

$\underline{I}$

$$\updownarrow$$

$$\begin{matrix} R-C-O^{\ominus} \\ || \\ R-C-O^{\ominus} \end{matrix} \quad \underline{I'} \qquad Na^{\oplus}$$

[7.21]

$$\begin{matrix} R-C-O^{\ominus} \\ || \\ R-C-O^{\ominus} \end{matrix} \xrightarrow{H^{\oplus}} \begin{matrix} R-C-OH \\ || \\ R-C-OH \end{matrix} \rightleftharpoons \begin{matrix} R-C=O \\ | \\ R-C-OH \end{matrix}$$

$\underline{K}$

[7.22]

$$\begin{matrix} R-C-OSi(CH_3)_3 \\ || \\ R-C-OSi(CH_3)_3 \end{matrix}$$

In cases where the two R-groups are different, hydrolysis of the bis-trimethylsilyl ether can lead to one of the two possible acyloin products.

The particularly high yield of the acyloin method in concentrated solution seems surprising at first glance. Prelog has explained this in terms of the reaction mechanism schematically represented in Figure 4. The two electrophilic carbon atoms at the extremities of the diester chain are first adsorbed onto the surface of the sodium covered with electrons. If the dimensions of the molecule permit, the electrophilic residues can approach each other by sliding along the metal surface, which finally results in cyclization. The sliding of the electrophilic groups once adsorbed, requires less energy than their desorption from the metal surface. Once the cyclization has taken place, the molecule no longer has electrophilic character and can detach itself from the metal. In fact, it will remain bound electrostatically because, after losing four electrons to the diester, the sodium will have acquired a positive charge and the diester will finally be transformed into a dianion of the type $\underline{I'}$. In this sense, the fourth phase of the scheme proposed by Prelog can possibly be criticized. All in all, however, after taking account of the thermodynamic analysis discussed earlier, it is possible to say that the sodium has the effect of considerably diminishing the entropic factor in this reaction and, for this reason, it increases significantly the rate constant k_c in equation 1.

Although the acyloin condensation gives high yields, the acyloins are fragile substances and their transformation into ketones is relatively difficult [7.23].

It is possible, as we have seen earlier, to proceed beforehand with catalytic dehydration on Al_2O_3 according to the procedure of Stoll. At the Roure Bertrand Dupont Research Center, we obtained excellent results by distilling the acyloins over H_3PO_4 under very high vacuum. The cyclenones thus obtained can then be hydrogenated into cyclanones.

[7.23]

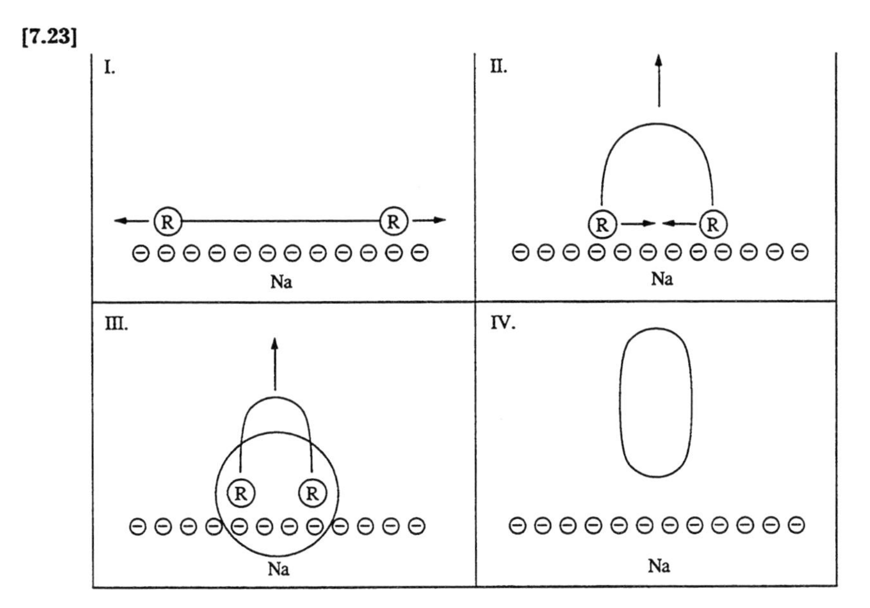

It is also possible to treat the acyloins by distilling them over zinc powder according to the old method of Baeyer. In this case, however, a mixture of the cyclanone and the corresponding hydrocarbon is always obtained. The method of Zinin gives medium yields; the acyloin is treated with hydrochloric acid in dioxan and then the resulting chloro-ester is reduced by zinc turnings.

In 1955, Cram and Cordon perfected a method for obtaining this conversion under good conditions. The acyloin, or better still, its acetate, is treated with 1,3-propandithiol, hydrochloric acid and zinc chloride in benzene solution. With either the acyloin or its acetate, a simple mercaptal is formed, the hydroxyl or acetyl groups having been reduced. The mercaptals obtained in this way can then be hydrolyzed to the ketones. They can also be desulfurated into macrocyclic hydrocarbons by Raney nickel [7.24].

In 1948, Stoll and co-workers, using the acyloin method, made the first total synthesis of civettone which was identical to natural civettone. The principle of this synthesis, which starts with oleic acid, is indicated in the following scheme [7.25].

Hydrobromic acid in anhydrous acetic acid gives two diastereomeric bromo acetates a + a'. Zinc reacts with this mixture to give the dioxolanes of cis- and trans-civettone. These two dioxolanes can be crystallized and even separated by fractional crystallization. Alternatively, the totality of the product can be used and transformed into the dioxolane of dehydrocivettone, which by semi-hydrogenation and hydrolysis yields only cis-civettone. This reaction sequence therefore assures the cis geometry of natural civettone.

In 1943, Hunsdiecker described a synthesis of civettone starting with aleuritic acid extracted from shellac. The civettone that he obtained was not the same as

[7.24]

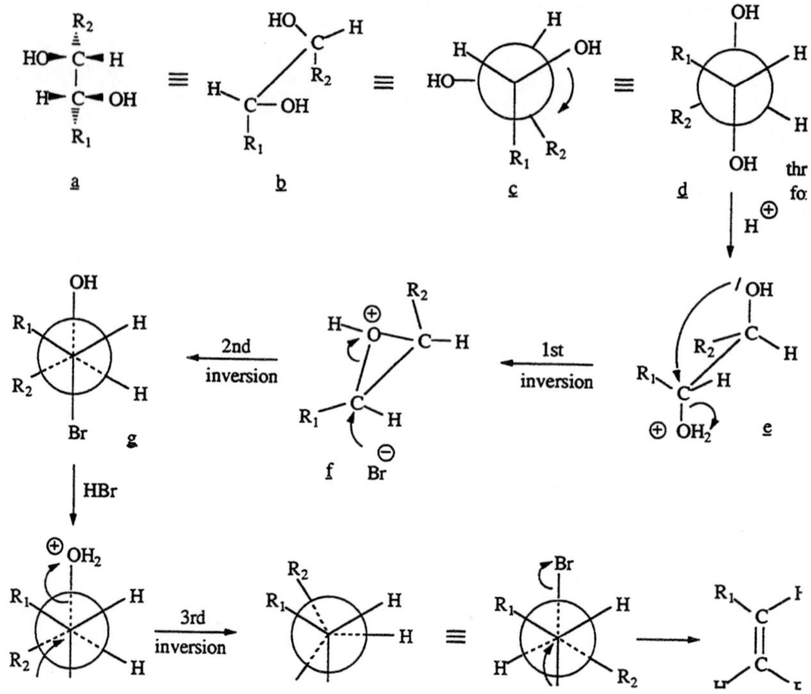

a ≡ b ≡ c ≡ d

$\xrightarrow{H^{\oplus}}$ e $\xrightarrow{\text{1st inversion}}$ f $\xrightarrow{\text{2nd inversion}}$ g $\xrightarrow{HBr}$ $\xrightarrow{\text{3rd inversion}}$

thr fo

[7.25]

$CH_3-(CH_2)_7-CH=CH-(CH_2)_7-COOC_2H_5 \xrightarrow{1)\,[O_3]\quad 2)\,[O]} HOCO-(CH_2)_7-COOC_2H_5$

$\xrightarrow{\text{Piria}} O=C[(CH_2)_7-COOC_2H_5]_2 \longrightarrow$ (ethylene ketal) $\longrightarrow$ (acyloin: $(CH_2)_7-C=O$ / $(CH_2)_7-CHOH$) $\longrightarrow$ (diol: $(CH_2)_7-CHOH$ / $(CH_2)_7-CHOH$) $\xrightarrow[AcOH]{HBr}$

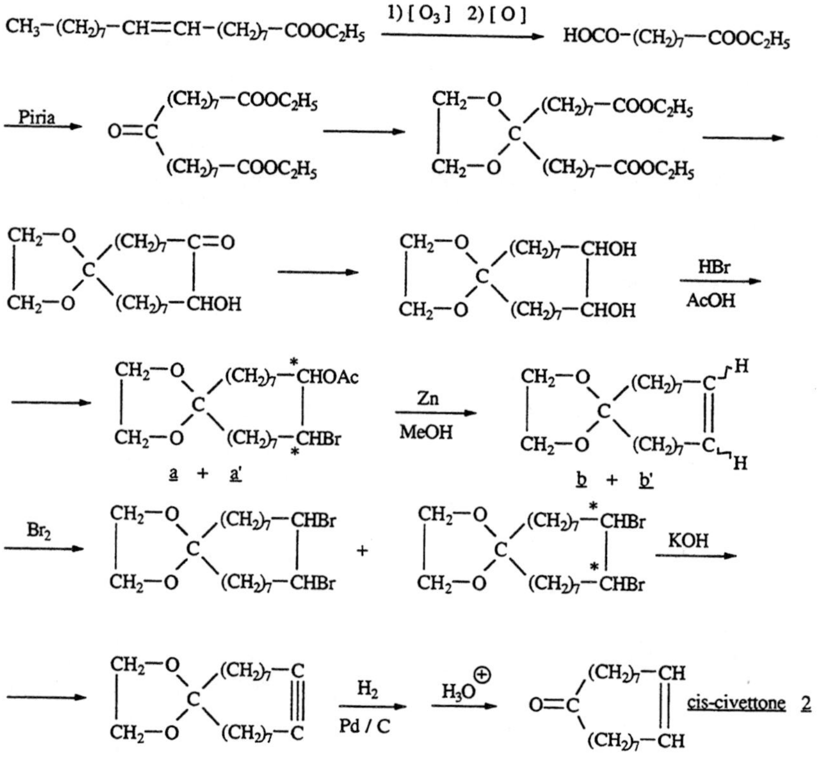

$\longrightarrow$ ($(CH_2)_7-\overset{*}{C}HOAc$ / $(CH_2)_7-\overset{*}{C}HBr$) a + a' $\xrightarrow[MeOH]{Zn}$ ($(CH_2)_7-CH=CH-(CH_2)_7$, trans) b + b'

$\xrightarrow{Br_2}$ ($(CH_2)_7-CHBr$ / $(CH_2)_7-CHBr$) + ($(CH_2)_7-\overset{*}{C}HBr$ / $(CH_2)_7-\overset{*}{C}HBr$) $\xrightarrow{KOH}$

$\longrightarrow$ ($(CH_2)_7-C\equiv C-(CH_2)_7$ ketal) $\xrightarrow[Pd/C]{H_2}$ $\xrightarrow{H_3O^{\oplus}}$ $O=C[(CH_2)_7-CH=CH-(CH_2)_7]$ cis-civettone 2

natural civettone. In fact, in 1943, the stereochemistry of aleuritic acid was not known and it was therefore impossible to predict the nature of the double bond in the derived product. Following the work of McGhie and collaborators in 1968, it is now known that natural aleuritic acid has the threo configuration and is racemic; it is therefore (±)-threo aleuritic acid A. This acid must exist mainly in conformation B. Previous researchers have also shown that anhydrous hydrobromic acid in hot acetic acid transforms threo aleuritic acid into (±)-erythro–9,10,16-tribromohexanoic acid C [7.26].

[7.26]

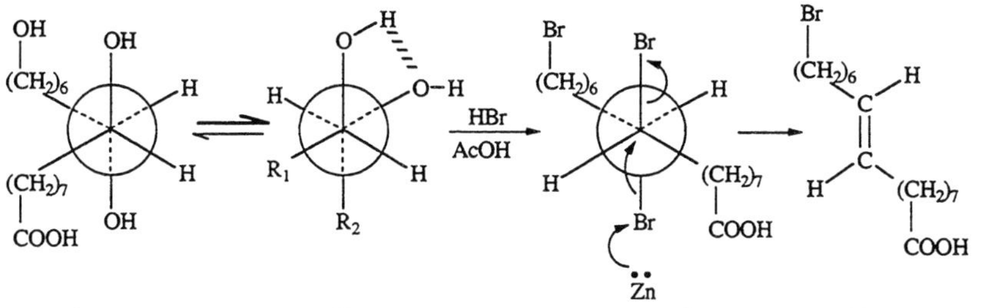

OH, OH, $(CH_2)_6$, H, H, $(CH_2)_7$, COOH, OH — A ⇌ B: O–H, H, O–H, R_1, H, R_2 — HBr / AcOH → C: Br, $(CH_2)_6$, Br, H, H, $(CH_2)_7$, COOH, Br, Zn → D: Br, $(CH_2)_6$, H, C=C, H, $(CH_2)_7$, COOH

A B C D

When treated with zinc, the compound B leads to the trans bromo ethylenic acid D.

$$Br\,(CH_2)_6\text{-}CH{=}CH\text{-}(CH_2)_7\text{-}COOH \quad \text{D}$$

t

This is converted into trans-civettone by the previously described method of Hunsdiecker [7.27].

[7.27]

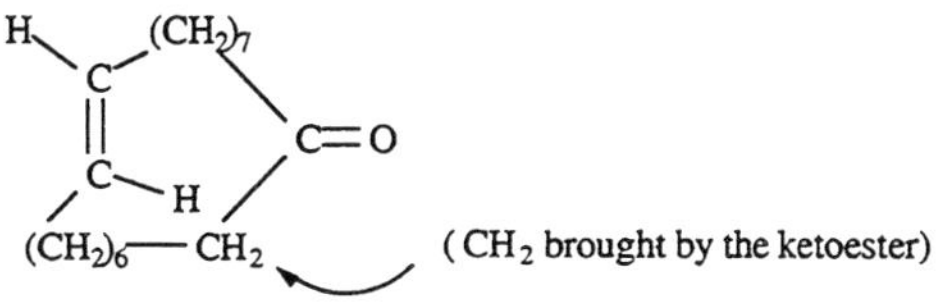

H, $(CH_2)_7$, C=C, C=O, H, $(CH_2)_6$—CH_2 (CH_2 brought by the ketoester)

Note: *A priori*, it seems surprising that threo aleuritic acid A should afford the erythro dibromo derivative C under the effect of hydrobromic acid. This apparently implies that only a single inversion of configuration or an uneven number of inversions can occur. The mechanism leading to C must therefore be more complex and must involve intermediates not isolated by McGhie's team.

The following scheme involving epoxide f as an intermediate explains the observed results, even though it has never been supported by any experimental proof [7.28].

The olefin E (trans), obtained when the erythro dibromo derivative j is treated with zinc, is the principal argument in favor of the threo configuration of aleuritic acid.

A more direct reaction sequence, which converts the acyloins into their cor-

[7.28]

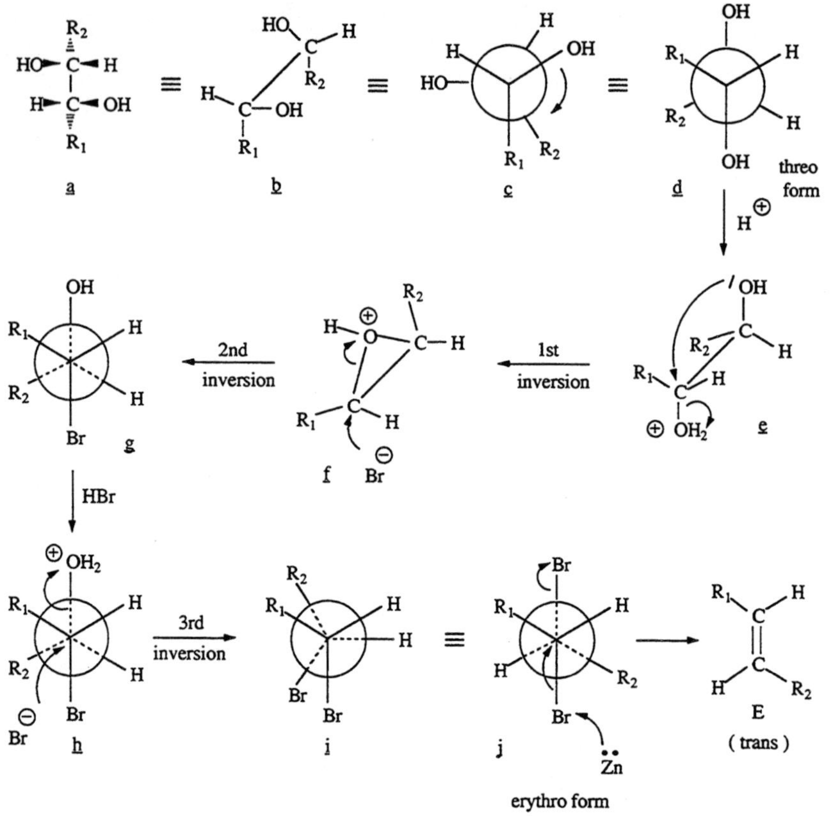

a ≡ b ≡ c ≡ d threo form

$H^{\oplus}$

e → 1st inversion → f → 2nd inversion → g

HBr

h → 3rd inversion → i ≡ j → E (trans)

erythro form

responding bisubstituted acetylenic derivatives, is outlined in the following scheme [7.29]

[7.29]

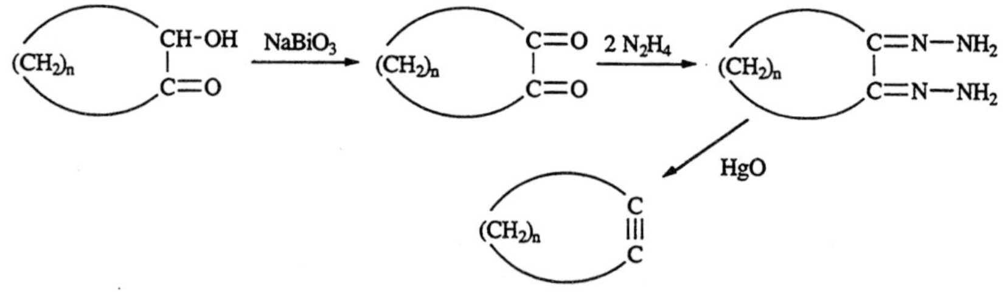

$(CH_2)_n$ CH-OH / C=O —NaBiO₃→ $(CH_2)_n$ C=O / C=O —2 N_2H_4→ $(CH_2)_n$ C=N—NH_2 / C=N—NH_2 —HgO→ $(CH_2)_n$ C≡C

While on the subject of acyloin syntheses, it should be pointed out that this method has also permitted Cram's group to attain a new class of macrocyclic compounds, the "paracyclophanes" [7.30]. It has even been possible to prepare paracyclophanes containing two aromatic nuclei [7.31].

[7.30]

$(CH_2)_m-CO_2CH_3$ … $(CH_2)_n-CO_2CH_3$ → $(CH_2)_m-C{=}O$ … $(CH_2)_n-CHOH$

m = 6 5 4 3 2 1
n = 6 5 4 4 5 6

[7.31]

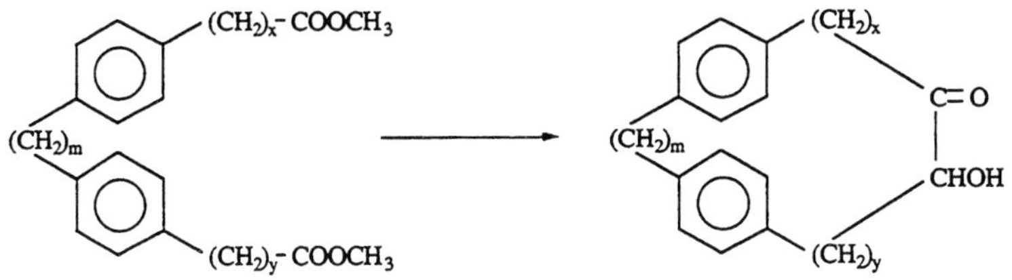

$(CH_2)_x-COOCH_3$, $(CH_2)_m$, $(CH_2)_y-COOCH_3$ → $(CH_2)_x$, $C{=}O$, $CHOH$, $(CH_2)_m$, $(CH_2)_y$

m = 1–6; x = 0–6; y = 1–5

F. Method of Goldfarb and Taits

A very interesting method of synthesizing macrocyclic ketones has been developed by a Russian group. It consists of replacing the benzene ring by a thiophene and is based on the desulfurizing reduction of thiophene derivatives using Raney nickel.

There exist two possible routes:

- bifunctional aliphatic compounds are prepared by desulfurizing reduction and the resulting compounds are then cyclized in the usual way, notably by the acyloin method.
- bi- or polycyclic compounds having a thiophene ring are first synthesized and then the sulfur atom is eliminated.

The following scheme represents the two routes [7.32]:

[7.32]

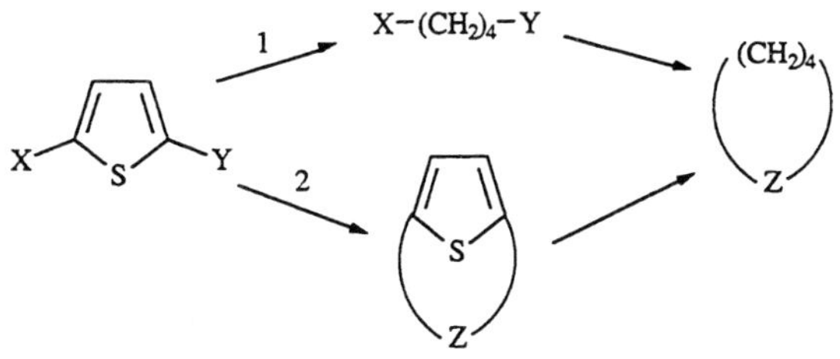

X, S, Y; 1 → $X-(CH_2)_4-Y$ → $(CH_2)_4$, Z; 2 → S, Z →

The second method is the more advantageous in that it generally gives higher yields of macrocyclic compounds when the same cyclization procedures are used. In fact, the presence of the rigid ring facilitates the cyclization by favoring the entropic effect (this effect is similar to the "Thorpe-Ingold effect"). Furthermore, the thiophene nucleus can be used for extending the chain by four units and for the incorporation of various substituents.

We will briefly examine some examples starting with the acyloin condensations. The conditions under which this cyclization is normally carried out cannot be used for the cyclization of thiophene esters. This is explicable first in terms of the sensitivity of the thiophene nucleus to sodium under the reaction conditions and second in terms of the deactivation of the surface of this metal by sulfur compounds derived from degradation of the thiophene ring.

If the reaction temperature is reduced to 50°–60°C, the yields of acyloins are greatly reduced because, under these conditions, sodium is solid and its surface is insufficiently renewed. It is possible to increase the yield by using a Na/K alloy—liquid under these conditions—but even in this case, yields remain modest when the thiophene nucleus is α,α'-bisubstituted.

When the bisubstituted derivatives are β,β' rather than α,α', however, yields increase substantially. Furthermore, if the nucleus is completely substituted, the yield of conversion can attain 100%. Examples are given in the following two schemes [7.33, 34].

[7.33]

1) $ClCO-(CH_2)_n-COOC_2H_5$ / $SnCl_4$; 2) NH_2-NH_2 / KOH; 3) CH_3OH / HCl → thiophene$-(CH_2)_n-COOCH_3$

n = 4, yield = 98%
n = 5, yield = 76%

$ClCO-(CH_2)_m-COOCH_3$ / $SnCl_4$ → (S) $(CH_2)_n-COOCH_3$, $CO-(CH_2)_m-COOCH_3$

n = m + 1 = 4, yield = 72.5%

NH_2-NH_2 / KOH → (S) $(CH_2)_n-COOH$, $(CH_2)_{m+1}-COOH$

1 + m = n = 4, yield = 97%

CH_3OH, HCl, yield = 87.5% → (S) $(CH_2)_n-COOCH_3$, $(CH_2)_m-COOCH_3$

K / Na, yield = 25–30% → (S) $(CH_2)_4-C{=}O$, $(CH_2)_4-CHOH$

[7.34]

CH_3, CH_3 (2,5-dimethylthiophene) → yield = 84% → $CO-(CH_2)_4-COOC_2H_5$ → yield = 89% → $(CH_2)_5-COOCH_3$

→ yield = 82% → $(CH_2)_5-COOCH_3$, $CO-(CH_2)_4-COOC_2H_5$ → yield = 95.5% → $(CH_2)_5-COOH$, $(CH_2)_5-COOH$

→ yield = 76.5% → $(CH_2)_5-COOCH_3$, $(CH_2)_5-COOCH_3$ → yield = 72% → $(CH_2)_5-C{=}O$, $(CH_2)_5-CHOH$

As noted, the acyloin condensation in the thiophene series is associated with serious problems when the nucleus is not completely substituted. It is possible to use another route, cyclization according to Hunsdiecker (i.e., intramolecular alkylation of ω-halo-ketoesters). The following sequence is an example of this approach [7.35].

The principle of large dilutions has also been employed for the cyclization of thiophene acid chlorides using aluminum chloride etherate in the presence of adsorbants with large surface areas [7.36].

The following table indicates the yields of ketones obtained from the acid chlorides, as a function of n.

n	Yield %	Ketones obtained
9	53.6	cyclotetradecanone
10	63.5	cyclopentadecanone or exaltone
11	62.5	cyclohexadecanone or dihydroambretolone
12	60.0	cycloheptadecanone or dihydrocivettone

G. Methods of Eschenmoser et al. and of Tanabe et al.

Two groups of researchers, one Swiss (Eschenmoser, Félix and Ohloff) and the other American (Tanabe, Growe and Dehn), discovered in 1967 a new type of fragmentation affecting the tosylhydrazones of α-epoxy ketones. This new reaction offers a route to, among others, exaltone and (d,l)-muscone. The starting material is cyclodecanone, transformed by the Stobbe reaction into 1,5-bicyclo [10.3.0] pentadecene–2-one. This reaction is summarized below [7.37]:

[7.35]

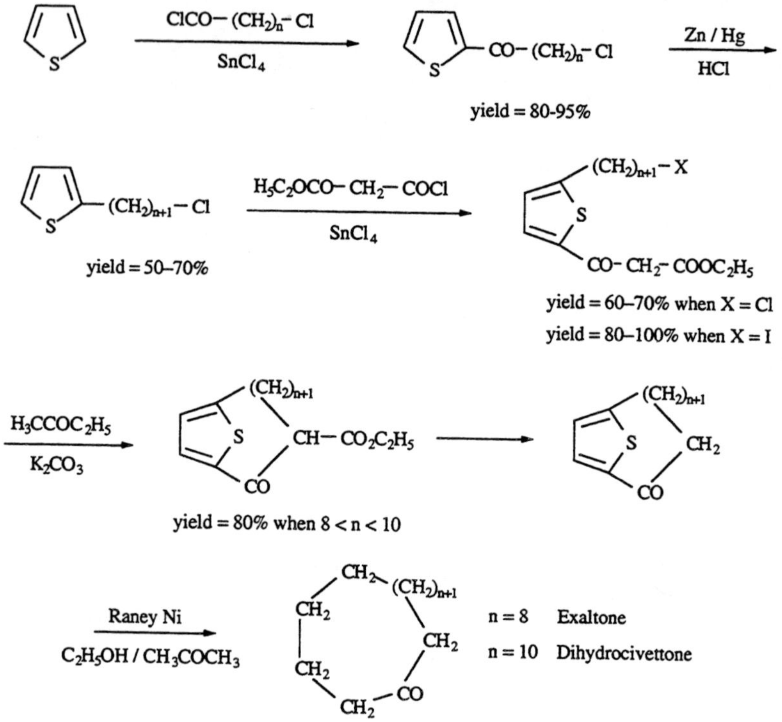

$ClCO-(CH_2)_n-Cl$ / $SnCl_4$ → $CO-(CH_2)_n-Cl$ → Zn/Hg / HCl

yield = 80-95%

$(CH_2)_{n+1}-Cl$ → $H_5C_2OCO-CH_2-COCl$ / $SnCl_4$ → $(CH_2)_{n+1}-X$, S, $CO-CH_2-COOC_2H_5$

yield = 50–70%

yield = 60–70% when X = Cl

yield = 80–100% when X = I

$H_3CCOC_2H_5$ / K_2CO_3 → $(CH_2)_{n+1}$, S, $CH-CO_2C_2H_5$, CO → $(CH_2)_{n+1}$, S, CH_2, CO

yield = 80% when 8 < n < 10

Raney Ni / C_2H_5OH / CH_3COCH_3 → CH_2, $(CH_2)_{n+1}$, CH_2, CH_2, CH_2, CH_2, CO

n = 8 Exaltone

n = 10 Dihydrocivettone

[7.36]

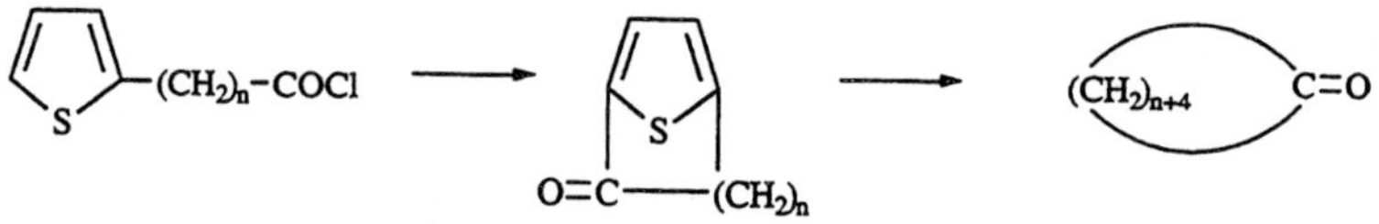

$(CH_2)_n-COCl$ → S, O=C, $(CH_2)_n$ → $(CH_2)_{n+4}$ C=O

[7.37]

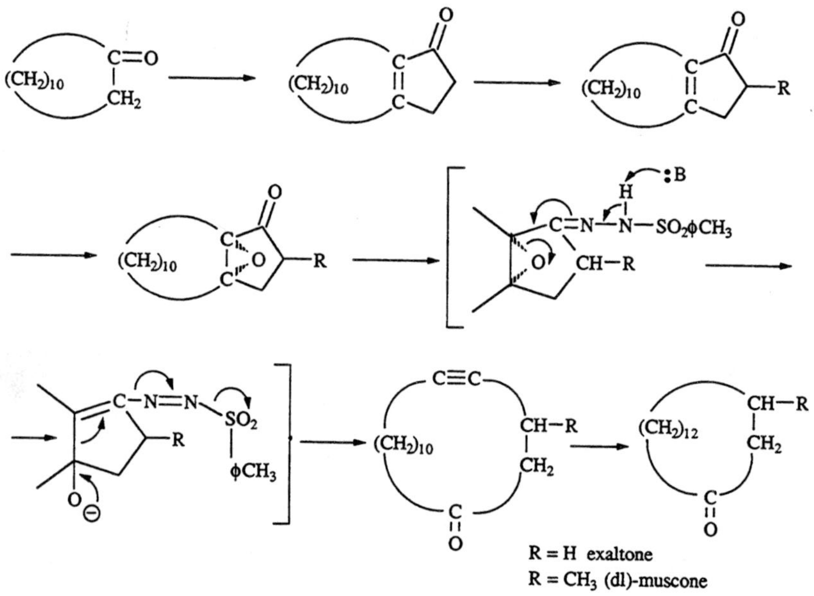

$(CH_2)_{10}$ C=O, CH_2 → $(CH_2)_{10}$ C, C, O → $(CH_2)_{10}$ C, C, O, R

→ $(CH_2)_{10}$ C, O, C, O, R → [:B, H, C=N–N–$SO_2\phi CH_3$, O, CH–R →

→ C–N=N–SO_2, ϕCH_3, R, O] → C≡C, $(CH_2)_{10}$, CH–R, CH_2, C, O → $(CH_2)_{12}$, CH–R, CH_2, C, O

R = H exaltone

R = CH_3 (dl)-muscone

The bicyclopentadecenone can then be reduced using the Wolff-Kishner reaction to form a mixture of 1,5-bicyclo [10.3.0] pentadecene (30%) and 1,2-bicyclo [10.3.0] pentadecene (70%). Ozonolysis of the former of these gives a diketone (5-oxo-exaltone, R = H). Selective reduction of one of the carbonyl groups leads to exaltone. When R = CH_3, muscone is obtained.

It is quite surprising that 5-oxo-exaltone has a strong musk odor whereas the 4-oxo- and the 6-oxo analogues are practically odorless. This specificity is in fact general to all of the macrocyclic diketones.

It should also be pointed out that starting with the mixture of bicyclo [10.3.0] pentadecenes obtained above, Biemann, Büchi and Walker synthesized muscopyridine. We have seen that this substance was isolated in 1946, but at that time it was only known that it was an α,α'-bisubstituted pyridine.

The synthesis shown earlier and undertaken on the basis of biogenetic considerations, has permitted correlation of the structures of the synthetic and the natural products. This synthesis is essentially based on the application of the Schmidt reaction with olefins. It should be recalled that the Schmidt reaction is known especially in the form of the reaction of hydrazoic acid on carbonyl derivatives in the presence of strong acids; however, if the antagonist is a carboxylic acid, the product will be an amine.

The Schmidt reaction has also been applied to olefins. In such cases, a substituted imine (ketimine) is obtained as the result of rearrangement. Protonation of the double bond is the first step of this reaction. The protonated olefin then undergoes nucleophilic attack from hydrazoic acid (which retains a strong nucleophilic potential even in strong acid). The last step is a rearrangement leading to an imine.

This mechanism is strongly supported by the following interesting example involving terpenic compounds. The action of hydrazoic acid on camphene in acidic media furnishes a mixture containing 50% α- and 25% β-N-dehydrocamphidine [7.38].

[7.38]

In obtaining products which result from a Wagner-Meerwein rearrangement, it is clear that there is first of all protonation of camphene and this is then followed by nucleophilic attack of HN_3 on the bornylium ion. It is also possible to conclude that the Wagner-Meerwein rearrangement is more rapid than the nucleophilic attack on the camphenylium cation. The last step leading to the N-dehydrocamphidines is a sort of heterocyclic Demjanov reaction.

Having discussed these examples, we can return now to the synthesis of muscopyridine involving 1,5-bicyclo [10.3.0] pentadecene and 1,2-bicyclo [10.3.0] pentadecene. The sequence, discovered in 1953, consists of a series of key reactions that acetylate the β-carbon of a 2-alkylpyridine as shown below [7.39].

[7.39]

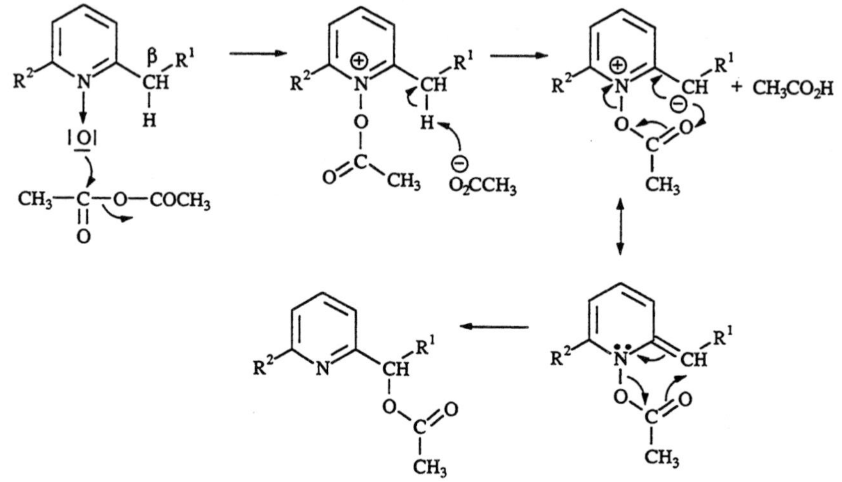

The action of HN_3 leads to a single addition product a, affording the imines b, c and d by ring extension (Demjanov type). When treated with palladium in refluxing methyl naphthalene, the imines b and c are dehydrogenated into the pyridines e and f; the imine d only gives resins. At this stage, it is necessary to separate the two pyridines e and f. Hydrogen peroxide converts pyridine e into pyridine oxide g, and when this is treated with refluxing acetic anhydride, the α-acetoxypyridine h is obtained.

Saponification of h and subsequent oxidation of the resulting secondary alcohol gives a ketone which is then methylated at its α-position using methyl iodide in the presence of potassium tertiary butylate. The carbonyl group is finally eliminated by a Wolff-Kischner reaction, thereby giving muscopyridine. This synthesis is summarized as follows [7.40]. Another more recent access route to (d,l)-muscopyridine is based on a completely different principle.

The condensation of 2,6-dihalopyridines and, in particular, dichloropyridines with α,ω-dimagnesium compounds in the presence of catalytic quantities of nickel-phosphine complexes leads to the formation of n-(2,6)-pyridinophanes. The catalysts used in this reaction are of the type: $Cl_2Ph_2P(CH_2)_nPh_2Ni(II)$ and, particularly, 1,3-dichloro-bis-(diphenylphosphine) propane nickel (II) or Cl_2(dppp) Ni in which n = 3. The reaction can be represented schematically as follows [7.41]. The yields as a function of the size of n are indicated in the table shown below.

n	6	7	8	9	10	12
R%	11	10	16	15	33	26

In order to obtain muscopyridine, it is therefore necessary to prepare 2-methyl-α,ω-dibromodecane [$Br\text{-}(CH_2)_8\text{-}CH(CH_3)\text{-}CH_2Br$], which is not particularly diffi-

[7.40]

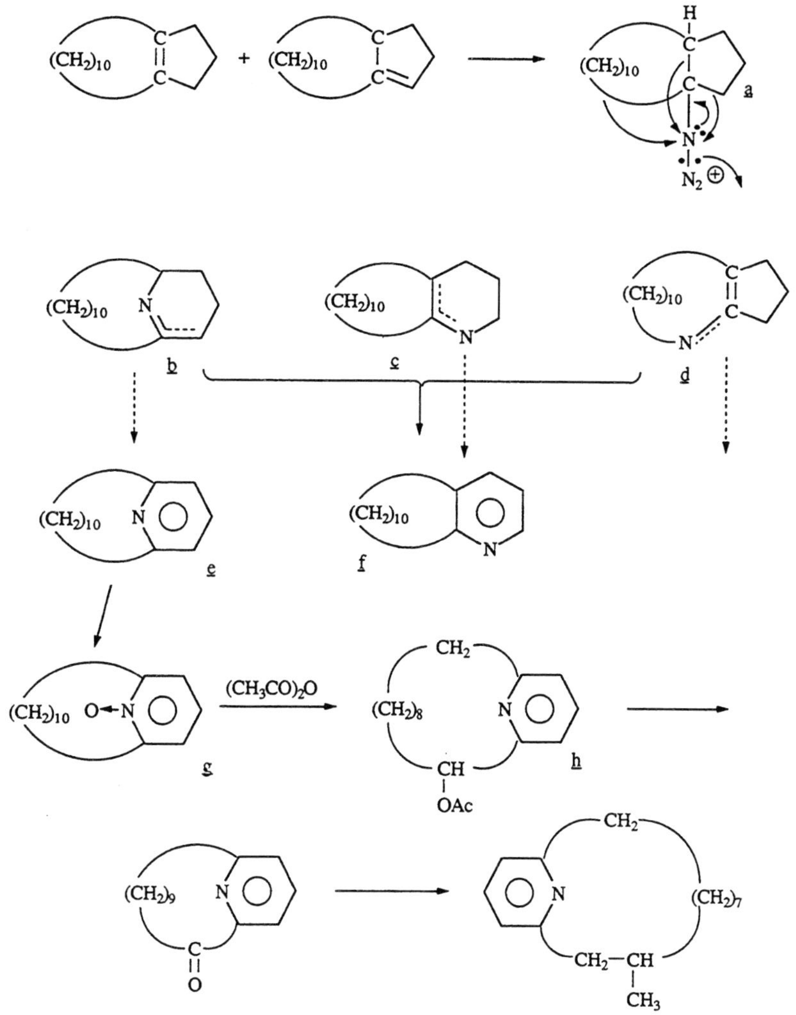

(CH2)10 C=C + (CH2)10 C–C= → (CH2)10 H–C–C–N–N2⊕ a

(CH2)10 N b (CH2)10 N c (CH2)10 C=C N d

(CH2)10 N e (CH2)10 N f

(CH2)10 O←N g —(CH3CO)2O→ CH2 (CH2)8 N CH OAc h →

(CH2)9 N C=O → CH2 N (CH2)7 CH2–CH CH3

[7.41]

Cl N Cl + XMg–(CH2)n–MgX —Cl2(dppp)Ni→ N (CH2)n

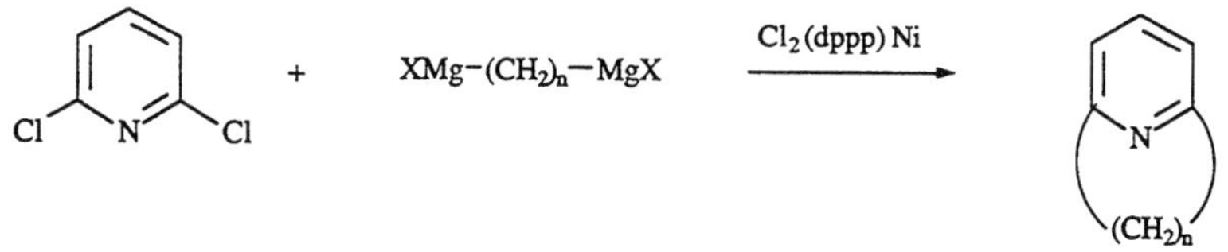

cult. The cyclization of the bis-magnesium derivative of this compound with 2,6-dichloropyridine in the presence of Cl_2(dppp) Ni gives d,2-muscopyridine with a yield of 20%.

Although this yield is relatively modest, this synthesis is far more simple and rapid than the preceding one, which requires many steps and multiple separations.

H. Methods which use cyclododecanone

Since the discovery by Wilke in 1959 of the cyclotrimerization of butadiene to give 1,5,9-cyclododecatriene which is transformable into cyclododecanone, this ketone has become a readily available and relatively cheap product. It has therefore been used as the starting material for many studies aimed at transforming it into higher macrocyclic ketones with musk odors such as muscone and exaltone.

All of these studies have necessitated the development of methods of ring homologation which allow the inclusion of one, two or three carbon atoms. In the course of our discussion of general methods of synthesizing macrocyclic ketones, we have, as an example, presented the synthesis of exaltone and of muscone starting from cyclododecanone using the method of Eschenmoser and colleagues (see section G). This was based on the homologation of the ring by three carbon atoms using the Stobbe reation.

We will now examine the principal methods of ring homologation.

1. *Homologation by one carbon atom*

a. The classic method for homologating carbocyclic ketones

This method involves the insertion of a methylene group by the reaction of diazomethane. The general reaction of diazoalkanes with carbonyl derivatives has been known since the end of the nineteenth century. However, it was essentially due to the research of Arndt, Mosettig and Meerwein and their co-workers between 1927 and 1930, that the utility of this reaction was developed. As we shall see, it is, unfortunately, far from being a specific reaction.

Diazomethane is one of those molecules which cannot be represented by a single structural formula. Using the method of mesomerism, it can be considered a resonance hybrid of several linear contributing structures. The linearity of the diazo group has been demonstrated using several physical methods, notably that of gas-phase electronic diffraction [7.42].

[7.42]

$$R_2\overset{\oplus}{C}-N=\overset{\ominus}{N}\ (\underline{a}) \longleftrightarrow R_2C=\overset{\oplus}{N}=\overset{\ominus}{N}\ (\underline{b}) \longleftrightarrow R_2\overset{\ominus}{C}-\overset{\oplus}{N}\equiv N\ (\underline{c})$$

The mechanism of the insertion reaction can be interpreted as a nucleophilic attack on the carbonyl group by the anionic carbon of diazomethane in the form of its contributing structure c which has the most "weight" in the resonance hybrid [7.43].

The doubly ionic intermediate B decomposes to another doubly ionic intermediate C which can rearrange in two different ways to give either the homologous ketone D or the oxirane E. Furthermore, the ketone D can react again with an excess of diazomethane. If the two R groups are not identical, isomeric homologous ketones can be obtained in proportions which reflect the migratory aptitude of each R group.

[7.43]

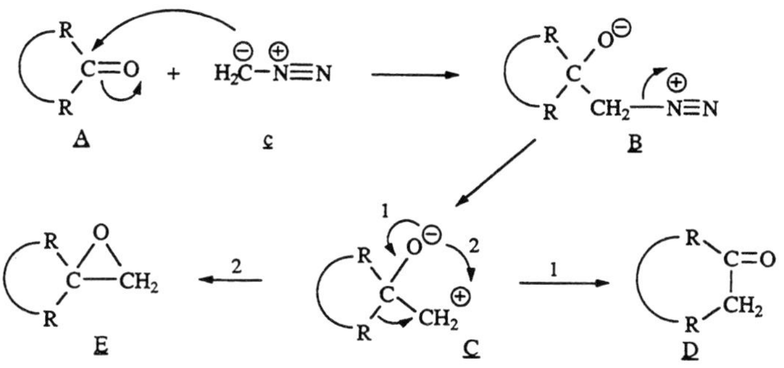

This reaction resembles the nitrous deamination of β-amino alcohols, introduced in 1907 by Wallach and developed especially by Tiffeneau and Tchoubar in 1937. It can be represented in the following way [7.44]:

[7.44]

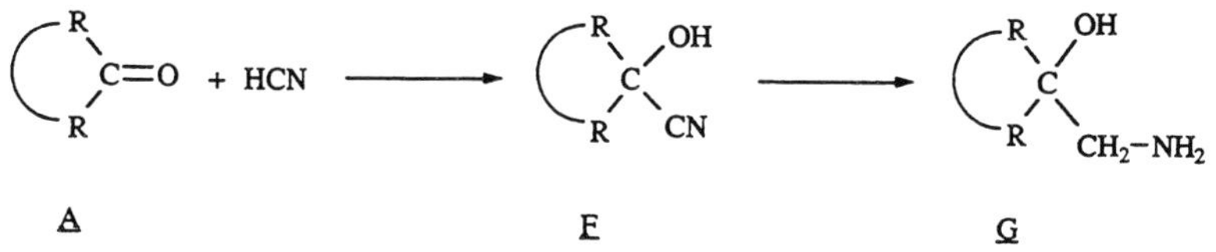

(These latter workers, however, obtained the β-amino alcohols by the action of pressured ammonia on the epoxide of the cycloalkene).

According to McKenzie and Richardson, nitrous acid reacts with the β-amino alcohol G to give the diazonium betaine B. On the other hand, Robinson proposed that the action of diazomethane on carbonyl derivatives is entirely concerted. It can be concluded that the mechanism of the two reactions still needs confirmation.

More recently in 1964, Tai and Warnoff indicated that Lewis acids catalyze the insertion reaction of ethyl diazoacetate on ketones. The Lewis acid used in this case is boron trifluoride. The reaction is carried out between –40° and –70°C, but yields of the homologated β-ketoesters still remain modest. It may be possible that BF_3 complexes with the oxyanion, thereby diminishing its nucleophilic potential so that the proportion of oxirane is lowered. Mock and Hartman revived this idea in 1970 using triethyloxonium fluoroborate as catalyst. The reaction, effected in methylene chloride at temperatures between 0° and 25°C, generally afforded yields in excess of 70% and often reached 90%. The only identifiable by-product (0–5%) was the corresponding glycidic ester J [7.45].

The β-ketoester H corresponding to the homologous ketone D can, in addition, be used to simply carry out substitution reactions in α to the carbonyl group.

The method of Mock and Hartman has been used more recently (1975) by Dreiding in a synthesis of cyclopentadecanone ("Exaltone") by three successive

[7.45]

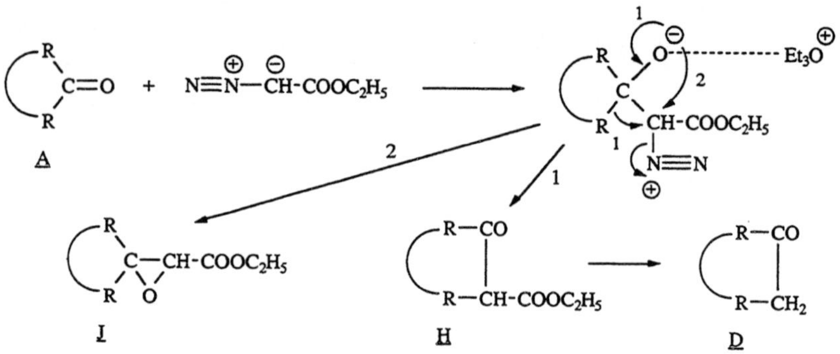

homologations with cyclododecanone as starting material. The overall yield was around 70%. Drieding has also used this reaction in a synthesis of racemic muscone. This will be discussed later on.

b. Method of Parham

This method involves adding dichlorocarbene to the enol ether of cyclododecanone; l-ethoxy–2,2-dichloro bicyclo [10.1.0] tridecane is obtained with a yield of 96%. This then undergoes ring enlargment under the action of refluxing pyridine to give a yield of 87% of 2-ethoxy–3-chloro–1,3-cyclotridecadiene. Potassium tertiary butylate in DMSO eliminates the chloride ion for a 76% yield of l-ethoxy-l-cyclotridecen–3-yne, which after hydrolysis and hydrogenation gives an almost quantitative yield of cyclotridecanone [7.46].

[7.46]

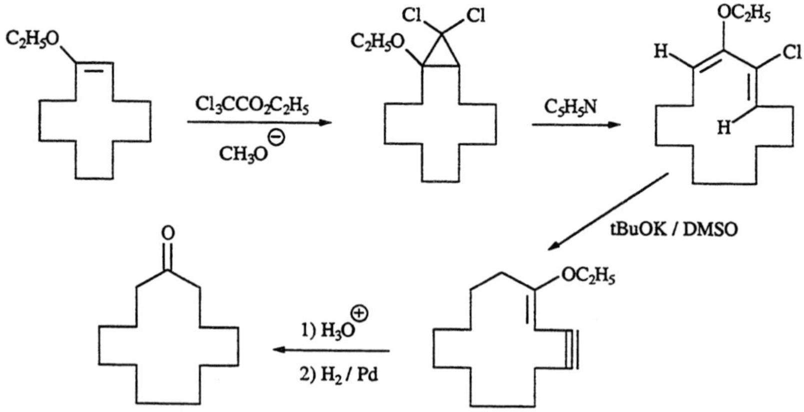

2. *Homologation by two carbon atoms*

a. Method of Brannock

The cycloaddition of ethyl propynoate to the enamine of cyclododecanone yields a bicyclic addition product with a cyclobutene structure which subsequently undergoes thermal opening in accord with the Woodward-Hoffmann rules. A yield of around 60% of the 14-membered conjugated dienic enamine is obtained. Hydrolysis in acidic media gives the β-keto ethylenic ester with a yield of around 90%. Saponification of this ester, followed by hydrogenation, affords cyclotetradecanone with a yield of 80% [7.47].

[7.47]

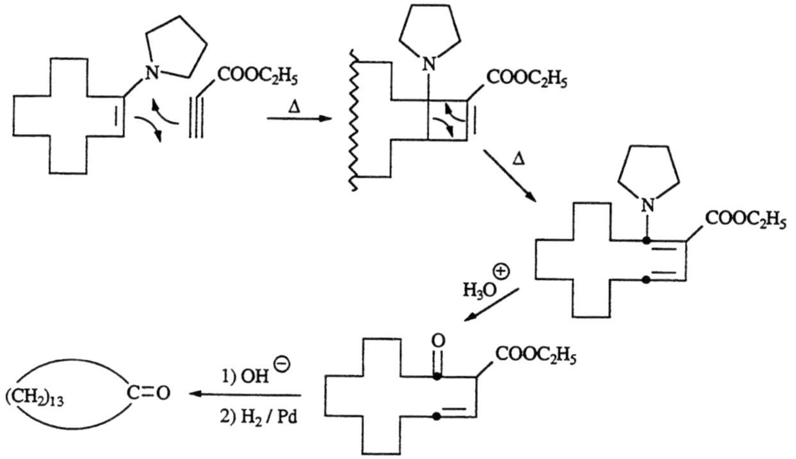

$COOC_2H_5$ Δ $COOC_2H_5$ Δ $COOC_2H_5$ $H_3O^{\oplus}$ $(CH_2)_{13}$ C=O 1) $OH^{\ominus}$ 2) H_2 / Pd O $COOC_2H_5$

If one uses the cycloaddition of ethyl tetrolate (ethyl 2-butynoate) to the previously obtained enamine of cyclotridecanone, the sequence shown here furnishes (d,l)-muscone [7.48].

[7.48]

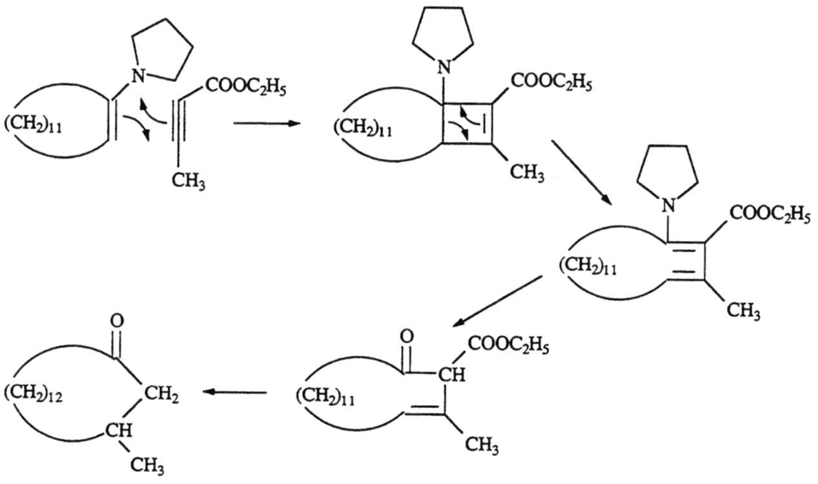

$(CH_2)_{11}$ $COOC_2H_5$ CH_3 $(CH_2)_{11}$ $COOC_2H_5$ CH_3 $(CH_2)_{11}$ $COOC_2H_5$ CH_3 $(CH_2)_{12}$ O CH_2 CH CH_3 $(CH_2)_{11}$ O $COOC_2H_5$ CH CH_3

A more recent Japanese patent (1975) claims to pass from the enol ether of cyclotetradecanone to muscone in only three steps. The first step is to add dichlorocarbene to this enol ether as in the method of Parham. In the second step, the addition product thus formed is treated with excess methyl lithium at –95°C during one hour followed by two hours at room temperature. Muscenone is obtained under these conditions in a yield of 80% and this is then hydrogenated to (d,l)-muscone [7.49].

[7.49]

$(CH_2)_{12}$ (cyclic enol ether, OC_2H_5) $\xrightarrow[NaOH]{Cl_3CH}$ $(CH_2)_{12}$ (dichlorocyclopropane, OC_2H_5, Cl, Cl) $\xrightarrow{CH_3Li}$ $(CH_2)_{12}$ (C=O, CH, =C–CH_3) $\xrightarrow{H_2}$ $(CH_2)_{12}$ (C=O, CH_2, CH–CH_3)

b. Method of Nozaki, Yamamoto and Mori

The principle of this method is far more classic. A 1,3-dihalopropane is reacted with the ketone or, even better, with the β-ketoester, obtained from prior carbethoxylation of the ketone. A bicyclic ketone having three more carbon atoms than the original monocyclic ketone is obtained.

This is based on an extension of the photochemical cleavage method of Gutsche. If this procedure is carried out in methanol, the intermediate ketene resulting from this cleavage provides directly a methyl ester possessing two supplementary ring carbons [7.50].

[7.50]

$(CH_2)_{n-2}$ (CH_2, C=O, CH_2) $\longrightarrow$ $(CH_2)_{n-2}$ (O=C, CH, CH) $\xrightarrow[CH_3OH]{h\nu}$ $\left[(CH_2)_{n+1}\ C{=}C{=}O \right]$

$\longrightarrow$ $(CH_2)_{n+1}$ CH–$COOCH_3$

A way is known for converting this methyl ester into the corresponding ketone using Barton's method of decarboxylation which converts the acid to its corresponding acetate. It seems more advantageous than the method of Hunsdiecker and its variants. It involves first treating the acid with lead tetraacetate (or mercuric acid) and iodine in benzene. This solution is irradiated using a tungsten lamp. The iodo derivative thus obtained is then acetolyzed using mercuric acetate in acetic acid [7.51].

[7.51]

$(CH_2)_{n+1}$ CH–COOH ⟶ $(CH_2)_{n+1}$ CH–I ⟶ $(CH_2)_{n+1}$ CH–OAc

Hydrolysis of the acetate, followed by oxidation, leads to the homologous ketone containing two supplementary carbon atoms.

This strategy has been used in the synthesis of (d,l)-muscone starting from cyclotridecanone and its derivative, 3-chloro–2-chloromethylpropene [7.52].

[7.52]

$(ClCH_2)_2C{=}CH_2$ ⟶ ; H_2, Pd / C ⟶ ; ⟶ $(CH_2)_{12}$, CH_3, CH, CH_2, C=O

The whole sequence involves eight steps and gives yields of 10–12% (d,l)-muscone relative to cyclotridecanone.

c. Method of Hünig and Hoch (1972)

This method uses the cycloaddition of ketene on the morpholino-enamine of a cycloalkanone [7.53].

As indicated earlier, Dreiding used this approach in a recent synthesis of racemic muscone. The initial cycloalkanone is cyclotridecanone obtained by the homologation of cyclododecanone using the method of Mock and Hartmann. This ketone is then treated according to the method of Hünig and Hoch. A yield of around 30% of 4,6-dodecamethylene–2-pyrone is obtained and the saponification of this furnishes a mixture of four 3-methyl–3-cyclopentadecenones with an overall yield of 96% (the intermediate is the carboxylate ion of a vinylogue of β-ketoacid). Hydrogenation of this mixture leads to racemic muscone. This reaction sequence can be summarized as follows [7.54].

3. Homologation by three carbon atoms

a. Fragmentation of the tosylhydrazones of α-epoxy ketones

This approach has already been discussed and illustrated.

b. Fragmentation of epoxysulfones

[7.53]

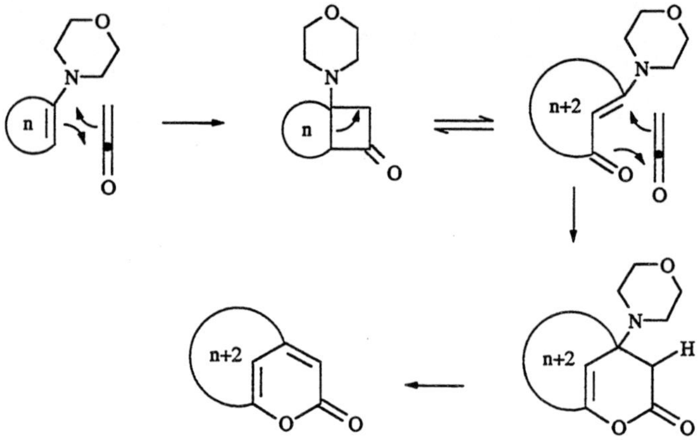

O N n O O N n O n+2 N O O O
n+2 O O n+2 N O H O O

[7.54]

4' 1' 3' 15 4 5 2 12' O 1 O OH ⊖
15 COO⊖ O
15 CH₃ O + 15 O CH₃ H
15 H CH₃ O + 15 H CH₃ O
H₂
15 CH—CH₃ CH₂ C O

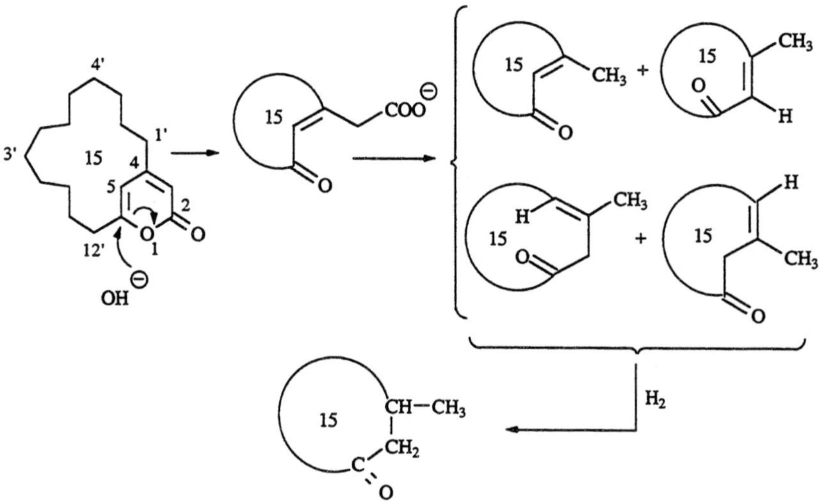

This new type of fragmentation was discovered in 1975 by Fischli. It can be simply summarized in the following way [7.55].

[7.55]

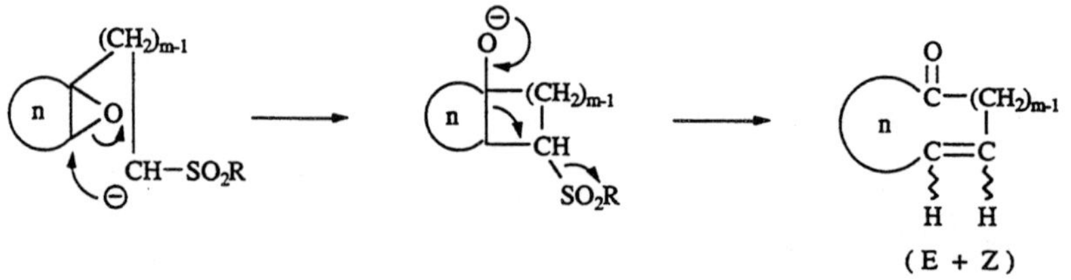

(CH₂)m-1 n O CH—SO₂R ⊖
⊖ O n (CH₂)m-1 CH SO₂R
O n C—(CH₂)m-1 C=C H H (E + Z)

Theoretically, this method could even permit homologation by more than three carbon atoms. It has, however, only been applied by its author to the synthesis of racemic muscone. This synthesis employs the following sequence [7.56].

[7.56]

The step indicated with an asterisk limits the overall yield of this synthesis and adds a further complication in that it is necessary to carry out fractional crystallization to isolate the desired isomer. Nonetheless, it can be considered that the two other rejected isomers could equilibrate to give, once more, a fraction of the desired isomer.

I. Method of Utimoto et al.

This method involves the synthesis of macrocyclic alkynones by cyclizing ω-trimethylsilylethynyl acid chlorides in the presence of sublimated aluminum chloride.

The reaction is carried out in dichloromethane solution and requires the technique of large dilutions. It can be used, in particular, for the synthesis of muscone and exaltone. It can be schematized as follows [7.57]:

[7.57]

$Me_3-Si-C\equiv C-(CH_2)_n-COCl$ $\xrightarrow{AlCl_3}$ cyclic $C(=O)-C\equiv C-(CH_2)_n$

A B

A vinylic carbocation is likely to intervene [7.58]. Table II indicates the yields of this reaction in relation to the initial ring size (n).

[7.58]

Table II

n	Cn + 3	Yield
8	C_{11}	46%
10	C_{13}	56%
12	C_{15}	65%

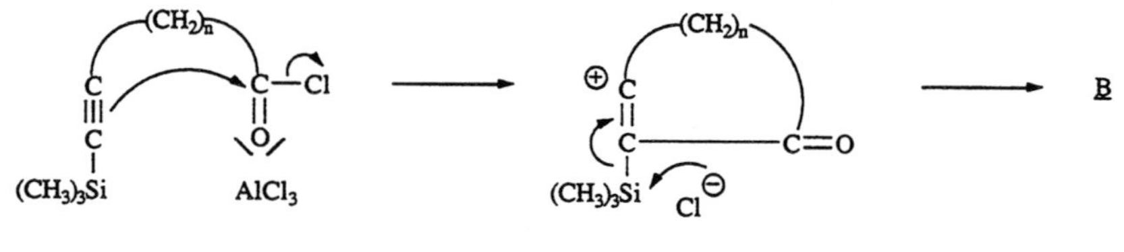

If the size of the ring is too small to permit triple bond formation, the reaction takes a different course. The vinylic carbonyl and not the trimethylsilyl group is attacked by the Cl^- ion. This occurs with normal-sized rings and leads to an α-silylated β-chlorovinyl ketone [7.59].

[7.59]

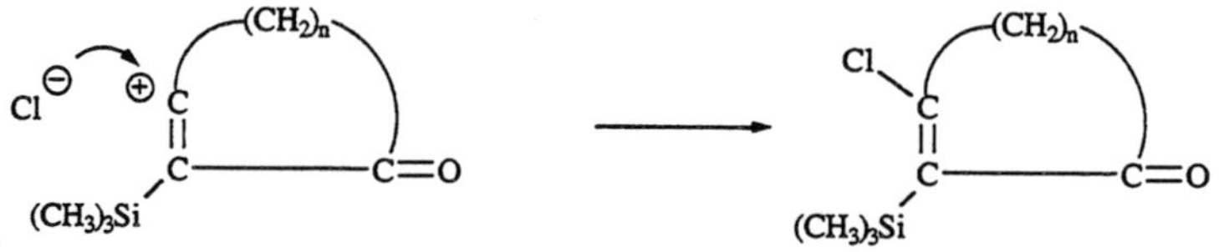

The yields of this reaction are indicated in Table III:

The synthesis of the C_{15} acetylenic acid chloride can be effected in different ways. For instance, condensation of l-bromododecane with propargyl alcohol gives the alcohol 1 with a yield of 83%. Condensation of π-butyl bromide with ω-undecynol, however, gives the alcohol 2 [7.60].

Treatment of the bisubstituted acetylenic alcohols 1 or 2 with an excess of the potassium salt of 3-aminopropylamide ($KNHCH_2$-CH_2-$CONH_2$) (KAPA) furnishes the desired acetylenic alcohol 3 with respective yields of 89% and 93%.

When the ω-acetylenic alcohol 3 is oxidized with chromic acid at 0°C (Jones reagent), it gives the corresponding acid 4 which can then react with oxalyl chloride to give the desired C_{15} acid chloride 5; the overall yield is 92% relative to the original acetylenic alcohol 3 [7.61].

Silylation of this acid chloride 5 is carried out by treating it in refluxing THF

Table III

n	Cn + 3	Yield
2	5	31%
3	6	74%
4	7	65%
5	8	40%

[7.60]

$$CH_3-(CH_2)_{11}-Br + HC{\equiv}C-CH_2OH \longrightarrow CH_3-(CH_2)_{11}-C{\equiv}C-CH_2OH \; (\underline{1})$$

$$CH_3-(CH_2)_3-Br + HC{\equiv}C-(CH_2)_9OH \longrightarrow CH_3-(CH_2)_3-C{\equiv}C-(CH_2)_9OH \; (\underline{2})$$

$$\underline{1} \text{ or } \underline{2} \xrightarrow{\text{KAPA}} HC{\equiv}C-(CH_2)_{12}-CH_2OH \; (\underline{3})$$

[7.61]

$$\underline{3} \longrightarrow HC{\equiv}C-(CH_2)_{12}-COOH \; (\underline{4}) \longrightarrow HC{\equiv}C-(CH_2)_{12}-COCl \; (\underline{5})$$

with 2 molar equivalents of ethyl magnesium bromide followed by an excess of trimethylsilyl chloride [7.62].

[7.62]

$$\underline{5} \longrightarrow Me_3-Si-C{\equiv}C-(CH_2)_{12}-COCl \; (\underline{6})$$

Treatment of the resulting compound 6 according to the sequence indicated below affords the cyclic alkynone 7 which can be hydrogenated to exaltone 8. Alternatively, this alkynone can be treated with lithium dimethyl cuprate to give a mixture of the compounds 9 and 9' which are transformed by hydrogenation into (d,l)-muscone [7.63].

[7.63]

$$\underline{6} \xrightarrow{AlCl_3} \text{cyclo}[C(=O)-C{\equiv}C-(CH_2)_{12}] \; (\underline{7})$$

$$\underline{7} \longrightarrow \text{cyclo}[(CH_2)_{14}\,C{=}O] \; (\underline{8})$$

$$\underline{7} \longrightarrow \text{cyclo}[C(=O)-CH{\cdots}C(CH_3){\cdots}CH-(CH_2)_{11}] \; (\underline{9} + \underline{9'}) \longrightarrow \text{dl-muscone}$$

Despite having these preceding processes at our disposal, at the present state of the art, apparently the most economical synthesis of (d,l)-muscone is that of Baker and colleagues, who employ nickel complexes.

It has been known since the work of Wilke that butadiene and Ni(0) give, under well defined conditions, the π-complex, dodecatrienyl-nickel 1. A solution of this complex in ether, when cooled to between –10° and –20°C, absorbs one equivalent of gaseous allene to give the bis-allyl-nickel 2 [7.64].

[7.64]

$CH_2{=}C{=}CH_2$

1

2

Wilke's group demonstrated that the complex 1 is also susceptible to insertion of isonitriles to give 11- and 13-membered cyclic imines. Baker and colleagues have extended this reaction to the bis-allyl-nickel 2. Thus, insertion of tertiary butyl isonitrile (added in excess to 2 at –20°C) gives a mixture of the compounds 3, 4, and 5 [7.65].

[7.65]

2 + tBuNC → 3 + 4 + 5

R—N

Acid hydrolysis of this mixture gives 5,9,13-cyclopentadecatrienone 6 along with the two unaltered hydrocarbons 3 and 4. These two impurities can be eliminated so that the pure ketone 6 will undergo catalytic hydrogenation to give (d,l)-muscone [7.66].

[7.66]

5 $\xrightarrow{H_3O^{\oplus}}$ 6 $\xrightarrow{H_2}$ dl-muscone

O

The yield of muscone calculated in relation to 1 is around 40% and the ratio of insertion of the isonitrile in the complex 2 is 10 to 1. This ratio can be reduced by the addition of ligands such as triphenyl phosphite.

We will finish this discussion on the synthesis of macrocyclic ketones by outlining the determination of the absolute configuration of natural muscone. This was achieved by correlation with (+)-R-citronellal, which already had a defined absolute configuration.

The following reaction sequence summarizes this determination and leads to the conclusion that the natural substance is (–)-R-muscone [7.67].

[7.67]

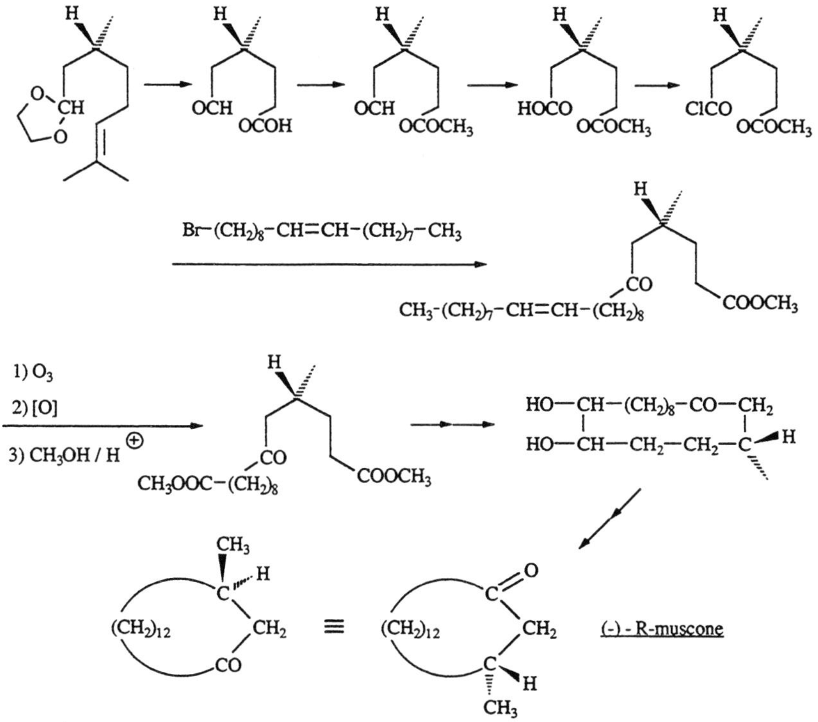

$Br-(CH_2)_8-CH{=}CH-(CH_2)_7-CH_3$

1) O_3
2) [O]
3) $CH_3OH / H^{\oplus}$

(–) - R-muscone

We have discussed syntheses of macrocyclic ketones at length partly for chronological reasons. These were the the first syntheses in this field to have been undertaken. They have also permitted us to discuss a certain number of original chemical reactions which have had an impact that has often gone beyond the specific synthetic goals for which they were originally intended.

Methods of Synthesis of Macrocyclic Lactones

We have just examined in detail various routes to the macrocyclic ketones. We will now study the principal methods for synthesizing macrocyclic lactones for which the economic importance is even greater. It transpires that of all macrocyclic substances of identical ring size, the lactones possess the finest odors.

A. Oxidation of the macrocyclic ketones by peracids

After having achieved the first syntheses of macrocyclic ketones, Ruzicka and Stoll showed, in 1928, that these lactones could be obtained from these ketones using Caro's acid (persulfuric acid); however, this reaction does not give good yields. In the so-called Baeyer-Villiger oxidation, the use of organic peracids such as trifluoroacetic acid is more effective in this regard. The overall reaction consists of inserting an oxygen atom into the bond $>C-C=O$. This actually results from a series of steps involving addition to the carbonyl followed by the migration of one of the groups onto the oxygen as represented here [7.68]:

[7.68]

$$R_2C=O \underset{}{\overset{H^{\oplus}}{\rightleftharpoons}} R_2C=OH^{\oplus} \xrightarrow{CH_3-CO-O-OH} R_2C(OH)-O^{\oplus}(H)-O-CO-CH_3$$

$$\rightleftharpoons R_2C(\ddot{O}H)-O^{\oplus}(H)-O-CO-CH_3 \xrightarrow{-\,CH_3CO_2H} R-C(=O^{\oplus}-H)-O-R \xrightarrow{-\,H^{\oplus}} R-C(=O)-O-R$$

B. Direct lactonization of ω-hydroxyacids

It is impossible to prepare macrocyclic lactones by conventional methods. The ω-hydroxyacids show a much too strong tendency to give linear polyesters by intermolecular polycondensation.

In 1934, Stoll and Rouvé achieved the first satisfactory direct lactonization of ω-hydroxyacids by employing the principle of large dilutions which has been discussed in detail with regard to the synthesis of macrocyclic ketones [7.69].

[7.69]

$$HOOC-(CH_2)_n-CH_2OH \longrightarrow \text{cyclo}[(CH_2)_n-C(=O)-O-CH_2]$$

C. Depolymerization of the polyesters

From 1929 onwards, Carothers (Dupont de Nemours) studied the polycondensation of hydroxyacids with the aim of preparing new fibers for textiles. This work ultimately led to the development of Nylon. With his collaborator Hill, Carothers prepared cyclic esters in 1933 and in 1935 by depolymerizing the corresponding polyesters, which was achieved initially with the polyesters of carbonic acid and oxalic acid.

This method was applied in 1936 to the preparation of macrocyclic lactones and has the advantage of being simpler in practice than the use of large dilutions. Today, it constitutes the principal approach to this class of compound.

The procedure involves first heating the α,ω-hydroxyacids to around 180°–200°C in the presence of catalysts so as to transform them into polyesters or "estolides."

$$_{n'}[OH(CH_2)_n\text{-}COOH] \rightarrow H[OCH_2\text{-}(CH_2)_n\text{-}CO]_{n'}OH + (n' - 1)H_2O$$

Second, the temperature is increased to 270°C under reduced pressure (1–2 mm Hg). Under these conditions, depolymerization occurs and a mixture of monomeric and dimeric lactones distills off. The dimer can be easily eliminated by rectification or by recrystallization in alcohol.

It can be demonstrated that larger cycles form with greater ease than medium-sized rings by examining the following table (extracted from work performed by Spanagel and Carothers in 1936).

The first line concerns depolymerization of 10-hydroxydecanoic acid (1); the second refers to that of 14-hydroxytetradecanoic acid (2).

Polyester	% of distillate	Monomer	Dimer	Monomer/ Dimer
1)	65	12	70	0.17/1
2)	77	90	6	15/1

It is instructive to examine the monomer/dimer ratio in this regard.

Various industrial versions of this method have been developed by several companies. The acid function of the hydroxyacid can be esterified (Givaudan) or both the alcohol and the acid functions can be esterified (Roure-Bertrand-Dupont).

In all of these syntheses, a major difficulty is often encountered in the preparation of the α,ω-hydroxyacids themselves.

Some of these compounds can be provided by nature. Indeed, in 1908, Bougault and Bourdier discovered α,ω-hydroxypalmitic acid (juniperic acid) and α,ω-hydroxylauric acid (sabinic acid) in the waxes covering the needles of certain conifers. We have also found α,ω-hydroxymyristic acid. The α,ω-diacid corresponding to juniperic acid (thapsic acid) also occurs in these waxes. Dimethyl thapsate can be readily prepared by the electrochemical condensation of dimethyl azelaate (dimethyl nonadioate) acccording to the method of Kolbe [7.70].

[7.70]

$$2\ COOCH_3\text{-}(CH_2)_7\text{-}COOH \longrightarrow H_3C{-}OCO{-}(CH_2)_{14}{-}COOCH_3 + 2\ CO_2\uparrow$$

We were able to transform this diester into juniperic acid passing through the intermediate thapsoin (corresponding acyloin) using the following sequence [7.71]:

Cyclization of methyl acetyljuniperate yields cyclohexadecanolide or dihydroambrettolide.

D. Method of Story (1968–1970)

More recently, Story realized a particularly original synthesis of cyclohexadecanolide involving a thermal or photochemical fragmentation reaction of the triperoxide of cyclohexanone [7.72].

[7.71]

COOCH$_3$ (CH$_2$)$_{14}$ COOCH$_3$ $\xrightarrow{Na}$ (CH$_2$)$_{14}$ CHOH C=O $\xrightarrow[Pb(OCOCH_3)_4]{CH_3OH}$ [(CH$_2$)$_{14}$ CHOH C—OH OMe] ⟶ CHO (CH$_2$)$_{14}$ COOCH$_3$ $\xrightarrow{H_2}$ CH$_2$OH (CH$_2$)$_{14}$ COOCH$_3$ $\xrightarrow{(CH_3CO)_2O}$ CH$_2$OCOCH$_3$ (CH$_2$)$_{14}$ COOCH$_3$

[7.72]

⟶ (CH$_2$)$_{15}$ C=O O + (CH$_2$)$_{15}$

This fragmentation can be interpreted in the following way: initial rupture of one of the -O-O- bonds leading to the formation of two alkoxy radicals which then undergo β-fragmentation resulting in the opening of two of the cyclohexane rings [7.73].

[7.73]

⟶ ⟶ (CH$_2$)$_{10}$ ⟶ etc.

Repetition of this process leads to the formation of cyclohexadecanolide and cyclopentadecane.

It is also possible to obtain the mixed peroxides of ketones. For example, if cyclopentanone is condensed with 1,1'-dihydroperoxydicyclohexyl in the presence of cupric sulfate, the 6,5,5-trimeric peroxide is obtained. Like the triperoxide, this substance undergoes a series of fragmentations leading to the formation of cyclopentadecanolide ("Exaltolide") along with cyclotetradecane [7.74].

This method necessitates a highly developed technology for the manipulation of peroxides, but, at the present time, it would seem to be the most economical method available.

[7.74]

$$\text{(dicyclohexylidene diperoxide)} + \text{cyclopentanone} \xrightarrow{CuSO_4} \text{(triperoxide)} \longrightarrow (CH_2)_{14}\text{–CO–O (lactone)} + (CH_2)_{14} \text{(cycloalkane)}$$

E. Intramolecular cyclization of α,ω-diacetylenic esters

Some preliminary explanation is called for to better understand this approach.

The coupling of terminal acetylenes was discovered in 1869 by Glaser and therefore carries the name "Glaser reaction." The acetylenes are oxidized in air in the presence of hydroxylamine with cuprous chloride as catalyst [7.75].

[7.75]

$$C_6H_5\text{-}C{\equiv}CH \xrightarrow[NH_2OH]{Cu_2Cl_2} \xrightarrow{air} [C_6H_5\text{-}C{\equiv}CCu] \longrightarrow C_6H_5\text{-}C{\equiv}C\text{—}C{\equiv}C\text{—}C_6H_5$$

The first important application of this coupling reaction was the historical synthesis of indigo dye by von Baeyer in 1883 in which he used potassium ferricyanide as catalyst [7.76].

[7.76]

$$o\text{-}NO_2C_6H_4\text{-}C{\equiv}C\text{-}Cu \longrightarrow o\text{-}NO_2C_6H_4\text{-}C{\equiv}C\text{-}C{\equiv}C\text{-}C_6H_4\text{-}o\text{-}NO_2 \xrightarrow[S(NH_4)_2]{H_2SO_4} \text{indigo}$$

Although von Baeyer attributed cis geometry to the structure of indigo, the true trans structure was demonstrated in 1928 from a study using X-rays. This reaction has been frequently used for the synthesis of natural products such as polyenes, vitamins, fatty acids, amino acids and sugars. Mixed Glaser couplings have been made, but the yields are not excellent. Specific disymmetrical coupling has been achieved by Cadiot and Chodkiewicz [7.77].

[7.77]

$$R\text{—}C{\equiv}CH + Br\text{—}C{\equiv}C\text{—}R' \xrightarrow[Base]{Cu^{\oplus}} R\text{—}C{\equiv}C\text{—}C{\equiv}C\text{—}R' + HBr$$

This is done by adding the acetylenic compound dissolved in a convenient solvent (H_2O, EtOH, C_6H_6) to a catalytic quantity of cuprous bromide or chloride dissolved in a primary amine. A small quantity of a reducing agent such as hydroxylamine hydrochloride is added in order to maintain constant the concentration of the cuprous ion. The bromoacetylenic derivative (prepared very easily by the action of sodium hypobromite on the second terminal acetylenic compound) is then poured into this vigorously stirred mixture [7.78].

[7.78]

$$R{-}C{\equiv}CH + Cu^{\oplus} \longrightarrow R{-}C{\equiv}CCu + H^{\oplus} \quad \text{(absorbed by the base)}$$

$$R{-}C{\equiv}CCu + Br{-}C{\equiv}C{-}R' \longrightarrow R{-}C{\equiv}C{-}C{\equiv}C{-}R' + BrCu \quad \text{(catalyst)}$$

The catalyst is used in small quantities (approximately 1%) in order to avoid coupling R'-C≡C-Br with itself.

Intramolecular oxidative coupling was described for the first time in 1956 by two independent groups of workers, one in England (Eglinton and Golbraith), the other in the United States (Sondheimer and Amiel). The method used by these groups was derived from the original method of Glaser; however, the cuprous salt and oxygen are replaced by cupric acetate and the reaction is carried out in homogeneous solution constituted by a mixture of benzene-pyridine or sometimes diethyl ether. One inconvenient aspect of this method is its reliance on the use of high dilutions. Under these conditions, however, yields can equal or exceed 90%.

This route presently permits the preparation of a large variety of macrocyclic compounds possessing up to 54-membered rings. These include cyclic saturated and unsaturated hydrocarbons, ketones, ethers, lactones, etc., according to the following general scheme [7.79]:

[7.79]

$$HC{\equiv}C{-}X{-}C{\equiv}CH \longrightarrow \text{cyclo-}[C{\equiv}C{-}X{-}C{\equiv}C]$$

This has permitted the group of Sondheimer to discover a new class of conjugated polyolefinic cyclic compound, the annulenes.

A synthesis of exaltolide involving intramolecular oxidative coupling of an α,ω-diacetylenic ester was proposed by McCrae in 1964. This is summarized in the following scheme [7.80]:

[7.80]

$$HC{\equiv}C{-}(CH_2)_8{-}COOH \xrightarrow{(COCl)_2} HC{\equiv}C{-}(CH_2)_8{-}COCl \xrightarrow[\text{pyridine}]{HC{\equiv}C{-}CH_2{-}CH_2OH}$$

$$HC{\equiv}C{-}(CH_2)_8{-}COO{-}(CH_2)_2{-}C{\equiv}CH \xrightarrow[Et_2O\ /\ C_6H_6]{Cu(OAc)_2\ /\ Pyr} \text{cyclo-}[CH_2{-}C{\equiv}C{-}C{\equiv}C{-}CH_2{-}CH_2{-}COO{-}(CH_2)_7]$$

$$\xrightarrow{H_2} \text{cyclo-}[(CH_2)_{14}{-}C({=}O){-}O]$$

The yield of the cyclization step is 90%. This method is certainly applicable to the synthesis of many other macrocyclic lactones. In this case, it has the advantage of employing 10-undecynoic acid, readily accessible from 10-undecenoic acid, which itself can be obtained from castor oil.

F. Fragmentation of hydroperoxides

It is also possible to synthesize exaltolide from cyclododecanone as starting material [7.81].

[7.81]

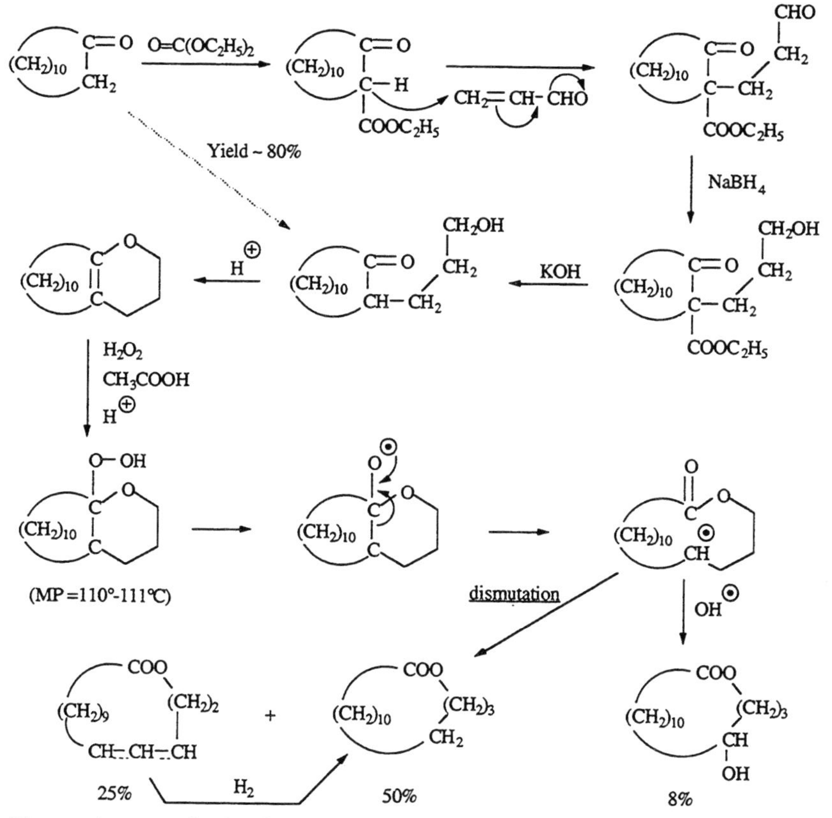

$(CH_2)_{10}$ C=O CH_2 → ($O{=}C(OC_2H_5)_2$) → $(CH_2)_{10}$ C=O, C–H, $COOC_2H_5$ + $CH_2{=}CH{-}CHO$ → $(CH_2)_{10}$ C=O, C–CH_2–CH_2–CHO, $COOC_2H_5$

Yield ~ 80%

$NaBH_4$

CH_2OH

KOH

$H^{\oplus}$

H_2O_2
CH_3COOH
$H^{\oplus}$

O–OH

(MP=110°-111°C)

dismutation

$OH^{\odot}$

COO $(CH_2)_9$ $(CH_2)_2$ CH=CH–CH + COO $(CH_2)_{10}$ $(CH_2)_3$ CH_2 ; COO $(CH_2)_{10}$ $(CH_2)_3$ CH OH

H_2

25% 50% 8%

The various methods already discussed all give access to different natural macrocyclic ketones and lactones. They can also be used to give numerous other substances which are entirely synthetic and which are, in some cases, more interesting when the economic aspects of the problem are considered, particularly the "quality-price" factor.

All of the substances with musk odor that have been cited up until now are characterized by their macrocyclic structure containing generally from 14 to 17 ring elements. This characteristic constitutes a "structure-odorant activity" relationship. In addition, besides their musk odor, the C_{14} compounds also possess a slightly moldy or camphoric olfactive character. The C_{16} and C_{17} compounds, in particular, manifest the "animal inflection."

On the basis of this "structure-odorant activity" relationship, macrocyclic compounds totally different from the natural products have been sought. As a result, two other groups of macrocyclic substances are presently known—the macrocyclic diesters and the lactone ethers or oxalactones.

Macrocyclic diesters

These are compounds of the type A. They are obtained by esterification of a diol and a diacid using the method of Carothers and Spanagel (see earlier). Ethylene brassylate (also called "astrotone" or "musk BRB"), in which m = 2 and n = 11, is obtained by this route from ethylene glycol and brassylic acid [7.82].

[7.82]

$$HO-(CH_2)_m-OH + HOCO-(CH_2)_n-COOH \xrightarrow{\text{polycondensation}}$$

$$\cdots\left[O-\overset{O}{\overset{\|}{C}}-(CH_2)_n-\overset{O}{\overset{\|}{C}}-O-(CH_2)_m\right]_x O-\overset{O}{\overset{\|}{C}}\cdots \xrightarrow{\text{depolymerization}}$$

$$\begin{array}{lcl} O & \text{———} & C{=}O \\ | & & | \\ (CH_2)_m & & (CH_2)_n \\ | & & | \\ O & \text{———} & C{=}O \end{array} \quad \underline{A}$$

Oxalactones

These are compounds of the type B and can be obtained by methods already explained (method of large dilutions, method of Carothers and Spanagel and its variants).

The general route for obtaining the monomer involves condensing an ω-halogenated acid with the sodium salt of a glycol. This salt is obtained by the reaction of the glycol with sodium dispersion in an inert solvent or by direct action of sodium on the glycol.

To cite an example, undecylenic acid obtained from the pyrolysis of castor oil, after passing through the intermediate ω-bromo undecylenic acid (raw material used for the manufacture of "Rilsan"), leads to several oxalactones having a musk odor.

When n = 4 (tetramethylene glycol), the product obtained has an intense musk odor. When n = 6 (hexamethylene glycol), the product has a 19-membered ring and an odor close to that of civettone [7.83].

A variant of the preparation of the oxalactones is to react ω-bromo-undecanol with acrylonitrile in the presence of "Triton B." Following a bromine-iodine exchange and subsequent hydrolysis of the nitrile function in acid, the resulting ω-iodoacid is cyclized by the Hunsdiecker-Erlbach process, employing, at the same time, the technique of large dilutions [7.84].

These macrocyclic substances do not however, retain exclusive rights to the

[7.83]

$$\begin{array}{l} (CH_2)_m\text{—}C{=}O \\ \;\;|\qquad\quad\;\; | \\ \;O \qquad\quad\;\; | \qquad \underline{B} \\ \;\;|\qquad\quad\;\; | \\ (CH_2)_n\text{—}O \end{array}$$

$$NaO{-}(CH_2)_n{-}OH + X{-}(CH_2)_m{-}COOH \longrightarrow HO{-}(CH_2)_n{-}O{-}(CH_2)_m{-}COOH \longrightarrow \underline{B}$$

[7.84]

$$Br{-}(CH_2)_{10}{-}CH_2OH + CH_2{=}CH{-}CN \longrightarrow Br{-}(CH_2)_{11}{-}O{-}(CH_2)_2{-}CN$$

$$\longrightarrow Br{-}(CH_2)_{11}{-}O{-}(CH_2)_2{-}COOH \longrightarrow I{-}(CH_2)_{11}{-}O{-}(CH_2)_2{-}COOH$$

$$\longrightarrow \begin{array}{l} (CH_2)_{11}{-}O{-}(CH_2)_2{-}C{=}O \\ |\qquad\qquad\qquad\qquad\;\; | \\ \text{———————————}O \end{array}$$

odor of musk. Another very important group also possesses this odor. These are the aromatic musks and they will be dealt with in the following section.

Aromatic Musks

The aromatic musks can be classed in two quite distinct groups: the aromatic nitro musks and the aromatic non-nitro musks.

As indicated in the introductory chapter of this book, Baur observed more than a century ago that certain nitrated aromatic compounds possessed a musk odor. This discovery has played an important role in the industrial research into new synthetic odorant compounds.

Since then, numerous studies have been undertaken on musk-smelling aromatic compounds and it is not possible to analyze them all here in detail. We will try, nonetheless, to point out the most essential points of interest.

A. Aromatic nitro musks

These compounds can also be grouped into three categories: the benzenic, indanic and tetralinic musks.

1. Benzenic nitro musks

The first musk synthesized from toluene (by Bauer) was 2-tertiary-butyl–6-methyl 1,3,5-trinitrobenzene 1. This, by the way, is called "Musk Baur" or "Musk B" or "Musk Toluene" or even "Tonkinol" and it was an immediate industrial success. Even before 1900, it was being sold in the form of a 10% mixture in acetanilide for abour $500 per kg, that is, about half the price of veritable musk tonkin! As time passed, however, this musk was gradually replaced by three other nitro derivatives also discovered by Baur: musk xylene 2, musk ketone 3 and musk ambrette 4. A fourth musk compound was added to these three 60 years later when Carpenter of the Givaudan company discovered musk tibetine 5. This chemist played an important role in the research of aromatic musk compounds.

In fact, the structures initially proposed by Baur for his compounds were inexact. It was not until the work of Tchichibabine in France in 1932 and Zeide and Dubinin in Russia in the same year that these were defined with certainty [7.85].

[7.85]

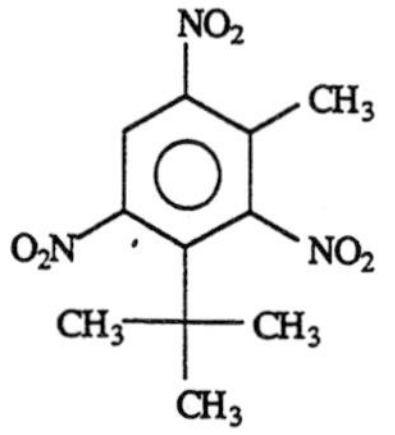

1 musk toluene

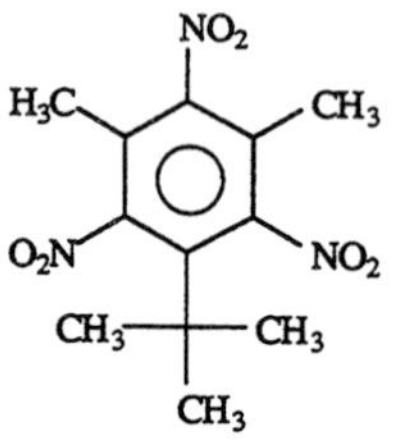

2 musk xylene

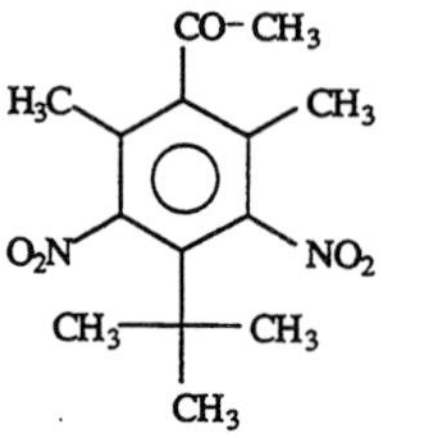

3 musk ketone

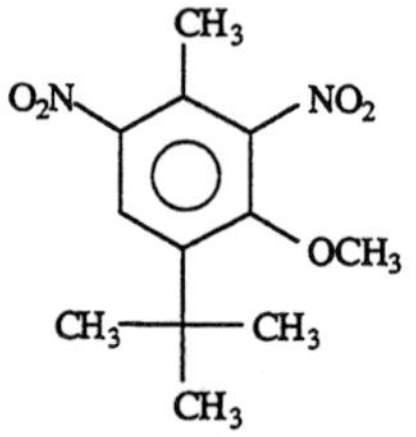

4 musk ambrette

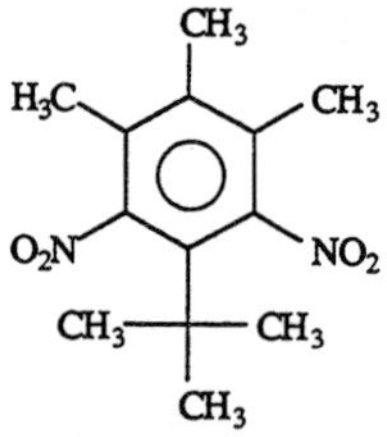

5 musk tibetene

The nitro musks are very crystalline substances which are only sparingly soluble in alcohol. For this reason, they can be purified by recrystallization in 90° alcohol.

Musk xylene:	MP = 112.5°–114.5°C (stable form)
	= 104°–106°C (metastable form)
Musk ambrette:	= 84°–86°C
Musk ketone:	= 134.5°–136°C
Musk tibetine:	= 136.5°C

Musk ketone is considered as having the closest odor to veritable musk.

In each group, the influence of the substituents on the musk odor has been studied in order to try to define a relationship between odor and molecular structure. The studies of Amoore, Beets, Wright and, more recently, Theimer and Davies can be cited as being among the most important research in this area [7.86].

When $R^1 = CH_3$ and $R^2 = -C_2H_5$ or Br and when $R^1 = H$ and $R^2 = CH_3$, the musk odor persists. If $R^1 = R^2 = -C_2H_5$, the musk odor disappears. When $R^2 = CH_3$ and R^1 = F,Cl, Br or I, the musk odor is very intense.

When R^1 = H and the tertiary butyl group is replaced by isoproyl or isobutyl

[7.86]

$R^1 = R^2 = CH_3$ = musk xylene

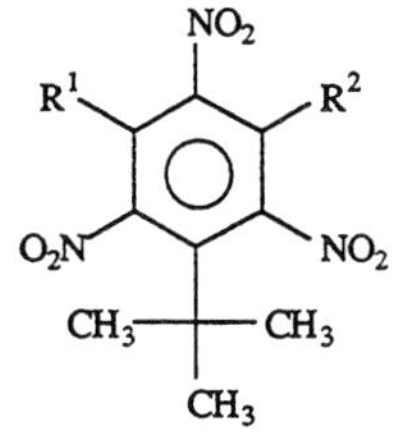

groups [CH_2-$CH(CH_3)_2$], the musk odor weakens noticably. If R^1 is replaced by an n-butyl group or an isopentyl group [CH_2-CH_2-$CH(CH_3)_2$], the musk odor disappears entirely [7.87].

[7.87]

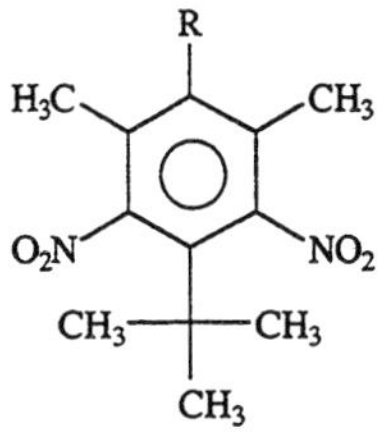

A

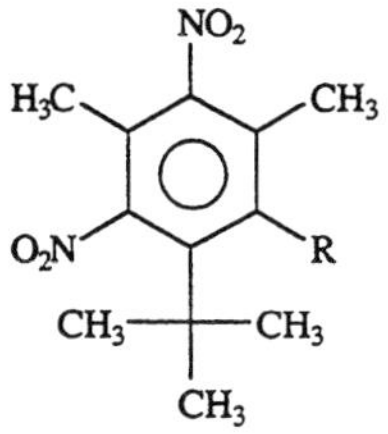

B

In 1954, Carpenter, Easter and Wood also studied the symmetrical compounds of group A as well as the unsymmetrical compounds of B. Musk ketone belongs to group A with R = $COCH_3$. In each group, when R = -$COCH_3$, -CN, -Br, -COCl or -$COOCH_3$, all compounds are found to have musk odors comparable in intensity and quality [7.88].

[7.88]

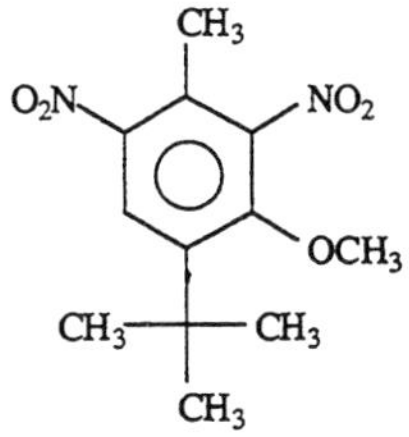

musk ambrette

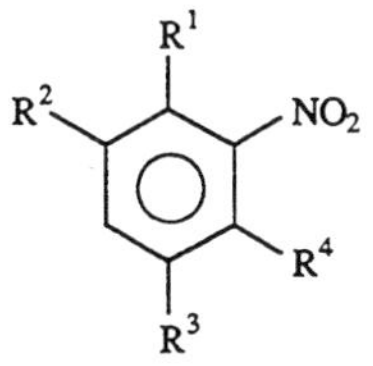

In Table IV, we have collected the olfactive characteristics of the twelve compounds belonging to this group. First, it can be said that homoambrette c possesses the most powerful musk odor. Unfortunately, this compound colors much more than musk ambrette.

Carpenter was the first to try to find a structure/odor relationship in this group of compounds and, as a result, he proposed the "ortho effect." According to this rule, compounds possessing a bulky group such as t-butyl or t-amyl or alkoxy in the ortho position, should possess a musk odor. This rule is substantiated by the

Table IV

Compound Substituents	a	b	c	d	e	f	g	h	i	j	k	l
R^1	H	CH_3	C_2H_5	H	H	H	OCH_3	CH_3	OCH_3	H	H	Br
R^2	NO_2	NO_2	NO_2	CHO	CHO	t-Bu	NO_2	NO_2	NO_2	$COCH_3$	$COCH_3$	NO_2
R^3	t-Bu	t-Bu	t-Bu	t-Bu	t-Bu	t-Bu	t-Bu	t-amyl	t-Bu	t-Bu	t-Bu	t-Bu
R^4	OCH_3	OCH_3	OCH_3	OCH_3	OC_2Hs	OCH_3	OCH_3	OCH_3	Br	OCH_3	OC_2H_5	OCH_3
Observations	<< norambrette >> << ambrette>><<homo ambrette>>											
Odor	m	m	m	m	m	m	m	m	m	nm	nm	nm

Remark: m = odor musk; nm = odor non musk

first eight compounds (a to h). However, compound i has a musk odor but does not obey this rule. Conversly, the last three compounds j, k and 1 satisfy this rule but have no musk odor at all!

Another interesting observation made by Carpenter was that the t-butyl radical is necessary for the presence of a musk odor. The same applies for the t-amyl radical. This is not, however, a sufficient condition in itself.

Beets tried to rationalize this structure/odor relationship by proposing "the rule of triplets," according to which the presence of three substituents, NO_2, OR and t-alkyl, is necessary in this group of ambrette-type musks for the presence of a musk odor. Unfortunately, compound 1, which complies with this requirement, does not present a musk odor. One is therefore tempted to conclude that the rule of Beets is valid on the condition that the substituent R_1 is not a halogen atom.

Synthesis of benzenic nitro musks

Musk xylene, musk ketone and musk tibetine are prepared from metaxylene. 7.89 summarizes these syntheses schematically. Musk ambrette is prepared from metacresol. This synthesis is shown in [7.90].

[7.89]

$(CH_3)_3CCl$

NO_3H / SO_4H_2

HCHO / HCl

$(CH_3CO)_2O$ / $AlCl_3$

Zn / H_2O

Musk xylene

Musk ketone

Musk tibetine

[7.90]

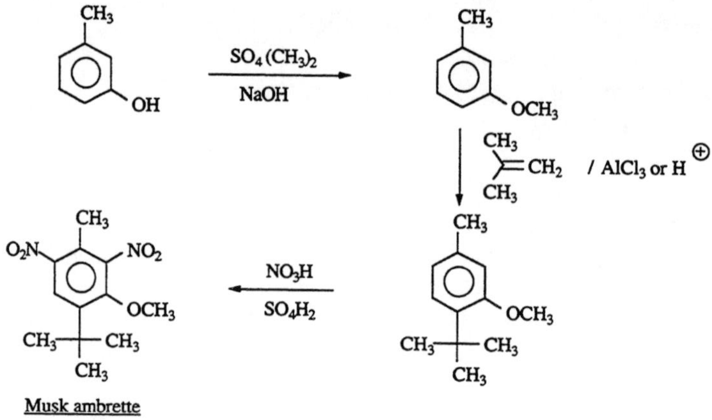

$SO_4(CH_3)_2$ / NaOH; $(CH_3)_2C{=}CH_2$ / $AlCl_3$ or $H^{\oplus}$; NO_3H / SO_4H_2

Musk ambrette

2. *Indanic nitro musks*

Another type of aromatic nitro musk was prepared by Barbier in 1932. The first step of his synthesis is to condense t-butyl alcohol or isobutylene with para-cymene in the presence of sulfuric acid.

Barbier thought that he had obtained 3-t-butyl–4-isopropyltoluene of which a subsequent nitration of this gave 2,6-dinitro–3-t-butyl–4-iso-propyltoluene which he called "Moskene."

About 20 years later, Carpenter tried to treat "Moskene" with nitric acid with the knowledge that isopropyl groups are readily replaced by nitro groups under these conditions. This reaction would therefore have given the original musk of Baur 1. However, he found that "Moskene" remained unaltered at the end of the reaction.

This observation gave cause to suspect the structure 7 and, in 1955, Grampoloff demonstrated that "Moskene" possessed an indanic structure 8 [7.91].

[7.91]

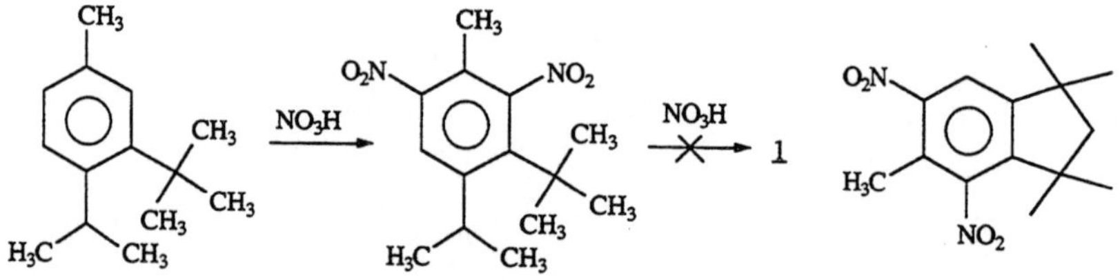

6 —NO_3H→ 7 —NO_3H ✗→ 1

8

In fact, the action of isobutylene *in excess* on paracymene gives an indanic derivative 9 after an overall reaction, summarized as follows [7.92]. It is therefore this intermediate 9 which on nitration, leads to "Moskene" 8.

The result of this reaction may seem surprizing. It should, however, be recalled that Ipatieff and co-workers had already shown from as early as 1948 that para-cymene acted like a "hydrogen donor" in the presence of olefins and in concentrat-

[7.92]

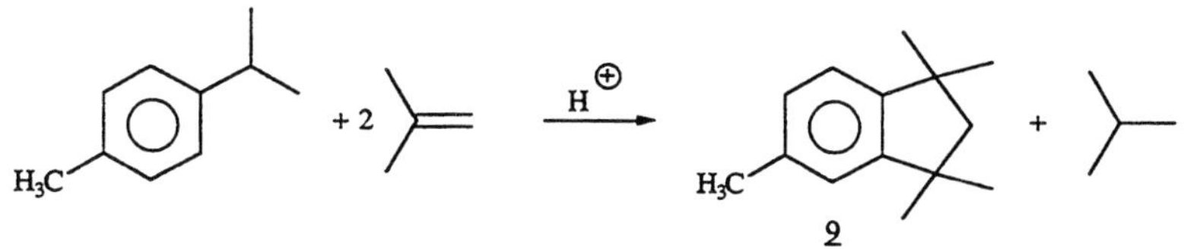

ed acidic media, with the olefin acting as a "hydrogen acceptor." Actually, when seen in this way, the Ipatieff reaction does not constitute a transfer of hydrogen, but rather a transfer of hydride (H^-) in acid media (H^+), which is highly unlikely.

It would seem more probable that there is a rupture of a C-H bond in para-cymene A with progressive transfer of the positive charge from the protonated olefin passing through the transition state E to give the carbocation B [7.93].

[7.93]

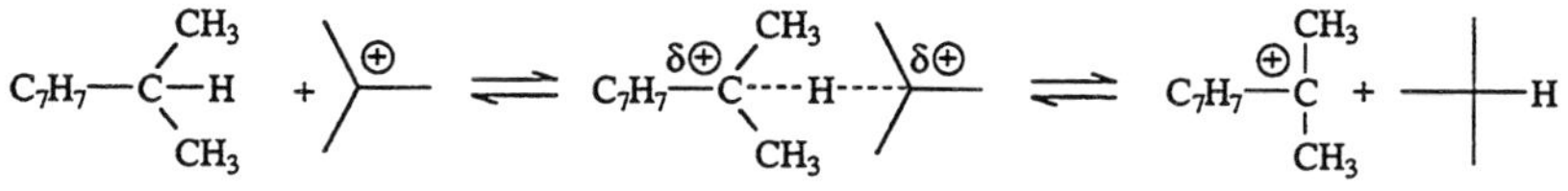

Moreover, this reaction is favored by a considerable enthalpic gain coming from the hybrid resonance stabilization of the carbocation B which will have four contributing structures a, b, c and d [7.94].

[7.94]

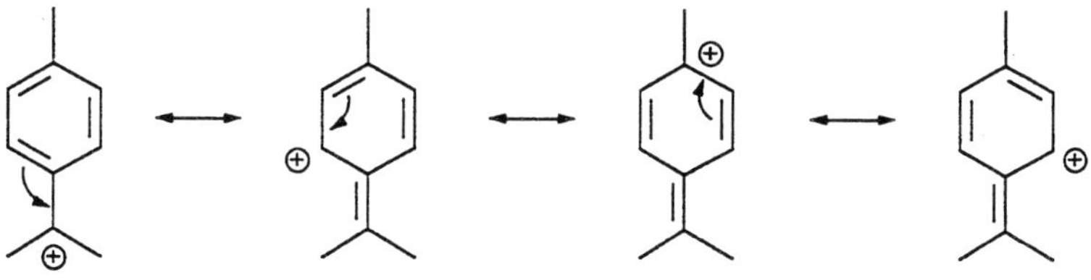

A second molecule of isobutylene then attacks this carbocation B to give a cationic intermediate C which cyclizes to give the indanic derivative 9 [7.95].

[7.95]

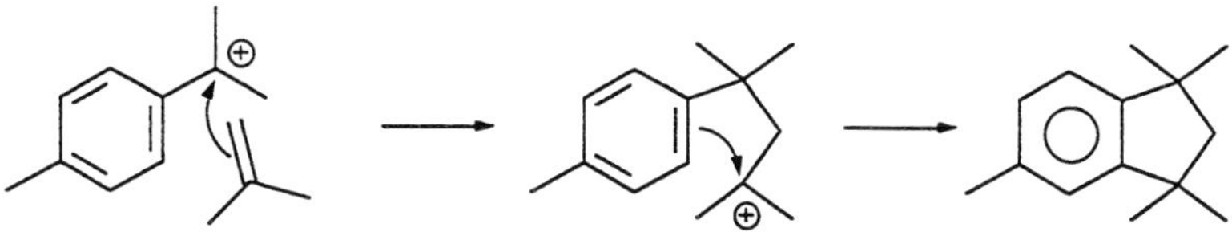

The same reaction has allowed the synthesis of a group of nitrated indane derivatives which all present musk odors. Furthermore, some of these compounds show interesting herbicidal properties [7.96].

[7.96]

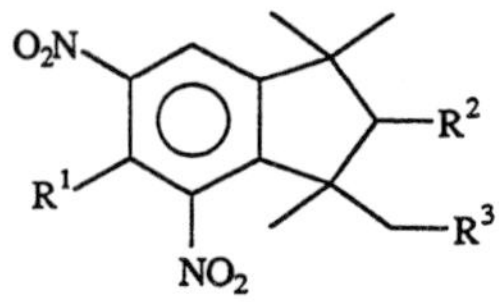

$R^1 = CH_3$, C_2H_5 , CH_3 , C_2H_5 , CH_3

$R^2 = CH_3$, H , C_2H_5 , CH_3 , CH_3

$R^3 =$ H , H , H , H , CH_3

3. Tetralinic nitro musks

The first two compounds in this series have been prepared from β-ionone (R = CH_3) and from normal β-methylionone (R = C_2H_5). Treatment of these ketones with iodine and heat affords the corresponding tetralinic hydrocarbons as the result of a dehydration-cyclization reaction. Nitric acid in concentrated sulfuric media gives two compounds of which the first (R = CH_3) has a medium-strength musk odor, while the second (R = C_2H_5) has a very powerful musk odor. Both of these compounds also have herbicidal properties [7.97].

[7.97]

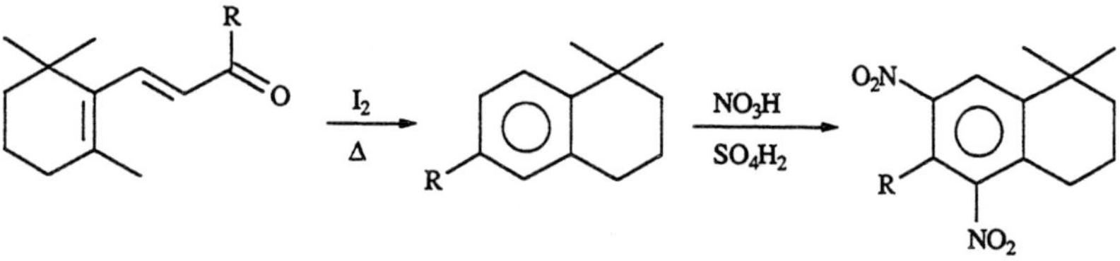

The preparation of several tetralinic hydrocarbons for use as substrates in the synthesis of different musk compounds has been the origin of many patented processes. The principal ones will be described here.

The 1,1,4,4-tetramethyl–6-alkyltetralins 10 are prepared by the condensation of alkylbenzenes with 2,5-dichloro–2,5-dimethylhexane in the presence of ferric chloride. This chlorinated derivative is prepared from dimethylhexyne diol which originates from the ethynylation of acetone [7.98].

[7.98]

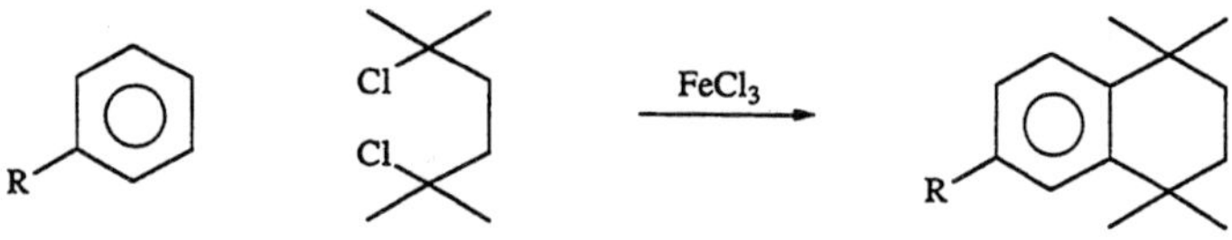

Paracymene reacts with dimethylbutene (or its equivalents) in acidic media and, as could be expected from what we have already seen, the product is 1,1,3,4,4,6-hexamethyltetralin 11 [7.99].

Other tetralins such as 12 can be prepared as shown here. The catalyst can be a cation exchange resin and the reaction temperature must be near to 100°C [7.100].

[7.99]

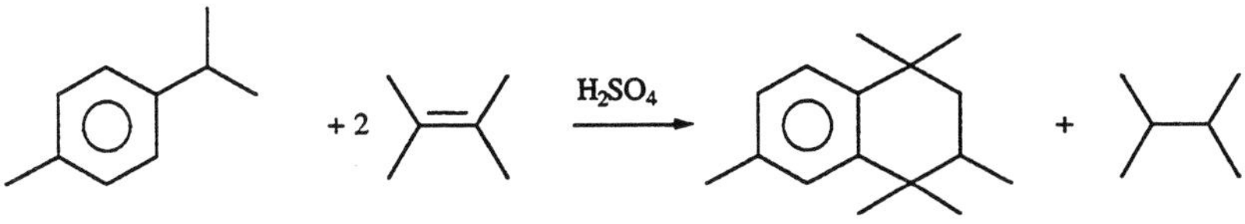

H_2SO_4

11

[7.100]

Resine⊕

R R

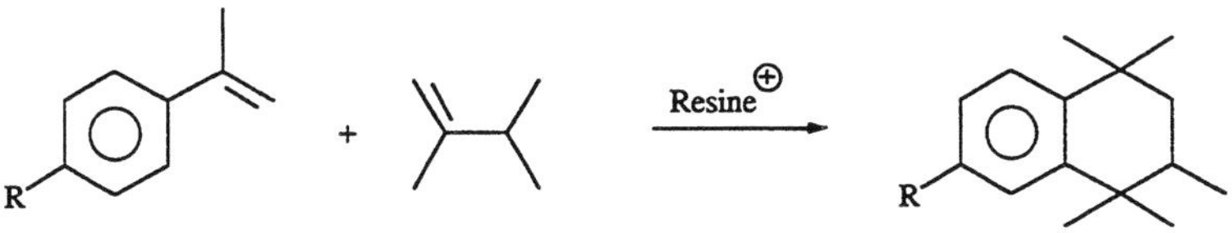

12

Nitration of these hydrocarbons makes available a whole series of substituted tetralinic nitro musks represented by the general formula 13 [7.101].

[7.101]

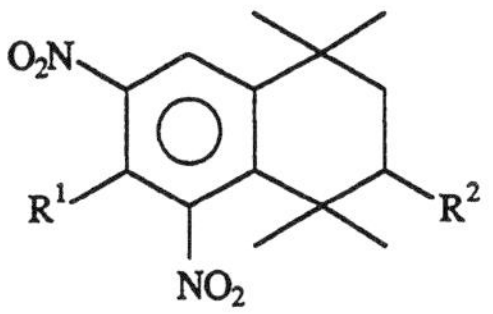

O_2N R^1 R^2 NO_2

13

$R^1 = CH_3$, C_2H_5, C_2H_5, CH_3, iC_3H_7, iC_3H_7, H, OCH_3, Cl, Br

$R^2 = CH_3$, CH_3, H, H, H, CH_3, CH_3, H, H, H

The first three of these are the most interesting in regard to both their power and their olfactive qualities, which can be compared to that of "Moskene."

B. Non-nitro aromatic musks

These compounds can also be grouped into five categories: the benzenic, indanic, tetralinic, tricyclic and isochromanic musks.

1. Benzenic non-nitro musks

The discovery that these aromatic compounds still had a musk odor despite the absence of any nitro group was made fortuitously by research workers at the Givaudan company in 1948. At that time, they were attempting the preparation of the nitro musk 15 according to the following reaction sequence [7.102].

They observed that the desired end product 15 was odorless. In contrast, however, the non-nitro derivative 14, which they named "Ambral," had a very fine and persistant musk odor. This observation opened the way for some very important research work, the most characteristic of which will be summarized here.

[7.102]

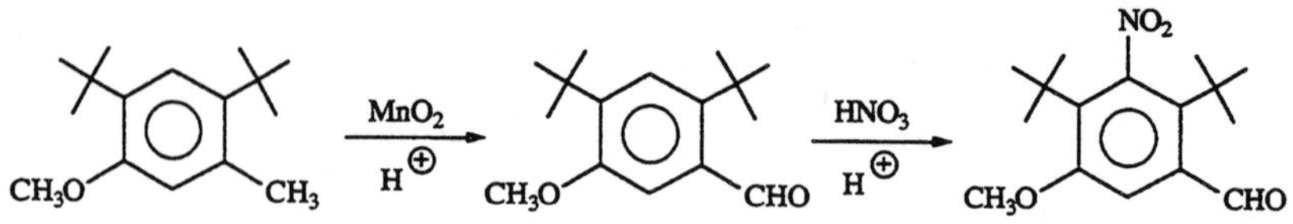

CH_3O CH_3 $\xrightarrow[H^{\oplus}]{MnO_2}$ CH_3O CHO $\xrightarrow[H^{\oplus}]{HNO_3}$ NO_2 CH_3O CHO

14 15

[7.103]

CH_3O CHO CHO CH_3 CN CH_3 NO_2 CH_3

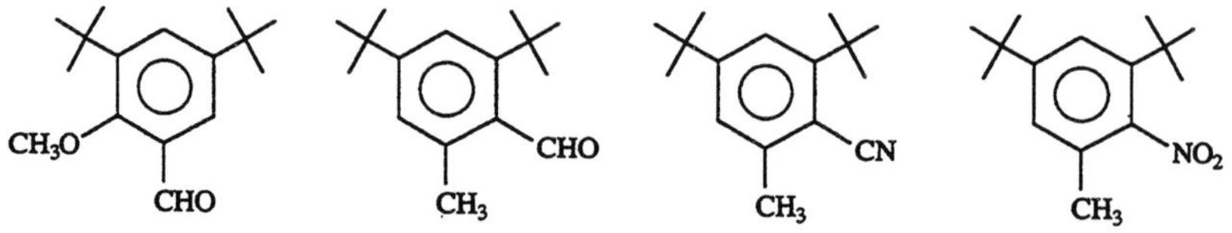

16 17 18 19

It should be pointed out in passing that replacing one of the meta-aldehyde groups with a methoxy and moving the other to the ortho position, as in structure 16, completely removes the odoriferous properties of the molecule. While compound 17 has a medium-strength musk odor, its analogue 18, which carries a nitrile function, only gives a slight musk odor on heating. The corresponding nitro musk 19 is completely odorless [7.103].

[7.103]

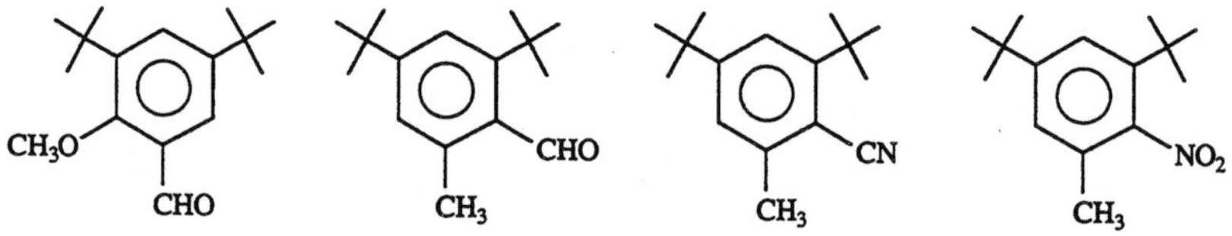

CH_3O CHO CHO CH_3 CN CH_3 NO_2 CH_3

16 17 18 19

Several members of the series of compounds with the general structure 20 have been prepared. Compounds in which R = H (20a) or R = CH_3 (20b) have a strong musk odor. If R = C_2H_5 (20c), the musk odor is strongly attenuated.

If the tertiary butyl groups are para rather than meta to each other in compound 20a, the musk odor disappears entirely and is replaced by a weak cuminic smell (as with compound 21). The separation of the aldehyde function from the aromatic nucleus also has a large influence. Compound 22 has only a weak musk odor and compound 23 is completely odorless. The same observation is made for analogues of compound 20b. When the acetyl group is displaced as in compound 24, no odor is apparent [7.104].

The syntheses of compounds 17, 20a, 20b and 20c from 3,5-ditertiary-butyl toluene, are summarized [7.105]. These musks have been used only sparingly in the perfumery industry, however.

[7.104]

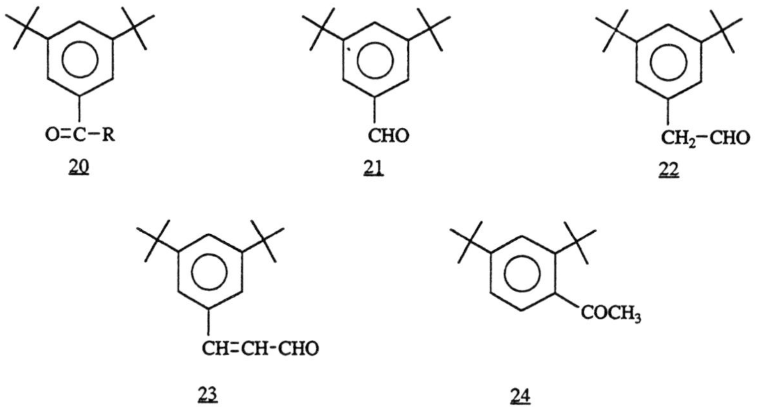

$O{=}C{-}R$

20

CHO

21

$CH_2{-}CHO$

22

CH=CH-CHO

23

$COCH_3$

24

[7.105]

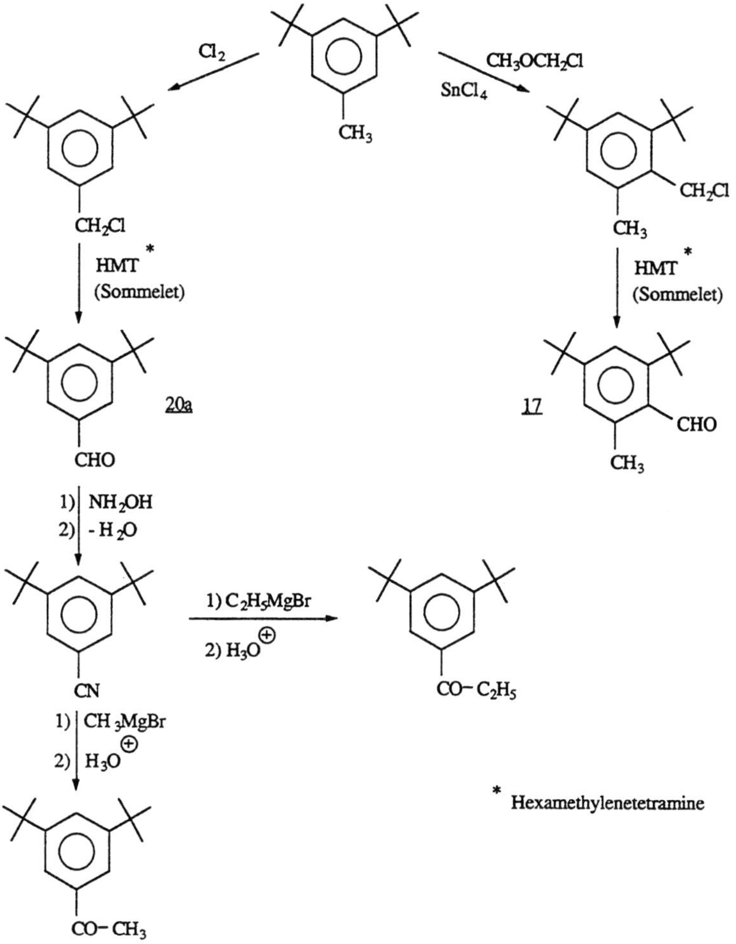

Cl_2

CH_3

CH_3OCH_2Cl

$SnCl_4$

CH_2Cl

HMT *

(Sommelet)

CH_2Cl

CH_3

HMT *

(Sommelet)

20a

CHO

17

CHO

CH_3

1) NH_2OH

2) $-H_2O$

CN

1) C_2H_5MgBr

2) $H_3O^{\oplus}$

$CO{-}C_2H_5$

1) CH_3MgBr

2) $H_3O^{\oplus}$

$CO{-}CH_3$

* Hexamethylenetetramine

It is worthwhile to briefly discuss the mechanism of the Sommelet reaction involved in the above sequences.

In 1913, when Sommelet heated benzylhalides with an excess of hexamethylenetetramine in water or in a mixture of water and alcohol, he discovered that a good yield of the corresponding aldehyde was obtained. The addition of alcohol to the reaction mixture permits a better contact between the halide and the aqueous phase. The other products formed in the course of this reaction are: ammonia, the mono-, di- and trimethylamines, as well as traces of the corresponding benzylamine.

The first phase of the reaction is the formation of a quaternary salt of hexamethyltetramine which does not need to be isolated. This salt is then hydrolyzed under the reaction conditions to give benzylamine. At the same stage, the excess of hexamethylenetetramine is hydrolyzed into ammonia and formaldehyde and these then react together to form methylenimine (CH_2=NH). A hydride ion is then transferred from benzylamine to protonated methylenimine, thereby affording the protonated benzylimine which hydrolyzes to the corresponding aldehyde. This sequence is summarized as follows [7.106].

[7.106]

CH_2 (quaternary hexamethylenetetramine salt) $\xrightarrow{H_2O}$ Ar—CH—NH_2 + CH_2=$NH_2^{\oplus}$ ⟶ Ar—CH=$NH_2^{\oplus}$ $\xrightarrow{H_2O}$ ArCHO + NH_3

CH_3—NH_2 ⟶ $(CH_3)_2NH$ ⟶ $(CH_3)_3N$

Transfer of hydride is completely justified in this case because the reaction media is neutral. In fact, this reaction also operates in diluted acetic acid providing the pH remains between 4 and 6. However, if the reaction is carried out in strongly acidic media, the transfer of hydride cannot occur and the reaction stops at the benzylamine stage. This is called the Delepine reaction.

The presence of electron-withdrawing substituents on the aromatic nucleus diminishes the yield of the Sommelet reaction. Ortho substituents also inhibit it. In such cases, it is preferable to use another reaction, the Kröhnke reaction, which is especially recommended for preparing fragile aromatic aldehydes.

2. *Indanic non-nitro musks*

After Carpenter had shown that non-nitro compounds could also have musk odors, much work was undertaken by different perfumery companies. In 1952, the Dutch firm, Polak's Frutal Works developed the first commercial non-nitro aromatic musk. At this time, the exact structure of this musk was not known and for this reason it was called "Phantolid." It was not until three years later following the work of Granpoloff on "Moskene," that the exact structure of "Phantolid" was demonstrated to correspond to the formula 25.

In 1958, Beets and co-investigators of the Dutch company, Polak and Schwarz, discovered another indanic musk which was commercialized under the name "Celestolide" and correspoonds to formula 26 [7.107].

[7.107]

25

26

The synthesis of "Phantolid" employs in its first phase, the condensation of paracymene with tertiaryamyl alcohol in concentrated sulfuric acid. The mechanism for the formation of the hydrocarbon 27 obtained from this is identical to that previously discussed for the hydrocarbon 9. Acetyl chloride in the presence of aluminum chloride reacts with the hydrocarbon 27 to give "Phantolid" [7.108].

[7.108]

+ 2 OH $\xrightarrow{H^{\oplus}}$ 27 + $\xrightarrow[AlCl_3]{CH_3COCl}$ 25

Before examining the synthesis of "Celestolide," a short discussion of the condensation of isoprene with different benzene hydrocarbons is appropriate. This reaction takes place in the presence of sulfuric acid at 95% and at temperatures between −10° and 0°C. The distribution of the two products 28 and 29 thus obtained is strongly influenced by the nature of the substituent R. This distribution, summarized in Table V, reflects the steric effect of this substituent [7.109].

[7.109]

R + $\xrightarrow{H^{\oplus}}$ R 28 + R 29

Table V

		Distribution (%)
a	R = CH_3	60 + 40
b	R = C_2H_5	70 +30
c	R = isoC_3H_7	80 + 20
d	R = tert-C_4H_9	98
		(Rdt = 70%)

The structure of the compound 28d has been unambiguously determined. Acetyl chloride reacts with this hydrocarbon in the presence of aluminum chloride 28 to give "Celestolide" 26 [7.110].

[7.110]

$CH_3{-}CO{-}CH_3$, $OH^{\ominus}$; 1) H_2 2) CH_3MgI; $H^{\oplus}$; CH_3COCl / $AlCl_3$ — 26, 28d

During the condensation of isoprene onto tertiary butylbenzene in acidic media, for example, the first phase is the protonation of isoprene which forms a delocalized carbocation. This cation is then attacked by the aromatic nucleus at position 1, thereby giving the more stable carbo-cation which then cyclizes to the indane 28d [7.111].

[7.111]

a — 28d — 26

Following the discovery of "Phantolid" and "Celestolide," many analogous compounds were prepared and their olfactive properties studied. These properties are summarized in Tables V and VI. In a general way, as represented here, the intensity of the musk odor decreases from left to right. In each table, the first compound corresponds respectively to "Phantolid" and to "Celestolide" [7.112].

Table VI: Intensity of musk odor as a function of substituents in structure 30. (M/m/0 = strong/weak/absent musk odor)

30	a	b	c	d	e	f	g	h	i	j	k	l	m
R^1	CH_3	CH_3	C_2H_5	C_2H_5	H	CH_3	CH_3	CH_3	H	CH_3	iPr	CH_3	CH_3
R^2	CH_3	CH_3	CH_3	CH_3	CH_3	CH_3	H	CH_3	CH_3	CH_3	CH_3	CH_3	CH_3
R^3	CH_3	C_2H_5	H	CH_3	CH_3	H	CH_3	H	H	CH_3	H	H	H
R^4	CH_3	CH_3	CH_3	CH_3	CH_3	$C_2H.$	CH_3	CH_3	CH_3	H	CH_3	C_3H_7	iBu
Observations	M	M	M	M	M	m	m	m	m	0	0	0	0

Table VII: Intensity of must odor as a function of substituents in structure 34. (VM/M/m/vm = very strong/strong/weak/very weak musk odor)

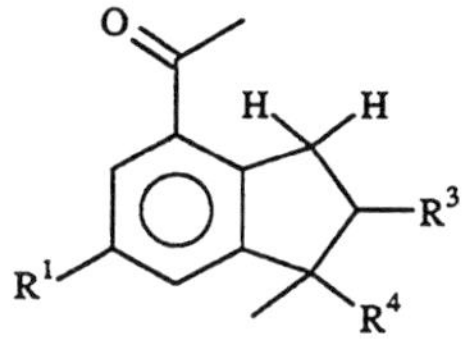

O H H R^3 34 R^1 R^4

34	a	b	c	d	e	f	g
R^1=	t-Bu	t-Bu	iPr	[illegible]	[illegible]	t-Bu	t-Bu
R^3=	H	H	H	H	H	H	CH_3
R^4=	CH_3	C_2H_5	CH_3	CH_3	CH_3	H	CH_3
Observations	VM	M	M	m	m	vm	vm

[7.112]

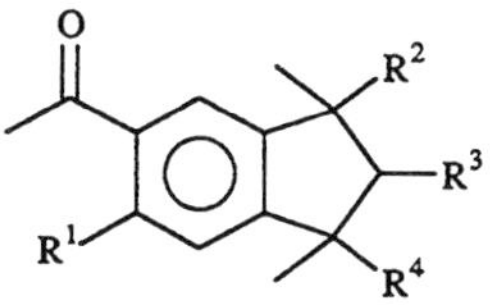

O R^2 R^3 30 R^1 R^4

If the acetyl group is replaced by a formyl group, as shown in the general formula 31, the compounds have musk odors that are near to those of the corresponding compounds with the general structure 30 and for which R_3 and R_4 are the same. Hpwever, replacing the acetyl group of "Phantolid" by a propionyl group gives a clearly less "musky" compound. Similarly, compounds 32 and 33 are only weakly "musky" [7.113].

[7.113]

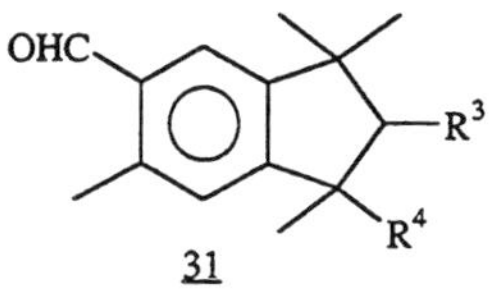

OHC R^3 R^4 31 O 32 O 33

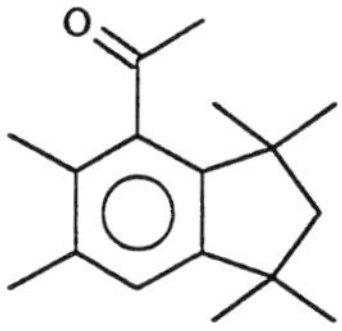

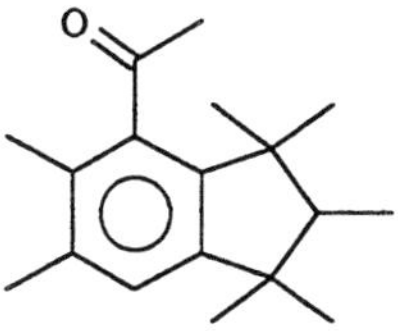

3. *Tetralinic non-nitro musks*

Research in this series has been even more considerable than in the preceding series. Consequently, two very important new musks were discovered, one by Polak's Frutal Works ("Tonalid" 35) and the other by Givaudan ("Versalide" 36).

The almost immediate commercial success of these substances, as well as that of the preceding two indanic derivatives, is due both to their excellent odor and to their fixative properties which are rather analogous to those of the macrocyclic musks. In addition, they are less costly than the macrocyclic musks. They are also colorless and, in contrast to the nitro musks, they are much more soluble in alcohol and do not produce coloration in the presence of other perfume ingredients or when exposed to light.

Following these discoveries, nearly 100 different tetralinic musks have been prepared. In the course of these syntheses, variations in both the aromatic nucleus and in the saturated ring have been explored [7.114].

[7.114]

O O

35 36

Variations in the aromatic nucleus

Three types of general structures can be defined: 37, 38 and 39. The synthesis of these different compounds always necessitates two steps. The first is to prepare an intermediate tetralinic hydrocarbon with the general structure 41. The second involves the introduction of the group R^2. The intermediate 41 is obtained by cycloalkylation of the aromatic hydrocarbon 40 with 2,5-dichloro–2,5-dimethylhexane, which has been prepared by chlorination of commercially available 2,5-dimethyl–2,5-hexanediol [7.115].

[7.115]

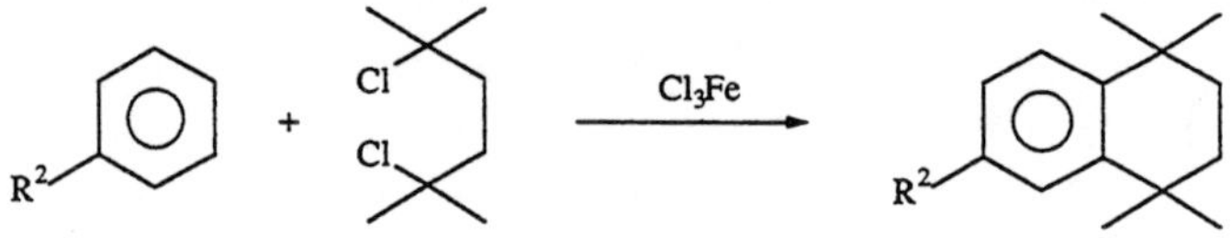

R^2 + Cl Cl —Cl_3Fe→ R^2

40 41

For the synthesis of "Versalide" (36 ≡ 37a), acetyl chloride reacts in the presence of aluminum chloride with the tetralin 41 in which R^2 is an ethyl group.

Tables VIII, IX, and X summarize the olfactive properties of a certain number of compounds belonging to the series generalized by structures 37, 38 and 39.

In Table VIII, the compounds 37c and 37d have very strong musk odors, but they are phototropic (in other words, they color strongly in light and decolorize again in darkness). Compound 37b was the first substance in this series to be synthesized; however, its olfactive properties are inferior to those of "Versalide."

The compounds in Tables IX and X are all practically odorless. The same applies to compounds 42–46 [7.116].

Condensation of an alkylbenzene (C_6H_5R) with 2,2,5,5-tetramethyl–3-furanone in the presence of aluminum chloride gives the β-tetralone 47 (in the form of a mixture of two products when R ≠ H).

When R = H, the resulting tetralone has a sweet camphoraceous odor. When R

Table VIII

37

37	R^1	R^2	Observations
a	Ac	Et	M
b	Ac	Me	M
c	CHO	Me	M
d	CHO	Me	M
e	Ac	i-Prop	vm
f	Ac	MeO	vm
g	Prop	Et	vm
h	Prop	Me	vm
i	Ac	H	O
j	CHO	MeO	O
k	Prop	i-Prop	O
l	i-But	Et	O
m	Ac	Br	O
n	Ac	CN	O
o	$CH = CH\text{-}COCH_3$	H	O
p	$CH(CH_3)CHO$	Et	O
q	Ac	F	
r	Ac	Cl	
s	CHO	H	

Notes:
Ac = CH_3CO;
i-But = >-CO
M = strong musk odor;
O = odorless;
Prop = CH_3CH_2CO;
vm =very weak odor;
37q and 37r have moldy odors and 37s has a slight amber odor.

Table IX

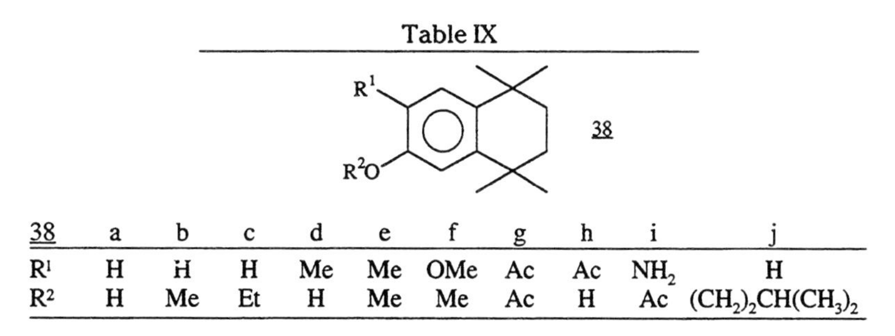

38	a	b	c	d	e	f	g	h	i	j
R^1	H	H	H	Me	Me	OMe	Ac	Ac	NH_2	H
R^2	H	Me	Et	H	Me	Me	Ac	H	Ac	$(CH_2)_2CH(CH_3)_2$

= CH_3, the odor is woody. The two acetylated derivatives 48 and 49 are odorless. Hydrogenation of the tetralone 47 gives an odorless tetralol 50. The acetylated dehydrated product 51 has a musk odor which is not very persistant while its homologue 52 is odorless [7.117].

Table X. Variations in the saturated ring

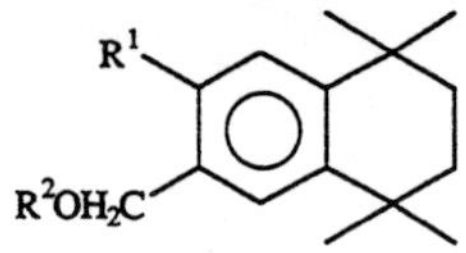

39

38	a	b	c	d	e	f
R¹	Me	Me	Et	Et	Ac	Me
R²	Me	Ac	Me	Ac	Me	$(CH_2)_2CH(CH_3)_2$

[7.116]

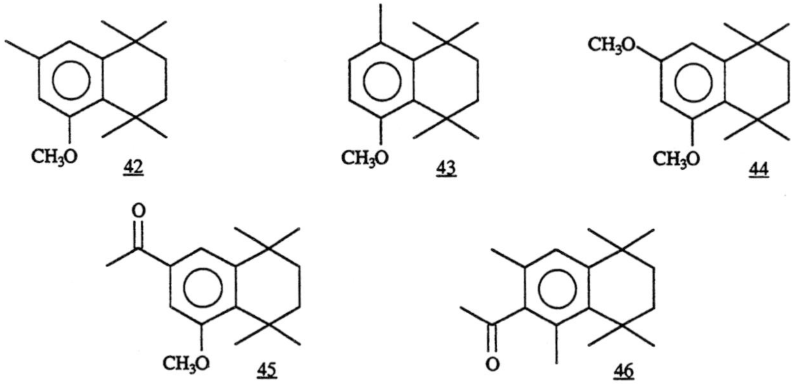

42 43 44 45 46

[7.117]

47 48 49 50 51 52

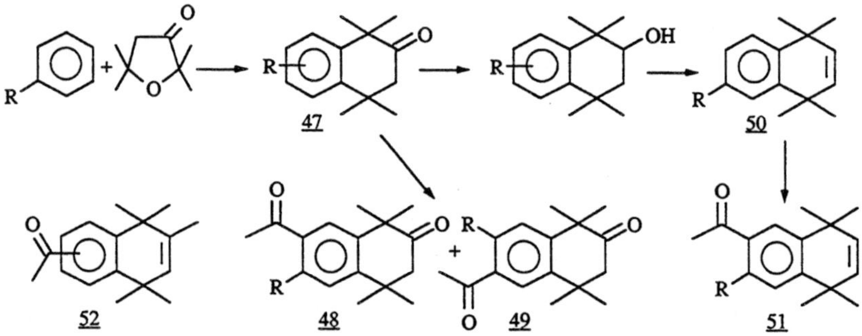

In [7.118] we have collected some of the procedures used for preparing the intermediate tetralins that have a methyl group in position 3. "Tonalid" 35 belongs to this group, but it can be prepared far more simply by using a process indicated earlier for the preparation of the hydrocarbon 11.

The compounds in this series are represented by the general formula 56 in Table XI. In this table, the six first compounds have strong musk odors, however, the first, 56a, has the most interesting odor: this is "Tonalid." The last seven

[7.118]

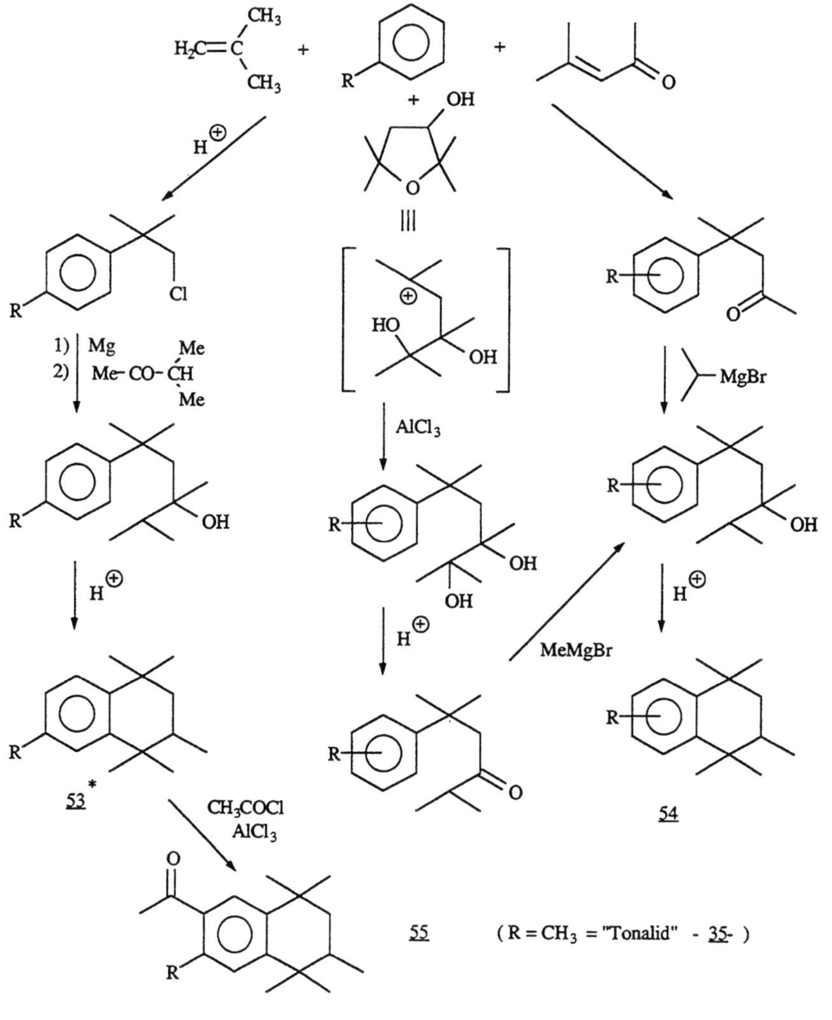

53*

54

55 (R = CH_3 = "Tonalid" - 35-)

(* For R = CH_3, 53 ≡ 11 § 3)

compounds in this table are odorless or possess non-musky odors. Compound 57 is also musky while 58 is non-musky and practically odorless [7.119].

4. *Tricyclic non-nitro musks*

The tricyclic musks can be grouped into three categories: hydrindacene derivatives; tetrahydronaphthindanone derivatives; and acenaphthene derivatives.

a. Hydrindacene derivatives

Condensation of meta bis-chloromethylbenzene with methallyl magnesium chloride yields a hydrocarbon which cyclizes in acidic media to form a mixture of two

Table XI

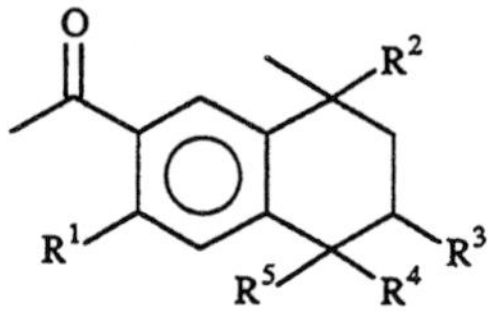

56

56	a	b	c	d	e	f	g	h	i	j	k	l	m
R^1=	Me	Me	Me	Me	Me	Me	Me	Et	Me	H	Me	Et	iPr
R^2=	Me	H	Me	H	Me	H	Me	Me	Me	Et	Et	Et	Et
R^3=	Me	H	H	H	H	Me	H	H	H	H	H	H	H
R^4=	Me	Me	Me	Me	Me	Me	H	H	iPr	Me	Me	Me	Me
R^5=	Me	Me	H	Et	Et	Me	H	H	H	Et	Et	Et	Et

[7.119]

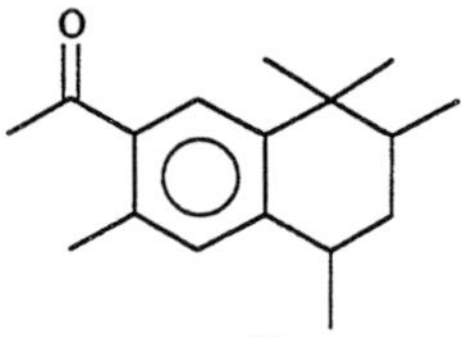

57

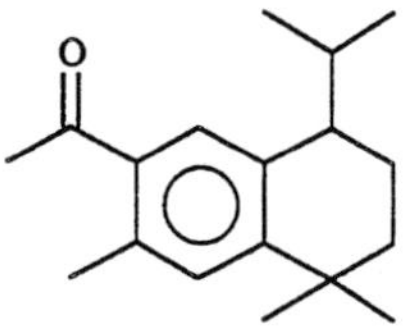

58

hydrindacenes. One is symmetrical (1,1,7,7-tetramethyl s-hydrindacene 59) and the other is asymmetrical (1,1,6,6-tetramethyl as-hydrindacene 60). The acetylated derivative of 59, with the structure 61, possesses a weak musk odor. However, the compound 62 has a medium but persistant musk odor, while 63 has a strong musk odor. In contrast, compound 64 has almost no musk odor and compound 65 is odorless.

The polyalkylhydrindacenones are also found in this category and several of these have strong musk odors. This is the case for the first four compounds in Table XII. The last two are odorless [7.120].

A second series of hydrindacenones with the general stucture 67 are shown in Table XIII. The first four compounds in this table also have strong musk odor while it is only weak for the last two.

b. Tetrahydronaphthindanone derivatives

Condensation of β-chloropropionic acid chloride with the polyalkyltetralin 68 in the presence of aluminum chloride leads to the chloroketone 69. Cyclization of this in acidic media with heating gives a tetrahydronaphthindanone 70 when $R^2 \neq$ H and another tetrahydronaphthindanone 71 when $R^2 = H$ [7.122].

In an analogous way, acrylic acid chloride reacts with the tetralin 68 in the presence of aluminum chloride to give the ketone 72. Cyclization of this ketone in acidic media gives tetrahydronaphthindanones which have either structure 73 (when $R^2 \neq$ H) or structure 74 (when $R^2 = H$) [7.122].

Table XII

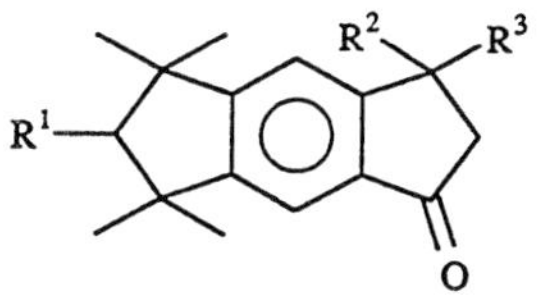

66

66	a	b	c	d	e	f
R^1 =	H	H	Me	Me	H	Me
R^2 =	H	Me	H	Me	Me	Me
R^3 =	H	H	H	H	Me	Me

[7.120]

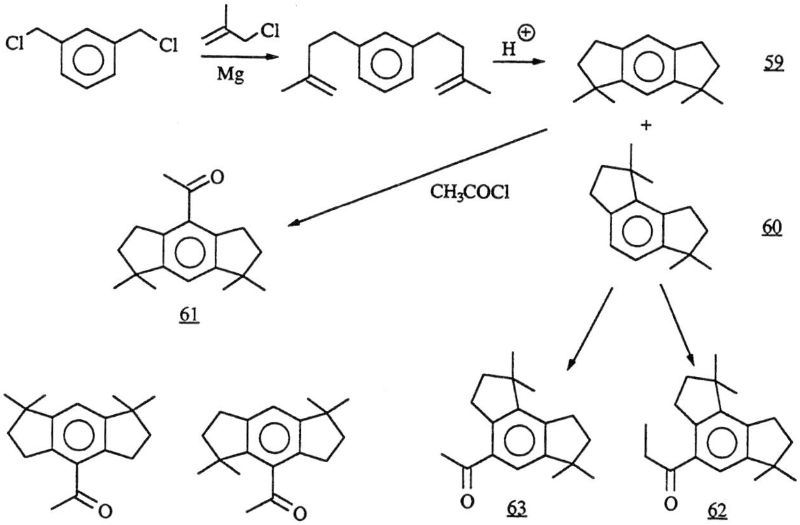

CH_3COCl

59 + 60 → 61; 63; 62

64 = 61

65

Table XIII

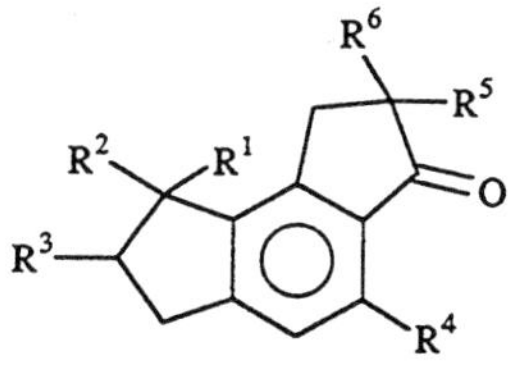

67

67	a	b	c	d	e	f
R^1 =	Me	Me	Me	Me	H	H
R^2 =	Me	Me	Me	Me	H	H
R^3 =	H	H	Me	Me	H	H
R^4 =	Me	Me	Me	Me	H	H
R^5 =	H	Me	H	Me	Me	Me
R^6 =	H	H	H	H	H	Me

[7.122]

The general formula 75 in Table XIV represents the group to which structures 70 and 73 belong. All of the compounds represented in this table have strong musk odors with the exception of the last, which is the weakest in this regard.

These compounds are listed in the order of increasing numbers of carbon atoms. It is quite remarkable and unforeseeable that compounds 75e and 75f have strong musk odors with 20 carbon atoms per molecule. In fact, these are probably the only odorant C_{20} compounds (molecular weight: 284). Generally, the upper limit for odorant compounds is around C_{18} and C_{19} depending on structure. It can be noted, however, that the structures of these compounds are extremely compact and possess very little conformational mobility, with the notable exception of 75g

Table X

75	a	b	c	d	e	f	g
R^1=	H	H	H	Me	Me	Me	H
R^2=	Me	Me	Et	Me	Me	Et	iPr
R^3=	H	Me	H	H	Me	H	H
C_n	18	19	19	19	20	20	20

for which the musk odor is indeed weaker (see also the general comments on musk substances in the section on macrocyclic musks).

In addition to these, the compounds with structures 76, 77 and 78 also have strong musk odors, whereas those with the structures 79, 80, 81, 82 and 83 are odorless [7.123].

[7.123]

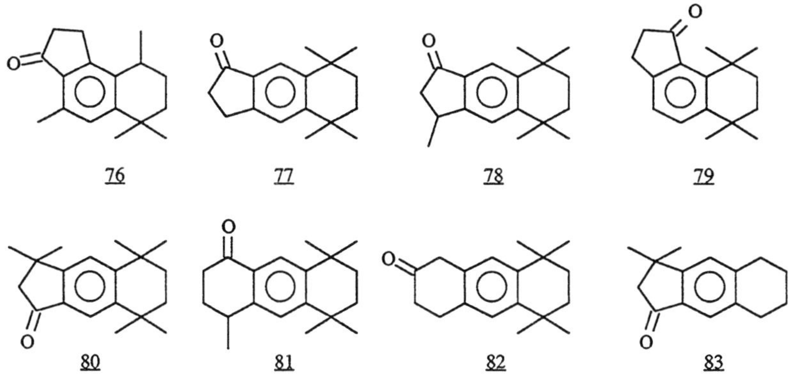

76 77 78 79

80 81 82 83

c. Acenaphthene Derivatives

We will only describe here one member of this series, compound 85, which is an acetylated derivative of acenaphthene 84 and has a strong musk odor [7.124].

[7.124]

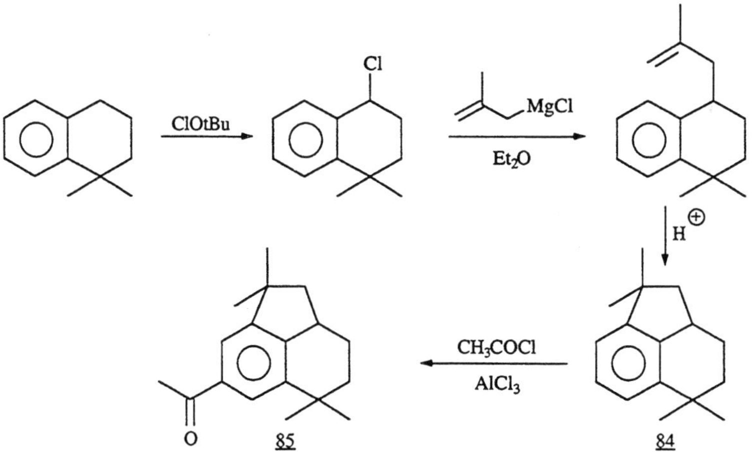

ClOtBu

MgCl

Et_2O

$H^{\oplus}$

CH_3COCl

$AlCl_3$

85 84

5. Isochromanic non-nitro musks

In 1967, Heeringa and Beets of International Flavors and Fragrances, Inc. discovered two new musks with isochromane structures. The first, "Galaxolide" 86, had an indane nucleus and the second, "Musk 89" 87, had a tetralin nucleus: These have a very powerful but refined musk odor with excellent tenacity. Moreover, they are stable in perfumes, soaps, and detergents and do not give rise to coloration of the bases of these products. Thanks to these properties, these two musks have enjoyed great commercial success, particularly "Galaxolide" 86. This product is commercialized under the name "Galaxolide 50" as a 70–75% solution in diethyl phthalate.

We have already discussed the methods used to prepare the indane and tetralin nuclei. We will therefore only summarize the syntheses of these two musks [7.125]. The synthesis of "Galaxolide" starts with α-methylstyrene and that of "Musk 89" starts with hydratropic alcohol.

[7.125]

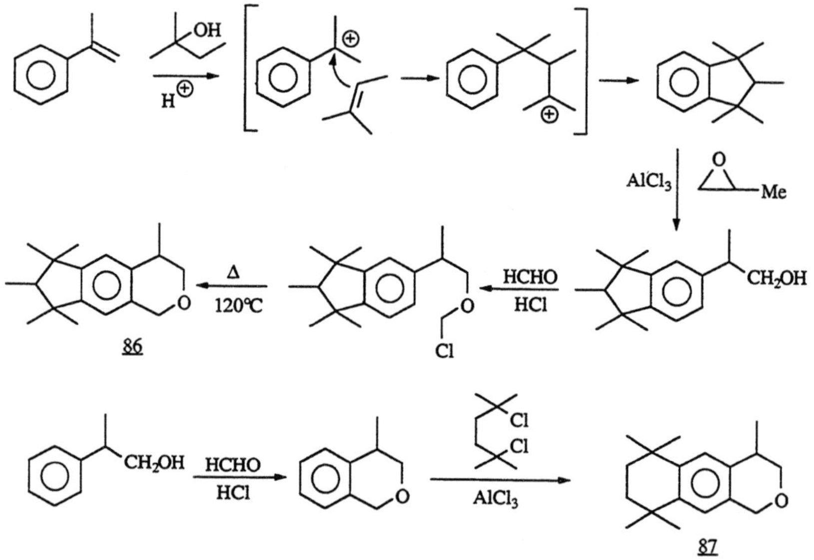

In addition to these two musks, several other isochromane compounds have been prepared. Table XV brings together most of the compounds with the general structure 88. These are analogous to "Galaxolide" in that they have an indane nucleus. In this table, 88a is "Galaxolide." The next three compounds also have a strong musk odor, although this is weaker for 88c. Compound 88f and compound 89, an isomer of 88e, are both odorless.

Table XVI brings together the compounds that have a tetralin nucleus in analogy to "Musk 89." These correspond to the general structure 90 and the first of these, 90a, is "Musk 89" itself. The next five substances also have strong musk odors. The compounds with structures 90f and 90g have a musk odor of medium intensity, while the last three are practically odorless. In this regard, it is interesting to note that 90j is a C_{20} compound ($C_{20}H_{30}O$ = 286). In addition, the compound 91 is practically odorless while its isomer 90b has a strong musk odor.

In addition to changing the position of the oxygen, replacement of this atom with nitrogen has also been tried. Three compounds 92, 93 and 94 have been pre-

Table XV

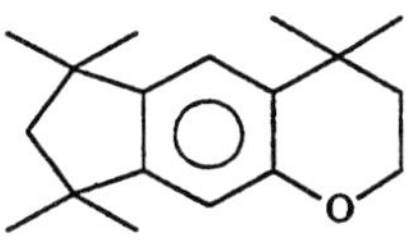

89

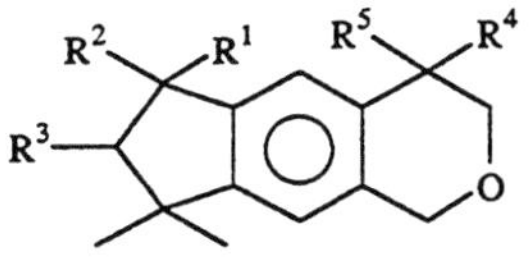

88

88	a	b	c	d	e	f
R^1	Me	Me	Me	Me	Me	H
R^2	Me	Me	Me	Me	Me	H
R^3	Me	H	H	H	H	H
R^4	Me	H	Me	Me	Me	Me
R^5	H	H	H	Me	Me	Me

pared with this in mind, but they all present only weak musk odors. Alternatively, the hydrogenated compounds 95, 96, 97 and 98, which are all derived from "Galaxolide" and "Musk 89," have, according to Beets, weak musk odors with a woody amber inflection [7.126].

More recently, a new tricyclic compound derived from dihydrocoumarin has been prepared. This compound is 1,1,4,4-tetramethyltetrahydro α-pyrone 99 and it is reported to have a musk odor [7.127].

Finally, in 1972, a group at Givaudan synthesized the first phenol to be found to have a musk odor. This compound, 5,7-diisopropyl–1,1-dimethyl–6-indanol 100, has been prepared by condensing isoprene with 2,6-diisopropylphenol in the presence of sulfuric or phosphoric acids [7.128].

Table XVI

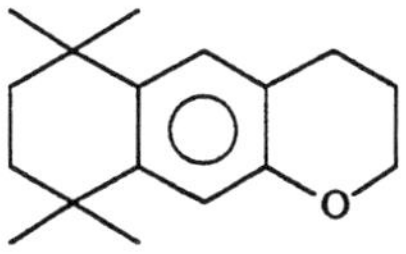

91

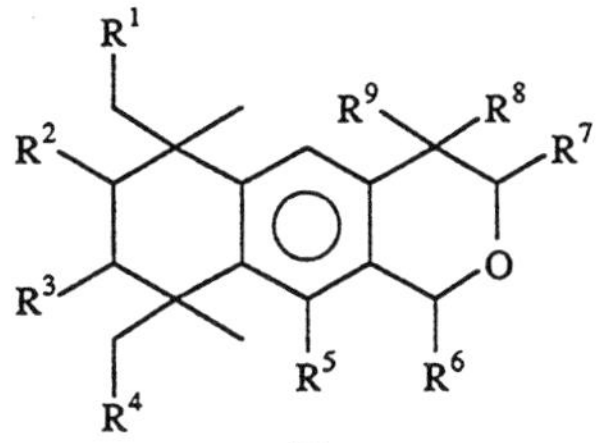

90

90	a	b	c	d	e	f	g	h	i	j
R^1	H	H	H	H	H	H	H	H	H	Me
R^2	H	H	H	Me	Me	H	H	H	H	H
R^3	H	H	H	Me	H	H	H	H	H	H
R^4	H	H	H	H	H	H	H	H	H	Me
R^5	H	H	Me	H	H	H	H	H	H	H
R^6	H	H	H	H	H	H	H	Me	H	H
R^7	H	H	H	H	H	H	H	H	Me	H
R^8	H	H	H	H	H	Me	H	H	H	H
R^9	Me	H	H	H	H	Me	Et	H	H	Me

[7.126]

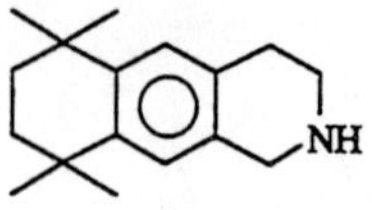

NH

92

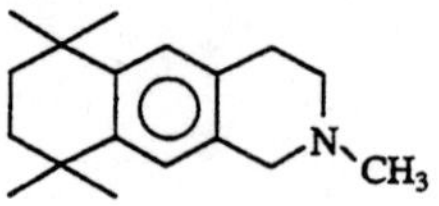

N CH$_3$

93

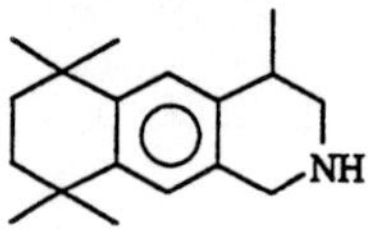

NH

94

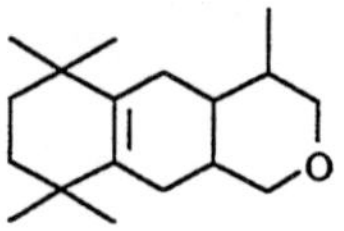

O

95

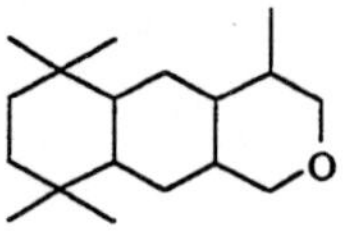

O

96

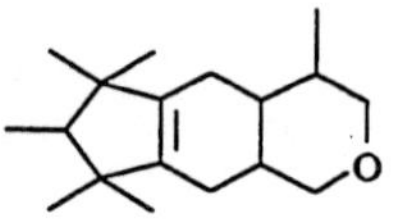

O

97

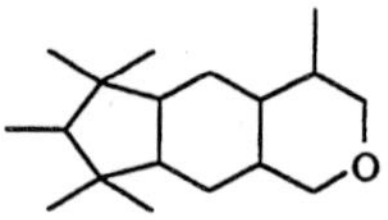

O

98

[7.127]

O O

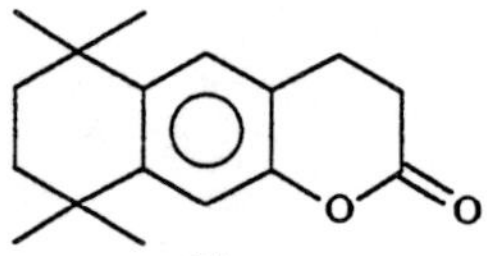

99

[7.128]

HO + $\xrightarrow{H^{\oplus}}$ HO

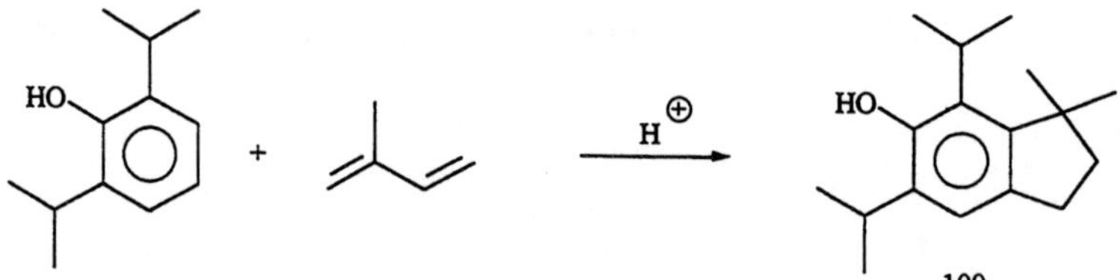

100

APPENDIX

1

The Concept of Prochirality

A. Introduction

It is possible to obtain a simplified structural description of a given compound by inspecting its symmetry properties, thus permitting an internal comparison of its component parts. Two types of stereoisomerism are shown in the following models.

1) Molecule 1 has second order symmetry in which a rotation of 180° around the axis indicated, superimposes $CH_{3(a)}$ to $CH_{3(b)}$. The same is true for the groups $CO_2H_{(a')}$ and $CO_2H_{(b')}$. [8.01]

[8.01]

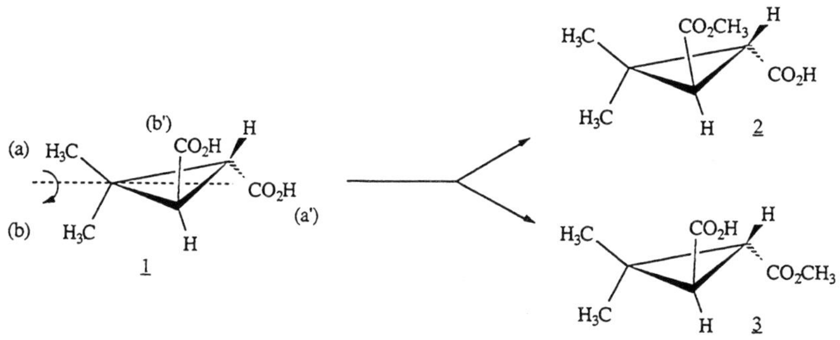

(b') (a) H_3C CO_2H H CO_2H (a') (b) H_3C H 1

H_3C CO_2CH_3 H CO_2H H_3C H 2

H_3C CO_2H H CO_2CH_3 H_3C H 3

It follows that the CO_2H groups are geometrically equivalent and have identical physicochemical properties; thus, a reagent which reacts on carboxylic acid groups will react with each fuctionality (a') and (b') at exactly the same rate to give only one product (i.e., 2 ≡ 3).

2) Alternatively, there is another class of molecule with a single plane of symmetry which has two regions (a) and (b) that are not strictly identical, defined as "enantiotopic," on opposite sides of this plane. For example, the two ester functions of dimethyl ethylmethylmalonate 4 are enantiotopic. [8.02]

Monosaponification of the ester function (b) affords mono-ester 5, whereas the same reaction on the ester function (a) gives its antipodal mono-ester 6. Monosaponification of diester 4 by a non-chiral base such as sodium hydroxide would give a racemic mixture of 5 and 6, the rate of attack on the ester functions (a)

[8.02]

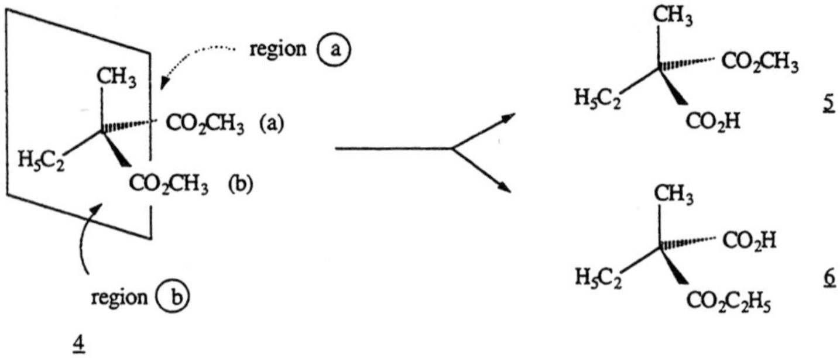

and (b) being the same. In contrast, a chiral (non-racemic) reagent is capable of distinguishing between the two enantiomeric products and thus the two enantiomeric regions. It should thus be theoretically possible, by employing an optically active base of defined configuration, to selectively hydrolyze either ester (a) or (b).

3) A third class is represented by 7, a chiral (non-racemic) compound which has no plane of symmetry. [8.03]

[8.03]

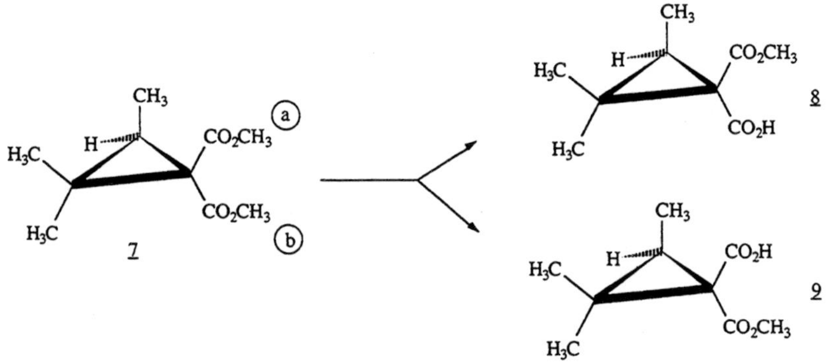

Here the groups $CO_2CH_{3(a)}$ and $CO_2CH_{3(b)}$ are clearly non-equivalent. Their environment is very different: $CO_2CH_{3(a)}$ is cis to the two CH_3 groups and $CO_2CH_{3(b)}$ is cis to an H atom and a CH_3 group. Monosaponification of $CO_2CH_{3(a)}$ gives 9, the diastereoisomer of mono-ester 8, which results from monosaponification of the other ester group $CO_2CH_{3(b)}$. For this reason, the two groups $CO_2CH_{3(a)}$ and $_{(b)}$ are termed "diastereotopic" and a non-chiral base such as sodium hydroxide would not attack them at the same rate, the ester function (b) reacting more rapidly than (a).

Although NMR spectroscopy cannot distinguish enantiotopic substituents, it can, however, make a distinction between two diastereotopic groups. Thus, in 7, the CH_3 groups of (a) and (b) exhibit distinct signals in their ^1H and ^{13}C NMR spectra. [8.04]

[8.04]

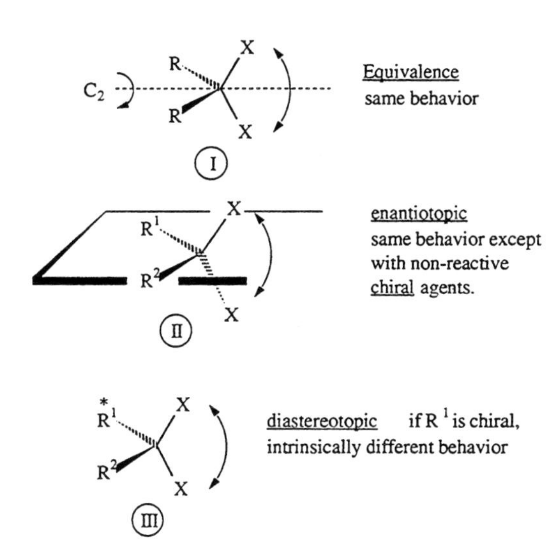

there is creation of an asymmetric carbon (non-racemic chiral center). Hanson proposed the term "prochiral" for every carbon atom $(R^1R^2)CX_2$ which is bonded to two identical groups X. Thus, there is a prochiral carbon in structures II and III. Stereoselective transformation of this carbon to an asymmetric carbon (R^1R^2) CXY in a preferred configuration (R or S) constitutes an asymmetric synthesis.

B. Nomenclature of prochirality

The necessity for a nomenclature of prochirality is self-evident. A careful observer is always capable of locating one of the two substituents X. Structure III contains an internal reference, the chiral group R^1, which facilitates analysis and in the cyclic compound 7, the ester function $CO_2CH_{3(b)}$ is readily located by its cis-stereochemical relationship to the H atom on the neighboring carbon.

Nevertheless, in acyclic compounds, the structure is more complex and it is easier to first choose an arbitrary configuration. For prochiral structures such as II, the substituents X can be distinguished by using an *external chirality*—an observer, for example. This observer, placing his feet on R^{*1} and his head on R^2, looking towards the prochiral carbon, will always see the X substituent which is above the plane on his left. On his right he will see the X substituent which is below this plane. [8.05]

Hanson proposed the terms "pro-R" and "pro-S" for the two X substituents which are enantiotopic (for II) and diastereotopic (for III). The convention is as follows: the group X will be pro-R if its replacement generates a carbon of configuration R when the replacing group is supposed to have priority over the other groups, using the Cahn, Ingold, and Prelog classification rule. [8.06]

[8.05]

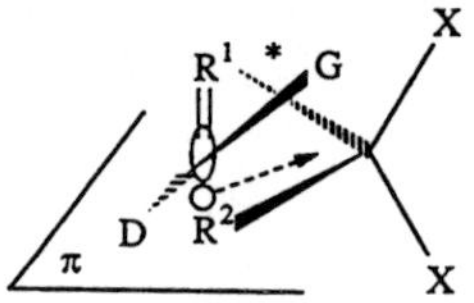

[8.06]

If for diester 4 $CO_2CH_{3(a)}$ is arbitrarily given priority over $CO_2CH_{3(b)}$, the classification order is the following: $CO_2CH_{3(a)} > CO_2CH_{3(b)} > C_2H_5 > CH_3$.

By looking along the the axis of the C-CH_3 bond and on the opposite side from CH_3, the foregoing order corresponds to an anticlockwise movement: $CO_2CH_{3(a)}$ is thus pro-S and $CO_2CH_{3(b)}$ is pro-R.

The concept of prochirality is immensely important in biochemistry because enzyme systems are capable of reacting stereospecifically with one of the two X substituents in a prochiral center.

A number of enzymatic reductions require a coenzyme derived from nicotinamide of partial structure: [8.07]

[8.07]

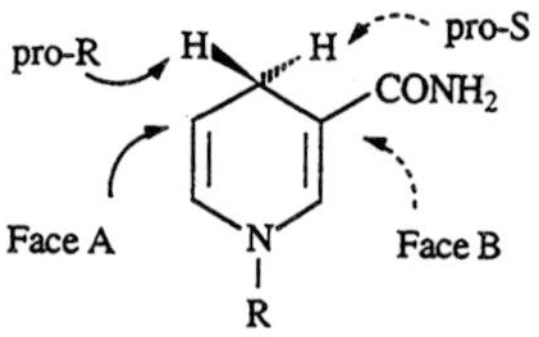

This compound is a hydride donor (H^-) using either the pro-R (H) or the pro-S (H) (defined using the preceding conventions). Depending on the enzyme systems employed, it is either the pro-R (H) (face A) or the pro-S (H) (face B) which is used in the enzymatic reaction. Enzymes capable of transferring the pro-R hydride belong to class A, whereas those which transfer the pro-S hydride are members of class B. [8.08]

[8.08]

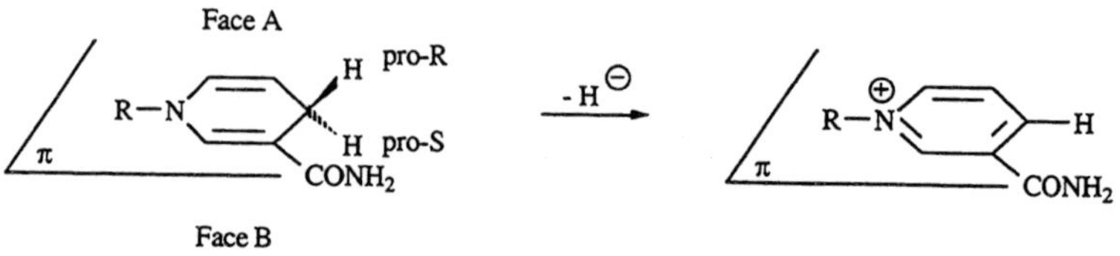

C. Face designation in trigonal systems

Related to the idea of prochirality is the problem of unambiguously describing the two faces of a trigonal system. The corresponding rules are particularly useful in the stereochemical description of reactions, enzymatic or otherwise, which affect a trigonal carbon atom.

1) Denomination of the faces of a system of the Yghi type

If the substituents of a trigonal atom are in descending order of priority (Cahn, Ingold, Prelog rule) in a clockwise sense, the face on which the observer is positioned is termed "re" (rectus); in the other case (anticlockwise sense), the face is designated "si" (sinister). [8.09]

[8.09]

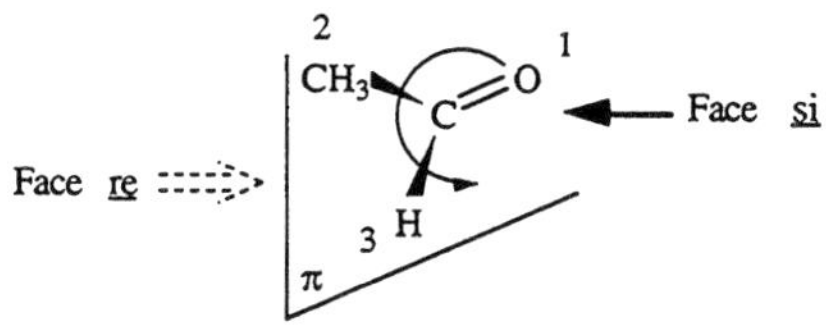

2 CH_3 C O 1 Face si

Face re 3 H π

2) System >Y=Z<

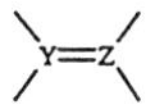

The faces corresponding to Y and Z are defined separately.

Examples: [8.10]

[8.10]

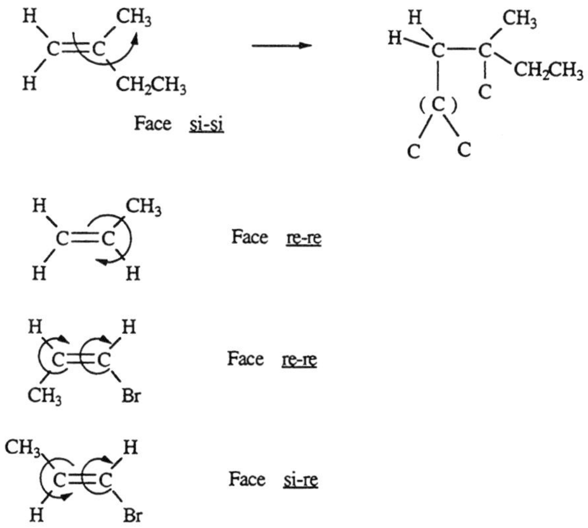

H, H, C=C, CH_3, CH_2CH_3 → H, H, C—C, CH_3, CH_2CH_3, C, (C), C C

Face si-si

H, H, C=C, CH_3, H — Face re-re

H, CH_3, C=C, H, Br — Face re-re

CH_3, H, C=C, H, Br — Face si-re

The utility of this system can be illustrated by taking the hydration of fumaric acid to give S-malic acid as an example. [8.11]

[8.11]

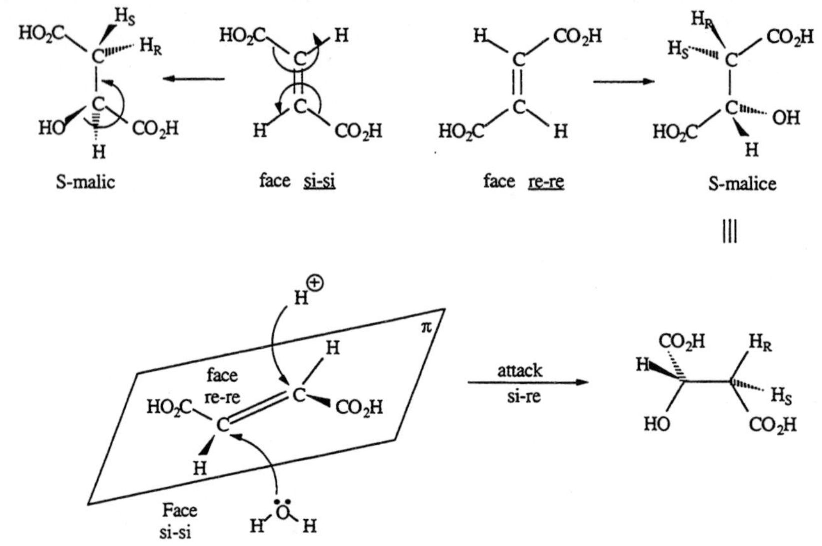

S-malic   face si-si   face re-re   S-malice

attack si-re

It can also be seen that the reduction by NADPH of oxaloacetic acid to give S-malic acid as a result of addition of hybride to the re face of the carbonyl.[8.12]

[8.12]

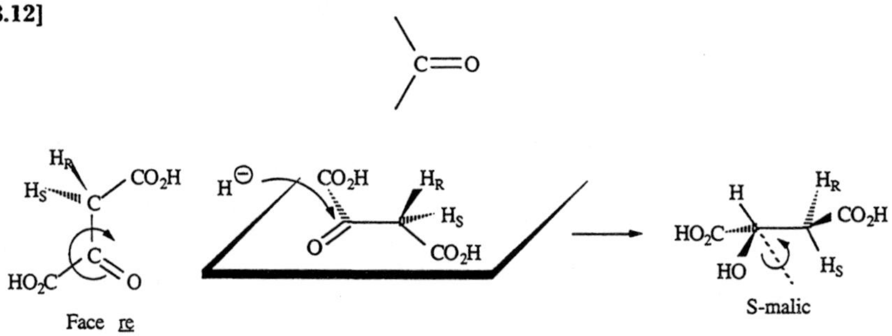

Face re   S-malic

These concepts have played a primordial role in the understanding of the stereochemistry and of the mechanisms of the biosynthesis of terpenoids. Elucidation of these mechanisms has allowed the "acetic" and the "isoprenic" hypotheses to be reconciled.

APPENDIX

2

The Wittig Reaction

This method of olefin synthesis was discovered in 1953 by Wittig and Giessler. It involves the addition of an alkylidenephosphorane 1 (in which R can be an aryl or and alkyl group) to a carbonyl compound 2, followed by elimination of the phosphine oxide 4 through an intermediate betaine 3 to give the olefin 5. [9.01]

[9.01]

$$R_3P{=}C(R_1)(R_2) + (R_3)(R_4)C{=}O \longrightarrow R_3\overset{\oplus}{P}{-}C(R_1)(R_2){-}C(R_3)(R_4){-}\overset{\ominus}{O} \longrightarrow R_3PO + (R_1)(R_2)C{=}C(R_3)(R_4)$$

1 2 3 4 5

The main advantage of this method is in the specific replacement of the carbonyl group with a carbon-carbon double bond without the formation of isomeric olefins.

Another advantage of this method is the mildness of the conditions employed. The reaction medium is generally basic and, as a consequence, this is the only method that can be applied to the synthesis of acid-sensitive olefins such as terpenoids, polyenes, carotenoids, vitamins, etc.

The Wittig reaction synthesis constitutes an example of the general elimination reaction. [9.02]

[9.02]

$$\overset{\oplus}{P}{-}X,\ \overset{\ominus}{O}{-}Y \longrightarrow \overset{\oplus}{P}{-}\overset{\ominus}{O} + X{=}Y$$

The phosphorus pentachloride-induced dehydration of amides to nitriles through the betaine 6 represents another example of this type of elimination. [9.03]

[9.03]

$$R{-}C({=}O){-}NH_2 + PCl_5 \longrightarrow R{-}C({=}O){-}N{=}PCl_3 \longleftrightarrow R{-}C{=}N(\overset{\ominus}{O})(\overset{\oplus}{P}Cl_3) \longrightarrow POCl_3 + R{-}C{\equiv}N$$

6

The existence of pentaalkylidenephosphoranes shows that phosphorus, unlike nitrogen, is capable of pentavalency by extending its valence layer to 10 electrons by including the d orbitals. These alkylidenephosphoranes can then be considered as resonance hybrids with two contributing structures: the ylid form 7a and the ylene form 7b. [9.04]

[9.04]

$$\underset{\underline{7a}}{R_3\overset{\oplus}{P}-\overset{\ominus}{C}(R_1)(R_2)} \longleftrightarrow \underset{\underline{7b}}{R_3P=C(R_1)(R_2)}$$

In order to prepare the Wittig reagent, the phosphonium salt must be prepared. It can be readily obtained from the organic halide and a trialkyl- or triarylphosphine. Triphenylphosphine $[(C_6H_5)_3P]$ is the most used. [9.05]

[9.05]

$$\phi_3P + R'X \longrightarrow \phi_3\overset{\oplus}{P}R'\ \overset{\ominus}{X}$$

The resulting salts are generally very stable, highly crystalline and have well defined melting points. They are then treated with a base which eliminates a proton and gives rise to a phosphorane. The choice of the operating conditions depends entirely on the nature of the desired ylid. [9.06]

[9.06]

$$\phi_3\overset{\oplus}{P}-CH(R_1)(R_2) + \overset{\ominus}{B} \rightleftharpoons \phi_3\overset{\oplus}{P}-\overset{\ominus}{C}(R_1)(R_2) \longleftrightarrow \phi_3P=C(R_1)(R_2) + HB$$

It can be said, in general, that the Wittig reaction with a carbonyl compound proceeds in two steps: the first involves formation of an intermediate which is either a "betaine" or its derivative "oxaphosphetane." The rate determining step can be either of the following:

- the formation of the intermediate, betaine or oxaphosphetane;
- the decomposition of this intermediate.

The free electron pair in the "ylid" form provides alkylidenephosphoranes with nucleophilic properties. [9.07]

[9.07]

betaine ⇌ oxaphosphetane

The betaine and oxaphosphetane forms can be in equilibrium and the accepted mechanism of the decomposition of these intermediates passes through a four-centered transition state which is already preformed in the oxaphosphetane. This is an example of a "syn"-elimination. In the case of the betaine, the charges remain separate. [9.08]

[9.08]

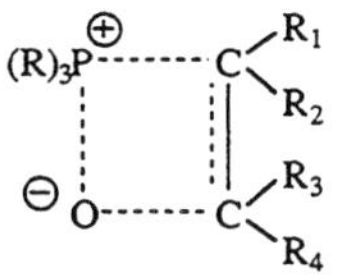

The rate of this step can be slowed by R substituents which diminish the electropositivity of phosphorus (by means of +I, +M effects or by hyperconjugation) and hence its affinity for the oxygen atom. It can also be accelerated by R_1 - R_4 substituents which can conjugate with the incipient double bond in the transition state. It is also clear that factors which stabilize the betaine will also impede its decomposition.

An example of this is seen in the reaction of very reactive phosphines such as $(MeO\Phi)_3P{=}CH_2$ with benzaldehyde. The resulting betaine is stable because the electropositivity of the phosphorus has been reduced to the extent that formation of the phosphine oxide cannot take place. This theoretically interesting example is of no interest in the area of synthesis, however. [9.09]

[9.09]

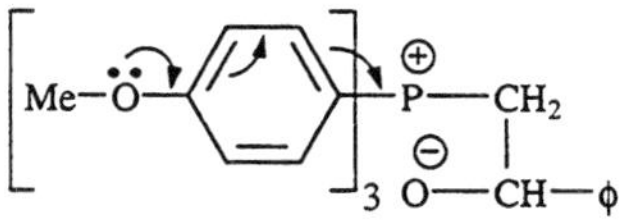

Conversely, the intermediate betaines can practically never be isolated during the synthesis of olefins involving the participation of stabilized phosphoranes such as $\Phi_3P{=}CH{-}CO{-}R'$. In this case, the first step is the slower of the two and is followed by a very rapid elimination in which the forming double bond is stabilized by conjugation with the carbonyl group. [9.10]

[9.10]

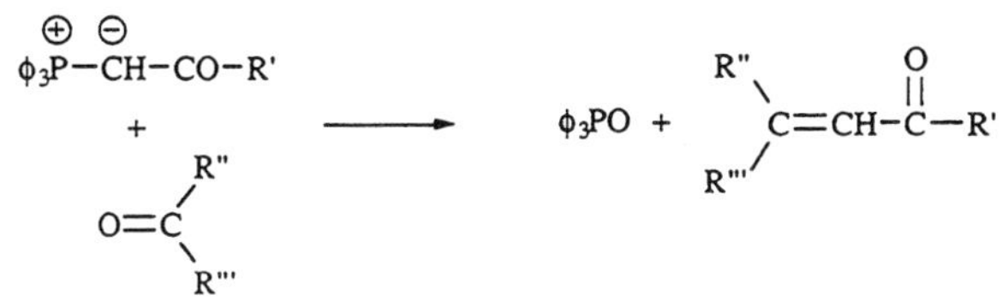

$$\overset{\oplus}{\phi_3P}-\overset{\ominus}{CH}-CO-R' + O{=}C(R'')(R''') \longrightarrow \phi_3PO + (R'')(R''')C{=}CH-\overset{O}{\overset{\|}{C}}-R'$$

However, in the case of very reactive unstabilized phosphoranes, the betaine is formed very rapidly and its decomposition constitutes the slower step. We see a concrete example of this latter case in the synthesis of lavandulol. The first step of this synthesis is to prepare the β-ketoester A starting with isoprene. The second step is a Wittig reaction. [9.11]

Methyltriphenylphosphonium iodide, obtained as a precipitant (MP: 176°–177°C) from the reaction of methyl iodide and triphenylphosphine in benzene, is converted by sodium amide in benzene suspension into a bright yellow solution of triphenylmethylenephosphorane which can be decanted onto this β-

[9.11]

$$\text{isoprene} + HCl \longrightarrow (CH_3)_2C{=}CH{-}CH_2Cl + CH_3{-}CO{-}CH_2{-}COOEt \longrightarrow \underline{A}$$

ketoester $\underline{A}$ at room temperature. As soon as this addition has been completed and stirring is halted, two layers are observed. The lower and smaller of the two is constitued by the betaine $\underline{B}$ which is insoluble in benzene. [9.12]

[9.12]

$$\phi_3\overset{\oplus}{P}CH_3\ \overset{\ominus}{I} + NH_2Na \longrightarrow \phi_3P{=}CH_2 + INa + NH_3\uparrow$$

$$\phi_3\overset{\oplus}{P}{-}CH_2{-}C(CH_3)(O^{\ominus}){-}CH(COOEt){-}CH_2{-}CH{=}C(CH_3)_2 \quad \underline{B}$$

These two layers persist even after 15 hours of stirring at room temperature. Homogenization is only observed after 6 hours reflux of the benzene during which the betaine has decomposed. [9.13]

[9.13]

$$\underline{B} \longrightarrow \phi_3P{=}O + \underline{C}\ (COOEt)$$

Note: The logic behind adding the phosphorane to the β-ketoester $\underline{A}$ is based on the existance in this substrate of two centers which are susceptible to nucleophilic attack: the ketone and the ethoxycarbonyl (COOEt). Addition of these two participants in the opposite order will lead, in part, to the following secondary reaction. [9.14]

In the last step of the synthesis, the dieneic ester $\underline{C}$ obtained from this Wittig reaction is reduced to lavandulol by the action of diisobutylaluminum hydride. [9.15]

A. The stereochemistry of olefins formed in the Wittig reaction

We have just seen how the nature of the phosphoranes (stabilized or not) can influence the kinetics of betaine formation. It will be seen that this can be critical in determining the stereochemistry of the resulting olefins. It should be noted

[9.14]

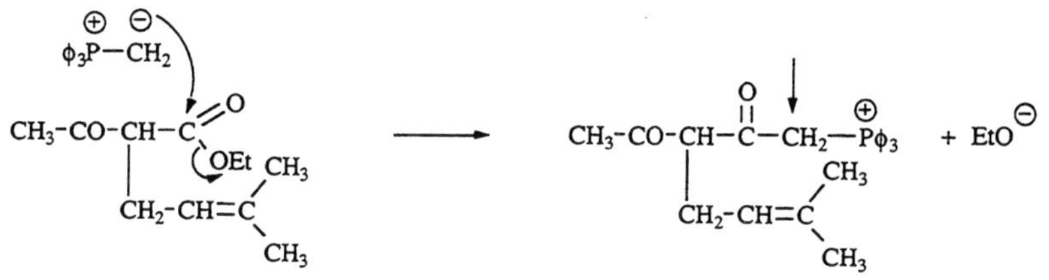

$$\Phi_3\overset{\oplus}{P}-\overset{\ominus}{C}H_2 + CH_3\text{-CO-CH}(CH_2\text{-CH=C}(CH_3)_2)\text{-C(=O)OEt} \longrightarrow CH_3\text{-CO-CH}(CH_2\text{-CH=C}(CH_3)_2)\text{-C(=O)-}CH_2\text{-}\overset{\oplus}{P}\Phi_3 + EtO^{\ominus}$$

[9.15]

COOEt + HAl(iBu)$_2$ ⟶ CH_2OH

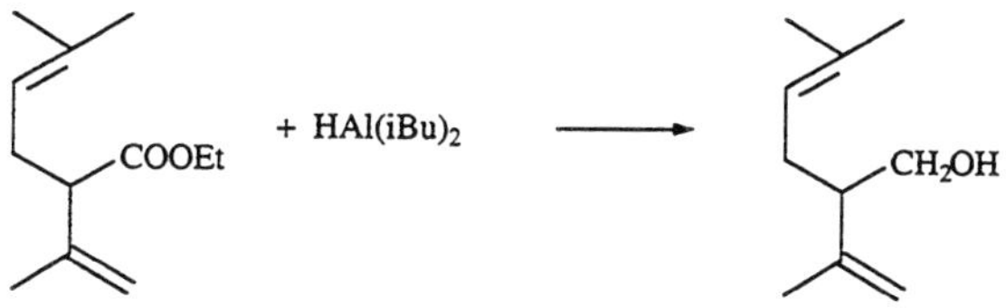

from the outset that the addition of a phosphorane $(R)_3P$=CH-R_1 on an aldehyde R_2-CHO can lead to the formation of two diastereoisomeric betaines 8a and 8b. [9.16]

[9.16]

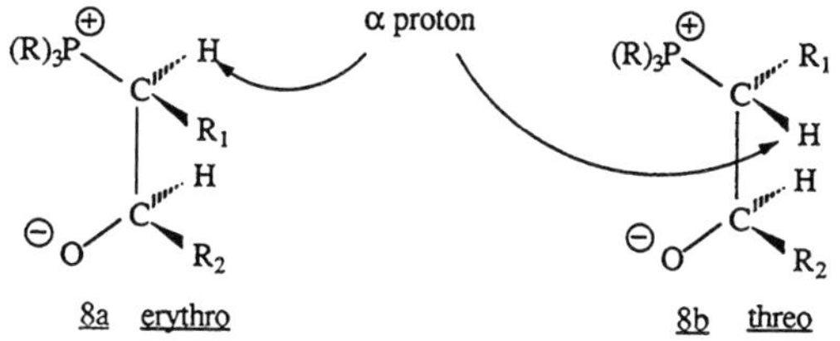

α proton

8a erythro   8b threo

Each of these betaines will then lead respectively to cis and trans olefins 10 and 11. One of the first questions posed is whether there is any reversibility in the formation of these betaines. [9.17]

[9.17]

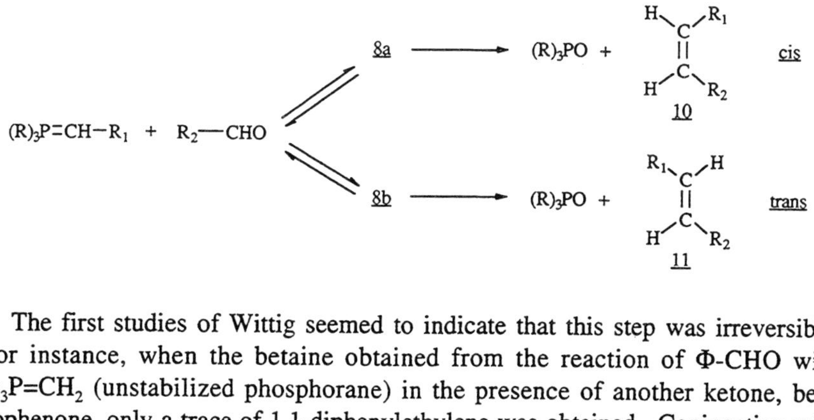

$$(R)_3P{=}CH{-}R_1 + R_2{-}CHO \rightleftharpoons 8a \longrightarrow (R)_3PO + \underset{\mathbf{10}}{H(R_1)C{=}C(H)R_2}\ \text{cis}$$

$$(R)_3P{=}CH{-}R_1 + R_2{-}CHO \rightleftharpoons 8b \longrightarrow (R)_3PO + \underset{\mathbf{11}}{R_1(H)C{=}C(H)R_2}\ \text{trans}$$

The first studies of Wittig seemed to indicate that this step was irreversible. For instance, when the betaine obtained from the reaction of Φ-CHO with $\Phi_3P=CH_2$ (unstabilized phosphorane) in the presence of another ketone, benzophenone, only a trace of 1,1-diphenylethylene was obtained. Conjugation with the phenyl group should accelerate somewhat the normally slow decomposition of betaines. This proves that if formation of the betaine A is indeed reversible, the equilibrium must be strongly displaced towards the right. [9.18]

[9.18]

$\phi_3P{=}CH_2 + \phi CHO \rightleftharpoons \phi_3\overset{\oplus}{P}{-}CH_2{-}CH(\phi){-}O^{\ominus}$ (A) $\longrightarrow \phi_3PO + \phi{-}CH{=}CH_2$

non-reactive phosphorane: $\phi_3P{=}CH_2$ + $\phi{-}CO{-}\phi$ $\not\longrightarrow$ $\phi_3\overset{\oplus}{P}{-}CH_2{-}C(\phi)_2{-}O^{\ominus}$ $\not\longrightarrow$ $(\phi)_2C{=}CH_2$

Yet in 1963, Spéziale and Bissing reported a case in which the betaine is formed reversibly. They treated ethyl trans-phenylglycidate with tributylphosphine in the presence of meta-chlorobenzaldehyde (cross reaction) and obtained two different products: ethyl cinnamate and ethyl meta-chlorocinnamate. This proves that formation of betaine B from the stabilized phosphorane C occurs reversibly.

Note: In this regard it should be pointed out that the corresponding reactions of the cis and trans 2-epoxybutanes with tributylphosphine occur with high stereo-specificity to give, respectively, trans and cis 2-butene. In this case, however, while taking account of the previously discussed results of Wittig, it is difficult to invoke the reversible formation of the betaine from the very reactive ethylenephosphorane. [9.19]

It is possible to foresee another way in which the betaines interconvert. This is by formation of the phosphorane 9 which results from the loss of an α proton. [9.20] The reversible formation of a betaine and its interconversion through the mediation of a common phosphorane, can occur simultaneously, although it is still difficult to distinguish between these two processes. These factors are important in explaining the ratio of the resulting olefins. It is, however, essential to distinguish between stabilized and unstabilized phosphoranes.

- When the starting phosphoranes are stabilized, dissociation of the betaine is favored and the mobility of the H in the α-position (e.g., when R_1=COOEt) is increased;
- When the starting phosphoranes are not stabilized, the rate of formation of the final olefin is slow compared with the rate of equilibration of the betaine;
- A third possibility can also be foreseen if the rates of formation of isomeric olefins from diastereoisomeric betaines are clearly different. Hence, when R_1 or R_2 can conjugate with the forming double bond, the trans-olefin is generally obtained.

This conjugation does indeed increase the stability of the transition state when R_1 (or R_2) is coplanar with the forming double bond and this coplanarity is more readily attained in the transition state leading to the trans-olefin. This third case is identical to the first when it concerns only R_1.

[9.19]

Bu_3P: + $\phi-CH-CH-CO_2Et$ (epoxide)

$Bu_3\overset{\oplus}{P}-CH-CO_2Et$* ⟶ $\phi-CH{=}CH-CO_2Et$

$^{\ominus}O-CH-\phi$

ϕCHO + $Bu_3P{=}CH-CO_2Et$ ($\delta+$, $\delta-$) ⇌ (** $mCl-\phi-CHO$) $Bu_3\overset{\oplus}{P}-CH-CO_2Et$ / $^{\ominus}O-CH-C_6H_4Cl$ ⟶ $mCl\phi-CH{=}CH-CO_2Et$

C phosphorane

cis ⟶ trans

trans ⟶ cis

$Bu_3P{=}CH-CH_3$ + $O{=}CH-CH_3$

* The same argument as above, which proves that this is more critical than the nature of the phosphorane.

** Less reactive phosphorane than that used by Wittig.

[9.20]

$R_3P{=}CH-R_1$ + R_2-CHO ⇌ 8a / 8b ⇌ 9

B. Stereochemistry of olefins formed from stabilized phosphoranes

In the absence of solvation effects, kinetic studies have shown that formation of the betaine is slow and reversible (Spéziale and Bissing, 1963). Under these conditions, there will be interconversion in favor of the more thermodynamically stable form. This conformation will be largely determined by electrostatic attraction between the P and O atoms, and it is reasonable to anticipate that the threo betaine 8b will be favored by steric factors. Under these conditions, however, decomposition of the betaine is generally rapid and although the trans-olefin is obtained preferentially, this selectivity is usually weak, particularly in aprotic dipolar solvents. [9.21]

[9.21]

R−CHO + (R')$_3$P=CHR"

threo

trans

erythro

cis

The Russian group of Shemyakin noted in 1963 that the proportion of cis-olefins increased with the polarity of the solvent. It also increased when nucleophilic Lewis bases were added to the reaction mixture. Yet, the decomposition of the betaine is slowed in apolar solvents to the extent that equilibration in favor of the threo isomer has a better chance to occur. In such cases, the reaction is more selective and the trans-olefin is obtained. It is also possible to reduce the rate of decomposition of the betaine by modifying the substituents on the phosphorus atom. Tricyclohexylphosphine, for instance, which is more weakly electrophilic than triphenylphosphine, will eliminate the phosphine oxide $[(C_6H_{11})_3P{=}O]$ more slowly and, as a result, will lead to the trans-olefin with greater selectivity.

Finally, if this reaction is carried out in protic solvents or in the presence of lithium salts, selectivity in favor of the cis-olefin increases.

C. Stereochemistry of olefins formed from non-stabilized phosphoranes

Addition with the carbonyl derivative will be rapid in cases such as this. Formation of the betaine is still reversible, but its decomposition is generally slow.

It will be recalled that the very existance of an equilibrium has been contested in these cases. The first studies of Wittig tended to favor the opposite hypothesis. It was not until 1965 that the reversibility of betaine formation was unequivocably demonstrated by Schlosser and Christmann. [9.22]

[9.22]

$\phi_3\overset{\oplus}{P}-\overset{\ominus}{C}H-CH_3$

When Φ-CHO is made to react with Φ_3P^+-C^-(H)-CH_3, a mixture of threo and erythro betaines is obtained. These can be protonated by the addition of HBr to give stable salts which can be separated into their threo and erythro forms. [9.23]

[9.23]

$$\phi\text{-CHO} + \overset{\oplus}{\phi_3P}-\overset{\ominus}{CH}-CH_3 \rightleftharpoons \overset{\oplus}{\phi_3P}-CH(CH_3)-\underset{t\,+\,e}{\overset{O^{\ominus}}{CH}}-\phi \underset{tBuOK}{\overset{HBr}{\rightleftharpoons}} \overset{\oplus}{\phi_3P}-CH(CH_3)-\underset{t\,+\,e}{\overset{OH}{CH}}-\phi$$

$$\xrightarrow{} CH_3-CH=CH-\phi$$

$$\downarrow (p)ClC_6H_4-CHO$$

$$\overset{\oplus}{\phi_3P}-CH(CH_3)-\overset{O^{\ominus}}{CH}-C_6H_4(p)Cl \longrightarrow CH_3-CH=CH-C_6H_4(p)Cl$$

When the erythro salt is treated with an equivalent of potassium tertiary butylate, the erythro betaine is regenerated and this is shown to requilibrate with its conformationally predominant threo isomer. Addition of another aldehyde such as para-chlorobenzaldehyde to this mixture gives para-chloropropene which shows that this equilibration is due to reversibility in the formation of betaine intermediates.

Another way to stabilize these betaines is to add LiBr so that the erythro and threo complexes C can be isolated. Their betaines can also be regenerated by the addition of t-BuOK. This procedure offers an extremely effective possibility of controlling the reaction. [9.24]

[9.24]

$$\left[\overset{\oplus}{\phi_3P}-CH(CH_3)-\overset{OLi}{CH}\right] Br^{\ominus} \qquad \underline{C}$$

Betaines obtained from non-stabilized phosphoranes almost completely lose their aptitude to form olefins in the presence of lithium salts, yet once t-BuOK has been added, olefins are formed in a matter of minutes.

If the ylid is generated in a medium completely devoid of salts and particularly when the solvent is apolar, the reaction will proceed under kinetic control to give (via the erythro betaine) the cis-olefin in a selectivity exceeding 90%. A synthesis of cis-jasmone (an important constituent of jasmin absolute) is a good example. [9.25]

In order to obtain trans-olefins from non-stabilized phosphoranes, Schlosser and co-workers developed the following procedure in 1966.

When the salt C is treated with BuLi, an anion is obtained. Although the original erythro complex inverts with difficulty, the erythro anion D obtained in this way equilibrates rapidly in favor of a predominant threo form. After protonation with t-BuOH and reactivation with t-BuOK, an almost purely trans-olefin is obtained via the threo betaine E. This is the "Wittig-Schlosser procedure." [9.26]

How can the preferential formation of the erythro betaine occur from non-stablized phosphoranes in the absence of salts when, at first glance, this form would

[9.25]

$\phi_3P{=}CH{-}C_2H_5$ + CHO → ... C_2H_5 H H → ... O O C_2H_5 H H → $OH^{\ominus}$ → H H O

cis-jasmone

[9.26]

C (e') —BuLi→ D (e') ⇌ D (t) —t-BuOH→ C (t) (protonation) —t-BuOK→ E (t) (reactivation) → trans-olefin

seem to be more sterically crowded than the threo betaine? Furthermore, it was reported in 1969 that selectivity during olefin formation is even higher when the starting carbonyl compound is itself sterically crowded. In order to explain these effects, Schneider (also in 1969) considered the importance of the configuration of the substituents around the phosphorus atom during the reaction. He postulated that in the case of non-stabilized phosphoranes, the phosphorus-oxygen bond formed before the carbon-carbon bond (nucleophilic attack on P^+ by O^- to form a dipolar ion).

Hence, the structural description A in which the oxygen is in the apical position and the ylid carbon is in the equatorial position (which is necessary for the formation of a bond with the aldehyde), represents a situation in which there is very little steric crowding for the aldehyde group R_1. [9.27]

The very bulky R_1 group is initially placed in the vertical plane of the carbenium ion in the farthest possible position from the substituents of the phosphorus atom. In this structure there is nonetheless an important steric interaction between the R_2 group and one of the phenyl substituents on the phosphorus atom. In order for the C-C bond and the oxaphosphetane intermediate to form, it is necessary for the O-C bond to rotate in one direction or another so that the plane of the carbenium ion ends up perpendicular to the orbitals of the ylid ion. This can happen in one of two following manners:

[9.27]

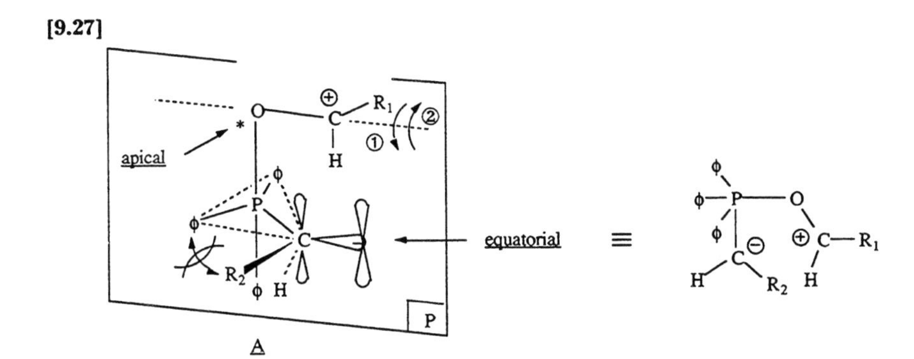

- the O-C bond can rotate in the sense indicated in diagram 1, resulting in the erythro configuration in which R_1 and R_2 are eclipsed—but only one of the Φ groups is sterically compressed;
- the O-C bond can rotate in the sense indicated in diagram 2, resulting in the threo configuration in which R_1 and R_2 are not eclipsed—but a very important steric compression develops between R_1 and another Φ group.

The erythro configuration is therefore the more stable energetically and leads to the cis-olefin.

In the presence of polar solvents or of salts which coordinate more rapidly with O^- than with P^+, the mechanism is not possible and the trans-olefin is obtained preferentially. [9.28]

[9.28]

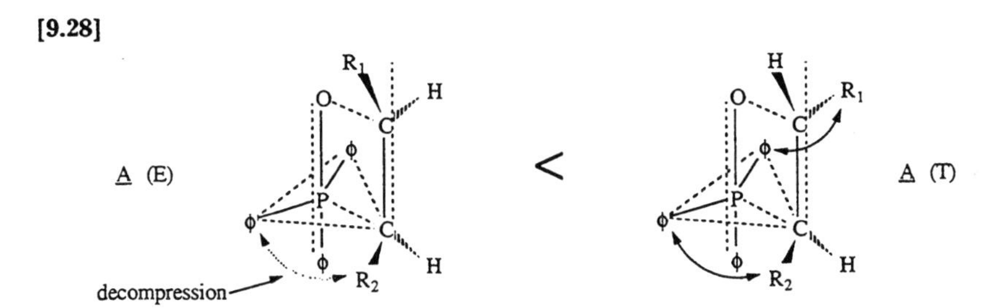

We have tried to summarize the principal properties of the Wittig reaction in the following table (Properties of the Wittig reaction).

D. Special conditions for employing the Wittig reaction

After having carefully studied the work of Schlosser and Christmann, Corey developed the "Wittig-Schlosser reaction" to a stereospecific synthesis of allyl alcohols. Indeed, if instead of protonating the betaine D(t) (see earlier), a second aldehyde R-CHO is added, an intermediate F(t) is obtained that transforms rapidly and stereospecifically into an allyl alcohol. The oxygen brought by the second

Properties of the Wittig reaction: Type of phosphoranes

Reaction conditions	"Selectivity" Stabilized	Non-stabilized
Polar solvents		
Aprotic	weak selectivity but trans formed principally	weak selectivity
Protic	increase in the selectivity in favor of cis-isomer	increase in the selectivity in favor of trans-isomer
Nonpolar solvents		
Without salts	high selectivity → trans	high selectivity → cis
With salts	increase in the selectivity in favor of cis-isomer	increase in the selectivity in favor of trans-isomer*

Reaction conditions	"Kinetics" Stabilized	Non-stabilized
First step: reversible betaine formation	slow	rapid
Second step: betaine decomposition	rapid	slow**

* Wittig-Schlosser procedure

** more rapid, however, if the reaction is carried out in the total absence of salts

aldehyde R-CHO links to the phosphorus and will eventually give triphenylphosphine oxide. [9.29]

[9.29]

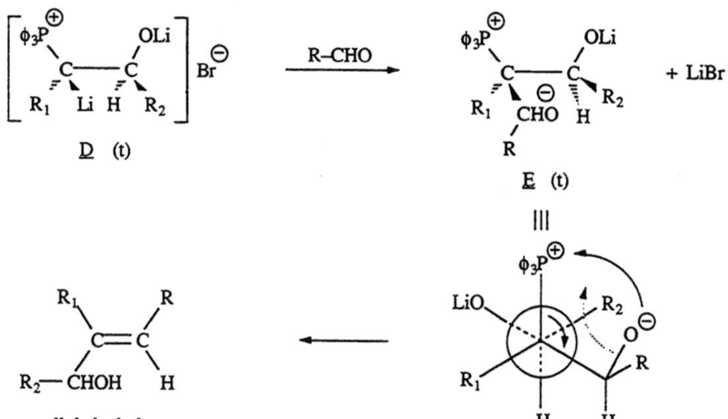

D (t) + R–CHO → E (t) + LiBr → allyl alcohol

Simple examination of the betaine formed in this reaction does not allow prediction of the relative positions of H and R. [9.30]

[9.30]

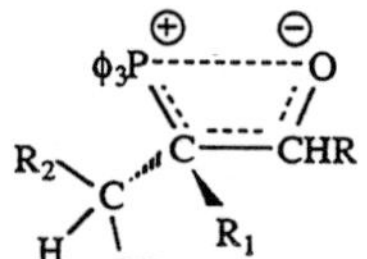

This special bonding of the second oxygen can also be explained by the hypothesis of Schneider which places formation of the P-O bond before that of the C-C bond and which, in this case, will predict formation of the di-ionic intermediate B. This can be transformed by rotation in the sense indicated in diagram 1 to give the oxetane B_1. Rotation in the sense shown in diagram 2 to give the oxetane B_2 does not occur because this would result in serious compression between the R group with the voluminous $-CHR_2OLi$ group. [9.31]

[9.31]

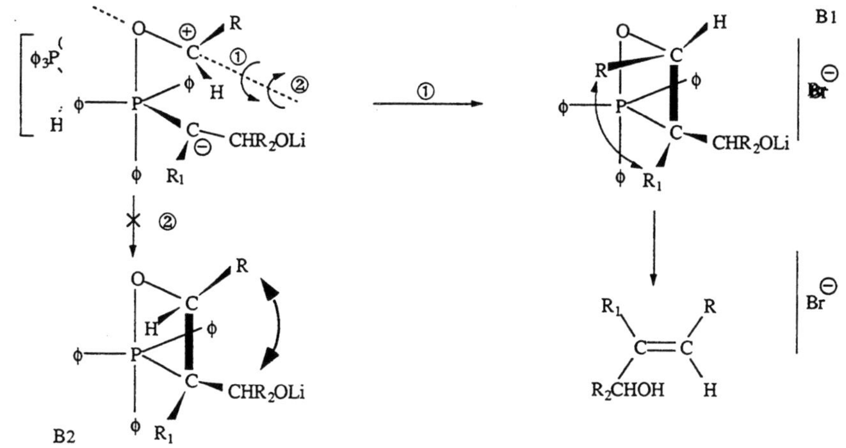

B1

B2

The results shown below are quite general:

Example [9.32]

[9.32]

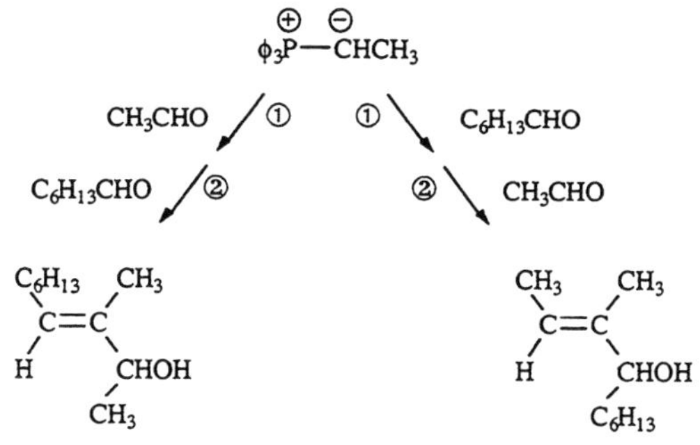

$$\phi_3\overset{\oplus}{P}-\overset{\ominus}{C}HCH_3$$

① CH_3CHO, ② $C_6H_{13}CHO$ → $(C_6H_{13})(H)C{=}C(CH_3)(CHOH\,CH_3)$

① $C_6H_{13}CHO$, ② CH_3CHO → $(CH_3)(H)C{=}C(CH_3)(CHOH\,C_6H_{13})$

In contrast with the cases just analyzed, the use of anhydrous para-formaldehyde $(CH_2O)_n$ gives principally allyl alcohol in which the oxygen eliminated in the phosphine comes from the former aldehyde used. In fact, the para-formaldehyde does not contain any free carbonyl groups.

Example [9.33]

[9.33]

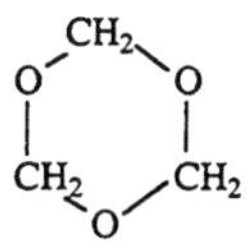

With this type of structure, initial formation of the C-O bond cannot occur. Para-formaldehyde behaves, therefore, more like an alkylating agent. [9.34]

[9.34]

$$\left[\begin{array}{c} \Phi_3\overset{\oplus}{P}\text{-}C(R_1)(Li)\text{—}C(H)(R_2)\text{-}OLi \end{array}\right] Br^{\ominus} \xrightarrow{(CH_2O)_n} \Phi_3\overset{\oplus}{P}\text{-}C(R_1)(CH_2(OCH_2)_{n-1}\text{—}O^{\ominus})\text{—}C(H)(R_2)\text{-}O^{\ominus} + LiBr$$

D (t)

$$\downarrow$$

$$R_1(CH_2OH)C{=}C(H)R_2$$

allyl alcohol

If $R_1 = CH_3(\Phi_3P{=}CH\text{-}CH_3$ for the initial ylid) and R_2 = tricycloekasantalal, a good yield of α-santalol is obtained. [9.35]

[9.35]

CHO → H, CH_3, CH_2OH

tricycloekasantalal α-santalol

E. Reactions related to the Wittig reaction

A serious handicap of the Wittig reaction is that, up until recently, triphenylphosphine was a relatively costly reagent. This incited studies into cheaper alternatives to this reaction. Among these, one of the more original reactions is that of Horner. It involves of using triethylphosphite (very cheap) instead of triphenylphosphine. Several examples of this reaction will be discussed, but beforehand, two other extremely important and more classic reactions will be described.

F. Michaelis-Arbuzov reaction

Organic halides react with triethylphosphite with the elimination of the ethyl halide to give a phosphonate. For the Horner reaction to proceed with this compound, it must contain a group A (electron-withdrawing) which activates the hydrogen atoms positioned α to the phosphorus. [9.36]

[9.36]

$$P(OEt)_3 + XCH_2A \xrightarrow{\Delta} O{=}P(OEt)_2\text{—}CH_2A + EtX\uparrow$$

A = -COOEt; -CHO; -R-C=O; -C≡N

G. Arbuzov-Razumov reaction

The phosphonate obtained in the preceding reaction can be treated with base resulting in a carbanion [9.37]

[9.37]

$$O{=}P(OEt)_2{-}\overset{\ominus}{C}H{-}A$$

susceptible to alkylation by another alkyl halide RX', thereby yielding a second substitued phosphonate: [9.38]

[9.38]

$$O{=}P(OEt)_2{-}\bar{C}(H){-} + RX \longrightarrow O{=}P(OEt)_2{-}CH(R)(A)$$

H. Horner reaction

If the phosphonate carbanion is treated with a carbonyl compound, a reaction occurs which is completely analogous to that of Wittig. This is the Horner reaction. [9.39]

[9.39]

$$(EtO)_2P({=}O){-}\overset{\ominus}{C}(R)(A) + O{=}C(R_1)(R_2) \longrightarrow [\text{four-membered P--C--C--O}^{\ominus}\ \text{intermediate}] \longrightarrow R(A)C{=}C(R_1)(R_2) + O{=}P(OEt)_2{-}O^{\ominus}$$

An olefin substituted with the activating group A is obtained along with a mixed phosphate.

I. Applications

1) Synthesis of isolavandulol [9.40]

2) Synthesis of isolavandulal [9.41]

3) Synthesis of α-cyclopropyl ketones [9.42]

This occurs as if the residue shown below had been attached to the double bond of an alkene R_1-CH=CH-R_2.

$$R{-}CO{-}CH\langle$$

J. Ylids obtained with atoms other than phosphorus

Phosphorus is not the only element capable of giving ylids. Sulfur gives them also. Their formation and their reactions have been particularly well studied by Corey.

[9.40]

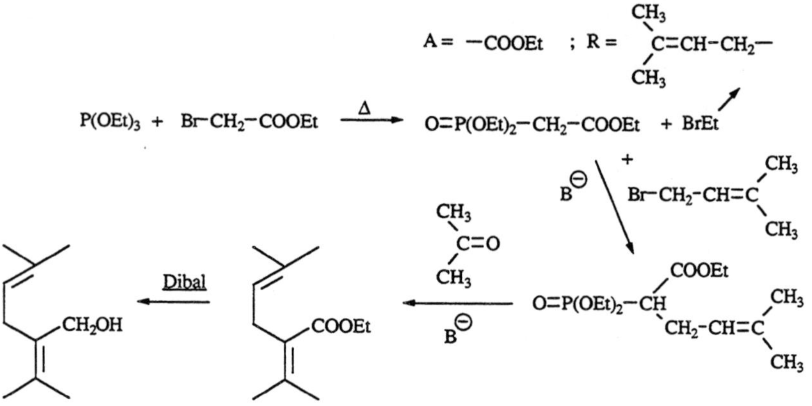

A = $-COOEt$; R = $(CH_3)_2C{=}CH{-}CH_2{-}$

$P(OEt)_3 + Br{-}CH_2{-}COOEt \xrightarrow{\Delta} O{=}P(OEt)_2{-}CH_2{-}COOEt + BrEt$

$+ Br{-}CH_2{-}CH{=}C(CH_3)_2$, $B^{\ominus}$ → $O{=}P(OEt)_2{-}CH(COOEt){-}CH_2{-}CH{=}C(CH_3)_2$

$(CH_3)_2C{=}O$, $B^{\ominus}$ → COOEt → Dibal → CH_2OH

isolavandulol

[9.41]

A = $-CN$; R = $(CH_3)_2C{=}CH{-}CH_2{-}$

$P(OEt)_3 + Br{-}CH_2{-}CN \longrightarrow O{=}P(OEt)_2{-}CH_2{-}CN + BrEt$

$B^{\ominus}$, R → $O{=}P(OEt)_2{-}CH(CN){-}CH_2{-}CH{=}C(CH_3)_2$

$(CH_3)_2C{=}O$, $B^{\ominus}$ → CN → Dibal → CHO

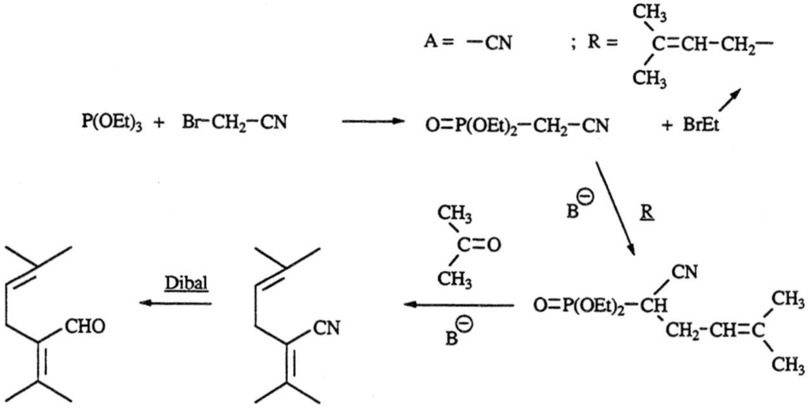

isolavandulal

1) Sulfoxides of the type $R_1{-}S(=O){-}R_2$

These sulfoxides are analogous to amine oxides $R_1R_2N{\rightarrow}O$ and like these they are bases. It is known that amine oxides react with alkyl halides to give salts of the type $R_1R_2RN^{\oplus}\ X^{\ominus}$

The same happens, for example, with dimethylsulfoxide (DMSO), which reacts

[9.42]

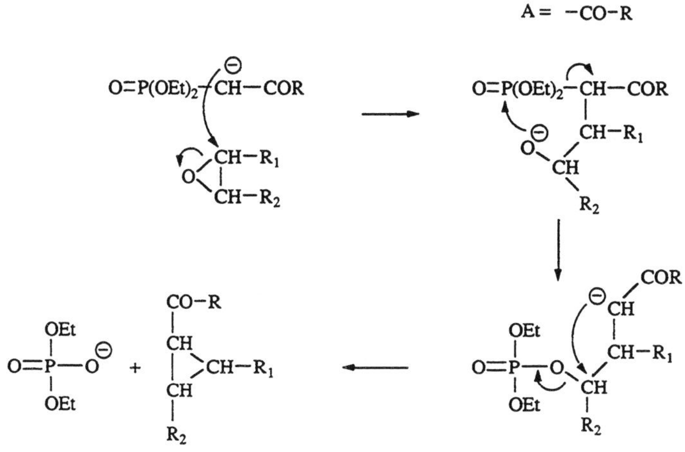

A = −CO−R

$O{=}P(OEt)_2{-}\overset{\ominus}{C}H{-}COR$ + epoxide $CH{-}R_1$ / $CH{-}R_2$ ⟶ $O{=}P(OEt)_2{-}CH{-}COR$ ($CH{-}R_1$, $O^{\ominus}$, CH, R_2)

⟶ $O{=}P(OEt)_2{-}O{-}CH(R_2){-}CH(R_1){-}\overset{\ominus}{C}H{-}COR$

⟶ $O{=}P(OEt)_2{-}O^{\ominus}$ + cyclopropane (CO−R, CH, CH−R_1, CH, R_2)

with CH_3I to give trimethylsulfoxonium iodide $(CH_3)_3$. $\overset{\oplus}{(CH_3)_3S{=}O}\ \overset{\ominus}{I}$

Example of this reaction:

Synthesis of oxiranes and cyclopropyl ketones [9.43]

[9.43]

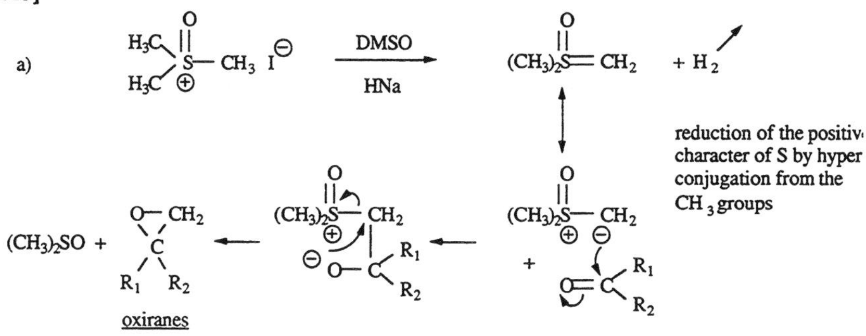

a) $(H_3C)_2\overset{\oplus}{S}(=O){-}CH_3\ I^{\ominus}$ $\xrightarrow[\text{HNa}]{\text{DMSO}}$ $(CH_3)_2S(=O){=}CH_2$ + H_2↗

↕

$(CH_3)_2\overset{\oplus}{S}(=O){-}\overset{\ominus}{C}H_2$ + $O{=}C(R_1)(R_2)$

reduction of the positiv character of S by hyper conjugation from the CH_3 groups

⟶ $(CH_3)_2\overset{\oplus}{S}(=O){-}CH_2{-}C(R_1)(R_2){-}O^{\ominus}$ ⟶ $(CH_3)_2SO$ + $O{-}CH_2{-}C(R_1)(R_2)$

oxiranes

[9.43]

b) $R{-}CH{=}CH{-}CO{-}R'$ + $(CH_3)_2\overset{\oplus}{S}(=O){-}\overset{\ominus}{C}H_2$ ⟶ $R{-}CH(CH_2{-}\overset{\oplus}{S}(=O)(CH_3)_2){-}\overset{\ominus}{C}H{-}CO{-}R'$

⟶ $(CH_3)_2SO$ + $R{-}CH{-}CH{-}CO{-}R'$ (bridged by CH_2)

α-cyclopropyl ketones

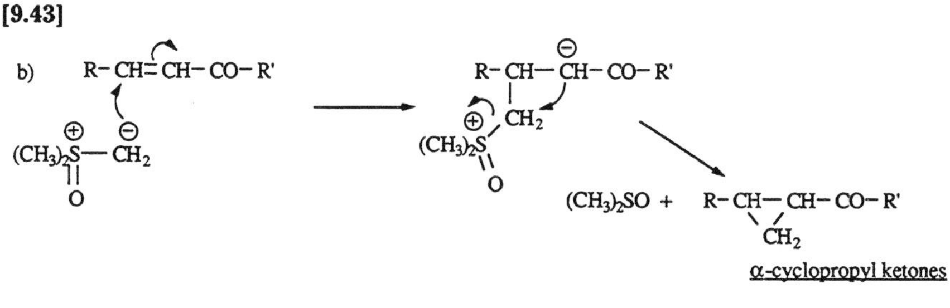

Synthesis of ketones from esters and the carbanion of DMSO (which plays the role of a base) [9.44]

[9.44]

$R-C(=O)-OR' + {}^{\ominus}CH_2-S(=O)-CH_3 \longrightarrow R-CO-CH_2-SO-CH_3 + R'O^{\ominus}$

$\downarrow CH_3-SO-CH_2^{\ominus}$ (which plays the role of base)

$R-CO-\overset{\ominus}{C}H-SO-CH_3 + CH_3-SO-CH_3$

$\downarrow H_2O$

β-ketosulfoxides $R-CO-CH_2-SO-CH_3 + OH^{\ominus}$

$\downarrow Al/Hg + H_2O$

$R-CO-CH_3$ methyl ketone

2) Ylids derived from dimethyl sulfide $(CH_3)_2S^+-CH_2$

These ylids react with derivatives of saturated or α,β-unsaturated carbonyl compounds to give oxiranes as exclusive products. [9.45]

[9.45]

$R_1R_2C=O + (CH_3)_2\overset{\oplus}{S}-\overset{\ominus}{C}H_2 \longrightarrow R_1R_2C(-O^{\ominus})-CH_2-\overset{\oplus}{S}(CH_3)_2 \longrightarrow R_1R_2C(-O-)CH_2$

3) Ylids of the type $\Phi_2S^+-2CH_3$

a) Synthesis of methyl chrysanthemate [9.46]

[9.46]

$CH_2=C(CH_3)-CH_2Cl + HC\equiv CH + MeOH \xrightarrow[MeO^{\ominus}]{Ni(CO)_4} (H_3C)_2C=CH-CH=CH-COOMe$ (trans)

Synthesis of Chuisoli

$+ \Phi_2\overset{\oplus}{S}-\overset{\ominus}{C}(CH_3)_2$

$\downarrow$

$(H_3C)_2C=CH-CH-CH-COOMe$ (cyclopropane ring with $C(CH_3)_2$; H, H)

b) Synthesis of aristone [9.47]

[9.47]

APPENDIX

3

The Chemistry of Enol Ethers

Enol ethers (or vinyl ethers) show reactivity which is intermediate between that of olefins and enamines. One of the free electron pairs on the oxygen atom and the π electrons of the double bond can at least partially overlap. For this reason, enol ethers can be considered resonance hybrids of two mesomeric forms. [10.01]

[10.01]

R—O—C=C ⟷ R—O⊕=C—C⊖

1 α β 2

The same is true of enamines: [10.02]

[10.02]

N—C=C ⟷ N⊕=C—C⊖

3 α β 4

In fact, the enammonium structure 4 is far more energetically favourable than the enoxonium structure 2, and therefore the contribution of 4 to the enamine hybrid is greater than that of 2 to the enol ether. This explains the intermediate reactivity.

Enol ethers are capable of undergoing three fundamentally different types of reaction:

1) Cationic polymerization in the the presence of Lewis acids. This type of reaction will not be discussed here, but its existence must be recognized in order to explain certain particularities of the reactivity of these substances.
2) Reactions with H-X compounds to give the carbonyl compounds from which the enol ethers are derived, or derivatives of these carbonyl compounds. Simple hydrolysis is an example of this. [10.03]

[10.03]

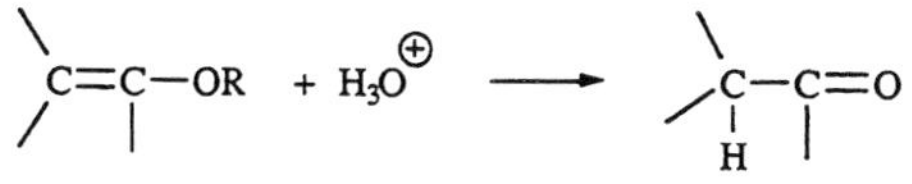

$$C{=}C{-}OR + H_3O^{\oplus} \longrightarrow H{-}C{-}C{=}O$$

Another example is found in their reactions with alcohols to give mixed acetals. This reaction has already been evoked on previous occasions in this text. [10.04]

[10.04]

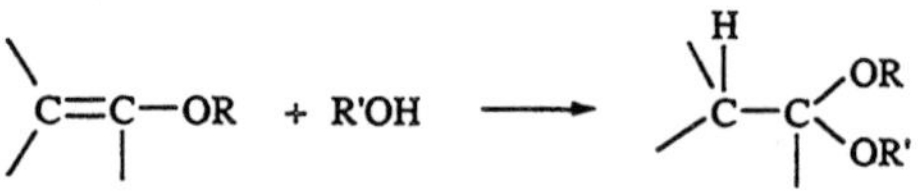

3) Reactions with compounds of the A-X type to form a bond between A and the β-carbon. These reactions lead to addition or substitution compounds. [10.05]

[10.05]

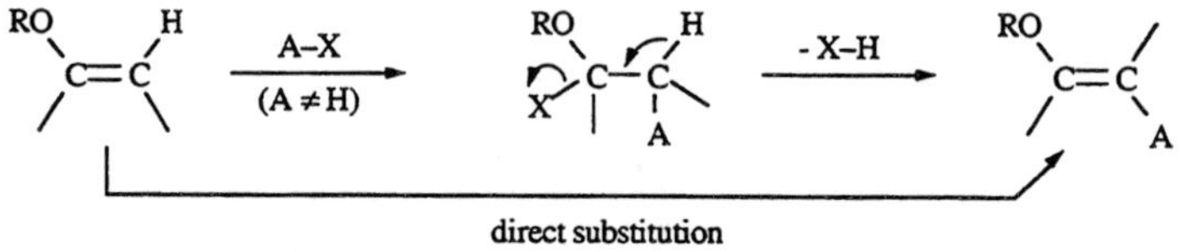

direct substitution

This third type of reaction is by far the most interesting and some examples will be discussed.

A. Action of boron or aluminum hydrides

(also similar to the second type of reaction with H-X compounds)

1. Dialkylborane hydrides [10.06]

[10.06]

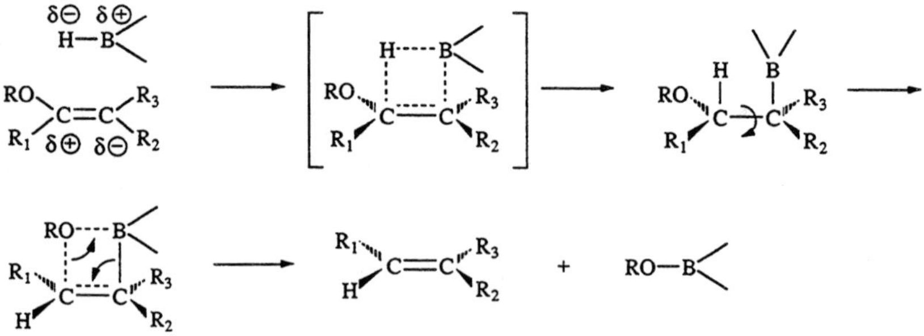

2. Dialkylaluminum hydrides [10.07]

In the case of dialkylborane hydrides, the addition and elimination steps proceed with cis-stereospecificity. In these two cases, the reactions proceed without catalysis and lead to olefins. They therefore permit easy passage from carbonyl

[10.07]

$$\underset{\delta\oplus\ \ \delta\ominus}{\overset{\delta\ominus\ \ \delta\oplus\ \ H-B<}{RO(R_1)C{=}C(R_3)R_2}} \longrightarrow \left[RO(R_1)C{=}C(R_3)R_2 \cdots H{-}B< \right] \longrightarrow RO(R_1)C(H){-}C(B)(R_3)R_2 \longrightarrow$$

derivatives (R^1-CH_2-CO-R^2) to olefins (R^1-CH=CH-R^2). [10.08]

[10.08]

$$R^1{-}CH_2{-}CO{-}R^2 \longrightarrow R^1{-}CH{=}\underset{R^2}{\underset{|}{C}}{-}OR' \longrightarrow R^1{-}CH{=}CH{-}R^2$$

If the carbonyl derivative is an aldehyde, this procedure yields a terminal olefin (R^2 = H).

$$R\text{-}CH_2\text{-}CHO \rightarrow R\text{-}CH{=}CH\text{-}OR' \rightarrow R\text{-}CH{=}CH_2$$

B. Action of organo-metallic compounds

1. Alkyl sodium compounds

Pentyl sodium, for example, can metallate enol ethers in the α- or the β-positions on the double bond. While the α-metallo derivative is stable under the reaction conditions, the β-metallo derivative rearranges immediately to an alcohol and an acetylenic derivative. [10.09]

[10.09]

$$\underset{\text{Na (stable)}}{CH_2{=}\overset{\alpha}{C}{-}OR} \longleftarrow CH_2{=}CH{-}OR + C_5H_{11}Na \longrightarrow \left[Na{-}\overset{\beta}{CH}{=}CH{-}OR \right]$$

$$\downarrow + CO_2 \qquad\qquad\qquad \downarrow \text{AmNa (immediate rearrangement)}$$

$$\underset{COONa}{CH_2{=}\underset{|}{C}{-}OR} \qquad NaOOC{-}C{\equiv}C{-}COONa \xleftarrow{CO_2} Na{-}C{\equiv}C{-}Na + RONa$$

When the enol ether is already β-substituted, α-metallation predominates while with α-substituted enol ethers, β-metallation is the only mode of reaction. [10.10]
Note: Alkyl lithium compounds, being less reactive than alkyl sodium compounds, do not react with enol ethers below –30°C. Even above this temperature, no defined product is obtained.

2. Organo-magnesium compounds

Grignard reagents have metal-carbon bonds in which the polarity is far weaker than in the preceding derivatives. As a result, they are not capable of metallating

[10.10]

$$\text{Et-}\overset{\beta}{\text{CH}}\text{=}\overset{\alpha}{\text{CH}}\text{-OR} \xrightarrow{\text{AmNa}} \text{Et-CH=}\underset{\text{Na}}{\text{C}}\text{-OR} \xrightarrow{CO_2} \text{Et-CH=C}\begin{smallmatrix}\text{OR}\\ \text{COONa}\end{smallmatrix}$$

β-substitution α-metallation

α-substitution β-metallation pentyne-4 ol-1

enol ethers. They do, however, form addition products. The negatively charged alkyl residue of the Grignard reagent can react with the α-carbon of the vinylic group to form:

- either an olefin (this is the case with enol ethers derived from aldehydes)
- or a ketone by cleavage of the ether bond (this is the case with enol ethers derived from ketones. In fact, the ketone is regenerated in such reactions). [10.11]

[10.11]

$$R^1\text{-CH=}\underset{R^2}{\text{C}}\text{-OR}^3 + \text{R'MgX}$$

R^2 = H, ex aldehyde
$R^2 \neq$ H, ex ketone

R^2 = H: $$R^1\text{-CH=}\underset{H}{\text{C}}\text{-OR}^3 + \text{R'-MgX} \longrightarrow R^1\text{-CH=CH-R'} + R^3\text{OMgX}$$

$R^2 \neq$ H: $$R^1\text{-CH=}\underset{R^2}{\text{C}}\text{-O-R}^3 + \text{XMg-R'} \longrightarrow R^1\text{-CH=}\underset{R^2}{\text{C}}\text{-OMgX} + \text{R'-R}^3$$

$$\downarrow$$

$$R^1\text{-CH}_2\text{-CO-R}^2$$

3. Organo-mercury compounds

These behave totally differently due to the clearly homopolar character of the carbon-mercury bond. Addition occurs exclusively in one direction, as the consequence of electrophilic addition. Mercury salts also add slowly, but with an excellent yield, to the double bond of enol ethers. [10.12]

[10.12]

$$R^1\text{-O-}\underset{R^2}{\text{C}}\text{=CH}_2 \xrightarrow[Et_2O]{Hg(OAc)_2} R^1\text{-O-}\overset{\text{OAc}}{\underset{R^2}{\text{C}}}\text{-CH}_2\text{-HgOAc}$$

The "acylals" obtained can be readily converted into "halo-mercury ketones" by the action of mercuric chloride and water.

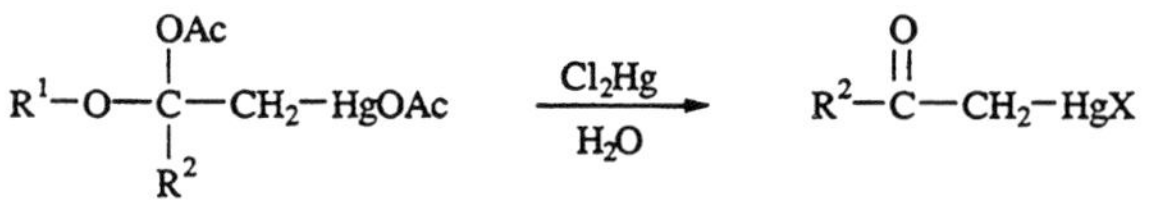

$$R^1{-}O{-}\underset{R^2}{\overset{OAc}{C}}{-}CH_2{-}HgOAc \xrightarrow[H_2O]{Cl_2Hg} R^2{-}\overset{O}{\overset{\|}{C}}{-}CH_2{-}HgX$$

C. Reactions of enol ethers leading to the formation of carbon-carbon bonds

The reaction of enol ethers with acetals is extremely important. It was observed for the first time in 1939 by Müller-Kunradi and Pieroh (BASF). [10.13]

[10.13]

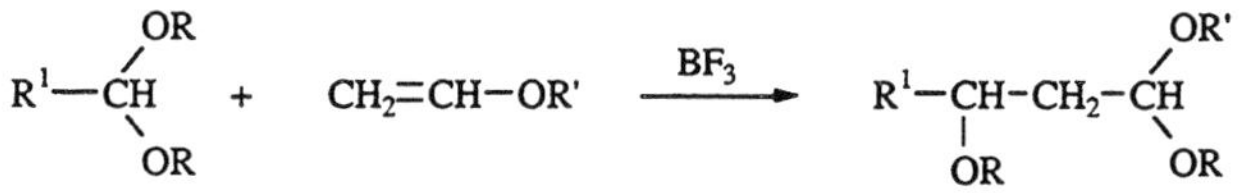

$$R^1{-}CH(OR)_2 + CH_2{=}CH{-}OR' \xrightarrow{BF_3} R^1{-}\underset{OR}{CH}{-}CH_2{-}CH(OR')(OR)$$

Isler and co-workers at Hoffmann-La Rôche have shown that the resulting adduct can be transformed into an α,β-unsaturated aldehyde with the loss of alcohol by treatment with an aqueous solution of acetic acid and sodium acetate. [10.14]

[10.14]

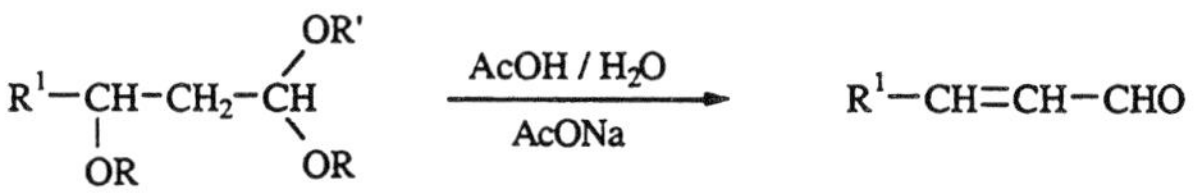

$$R^1{-}\underset{OR}{CH}{-}CH_2{-}CH(OR')(OR) \xrightarrow[AcONa]{AcOH\,/\,H_2O} R^1{-}CH{=}CH{-}CHO$$

This reaction provides an extremely useful complement to the aldol condensation and is quite superior to it in some regards due to its specificity. The example shown above is analogous to the aldol condensation R_1-CHO + CH_3-CHO → R_1-CH=CH-CHO. It represents an extremely powerful tool for extending a chain by two carbon atoms, particularly when an α,β-ethylenic aldehyde is used. [10.15]

[10.15]

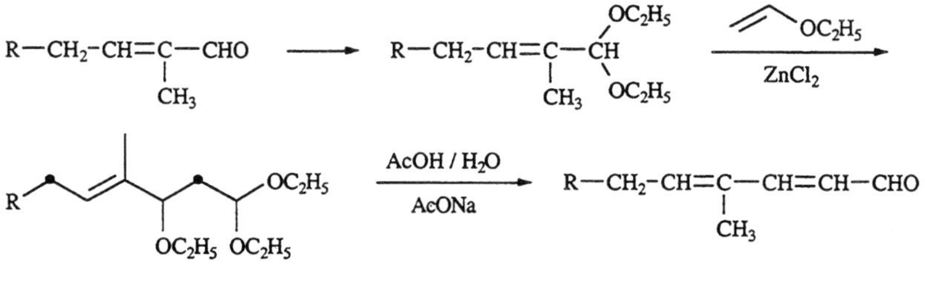

$$R{-}CH_2{-}CH{=}\underset{CH_3}{C}{-}CHO \longrightarrow R{-}CH_2{-}CH{=}\underset{CH_3}{C}{-}CH(OC_2H_5)_2 \xrightarrow[ZnCl_2]{CH_2{=}CH{-}OC_2H_5}$$

$$R{-}CH_2{-}CH{=}C(CH_3){-}\underset{OC_2H_5}{CH}{-}CH_2{-}CH(OC_2H_5)_2 \xrightarrow[AcONa]{AcOH\,/\,H_2O} R{-}CH_2{-}CH{=}\underset{CH_3}{C}{-}CH{=}CH{-}CHO$$

The reaction of α,β-ethylenic aldehydes with enol ethers is often used in the synthesis of polyenes and has held pride of place in the synthesis of carotenoids. For instance, one of the industrial syntheses of β-carotene involves the condensations of ethylvinyl ether and methylpropenyl ether with acetals. This subject will be discussed later in more detail.

Mechanism of the reaction [10.16]

[10.16]

$$CH_3-CH(OR)-OR + BF_3 \rightleftharpoons CH_3-CH(OR)-\overset{\delta\oplus}{O}(R)-\overset{\delta\ominus}{BF_3} \rightleftharpoons \underbrace{CH_3-\overset{\oplus}{CH}-OR + R-O-\overset{\ominus}{BF_3}}_{\underline{D}}$$

$$CH_3-\overset{\oplus}{CH}-OR + CH_2{=}CH-OR' \rightleftharpoons \underset{\underline{A}}{CH_3-CH(OR)-CH_2-\overset{\oplus}{CH}-OR'} + CH_2{=}CH-OR'$$

a: $\underline{A} + R-O\overset{\ominus}{B}F_3 \rightleftharpoons \underset{\underline{B}}{CH_3-CH(OR)-CH_2-CH(OR')(OR)} + BF_3$

b: $\underline{A} + CH_2{=}CH-OR' \rightarrow \underset{\underline{C}}{CH_3-CH(OR)-CH_2-CH(OR)-CH_2-\overset{\oplus}{CH}-OR'} + R-O\overset{\ominus}{B}F_3$

Nucleophilic addition of the vinyl ether with the intermediate complex of the acetal and BF_3 gives the carbocation A which can combine with the RO-BF_3 anion by process α to give the product B with regeneration of BF_3. Alternatively, the cation A can undergo a second nucleophilic attack from the vinyl ether to give C by the process b. This cation C can either combine with the RO^- anion D or be attacked by a third molecule of the vinyl ether. Repetition of this last alternative pathway constitutes the polymerization phase, which, incidentally, Müller-Kunradi and Pieroh were seeking in the first place.

In the context of chemical synthesis, however, process a is obviously the most interesting. It is favored by the following factors:

1. a high ratio (r) of acetal to enol ether;
2. greater reactivity of the enol ether with D than with A.

Comparison of the reactions of ketals with those of acetals of aromatic aldehydes indicates the important influence of the reactivity of the intermediate cation. Ketals give practically no products of 1:1 addition even when the ratio (r) is greater than 1. This is because the cation F is more reactive than E and reacts immediately with another molecule of the enol ether. [10.17]

[10.17]

$$R^1_2C(OR)_2 + BF_3 \rightleftharpoons \underset{\underline{E}}{R^1_2\overset{\oplus}{C}-OR} + RO\overset{\ominus}{B}F_3 \xrightarrow{CH_2=CH-OR'} \underset{\underline{F}}{R^1_2C(OR)-CH_2-\overset{\oplus}{CH}OR'} + RO\overset{\ominus}{B}F_3$$

$$\underline{F} \xrightarrow{CH_2=CH-OR'} R^1_2C(OR)-CH_2-CH(OR')-CH_2-\overset{\oplus}{CH}-OR' + RO\overset{\ominus}{B}F_3$$

It has recently been shown that:

1. montmorillonite clay K-10 absorbs methyl and ethyl orthoformates and readily gives yields of about 90–100% of the corresponding acetals with their carbonyl derivatives;
2. vinyl ethers in the presence of catalytic quantities of this clay, react readily with acetals or ketals to give rise to alkoxyalkanes with yields from 50 to 95%.

This catalyst is clearly superior to $BF_3 \cdot Et_2O \cdot FeCl_3$ and avoids completely the polymerization side reaction. This is performed in the absence of solvent. At the end of the reaction, the only manipulation required is for the catalyst to be filtered.

Acetals of aromatic aldehydes give only those products which arise from 1:1 addition even when the ratio (r) is equal to 1. This is due to the greater reactivity of G compared with that of H. [10.18]

[10.18]

$$\phi-CH(OR)_2 + BF_3 \rightleftharpoons \underset{\underline{G}}{\phi-\overset{\oplus}{C}HOR} + RO\overset{\ominus}{B}F_3 \xrightarrow{CH_2=CH-OR'} \underset{\underline{H}}{\phi-\underset{OR}{CH}-CH_2-\overset{\oplus}{C}H-OR'} + RO\overset{\ominus}{B}F_3$$

The reactivity of acetals with enol ethers increases in the following order:

saturated acetals < aromatic acetals < α,β-ethylenic acetals.

As the catalysts used in these reactions can also cause polymerization of enol ethers, their ease of addition to acetals is also influenced by this tendency which increases in the following order: [10.19]

[10.19]

$$\underset{\text{not substituted in the }\alpha\text{ position}}{R-CH=CH-OR'} < \underset{\text{substituted in the }\alpha\text{ position}}{R-CH=\underset{R_1}{C}H-OR'} < \underset{\text{1-alkoxydienes}}{R-CH=CH-CH=CH-OR'}$$

Thus, fully saturated acetals, the least reactive acetals (and *a fortiori* the others), combine normally with enol ethers which are not substituted in the α-position. In contrast, polymerization predominates when they react with the other types of enol ethers.

The most reactive aromatic acetals or the very reactive α,β-ethylenic acetals combine even with alkoxydienes as well as with all of the other types of vinyl ethers.

Alkoxydienes have, for instance, been used as enol ethers in the synthesis of the acetal of ethoxycitral. [10.20]

A 1% solution of phosphoric acid in ethanol transforms this acetal of ethoxycitral into 5-ethoxycitral, which can be subsequently converted into dehydrocitral by treatment with a trace of para-toluenesulfonic acid. [10.21]

[10.20]

$$(H_3C)_2C{=}CH{-}CH(OR)_2 + CH_2{=}C(CH_3){-}CH{=}CH{-}OR' \longrightarrow (H_3C)_2C{=}CH{-}CH(OR){-}CH_2{-}C(CH_3){=}CH{-}CH(OR')(OR)$$

[10.21]

OEt, CH, OEt, EtO — PO_4H_3 → CHO, EtO — pTsOH, ϕCH_3 → CHO

Advantages of this type of condensation over the aldol condensation

If acetaldehyde and propionaldehyde are condensed under alkaline conditions (aldol condensation), four products are obtained (2+2, 2+3, 3+2, 3+3). [10.22]

[10.22]

$$CH_3{-}CHO + CH_3{-}CH_2{-}CHO \longrightarrow$$

$CH_3{-}CH{=}CH{-}CHO$ (2 + 2) a

$CH_3{-}CH{=}C(CH_3){-}CHO$ (2 + 3) b

$CH_3{-}CH_2{-}CH{=}CH{-}CHO$ (3 + 2) c

$CH_3{-}CH_2{-}CH{=}C(CH_3){-}CHO$ (3 + 3) d

It is feasible in practice to favor formation of one of these four possible products (i.e., the adduct b from tiglic and angelic aldehydes) by modifying the proportions of the two aldehydes and by studying the influence of the reaction conditions.

However, enol ethers can only act as nucleophiles and acetals can only act as electrophiles. As an example, we can cite the case of the addition of ethylvinyl ether to the diethyl acetal of propionaldehyde which only gives a single product (3 + 2). [10.23]

[10.23]

$$CH_3{-}CH_2{-}CH(OEt)_2 + CH_2{=}CH{-}OEt \longrightarrow CH_3{-}CH_2{-}CH(OEt){-}CH_2{-}CH(OEt)_2 \longrightarrow CH_3{-}CH_2{-}CH{=}CH{-}CHO$$

The same can be seen with the addition of ethylpropenyl ether to the diethyl acetal of acetaldehyde (2 + 3). [10.24]

It can, however, be noted that mixtures of cis- and trans-isomers are obtained. In all of the reactions discussed here, the elimination of the alkoxy group (ethoxy, in particular) is carried out in acidic media. In contrast, if trialkoxy derivatives

[10.24]

$$CH_3\text{-}CH(OEt)_2 + CH(CH_3)\text{=}CH\text{-}OEt \longrightarrow CH_3\text{-}CH(OEt)\text{-}CH(CH_3)\text{-}CH(OEt)_2 \xrightarrow{H^{\oplus}} CH_3\text{-}CH\text{=}C(CH_3)\text{-}CHO$$

are pyrolyzed, a completely different reaction, leading to a cis + trans mixture of an alkoxydiene, occurs. [10.25]

[10.25]

$$CH_3\text{-}CH(OEt)\text{-}CH(CH_3)\text{-}CH(OEt)_2 \xrightarrow[(cat.)]{\Delta} CH_2\text{=}CH\text{-}C(CH_3)\text{=}CH\text{-}OEt$$

D. Rearrangements of enol ethers

These can either be acid-catalyzed (the more important cases) or thermally induced.

1. Acid-catalyzed rearrangements [10.26]

[10.26]

$$C_6H_5\text{-}O\text{-}C(CH_3)\text{=}CH_2 \underset{}{\overset{H^{\oplus}}{\rightleftharpoons}} C_6H_5\text{-}\overset{\oplus}{O}\text{=}C(CH_3)_2 \longrightarrow \text{(cyclohexadienone, H, } \overset{\oplus}{C}(CH_3)_2) \xrightarrow{-H^{\oplus}} o\text{-}HO\text{-}C_6H_4\text{-}C(CH_3)\text{=}CH_2$$

The exact mechanism of this rearrangement is not known. However, as the ortho and not the para derivative is formed, it is probable that this transformation proceeds by electrophilic substitution of the first protonated species.

2. Thermal rearrangements of alkylvinyl ethers

Normal ethers of this type rearrange to give aldehydes or ketones in the presence of dehydration catalysts such as the oxides of aluminum or zirconium. This reaction occurs between 200° and 300°C. [10.27]

[10.27]

$$CH_2\text{=}C(R')\text{-}OR \longrightarrow R\text{-}CH_2\text{-}C(\text{=}O)\text{-}R'$$

It is possible to envisage a a four-centered cyclic mechanism. The α-alkoxystyrenes isomerize, for example, by heating above 200°C, to give alkylphenyl ketones. [10.28]

[10.28]

$$H_2C=C(\phi)-O-R \longrightarrow R-CH_2-C(=O)-\phi \quad \Delta$$

However, when an initiator of free radical reactions (such as azaisobutyronitrile) is present, the same reaction occurs at temperatures as low as 70°C with yields of about 90% being obtained. In this case, it is clearly a free radical mechanism which operates (X· = radical initiator). [10.29]

[10.29]

$$\phi-C(OR)=CH_2 + X^{\bullet} \longrightarrow \phi-\overset{\bullet}{C}(OR)-CH_2-X \longrightarrow \phi-C(=O)-CH_2-X + R^{\bullet}$$

$$\phi-C(OR)=CH_2 + R^{\bullet} \longrightarrow \phi-\overset{\bullet}{C}(OR)-CH_2-R \longrightarrow \phi-C(=O)-CH_2-R + R^{\bullet} \quad \Delta$$

chain radical reactions

Chain termination reactions :

$$2X^{\bullet} \longrightarrow X-X \;;\; X^{\bullet} + R^{\bullet} \longrightarrow XR \;;\; 2R^{\bullet} \longrightarrow R-R$$

3.Thermal rearrangements of allylic ethers of enols

This reaction is very important, as well as the synthesis of these allylenol ethers which was discovered in 1912 by Claisen.

This rearrangement occurs far more readily than that of normal alkylenol ethers and it is favored by the presence of substituents at the α-position of the vinyl group (Hurd and Pollack, 1938). [10.30]

[10.30]

Δ → α, β, γ

R = H	t = 225°C
$R = CH_3$	t = 210°C
R = ϕ	t = 175°C

This is a stricly monomolecular reaction and the parameters of activation all point to the operation of a cyclic transition state. The intramolecular nature of the transfer resulting from such a mechanism has been confirmed by labeling studies using ^{14}C and by the observation of stereospecific bridgehead alkylation.

A particularly elegant confirmation of the cyclic mechanism has been provided by the rearrangement of an optically active allylvinyl ether to give an aldehyde which retains this activity. [10.31] The new C-C bond is formed cis to the old C-O bond which is broken. Bridgehead alkylation: stereopecific reactions [10.32]

[10.31]

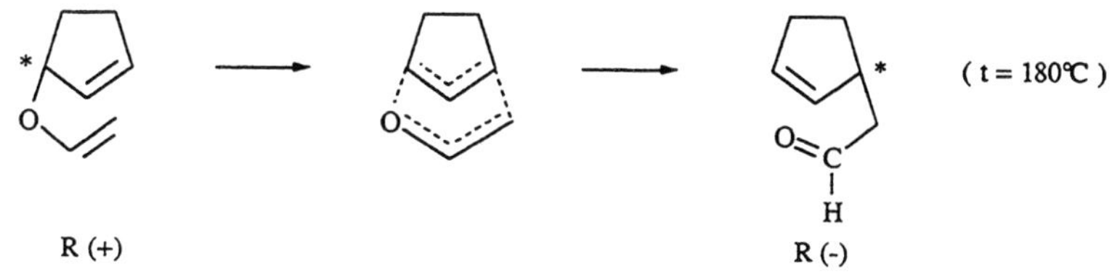

R (+) → → R (-) (t = 180°C)

[10.32]

cholesterol + CH_2=CH-OEt —Hg(OAc)$_2$→ —Δ→ CHO

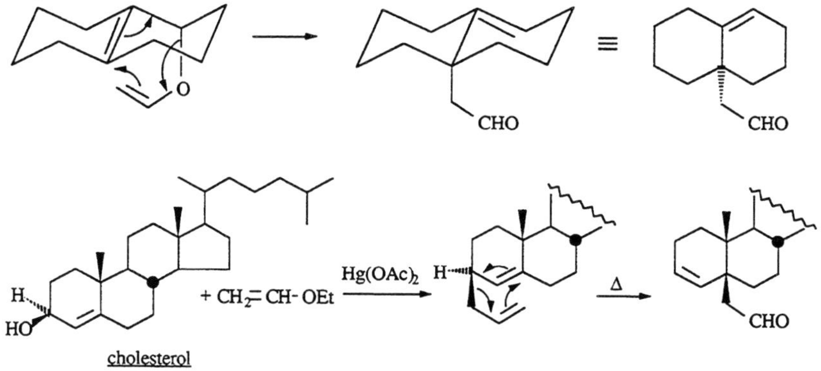

Example: Brannock reaction: [10.33]

[10.33]

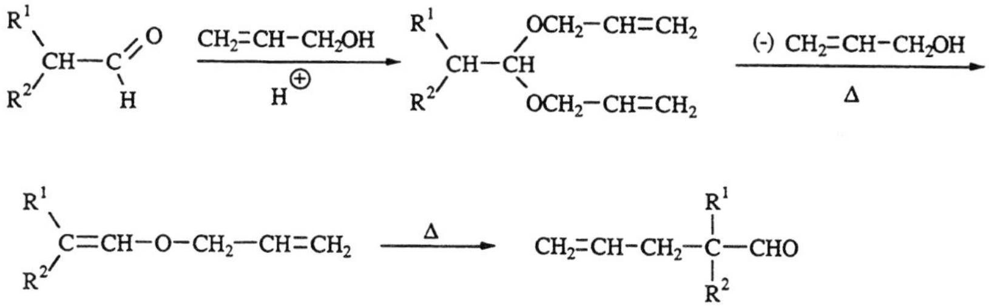

R^1R^2CH–CHO —CH_2=CH-CH_2OH, $H^{\oplus}$→ R^1R^2CH–CH(OCH_2–CH=CH_2)$_2$ —(-) CH_2=CH-CH_2OH, Δ→

R^1R^2C=CH–O–CH_2–CH=CH_2 —Δ→ CH_2=CH-CH_2–C(R^1)(R^2)–CHO

E. Examples of the application of these types of reactions

1. Partial synthesis of diterpenes [10.34]

2. Total synthesis of Kaurène (diterpene) [10.35]

The two ketones must be separated, keeping only the first. [10.36]

3. Synthesis of tagetones and dihydrotagetone [10.37]

4. Addition of a functionalized isoprenic unit to an allylic alcohol

[10.34]

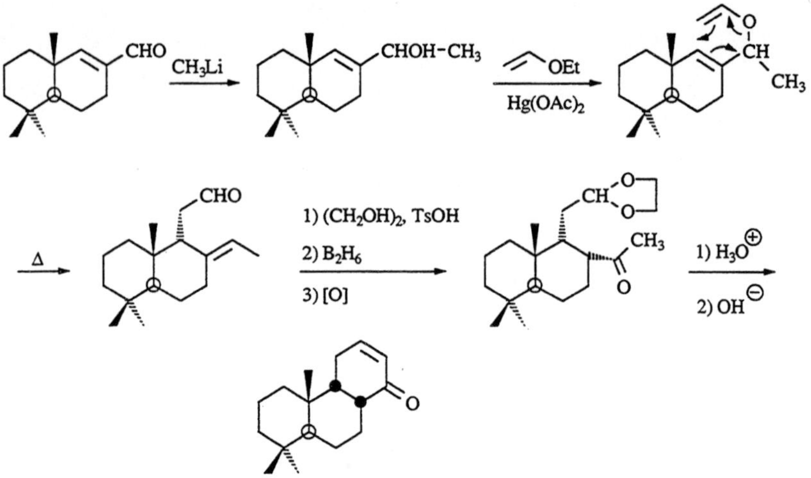

CHO $\xrightarrow{CH_3Li}$ CHOH−CH_3 $\xrightarrow[Hg(OAc)_2]{\text{OEt}}$ O, CH, CH_3

$\xrightarrow{\Delta}$ CHO $\xrightarrow[\text{3) [O]}]{\text{1) }(CH_2OH)_2\text{, TsOH; 2) }B_2H_6}$ CH, O, O, CH_3, O $\xrightarrow[\text{2) }OH^{\ominus}]{\text{1) }H_3O^{\oplus}}$ O

[10.35]

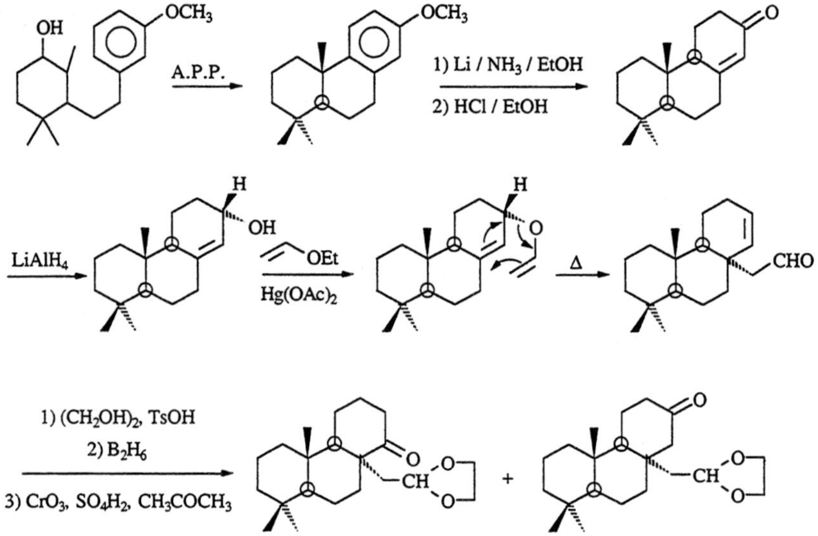

OH, OCH_3 $\xrightarrow{\text{A.P.P.}}$ OCH_3 $\xrightarrow[\text{2) HCl / EtOH}]{\text{1) Li / }NH_3\text{ / EtOH}}$ O

$\xrightarrow{LiAlH_4}$ H, OH $\xrightarrow[Hg(OAc)_2]{\text{OEt}}$ H, O $\xrightarrow{\Delta}$ CHO

$\xrightarrow[\text{3) }CrO_3\text{, }SO_4H_2\text{, }CH_3COCH_3]{\text{1) }(CH_2OH)_2\text{, TsOH; 2) }B_2H_6}$ O, CH, O, O + O, CH, O, O

[10.36]

O, CH, O, O $\xrightarrow[\text{2) }\Delta]{\text{1) HCl / }H_2O}$ OH, O, H $\xrightarrow[\text{2) }\phi_3P{=}CH_2]{\text{1) }CrO_3\text{, }H^{\oplus}\text{, }CH_3COCH_3}$

O, CH_2 $\xrightarrow{\text{Wolff-Kischner}}$ CH_2

Kaurene

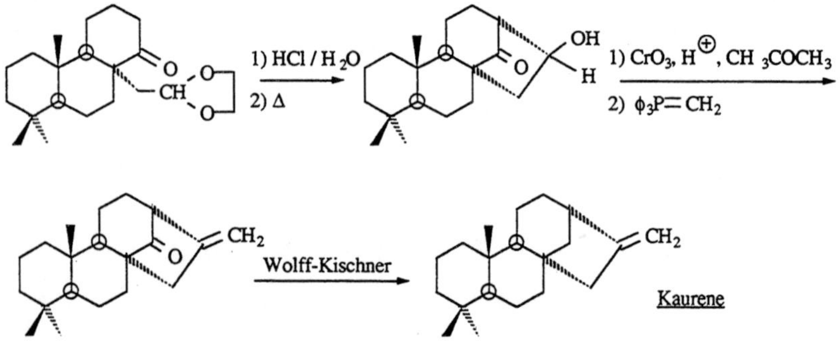

[10.37]

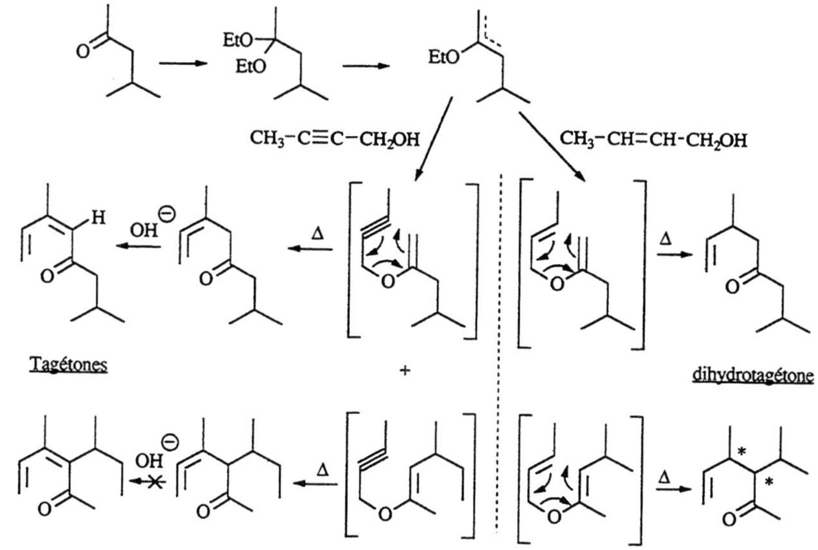

O EtO EtO EtO

$CH_3-C\equiv C-CH_2OH$ $CH_3-CH=CH-CH_2OH$

H $OH^{\ominus}$ O O Δ O O Δ O

Tagétones + dihydrotagétone

$OH^{\ominus}$ O O Δ O O Δ O * *

a. Principle: [10.38]

[10.38]

R^1 R^3 CH_3 allyle vinyle R^1 R^3 CH_3

$C=C-CH_2OH$ + $EtO-CH=C-CH=CH_2$ ⟶ $C=C-CH_2-O-CH=C-CH=CH_2$

R^2 A B R^2

R^2 R^3 $R^1-C=C$ CH_2 H_3C $CH-O$ ⟶ R^2 R^3 R^1-C-C C CH_2 H_3C $CH=O$ ≡ R^2 R^3 R^1-C-C H_3C-C CH_2 CHO CH_2 ⟶ R^2 R^3 $R^1-C=C$ H_3C-C CH_2 CHO CH_2

Claisen-type rearrangement

Cope-type rearrangement

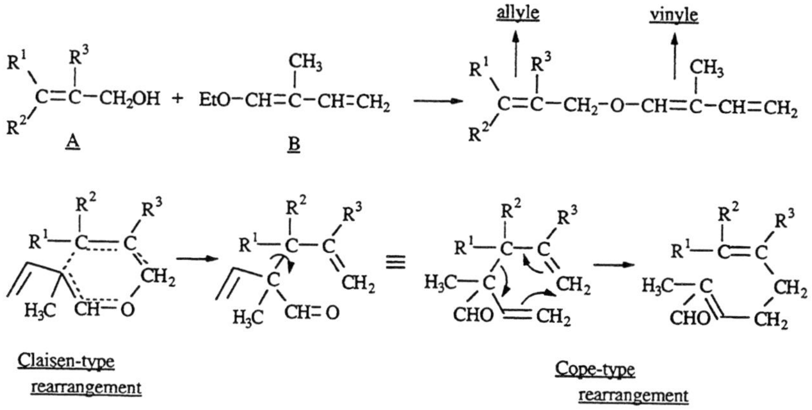

b. Example: Synthesis of β-sinesal

- Preparation of unit B [10.39]
- Preparation of sinesal [10.40]

c. Example

- Synthesis of citral [10.41]
- Synthesis of citral (BASF) [10.42]

[10.39]

$$CH_3-CH(OEt)_2 + BF_3 \longrightarrow CH_3-\overset{\oplus}{C}H-OEt + Et\overset{\ominus}{O}BF_3$$

$$CH_3-CH=CH-OEt$$

$$\downarrow$$

$$CH_3-CH(OEt)-CH(CH_3)-\overset{\oplus}{C}H-OEt$$

$$\xrightarrow{Et\overset{\ominus}{O}BF_3} BF_3 + CH_3-\overset{*}{C}H(OEt)-\overset{*}{C}H(CH_3)-CH(OEt)_2$$

$$\downarrow \Delta \text{ (catalyst)}$$

$$CH_2=CH-C(CH_3)=CH(OEt) + CH_2=CH-C(CH_3)=CH(OEt)$$

[10.40]

1) SeO_2

2) $LiAlH_4$

OEt

CH_2OH

Myrcene 1

CHO

H

O

H

CHO

Sinensal

[10.41]

OH

Pauling

Vanadates

CHO

OMe

CH

OMe

OMe

H

+ H

H

OMe

H

$Hg(OAc)_2$

CH_2OH

+

CH_2Cl

O

CHO

O

H

CHO

Citral

[10.42]

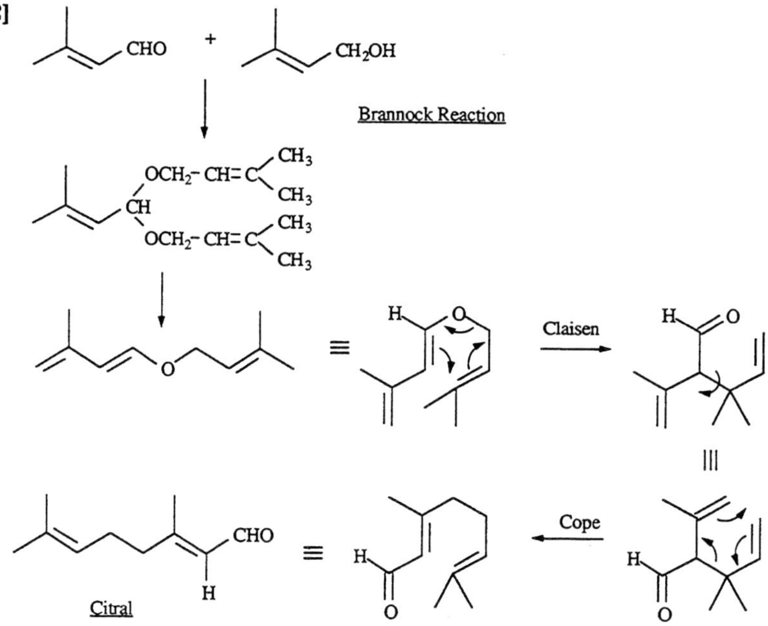

CHO + CH_2OH

Brannock Reaction

$OCH_2-CH=C(CH_3)_2$

CH

$OCH_2-CH=C(CH_3)_2$

H O ≡ Claisen H O

|||

Citral CHO H ≡ H O Cope H O

APPENDIX

4

Functionalization of Unactivated Carbon Atoms: Generalities and Biomimetic Reactions

Reactions which permit the functionalization of unactivated carbon atoms in molecules proceed principally by free radical mechanisms. This generally implies that the molecule is saturated and that free radicals attack either a methine (→ CH), a methylene (>CH_2) or a methyl (-CH_3), preferably in an unactivated position. This type of reaction generally leads, of course, to mixtures and to polysubstituted products. Indeed, for a long time, the area of free radical chemistry was considered to be a non-selective form of chemistry and was associated with the production of tars and resins. What then, are the necesary conditions for selectively functionalizing any chosen unactivated carbon atom with acceptable yields?

An initial solution to this problem arose from the observations of Hofmann in 1878 while he was attempting to determinine the structure of piperidine.

Hofmann discovered that the treatment of (+) –2-propyl-N-bromopiperidine with hot sulfuric acid gave a tertiary amine. Its structure was subsequently determined in 1899 by Lellmann to be (+)-octahydroindolizine: [11.01]

[11.01]

These observations slumbered for 20 years before Löffler and Freytag developed this reaction in 1909 as a general method for synthesizing pyrrolidines from secondary amines. Thus, it has become known as the "Hofmann-Löffler-Freytag reaction" (H.L.F.); it can be represented schematically as follows (X = halogen): [11.02]

[11.02]

$$R-CH_2-(CH_2)_3-N(X)-R' \longrightarrow R-CH_2-(CH_2)_3-N-R' \text{ (ring)}$$

It was not until 40 years later in 1950 that Wanzonek and co-workers demonstrated that this reaction proceeds by a free radical mechanism. They suggested that the N-chloramine 1 gives first the salt 2 with the aid of sulfuric acid and then the N-Cl bond undergoes homolytic cleavage under the influence of the heat or light or any other free radical initiator. The ammonium radical obtained 3 (radical cation) abstracts a hydrogen from the δ-carbon affording the radical 4 which recombines with the chlorine radical just generated, to give the δ-chloro ammonium salt 5. [11.03]

[11.03]

Therefore, the carbon in position δ with regard to the amino group has been specifically functionalized by a chlorine atom. As a rule, the chloro derivative 5 is not isolated, but is directly treated in alkaline solution to give the pyrrolidine 6. Apart from using heat or light, initiators such as metal salts are used the most frequently. Ferrous salts provide, by a redox process, an electron to the ammonium salt 2 to give 3 together with the corresponding ferric salt. Chlorine is liberated as the anion and not as the free radical. [11.04]

[11.04]

$$Fe^{2+} + R_2\overset{\oplus}{N}HCl \longrightarrow Fe^{3+} + R_2\overset{\oplus\odot}{N}H + Cl^{\ominus}$$

In summary, the H.L.F. reaction essentially consists of a specific intramolecular halogenation of an unactivated carbon atom by a halo-amine.

The question arises as to the possibility of carying out this reaction on an intermolecular basis. This is indeed feasible. To illustrate this, a short analysis of the 1964 publications of the Italian group of Minisci and Galli can be given. They have shown that the action of halo-amines on saturated aliphatic compounds (containing a polar group such as a halogen, an ester or an ether) gives a derivative with highly specific chlorination at the ω=1 position. The reaction is generally intitiated by salts of metals which can exist in different oxidation states. In numerous cases, for compounds containing polar groups, it is possible to obtain 70–90% of the ω=1 halogenated isomer. For example, in the chlorination of methyl hexanoate with N-chlorodimethylamine in the presence of ferrous sulfate, a yield of 80% of mono-halogenated products is obtained with the following isomer distribution:

$$\underset{(5.7)}{CH_3}\text{—}\underset{(87.3)}{CH_2}\text{—}\underset{(6.3)}{CH_2}\text{—}\underset{(0.7)}{CH_2}\text{-}CH_3\text{-}COOCH_3 \qquad \Sigma = 100\%$$

The same reaction carried out in the presence of cuprous chloride gives 89.4% of the ω=1 isomer. This "ω=1 effect" can be further improved if a sterically hindered amine such as N-chlorodiisobutylamine is used. This obtains yields up to 90% of the ω=1 chloro derivative. If the aliphatic substrate does not contain a polar group, the ω=1 effect can still be noticed. The reaction of N-chlorodiisobutylamine on n-heptane gives the following distribution of mono-chloro derivatives:

$$\underset{(1.3)}{CH_3}\text{—}\underset{(\underline{64.4})}{CH_2}\text{—}\underset{(22.9)}{CH_2}\text{—}\underset{(11.3)}{CH_2}\text{—}CH_{2\pi}\text{—}CH_3 \qquad \Sigma = 99.9\%$$

Another aspect of this process of intermolecular halogenation is that even positions which are greatly distant can still be selectively halogenated. The reaction of N-chlorodiisobutylamine with methyl decanoate gives the following distribution:

$$\underset{(1.4)}{CH_3}\text{—}\underset{(\underline{57.7})}{CH_2}\text{—}\underset{(19.1)}{CH_2}\text{—}\underset{(13.1)}{CH_2}\text{—}\underset{(6.7)}{CH_2}\text{—}\underset{(1.7)}{CH_2}\text{—}(CH_2)_3\text{—}COOCH_3 \qquad \Sigma = 99.7\%$$

The mechanism proposed by Minisci and his colleagues is essentially the same as that proposed for the metal ion-catalyzed H.L.F. reaction; the only difference being that hydrogen abstraction takes place intermolecularly and not intramolecularly. This does not, however, explain the ω=1 effect!

This selectivity can be partly explained on the basis of the inductive effect of the polar group. The di-isomopyl radical cation $(i\text{–}Pr)_2HN^{+\cdot}$ is very electrophilic and will tend to react as far away as possible from groups such as esters for example. This, however, is certainly not the only explanation because the ω = 1 effect exerts itself in the reaction of simple alkanes (although less pronounced as in the case of heptane just described). Minisci also considered the intervention of steric effects which certainly intervene in the case of the two following photochemical reactions: [11.05]

[11.05]

		CH_3	CH_2	CH_2	CH_2	CH_2—COOH
$Cl_2(CCl_4)$	$\xrightarrow{h\nu}$	18	<u>37</u>	18	13	4
⊕ $(iPr)_2NHCl$ / SO_4H_2 / AcOH	$\xrightarrow{h\nu}$	1	<u>93</u>	0	0	0

This explanation is not entirely satisfactory and the profound reason for this phenomenon still seems to escape discovery.

Without diverging too far from this subject, it can be pointed out that if the substrate contains, for example, δ-ε unsaturation, hydrogen-abstraction leading to a substitution product will not occur. Instead, an addition reaction takes place and a carbon-nitrogen bond is directly formed. [11.06]

From 1909 onwards, Löffler applied the results of his studies to the synthesis of natural products such as nicotine. [11.07]

[11.06]

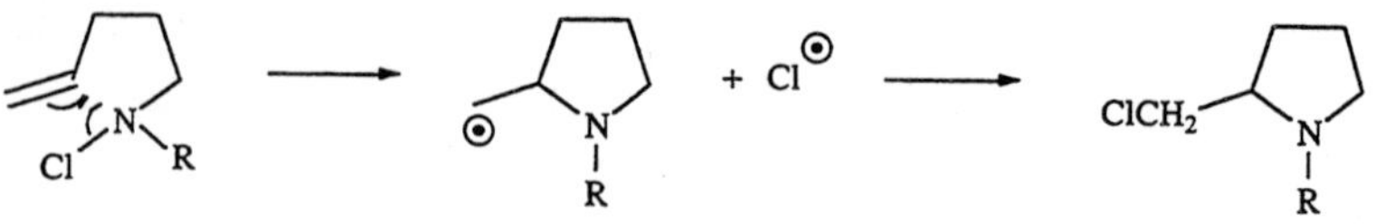

[11.07]

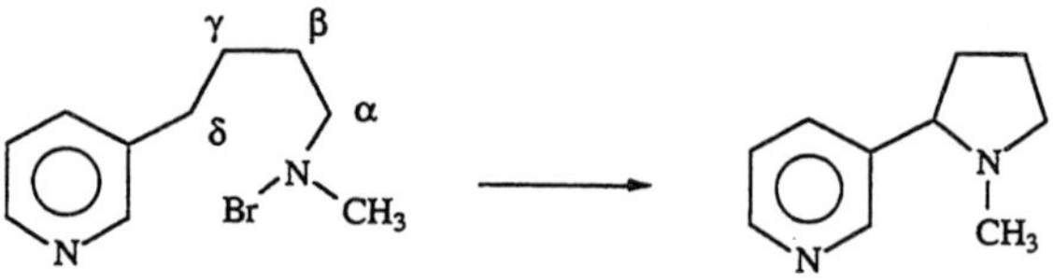

It was not until 50 years later, however, that this reaction was employed once again in the synthesis of natural products.

In fact, from 1958 onwards, several solutions to the problem of selectively functionalizing chosen unactivated carbon atoms in complex molecules were found in work on steroid chemistry.

The tertiary methyl groups at C_{18} and C_{19} in the steroid skeleton were for a long time considered to be inaccessible to chemical attack by common reagents. It was known, however, that these methyls possess a propensity to migrate to different parts of the skeleton or even to be eliminated during dehydrogenation reactions under the effect of agents such as sulfur or selenium, for example.

The problem of synthesizing aldosterone provided the necessary stimulus for developing different methods for attacking the tertiary C_{18} methyl; these methods are also applicable to the other tertiary methyl, C_{19}. [11.08]

[11.08]

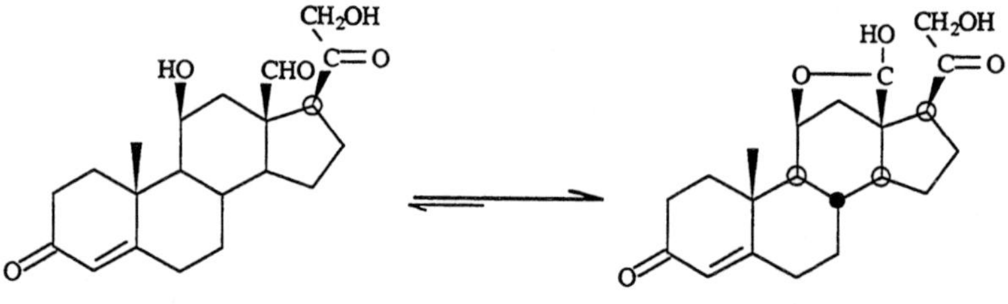

(aldol form) *aldosterone* (hemiacetal form)

Due to the ingenuity of the groups of Corey, Jeger, Arigoni and Barton, solutions to these problems were found. The first groups used the H.L.F. reaction to functionalize the C_{18} methyl according to the following scheme. [11.09]

Alkaline treatment of the C_{18} chloamine derivative gave access to a new class of compounds—the "conanine" alkaloids which are found in the *Holarrhena* species.

In 1959, Corey also used this reaction on a terpene model, camphidine (1,8,8-trimethyl–3-azabicyclo [3.2.1] octane), which was cyclized to give 1,8-

[11.09]

dimethyl–3,8-endomethylene–3-azabicyclo [3.2.1] octane with a yield approaching 70%. [11.10]

[11.10]

In 1959 as well, the group of Jeger and Arigoni showed that lead tetraacetate is also capable of reacting with steroids having hydroxyl functions at C_{20} to form a tetrahydrofuranyl ring incorporating C_{18}. This reaction will be discussed later. [11.11]

[11.11]

In addition to the oxidation reaction using lead tetraacetate, Barton proposed two other functionalization methods.

The former of these, developed in 1960, concerns the photochemical decomposition of alkyl nitrites. Since then it has acquired the name "Barton reaction" and involves, as in the previously discussed cases, the abstraction of a hydrogen at the δ position by an alkoxyl radical through a six-centered transition state. This reaction can be schematized in the following manner. [11.12]

Thus, Barton was able to functionalize the methyl at C_{19} of 6β-hydroxycholestanyl–3β-acetate. [11.13] He was also able to perform the first synthesis of aldosterone acetate (see earlier) by functionalizing the C_{18} methyl of corticosterone acetate. [11.14]

In the course of this work, Barton realized that derivatives of the type ROX (X being a halogen) should lead by homolytic decomposition to intramolecular halogenation of an unactivated but conveniently placed carbon.

A similar conclusion had already established precedents. In 1956, Cairns had shown that the thermal decomposition of 1-methylcyclopentyl hypochlorite led to 1-chloro–5-hexanone. [11.15] This is an example of a β-fragmentation reaction.

[11.12]

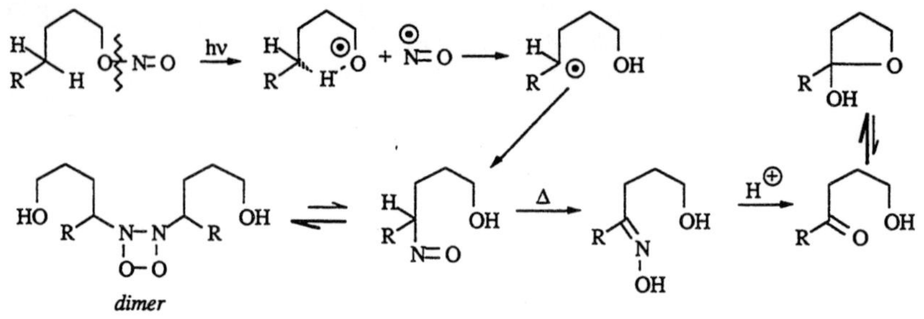

[11.13]

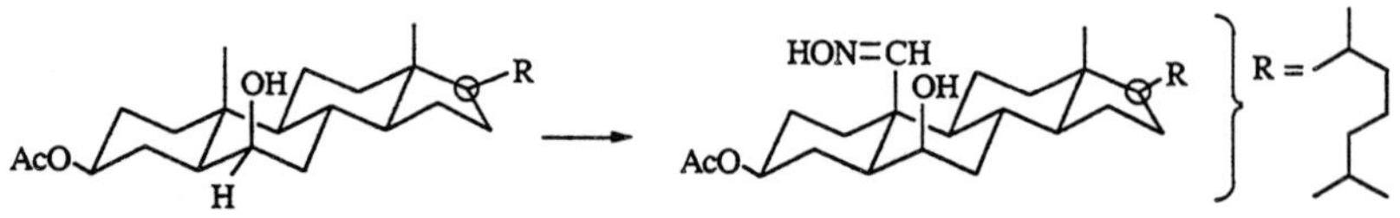

[11.14]

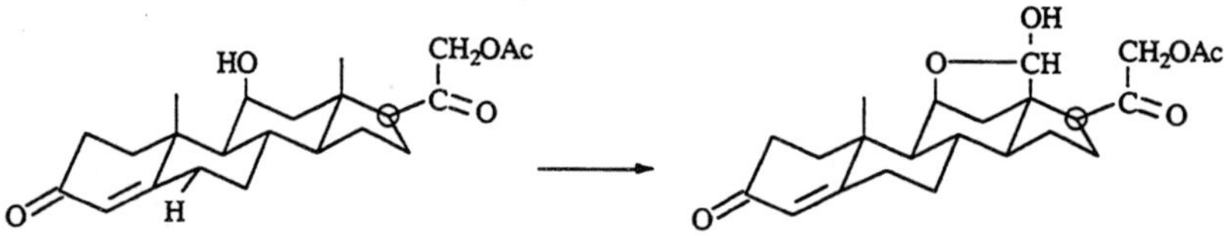

[11.15]

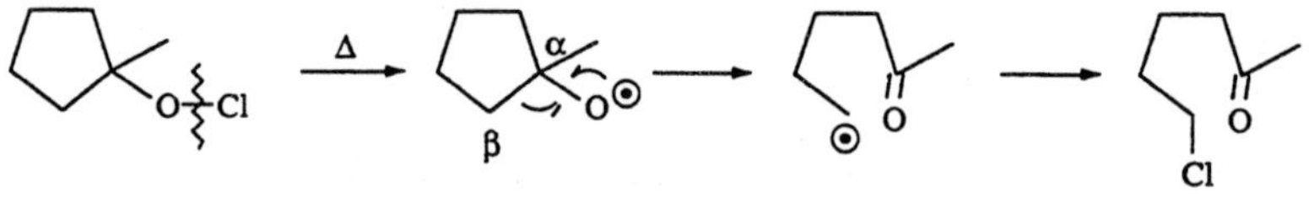

From 1961 onwards, Barton's hypothesis was developed further by several groups of research workers, but he provided the first two examples.

1. Functionalization at C_{19} (yield approximately 50%) [11.16]

[11.16]

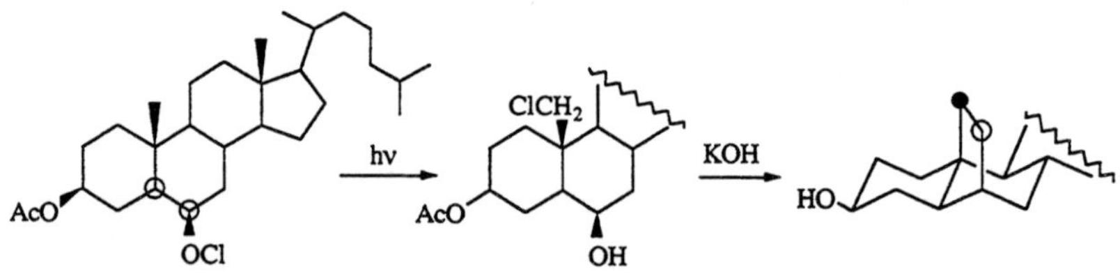

2. Functionalization at C_{18} [11.17]

We have just indicated very briefly the principal reactions which allow functionalization of unactivated carbon atoms situated at "position δ" with regard to

[11.17]

ClO

1) hν 2) KOH 3)Ac_2O

AcO AcO

an amine or hydroxyl group. These diverse reactions can be summarized by the following scheme. [11.18]

[11.18]

$Y = N^{\oplus}$	X = Cl, Br
Y = O	X = [$-Pb(OAc)_3$], –NO, Cl, Br, I
Y = N	X = I

The first case is the Hofmann-Löffler-Freytag reaction, which will not be discussed any further. The second incorporates several processes, including the Barton reaction. Some examples in the field of terpenes will now be given.

1. Synthesis of 8,14-cedranolide and 8,14-cedranoxide.

The 8,14-cedranoxide is a natural product which was first isolated in 1961 by Runeberg from *Juniperus fœtidissima* Willd. Its structure was established in 1968 by the Swedish group of Erdtman. Its synthesis was carried out simultaneously by the same group by oxidizing cedrol using lead tetraacetate in the presence of iodine. We have demonstrated elsewhere that this oxidation takes place even in the presence of iodine alone, whereas lead tetraacetate alone does not alter cedrol at all. The conjugated action of $Pb(OAc)_4$ and iodine generates hypoiodous acid, which gives a higher yield.

Another route involving bromine and mercuric oxide has been patented by I.F.F. Roure-Bertrand Dupont has prepared both compounds, the lactone and the oxide, starting with cedrol and using the Barton reaction. The common intermediate which permits access to both of these compounds is the hydroxy-aldehyde which is stable in the form of its internal hemi-acetal 4. [11.19]

When the crude product from these reactions is subjected to chromic acid oxidation, 8,14-cedranolide 5 is obtained with a yield of 60%. This substance can be reduced in a number of different ways, notably by the mixture $LiAlH_4/Et_2O$, BF_3/THF, to obtain 8,14-cedranoxide 6 with a yield of around 70%.

[11.19]

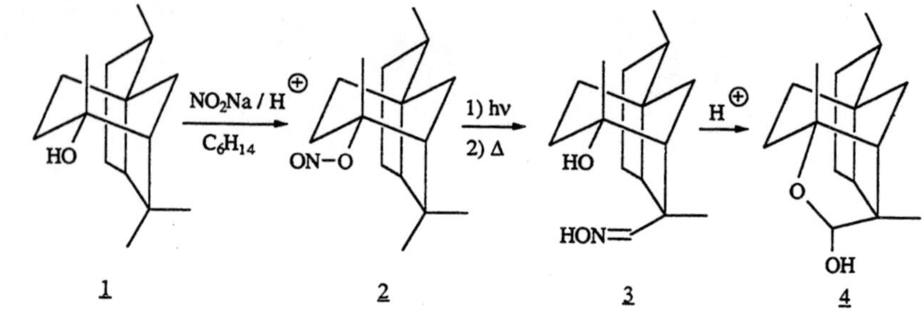

2. Functionalization of α-caryophyllenic alcohol at position 3.

In July 1969, Apollo–11 landed on the moon. Simultaneously and independently, the groups of Nickon and Sim demonstrated that α-caryophyllenic alcohol 2 (which had been long considered as the product of isomerization of caryophyllene—hence, its name) originated in fact from the acid-catalyzed rearrangement of humulene. [11.20]

[11.20]

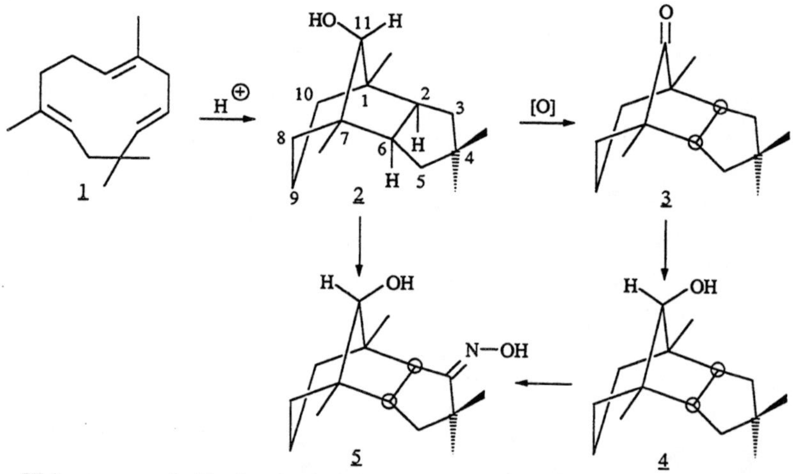

This α-caryophyllenic alcohol 2 contains a secondary hydroxyl at position 11 and possesses a symmetry similar to that of Apollo-11. This coincidence led Nickon to call this alcohol apollan–11-ol. [11.21]

[11.21]

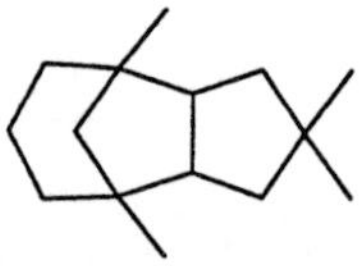

This alcohol as well as its epimer 4, when subjected to the Barton reaction, both give the same hydroxylamino alcohol 5 with excellent yields.

Although it is easy to see how 5 could be obtained from 4, it is not so easy to see how this could happen with apollan–11-ol. One explanation could be as follows. The alcohol 2 gives an alkoxyl radical a which is attached to a carbon that is part of a five-membered ring and which is adjacent to two tertiary centers. These two factors will facilitate β-fragmentation to give the radical b, which recyclizes producing epimerization to c. This is an identical radical to that which is obtained directly from 4. [11.22]

[11.22]

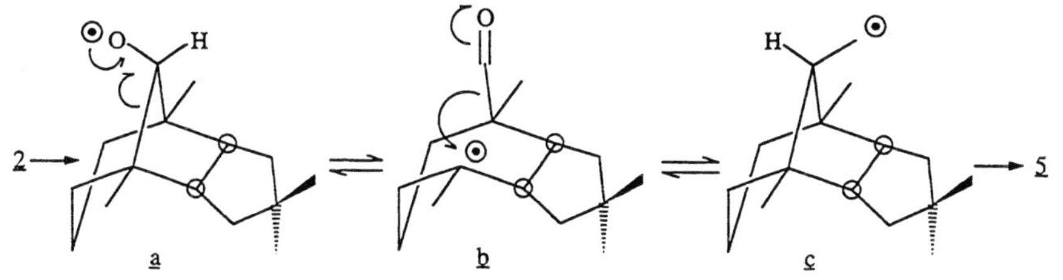

2 → a ⇌ b ⇌ c → 5

3. Synthesis of (–)-α-cis- and (+)-β-cis-bergamotenes

These syntheses involve the preparation of intermediates of the type A. [11.23]

[11.23]

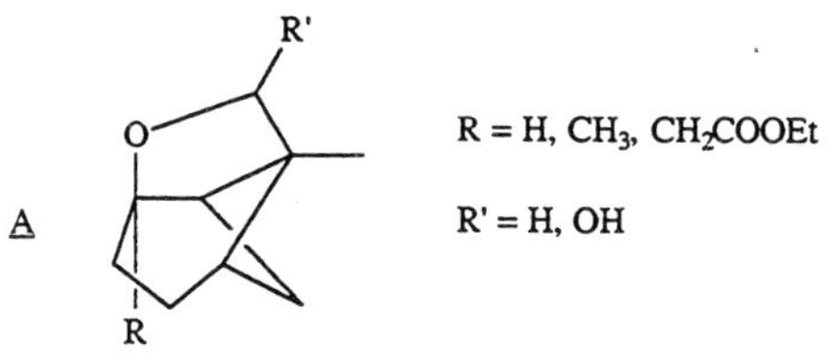

A

R = H, CH_3, CH_2COOEt

R' = H, OH

They can be obtained from nopinone, which itself results from the ozonolysis of β-pinene. A schematized example of the synthesis of (–)-α-cis-bergamotene 9 (isolated from the oil of opoponax) is given here: (A = 6: R = CH_3, R' = OH) [11.24]

The methyl group at position 8 is thus transformed into a primary alcohol with a yield of around 40% starting with a tertiary pinanol 3. This provides the possibility of attaching the remaining isoprenic unit in order to obtain (–)-α-cis-bergamotene 9. [11.25]

The synthesis of (+)-β-cis-bergamotene 14 (isolated from valerian oil) can be summarized by the following scheme. (A = 11: RCH_2COOEt, R' = H). O. Wallach had much earlier prepared compound 10 by using a Reformatsky reaction on nopinone. In this case, a hypohalogenite has been used. It can be prepared by the action of bromine and mercuric oxide with the result that the hetercyclization product 11 is obtained with a yield of 80%. [11.26]

In all of these reactions, the formation of an alkoxyl radical RO· is always involved in the first step. From thereon its course of reaction will vary depending on which reagent is used: lead tetraacetate or the mixture of bromine and mercuric oxide. In the latter case, it is known that a bromohydrin intermediate forms,

[11.24]

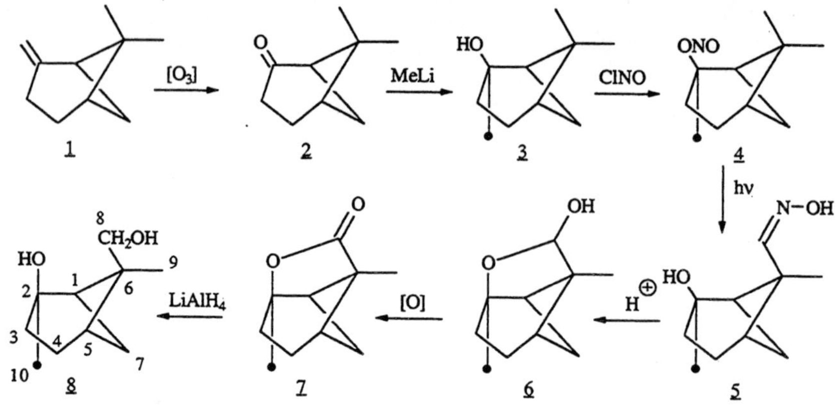

$[O_3]$ MeLi ClNO hν $H^{\oplus}$ [O] $LiAlH_4$

[11.25]

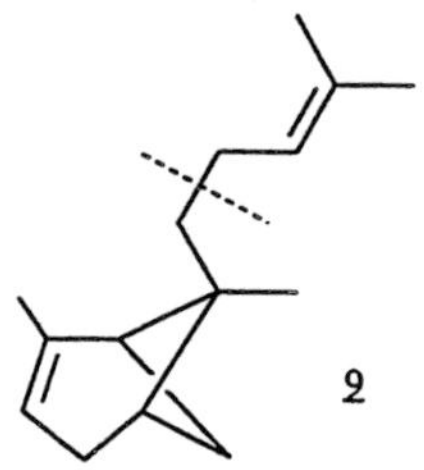

[11.26]

2 $BrCH_2COOEt$ / Zn → 10 → HgO / Br → 11 → Hunsdiecker / Kochi → 12

12 Na / OMe–$(CH_2)_2$–OMe → 13 → 14

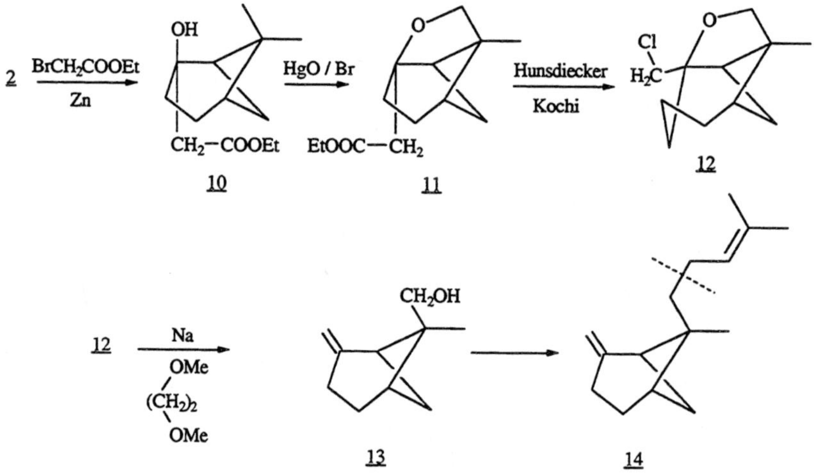

whereas in the former case, heterocyclization (abstraction of the δ-hydrogen) proceeds directly from the alkoxyl radical.

A very complete study on the role and the nature of the intermediates participating in these reactions has recently been carried out by Brun and Waegell. This study used alcohols with rigid structures thereby avoiding complications due to conformational variations. Despite this considerable work, it is still not possible to conclude whether the mechanisms of these reactions are entirely free radical or not. It is interesting to note that according to the reagents used, the behavior of certain alcohols varies enormously.

It has already been shown that the HgO/Br_2 mixture converts cedrol into 8,14-cedranoxide, while lead tetraacetate is unreactive. Neoisocedranol 1 (described by us in 1966), behaves quite differently. In fact, this alcohol, when treated with lead tetraacetate in refluxing benzene, furnishes a mixture of four products. The product of heterocyclization 2 is formed with a yield of 77% together with 15% of α-cedrene 3 (from dehydration of the starting alcohol), 3% of cedranone 4, as well as traces of neocedranyl acetate 5.

However, when neoisocedranol is treated with the HgO/Br_2 mixture, no heterocyclization product (2) is formed. Instead, cedranone 4 is produced with a yield of 64% along with α-cedrene 3 with a yield of 33%. [11.27]

[11.27]

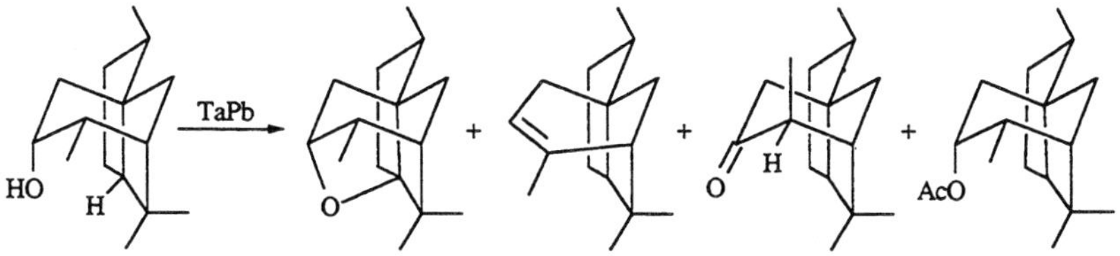

1 | 2 77% | 3 15% 33% | 4 3% 64% | 5

The terpene and steroid fields obviously have no exclusivity over these reactions. In 1961, the group of Mihailovic extended the reaction of lead tetraacetate to the aliphatic alcohols. He observed the formation of tetrahydrofuranyl derivatives with yields around 50%; traces of tetrahydropyranyl derivatives were also formed. [11.28]

[11.28]

In 1965, Cope extended this reaction to medium-sized rings. These seemed to constitute substrates of choice, because they readily undergo trans-annular reactions. In this way, cyclooctanol gives a mixture of four products, among which heterocyclization products constitute 37% (36% tetrahydrofuranyl derivative and 1% tetrahydropyranyl derivative). Cyclooctanone (23%) and cyclooctanyl acetate (17%) are also formed in the same reaction. Unreacted cyclooctanol (23%) is also recuperated. In fact, lead tetraacetate is acting here in part as an acetylating agent; the lead atom is in its maximum oxidation state in this reagent.

In the example shown here, two side reactions compete with the heterocyclization reaction (acetylation and oxidation). They both proceed at similar rates, but their mechanisms are quite different. [11.29]

[11.29]

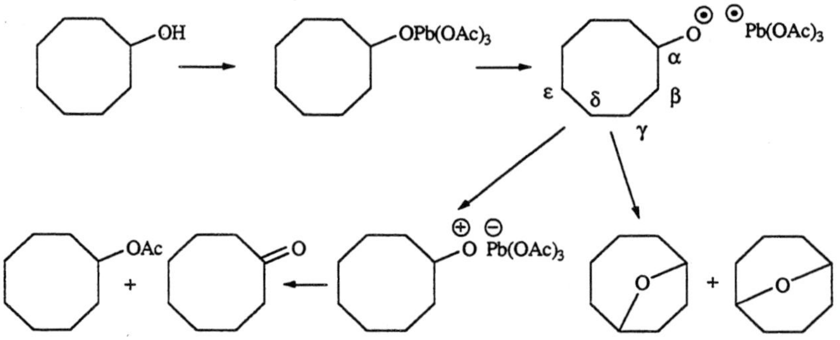

Formation of tetrahydrofuranyl derivatives is a characteristic of heterocyclization reactions. As seen earlier, the principal reason for this is in the existance of a six-membered transition state which can evolve in two different ways.

Furthermore, the results of studies carried out on conformationally immobile molecules indicate another extremely important factor in these heterocyclization reactions. If such a reaction is to occur, the distance between the oxygen atom and the δ-carbon must be between 2.4 and 2.7 Å. This is the case with all of the compounds discussed so far. In contrast, as shown by Partch, the oxidation of isoborneol (in which the distance is 2.9 Å) gives no trace of the heterocyclization product, but instead gives an 80% yield of the product of β-fragmentation. [11.30]

[11.30]

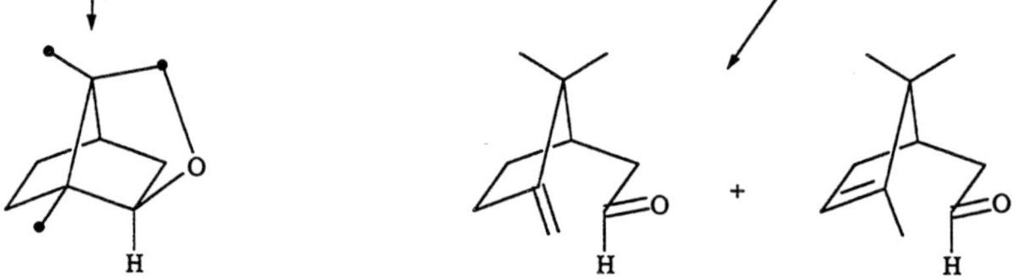

This proximity requirement is extremely important. It provides, for example, an explanation for the formation of tetrahydropyranyl derivatives. For instance, bicyclo [3.3.1.] nonan–3-ol exists in a double chair conformation in which the distance between the oxygen at position 3 and the carbon at position 7 is only 1.7 Å. Cyclization occurs between these two atoms, thus obtaining an analogue of adamantane with an 86% yield. [11.31]

[11.31]

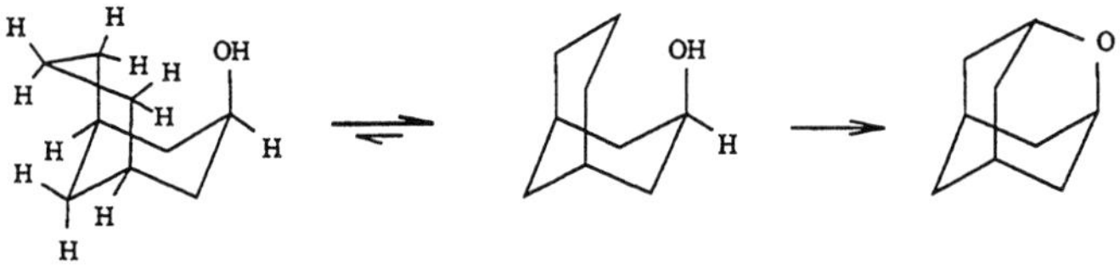

Among the examples cited so far, functionalization is usually obtained through heterocyclization between the hydroxyl group and the δ or ε carbon. For the needs of further synthetic elaboration, however, it is frequently desirable to "liberate" this position. Treatment with the mixture $Ac_2O/BF_3{\cdot}Et_2O$ will generally lead to opening of the heterocyclic ring.

For example: [11.32]

[11.32]

This reagent, however, cannot be used when the heterocyclic compound is susceptible to rearrangement. An example of this can be seen with the compounds resulting from the cyclization of pinanol. In cases such as this, acetic anhydride in the presence of pyridinium hydrochloride can be used. A product is thus obtained which has not undergone rearrangement and which can be used in the synthesis of α-bergamotene. [11.33]

[11.33]

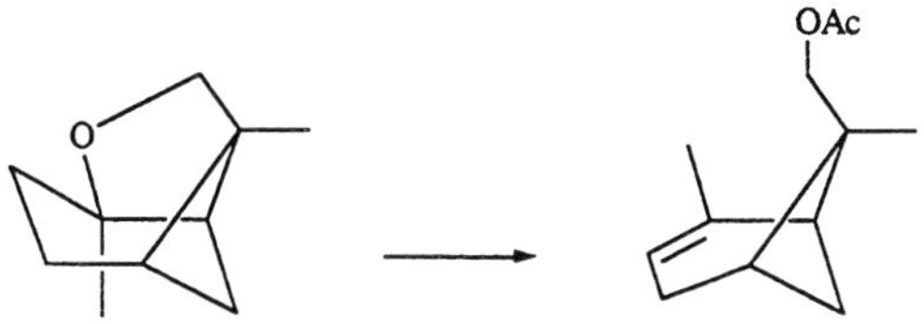

Problems due to ring opening are sometimes greatly facilitated when the final product is a γ- or a δ-lactone.

An example which corresponds to the third case given earlier (Y = N and X = I), when applied to amides, can be summarized as follows. [11.34] When these conditions are applied to stearinamide, a γ-lactone is obtained with a yield of 60%. This then gives 4-hydroxystearic acid on saponification. [11.35]

The reactions just discussed all lead to the functionalization of carbon atoms which are situated δ- or ε- to a hydroxyl or amino group which is initially present in the starting material.

[11.34]

H NH_2 → H NH I → H ⊙NH → ⊙ NH_2 → I NH_2 $\xrightarrow{OH^{\ominus}}$ O

[11.35]

$$CH_3-(CH_2)_{13}-CH_2-CH_2-CH_2-CONH_2 \longrightarrow CH_3-(CH_2)_{13}-\text{(lactone ring)}=O \longrightarrow$$

$$CH_3-(CH_2)_{13}-CHOH-(CH_2)_2-COOH$$

From 1969 onwards, Breslow sought to generalize this type of reaction in different ways. The idea was to try to imitate a certain number of factors involved in enzymatic catalysis such as the formation of an enzyme-substrate complex and the subsequent functionalization of a particular site in the substrate by a catalytic group conveniently placed in the part which simulates the enzyme.

It is known that enzymatic reactions proceed at high speed and, more importantly, show remarkable selectivity. Thus, we see that *Mycobacterium phlei* converts stearic acid into oleic acid. In the industrial manufacture of the corticosteroids, the C-ring is hydroxylized by microbiological fermentation using *Rhizopus nigricans.* [11.36]

[11.36]

$COCH_3$ O $\xrightarrow{R.n}$ HO C $COCH_3$ O ⟶ cortisone

The enzymatic oxidation of camphor by *Pseudomonas putida* results in 5-hydroxycamphor. [11.37]

[11.37]

$$\text{O} + O_2 + 2e^{\ominus} + 2H^{\oplus} \longrightarrow \text{HO} \ldots \text{O} + H_2O$$

In this case, the enzyme which intervenes in this process has been isolated and crystallized. It is cytochrome P450 for which the principal element is a hemoprotein. These cytochromes have been the subjects of several studies. They participate in numerous hydroxylation reactions in microorganisms and higher species.

Breslow has expressed the opinion that there is no special reason to limit the transition state to a six-membered ring as described earlier. There would seem to be an entropic advantage in larger cyclic transition states if the important parts of the molecule are rigid. Breslow has defined distant oxidation as a process in which a rigid reagent is attached to the substrate molecule in a way which leads to oxidation at a center relatively far from the point of attachment of the reagent. Breslow initially used derivatives of benzophenone as rigid reagents. The carbonyl group of this ketone readily gives a triplet excited state on ultraviolet irradiation. A general scheme of this type of reaction follows. [11.38]

[11.38]

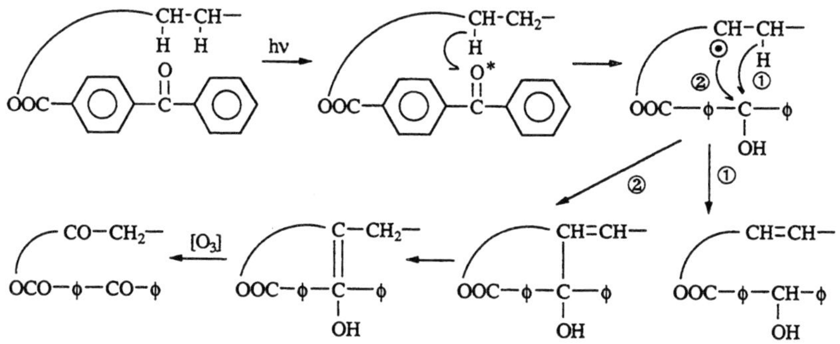

[11.39]

When applied to 3-cholestanol, this procedure produces abstraction of the 14-α hydrogen. On saponification, 3-cholesten–14-ol is obtained with a yield of 55%. [11.39]

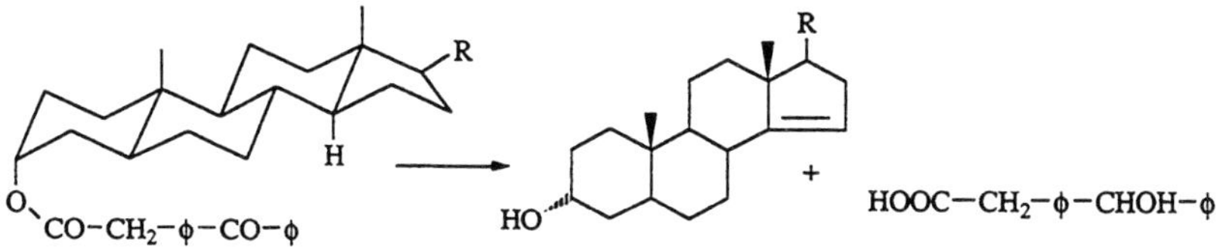

Another use of the Barton reaction provides the following example in which a derivative of acetophenone acts as the rigid reagent. In this case, the yields are much lower, although a substantial proportion of the unreacted substrate can be recovered. The 14-nitroso derivative is specifically obtained. [11.40]

[11.40]

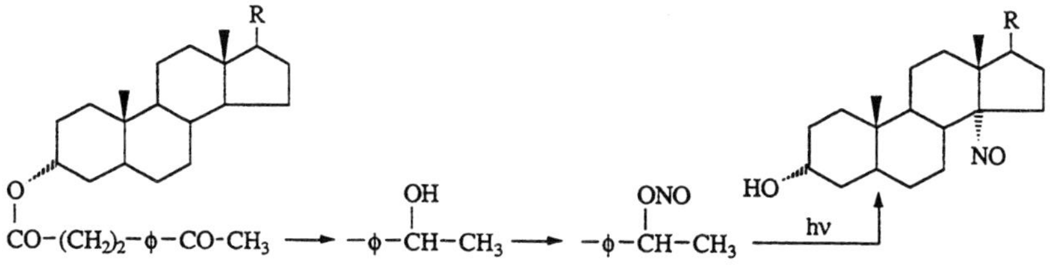

When reactions using benzophene carboxylic acid are applied to flexible molecules such as long-chain fatty alcohols, a rather random distribution of functionalized products is observed.

In 1974, Breslow's group developed a far more simple method of functionalizing steroids, employing a free radical mechanism. It will be recalled that an aryl iodide (ArI), when treated with chlorine, gives the $ArICl_2$ complex, which when irradiated with an ordinary lamp will give the ArICl- radical. This is susceptible to a reaction in which a chlorine atom ($Cl^{\cdot}$) is transferred to another aryl iodide. This chlorine transfer can be considerably simplified by using sulfuryl chloride in refluxing carbon tetrachloride in the presence of benzoyl peroxide. An example of this is provided by the ester of 3-cholestanol and meta-iodobenzoic acid. When this is treated as described earlier followed by saponification and acetylation, an 80% yield of the acetate of α-3-$\Delta^{9,11}$-cholestenol is obtained. [11.41]

[11.41]

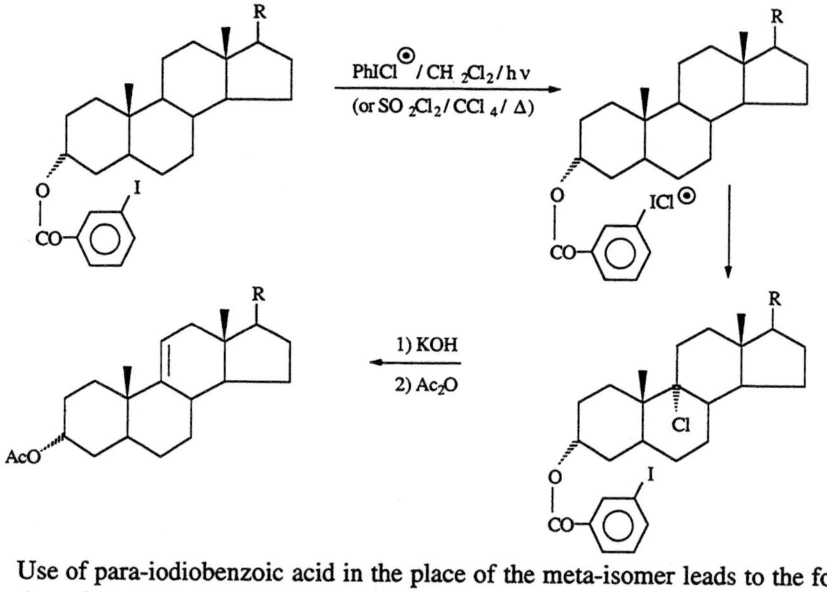

Use of para-iodiobenzoic acid in the place of the meta-isomer leads to the formation of the acetate of α-3-$\Delta^{14,15}$-cholestenol.

The key step in the synthesis of dihydrocortisone is the functionalization of ring C at position 11. To date, this has been carried out using microorganisms. Functionalization using meta-iodobenzoic acid does, however, offer a very attractive alternative!

Finally, it should be pointed out that the high yields indicated earlier can only be obtained when the principle of large dilutions originally enunciated by Rüggli is used. This allows intramolecular reactions to predominate to the detriment of competing intermolecular processes.

APPENDIX

5

Methods for Identifying Natural and Synthetic Organic Compounds

Chemically defined natural and synthetic compounds and ultimately their mixtures can, at the present time, be identified by several methods. Some are based on radioactivity (scintillation counter or Geiger-Müller counter); others are based on isotopic analysis using mass spectroscopy. A third and more recent method employs deuterium nuclear magnetic resonance spectroscopy (^2H NMR).

A. Methods based on radioactivity

Radioactive carbon (^{14}C) has a half-life of 5,730 years. Fossil products such as petroleum and coal contain practically none of this isotope at all. As a consequence, all compounds obtained from these sources show virtually no radioactivity.

However, compounds obtained from living plants have well-defined levels of ^{14}C which correspond to that displayed by atmospheric carbon dioxide.

Prior to 1950, the level of ^{14}C in atmospheric CO_2 was approximately 13.56 dpm.g^{-1}·C. As a result of nuclear tests, this level reached 26 dpm.g^{-1}·C in 1964. It has now stabilized to around 18 dpm.g^{-1}·C.

These observations have allowed us to distinguish between compounds which have been prepared from fossil and those from modern natural sources. Examples of the former are linalool, citral, geraniol, and the like, which are derived from processes we have baptized "acetylenic syntheses" (Hoffmann-la Rôche, BASF, Kuraray). Yet, products resulting from syntheses involving the pinenes (Glidden, Union Camp.) cannot be identified by this method.

Example of Citral:	^{14}C dpm.g^{-1}·C
Lemongrass	20.1 ± 0.4
Litsea cubeba	17.6 ± 0.4
Citral ex: dehydrolinalool	0.25
Citral ex: pinenes	18 to 20

B. Methods based on isotopic analysis using mass spectroscopy

In addition to radioactive ^{14}C, all organic compounds contain two other isotopes: ^{12}C, which clearly predominates, and ^{13}C. They also contain hydrogen with a mass of 1 (^{1}H) which predominates over its isotope with a mass of 2 (^{2}H or deuterium). When oxygenated, they will also contain three oxygen isotopes: ^{16}O (major), ^{17}O and ^{18}O.

1. Analysis of carbon isotopes

This involves determining the $^{13}C/^{12}C$ ratio by mass spectrometry of the CO_2 obtained from total combustion of the test material according to a well defined experimental protocol. The results are expressed in parts per thousand ($^{0}/_{00}$) of the observed difference with a standard called PDB. [12.01]

[12.01]

$$\delta\, ^{13}C = \left[\frac{^{13}C/\,^{12}C\ (\text{sample})}{^{13}C/\,^{12}C\ (\text{PDB})} - 1 \right] \times 1{,}000$$

The standard, PDB, is a fossil belemnite of the Cretaceous period from the Peedee rock formation in South Carolina. This belemnite (*Belemitilla americana*) was a elongated mollusk resembling the modern squid which had a carbonate shell. When treated with undiluted phosphoric acid, it liberates the standardizing carbon dioxide.

It has been shown that radiocarbon analysis of ^{14}C does not allow a distinction to be made in all cases. It is interesting to see how the $^{13}C/^{12}C$ ratio can help with this same example using citral.

Origin of citral	$\delta^{13}C$ $^{0}/_{00}$ PDB
Lemongrass	–11.6
Litsea cubeba	–26.7
Synthetics	–27.4

Here again it can be seen that the $^{13}C/^{12}C$ ratio does not permit distinction between synthetic citral and *Litsea cubeba.* Distinction is, however, possible between these citrals and the lemongrass product. The reason for this is far more complex, residing in the fundamental difference between the biosynthesis of compounds in lemongrass and in *Litsea cubeba* and between these and the synthetic material.

As there is, of course, no biosynthesis without photosynthesis, the difference is situated at the level of photosynthesis.

Photosynthesis is the use of solar energy by green plants and other organisms. Thanks to chlorophyll, these species manufacture from CO_2 and water the compounds which are necessary for their life—and ours—in particular sugars, proteins and fats.

De Saussure was the first to indicate that the global reaction of photosynthesis can be written:

$$CO_2 + H_2O \xrightarrow{h\nu} (CH_2O)_n + O_2$$

The studies of Calvin and his group, between 1946 and 1960, provided detailed understanding of the different steps responsible for the conversion of CO_2 into organic compounds in monocellular green algae such as *Chlorella pyrenoidosa* and *Scenedesmus obiquus.* They demonstrated that these mechanisms are the same in several higher plants and in bacteria. These studies resulted in the attribution of the 1961 Nobel prize in chemistry to Calvin. Since then, it has been demonstrated that nearly all plants fix carbon dioxide by the "Calvin cycle."

It would seem, however, that this neglects the fact that between 1953 and 1960, some researchers reported that certain plants were capable of fixing CO_2 at a rate two times faster than other plants. This is notably the case with sugarcane.

Between 1965 and 1970, Hatch and Slack re-examined these results, particularly those of Kortschak and co-workers. They showed that CO_2 fixation could also follow a different route than that demonstrated by Calvin. They called this the "C_4 cycle," whereas the Calvin cycle is a C_3 cycle. The terms C_3 and C_4 are derived from the first intermediates involved in photosynthesis.

It is now known that plants use both cycles, but differ in their relative proportions. C_3 plants use about 5% of the C_4 cycle and C_4 plants use about 20% of the C_3 cycle, but only to fix "internal" CO_2 and not atmospheric CO_2. But, most important is that the C_4 plants have enzymes which discriminate between the different carbon isotopes. The result is that C_4 plants are enriched in ^{13}C. This is the case for maize, sugarcane, sorghum, lemongrass, etc. This explains the differences seen in the second table of citral analyses between lemongrass and *Litsea cubeba.* The latter uses the C_3 cycle of Calvin as do the plants which are the ultimate sources of the synthetic citral.

There are, however, additional complications. Studies on the use of the C_4 cycle in tropical plants have led research workers to study this cycle in a very special group of plants, the crassulaceae. They are known to possess a special metabolism called "crassulacean acid mechanism" or CAM. These plants contain special enzymes which permit them to fix atmospheric CO_2 during the night by the C_4 cycle.

Recent studies have shown that certain tropical plants such as vanilla (*Vanilla planiflora*) use this type of fixation—that is, they fix atmospheric CO_2 at night by a C_4 mechanism and during the day by the Calvin (C_3) cycle. These studies allow us to distinguish between natural vanillin and those of synthetic origins.

Ratio of $^{13}C/^{12}C$ in different vanillins

Origins	*$\delta^{13}C$ ‰ PDB*
Madagascar	−21.0
Mexico	−20.8
Java	−20.9
Tahiti	−20.8
ex: guaiacol	−31.5
ex: lignin	−27.8

2.Isotopic analysis of oxygen and hydrogen

These analyses were initially developed with the aim of differentiating between natural fruit juices and those obtained by dilution of juice concentrates. The water in orange juice is, in fact, highly enriched in heavy isotopes compared with rain water.

As in the case of the $^{13}C/^{12}C$ isotopic analyses, it is necessary to carry out $^{18}O/^{16}O$ and $^{2}H/^{1}H$ analyses using some kind of standardized water. Standard Mean Ocean Water (SMOW) as defined by Craig is therefore used.

The results are expressed as the difference in parts per thousand of ^{18}O from the standard $^{18}O/^{16}O$ ratio of a sample of the standard SMOW. [12.02]

[12.02]

$$^{18}O = \left[\frac{^{18}O/^{16}O\ (sample)}{^{18}O/^{16}O\ (SMOW)} - 1 \right] \times 1{,}000$$

$$^{2}H = \left[\frac{D/H\ (sample)}{D/H\ (SMOW)} - 1 \right] \times 1{,}000$$

$D = {}^{2}H$ (deuterium)

Without going into detail, it is possible to say that this type of isotopic analysis provides a very reliable distinction between the different types of fruit juice.

As we have seen, the $^{13}C/^{12}C$ ratio is only helpful in a limited number of cases. It would be interesting to know whether the $^{2}H/^{1}H$ ratio can be more selective. To answer this, it is first of all important to study the D/H ratio in different classes of natural products to see how significant the isotope selection has been in the course of the biosynthesis of volatile compounds. This study has been limited to terpenoid and phenylpropanoid compounds. Results are shown below.

Origin	δD ‰ SMOW
Terpenoid	$-300 < \delta D < -250$
Phenylpropenoid	$\delta D > -100$

These originate from differences between the biosyntheses of these two classes of substance. The phenylpropanoid derivatives are formed by the shikimic acid pathway from the sugar reserves of plants. As a result, it is observed that the δD

Origin	δD ⁰/₀₀ SMOW
Linalool	
Thymus vulgaris (France)	–257
Coriandrum sativum (France)	–269
Petitgrain (Paraguay)	–244
Synthetic	–170
Citral	
Citrus aurantifolia	–258
Cympbopogon citratus (Lemongrass)	–276
Litsea cubeba	–252
Synthetic	–174
Menthol	
Mentha piperata (France)	–394
Synthetic (Europe)	–196
Synthetic (USA)	–242

value of phenylpropanoids and sugars are of the same order of magnitude. Terpenic compounds are very poor in deuterium compared with sugars. This is also the case with lipid products.

The reason for these low deuterium levels is linked principally to the mechanism of isotopic fractionation during the biosynthesis of acetyl coenzyme A, which is the direct precursor of terpenoid compounds.

We have just seen that examination of the D/H ratio provides a sensitive means of distinguishing between different metabolic routes. The question remains as to its utility in distinguishing between natural and synthetic compounds. The results shown here demonstrate that the response to this is affirmative.

Measurement of the D/H ratio has also been applied to distinguishing between natural and synthetic anethole:

Although this result is useful for distinguishing between the different natural anetholes on one hand and the synthetic quality on the other hand, this might a priori seem to be surprising in itself. This is because the synthetic quality is obtained by isomerizing estragole which, like anethole, is a phenylpropanoic derivative. If the δD is measured for estragole isolated from *Foeniculum vulgare* var. vulgare, a value of –70 ⁰/₀₀ SMOW is observed. This can be compared with the value of –91% for anethole from the same source.

Although there is a difference between these two, this is not sufficient to explain the difference between the δD values of natural and synthetic anethole. In fact, two questions are raised:

Origin	δD ⁰/₀₀ SMOW
Foeniculum vulgare var. dulce (France)	–86
Foeniculum vulgare var. vulgare (France)	–91
Illicium verum (China)	–96
Illicium verum (North Vietnam)	–84
Synthetic (USA)	–45

- Why is estragole enriched in deuterium compared with anethole from the same plant?
- Why is this enrichment even greater for synthetic anethole?

The first question has been almost answered. It has recently been shown that the biosynthesis of these two compounds is different. In fact, the first steps are common to both. The divergence manifests itself at the level of the cinnamic acid intermediate. [12.03]

[12.03]

D-glucose ⟶ shikimic acid ⟶ L.phenylalanine ⟶ cinnamic acid ⟶ Anethole / Estragole

In the case of anethole, the three carbon atoms in the lateral chain of cinnamic acid are conserved. In contrast, the biosynthesis of estragole proceeds by decarboxylation of cinnamic acid and the carbon lost in the form of CO_2 is subsequently reintroduced by methionine with the aid of an enzyme (methyl methionine transferase). It would seem that this reintroduction of carbon is the cause of the difference in deuterium levels found in these two compounds (–91 and –70 $^0/_{00}$).

With regard to the second question, its solution resides in the mechanism by which estragole isomerizes in basic aqueous or hydroalcoholic solution. This can be summarized in the following way. [12.04]

[12.04]

OCH3 — B⊖ H–CH–CH=CH2 (Estragole) ⇌ OCH3 — CH⊖–CH–CH2 H–OH ⇌ OCH3 — H–C=C–H, CH3 (Anethole)

In fact, this is not so simple. It should be recalled that the average concentration of the natural abundance of deuterium is .015%. This deuterium is spread over the different carbon sites in the estragole molecule, which can be schematically described as follows. [12.05]

[12.05]

$$\mathrm{Ar{-}\overset{1}{C}H_2{-}\overset{2}{C}H{=}\overset{3}{C}H_2}$$

If we only consider the three carbons of the lateral chain, it can be seen that they constitute six series of different cases. In the first case, there is no deuterium at all. In the second, there is one atom of deuterium (3 compounds). In the third, there are two atoms of deuterium (3 compounds). In the fourth, there are three atoms (3 compounds) and in the fifth, there are four (3 compounds). In the sixth, all hydrogens are replaced by deuterium (1 compound). In all, there are 16 different molecules among which 6 contain two hydrogens on carbon number 1 and another 6 which only contain 1 hydrogen at this position. Without going into detailed calculations, these considerations provide an understanding of why the hydrogen atoms attached to carbon number 1 are far more rapidly transferred than deuterium atoms. This implies arguments based on the concentration of deuterium in these molecules. In any case, the water used in the base catalyzed rearrangement has a δD which varies between –40 and +40 ‰ SMOW (because rain water condenses normally at atmospheric temperatures between 10° and 20°C), which will also increase the amount of deuterium on carbon 3.

These different considerations provide a better understanding of the deuterium enrichment seen in the synthetic qualities of anethole.

C. Method based on deuterium nuclear magnetic resonance spectroscopy—^{2}H NMR

It has just been mentioned that the natural abundance of deuterium is around .015% that of hydrogen. Despite this low concentration, the development of NMR spectometers employing superconducting magnets has led to the relatively easy determination of NMR signals from deuterium. This method has been used to analyze numerous natural and synthetic compounds. When applied to anethole, for example, it shows six different resonance signals. [12.06]

[12.06]

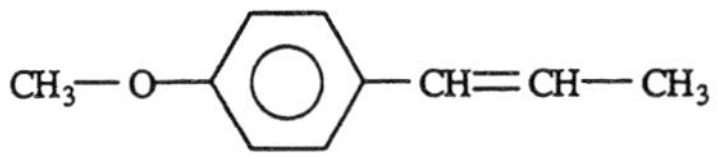

CH_3—O—C_6H_4—CH=CH—CH_3

1	2	3	4	5	6
3.75	6.80	7.23	6.28	6.08	1.83

The intensitiy and surface area intergrals of these signals are proportional to the abundance of deuterium at the six sites. This therefore gives us access to the intramolecular distribution of this isotope, which has, in fact, been shown to be completely heterogeneous.

Some sites in the molecule have deuterium levels which are far higher than others. What is remarkable, however, is that this internal distribution varies according to the origin of the anethole.

In anethole obtained from fenil or star anise oils, the highest level of deuterium is found on carbons 2 and 3 of the aromatic nucleus. In anethole produced by total synthesis (e.g., petroleum), the terminal methyl (1) is richest in deuterium. In anethole derived from the isomerization of estragole, the propenyl group contains the highest level (particularly the =CH_2 at position 6).

In summary, it can be noted that the results obtained with deuterium NMR confirm remarkably well the explanation just given for the isomerization of estragole into anethole in basic solution. This new analytical method therefore presents much potential for the identification of the origin of anethole.

APPENDIX

6

Summary of Some Physico-chemical Methods Used in the Analysis of Natural Odorant Products

A. Introduction

Analysis of natural odorant products is characterized by a certain number of particularities:

- large numbers of constituents reaching up to several hundred;
- large structural diversity. Near to 2,000 different mono- and sesquiterpenic and phenylpropanoid compounds are known, not to mention numerous "exotic" compounds which will be discussed later on;
- a large number of compounds are present at extremely low concentrations. It is not exceptional to find that 1–2% of a natural product contains more than 100 different compounds. These minor compounds are generally characteristic of the olfactive (or gustative) properties of the natural product. They are the "exotic" substances mentioned above. They can show great structural diversity, such as unsaturated or polyunsaturated hydrocarbons, thioesters, mercaptans, norterpenic and norsesquiterpenic compounds derived from the degradation of terpenoids and carotenoids, phenols, fatty acids and their derivatives, heterocycles such as pyridines, pyrazines, thiazoles, furans, etc. Among the last group, one can note, for example, 2-isobutyl-3-methoxypyrazine found in galbanum oil which has an olfactive threshold in water of 0.002 ppb (2×10^{-12} which is equivalent to 1 g in 500,000 metric tons of water)! A far more well known compound, β-ionone, has an olfactive threshold of the same order—.007 ppb (7×10^{-12} or .1 g in 143,000 metric tons of water)!

Concentrations of these minor compounds in natural odorant products can decrease from 10^{-3} to 10^{-9} and sometimes further. It is therefore easy to comprehend the olfactive role played by these compounds in the overall odor of the natural product. The odorant value of a compound can be defined in the following way.

$$\text{Odorant value} = \frac{\text{Concentration in the natural odorant product}}{\text{Olfactive threshold of the compound}}$$

The prodigious development of chromatographic and spectroscopic methods over the last 20 years has produced a veritable revolution in our intimate knowledge of natural products. Nevertheless, in numerous cases, this knowledge is still insufficient. However, there is every hope that this will improve, particularly as there is no reason to believe that this technical evolution will not continue. Either chemical or physical methods can be employed in the analysis of natural odorant products.

B. Chemical separation techniques

These are obviously the oldest. They carry the risk of sometimes leading to artifacts, therefore, they will only be discussed priefly here.

Extraction of a natural product with an alkaline or acidic aqueous solution, preferably using countercurrent techniques, allows isolation of, respectively, acids and phenols on one hand and basic complexes on the other. Carbonyl compounds can be extracted by transformation into water-soluble hydrazonium salts.

Esterification by reaction with anhydrides of dibasic acids such as phthalic acid gives derivatives having water-soluble sodium salts, thus allowing extraction of primary or secondary alcohols. An extremely mild method of extracting the totality of the alcohols is to exchange with a trialkyl borate for which the liberated alcohol has an appropriately low boiling point. After distillation of this alcohol, the remaining trialkyl borates can be readily hydrolized in hot water to regenerate the extracted alcohols.

Chemical methods generally suffer from the inconvenience of requiring large quantities of natural product. In any case, they suffer from an additional inconvenience in that they only separate constituents according to their chemical functionalities. It is still necessary to use physical methods to separate constitents sharing the same functional groups.

C. Physical separation methods

These employ the following properties of the constituents which are to be separated: vapor pressure, polarity and solubility. These properties lead to the use of distillation and a whole arsenal of chromatographic methods.

The high complexity of the mixtures being studied requires the use of preliminary simplifications. In this way, fractional distillation under reduced pressure (sometimes in a current of CO_2) permits initial separation of light constituents from those which only have a weak vapor pressure. In any case, constituents of the latter group could not be distilled without danger of thermally-induced structural modification. It is always a question of judgment as to how the distillation will be carried out.

According to the available quantity of material and the difficulty of the intended separation, either packed columns or spinning band columns can be used. The latter provides extremely efficient separation of small quantities of products (up to 200 theoretical plates).

Adsorption chromatography was discovered by Tswett in 1903, forgotten and then re-employed by Dhéré in the 1920s, forgotten again and definitively drawn from the ashes by Khun, Lederer and Brockmanne in 1931. It presently finds universal use in the separation of complex mixtures and the technique generally employed is that of the open column at atmospheric pressure. Since some time, this technique has also been used under a pressure of about 5 to 10 bars. This considerably increases the efficiency and the speed of the analysis especially when adsorbants with fine granulometric properties (e.g., from 10 to 15 microns) are used. In addition, these separations are greatly facilitated when a differential refractrometry detector is used. By increasing the speed of analysis, another fundamental advantage is that it diminishes the risk of artifact formation.

In order to display odorant properties, an organic compound must have a sufficiently high vapor pressure to enable a certain number of molecules to reach the olfactory mucous membranes. It is therefore evident that the technique of choice for the analysis of natural products is gas chromatography. Chemical techniques such as silylation, which increases the vapor pressure of very polar compounds such as alcohols, acids, amines, etc., can also be useful to the analyst. Furthermore, when account is taken of the enormous complexity and fragility of the products studied, the use of glass or melted silica capilliary columns is highly preferred.

When quantification of the levels of one or several compounds in a mixture is to be performed, it is indispensible to separate these compounds from their partners. Thus, capilliary columns are particularly well adapted to these studies providing precautions are taken during injection.

The use of detectors which are specific to sulfur or nitrogen provides important information on the presence of traces of compounds containing these elements (e.g., in galbanum, patchouly, sclary sage oils and in green pea and paprika flavors, etc.).

With regard to sulfur, an FPD detector can be rigorously specific. This is a flame photometer that analyzes the characteristic radiation emitted whem chemiluminescent reactions occur due to the passage of sulfur compounds in air-hydrogen flames. The sensitivity of these detectors is approximately 10^{-10}gS.sec^{-1}.

For nitrogenous compounds, the NPD detector is highly selective. This is a thermoionic detector whose actual mechanism of action is not well understood. Compared to hydrocarbons, for example, its selectivity to nitrogenous compounds is increased by a factor of 10^4. Yet, its sensitivity (largely a function of the hydrogen flow rate) can reach 10^{-13}gN.sec^{-1}. This is more sensitive than a flame ionization detector (FID) which is about 10^{-12}gC.sec^{-1}.

HPLC, however, is used more and more for separating and especially for quantifying compounds with a high boiling point such as indole and methyl anthranilate in some floral absolutes (orange flower, jasmin, etc.), 7-methoxycoumarin or herniarin in the oils of the labiates, of artemisia, of the Umbelliferæ, of the Rutacæ . . . (from 2 to 75,000 ppm!), vanillin in vanilla extracts, furocoumarins in citrus oils and so on.

Having summarized these analytical techniques, it is important that we distinguish between their uses for identification and for characterization.

Identification

a. Direct

This can be carried out by coupled GC/MS followed by coupling to a computer. In principle, it is then only necessary to compare the chromatographic and spectrometric findings with equivalent data stored in the computer memory. In fact, the coupling of a gas chromatograph with a mass spectrometer poses numerous problems that will not be discussed here. Nonetheless, these problems are partially simplified when using capilliary columns—the only type of column of interest here. It permits the direct introduction of the effluent gases into the ion source of the mass spectrometer. In fact, the average flow rate of the carrier gas at the exit of the column is roughly 2 ml. min^{-1} (expressed at atmospheric pressure); this permits a feasible evacuation rate for the pump which must maintain a minimum pressure of 10^{-4} torr at the ion source. Thus, with regard to maintaining the required partial vacuum, the problem is resolved. However, another fundamental problem remains—the sensitivity of the spectrometer. Without lingering over the details of the calculations involved, when an enriched product is analyzed, it is indispensable for the spectrometer to be able to detect a maximum concentration in the range of ng.sec^{-1}. It is therefore often necessary, even with capilliary columns, to use an interface which partially removes the carrier gas.

b. Indirect

Infrared spectroscopy remains the criterion of choice for identifying known organic compounds. It is sometimes necessary to use the technique of "trapping" when separating compounds to be examined in this way.

Techniques of microtrapping have to date, only been developed for chromatography using packed columns, which implies that prior enrichment is necessary. However, the development of Fourier Transform IR and NMR spectroscopic techniques has permitted certain workers specialized in natural product analyses to develop trapping systems at the exits of capilliary columns. This is notably the case in our laboratory. Very interesting work on this technique has recently been published by Roeraade and Enzell of the Research Department of the Swedish Tobacco Company in Stockholm. This system finds particular applicability in the analysis of tobacco smoke.

In addition to these problems of identification, there is an additional challenge in determining the origin of odorant compounds and flavors using mass spectrometry. It is known that it is possible to distinguish between a compound of natural origin and one obtained by synthesis from fossil products such as petroleum. As discussed, the latter compounds have reduced levels of the ^{14}C isotope. It is, however, impossible to use this technique to differentiate between chemically identical compounds which result from biosynthesis in a plant organism on one hand and from chemical synthesis from natural products on the other hand (e.g., such as pinenes or compounds obtained from degradation of lignin). It is possible to synthesize the near totality of the oxygenated monoterpenes from the pinenes. Products from the degradation of lignins such as methylchavicol and guaiacol are used as intermediates in the synthesis of anethole and vanillin.

By carrying out determinations of the deuterium/hydrogen (D/H) or $^{13}C/^{12}C$ isotopic ratios on these compounds (using internationally accepted standard samples), it is possible to obtain precise indications as to their origin. The D/H ratio also permits distinction between compounds of natural or synthetic origin. The $^{13}C/^{12}C$ ratio can, in certain cases, indicate which biosynthetic process has been employed in the plant species from which it is obtained.

Characterization

When no reference infrared spectra are available to permit identification of compounds which have been isolated, it is necessary to characterize the compound.

It is certain that detailed examination of infrared, ultraviolet, NMR and mass spectra will provide a lot of information. This may be sufficient to permit the proposition of a definitive structure or, at least, highly probable structural elements.

The exact structure can be confirmed in only one of two ways:

- by total synthesis, which is always long and not always easy given the structural complexity of natural products
- by correlation with a known natural compound. These correlations can be made in either of two ways:
  —by transforming a known compound in a small number of unambiguous reactions to arrive at the same compound as that which is being characterized (this is preferable because it does not involve use of this latter compound);
  —by transforming the compound which is being characterized. To do this, the isolated compound is transformed in a small number of unambiguous reactions to give a known compound. It is also possible to transform the isolated compound into a monocrystal and then determine its structure by radiocrystallography. This will generally allow determination of its absolute configuration.

Thioesters of galbanum

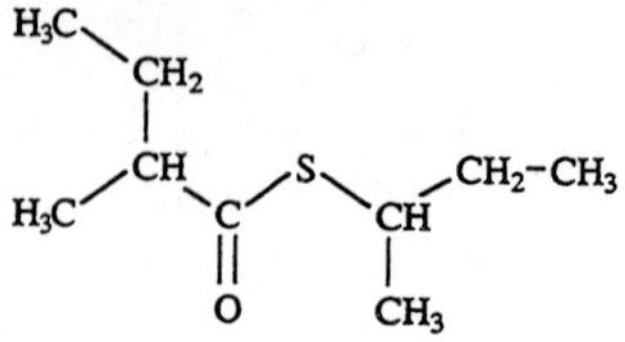

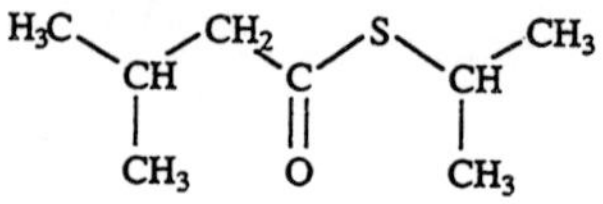

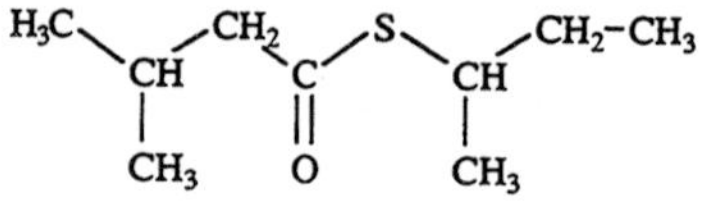

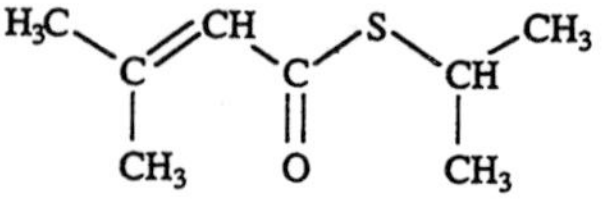

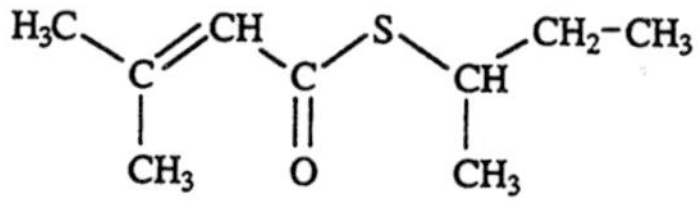

Pyrazines

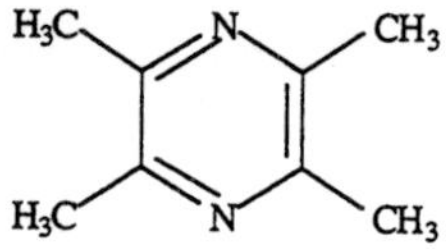

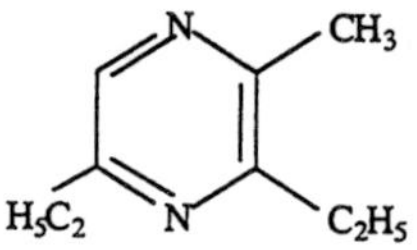

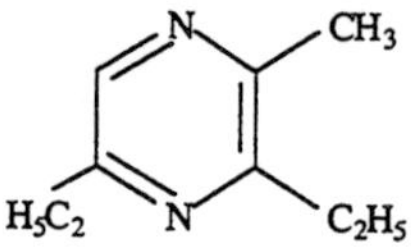

Galbanum of 1 to 7

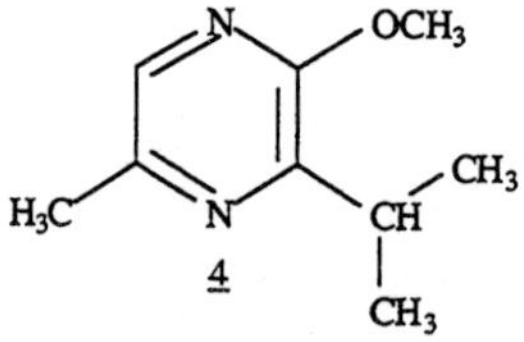

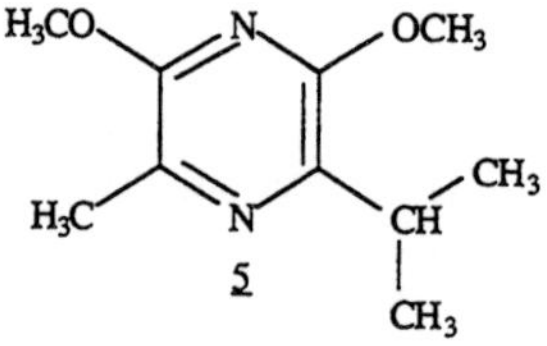

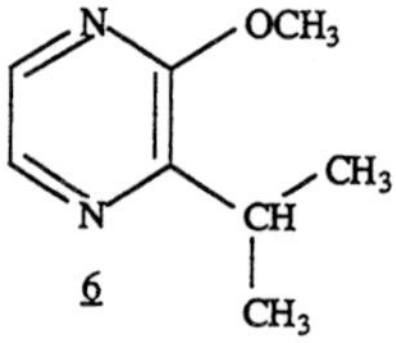

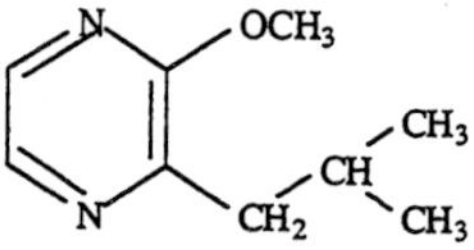

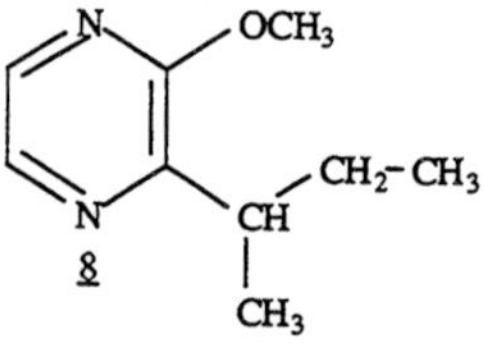

Bellpepper (*Capsicum annum* 7)

Pisum sativum 6 (~10^{-11}), 7 (~10^{-11}), 8 (~10^{-12})

Rayon x structure

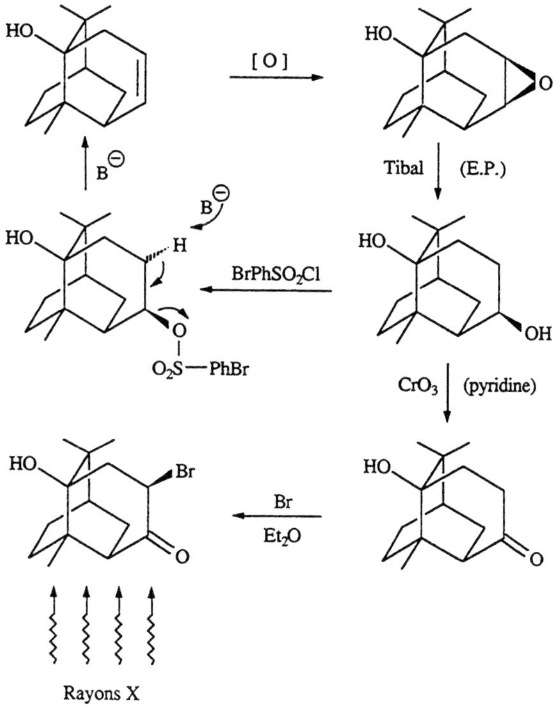

HO

[O]

HO

O

$B^{\ominus}$

Tibal (E.P.)

HO

$B^{\ominus}$

H

$BrPhSO_2Cl$

HO

OH

O

O_2S—PhBr

CrO_3 (pyridine)

HO

Br

Br

Et_2O

HO

O

O

Rayons X

Fatty acid compounds

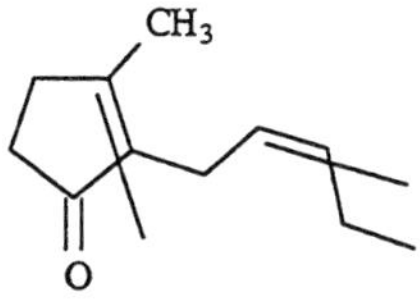

CH_3

O

cis-jasmone

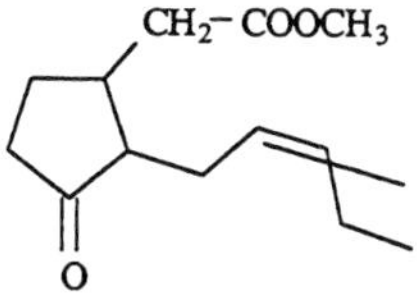

CH_2– $COOCH_3$

O

methyl cis-jasmonate

Nortepenic compounds

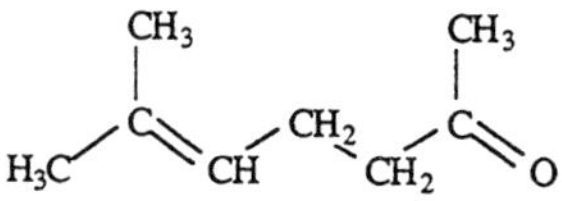

CH_3 CH_3

H_3C C CH CH_2 CH_2 C O

Methyl heptenone

$C_8H_{14}O$

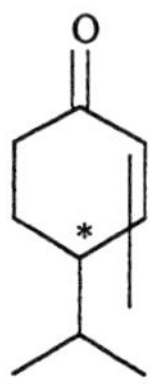

O

*

Cryptone

$C_9H_{14}O$

Pyridines

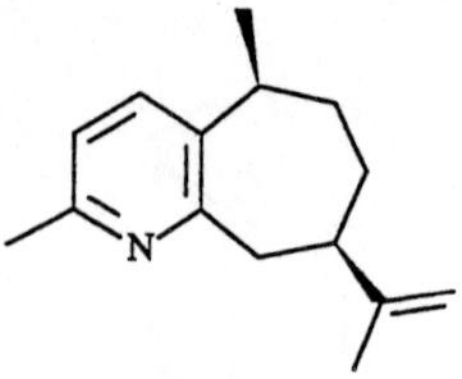

Gaïapyridine

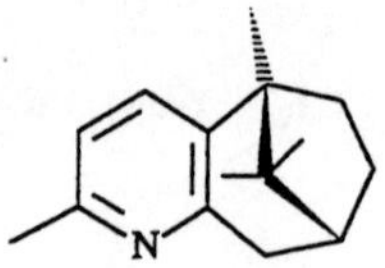

Patchoulipyridine

Norsesquiterpene compounds

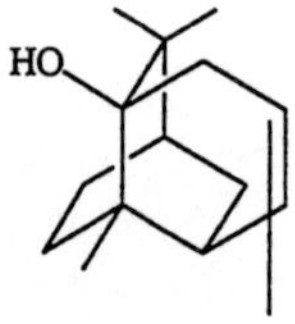

Norpatchoulenol
$C_{14}H_{22}O$

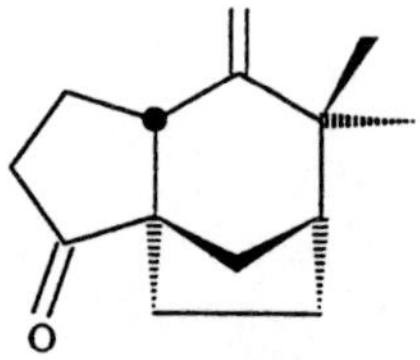

Khusimone
$C_{14}H_{20}O$

1, 3, 5-Undecatrienes from galbanum

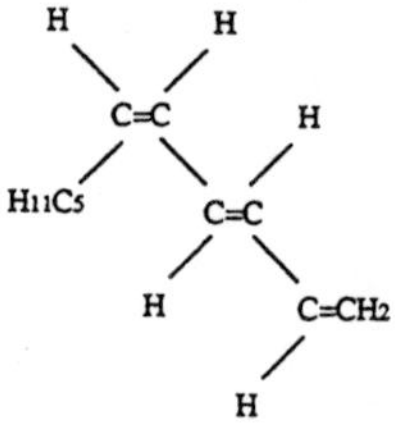

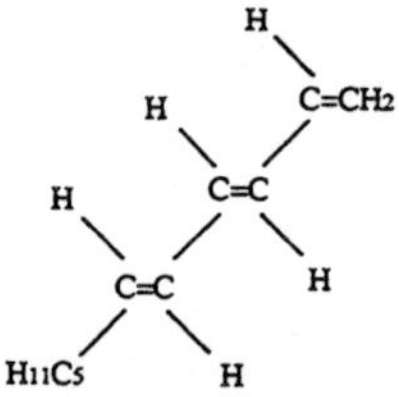

Compounds derived from carotenoids

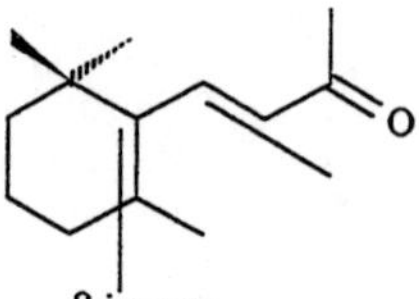

β-ionone

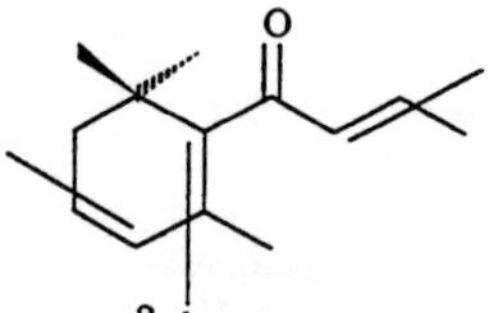

β-damascenone

Bibliography

General Works on Organic Synthesis

Aliphatic and Related Natural Product Chemistry, The Chemical Society, Burlington House, London; Vol. 1: *A Review of the Literature Published during 1976 and 1977,* 308 pages; Vol. 2: *A Review of the Literature Published during 1978 and 1979,* 265 pages; Vol. 3: *A Review of the Literature Published during 1980 and 1981,* 363 pages.

Allais A., Mathieu J., Petit A., Poirier P. and Velluz L., *Substances Naturelles de Synthèse:* Préparations et Méthodes de Laboratoires, 10 volumes, Masson & Cie, Pans.

Anand N., Bindra J. S. and Ranganathan S., *Art in Organic Synthesis,* Holden Day Inc., San Francisco, London, Amsterdam, 1970, 415 pages.

Annual Reports in Organic Synthesis—1970, edited by J. McMurry and R. Bryan Miller, Academic Press, New York, London, 1971, 339 pages.

Annual Reports in Organic Synthesis—1971, edited by J. McMurry and R. Bryan Miller, Academic Press, New York, London, 1972, 347 pages.

Annual Reports in Organic Synthesis—1972, edited by J. McMurry and R. Bryan Miller, Academic Press, New York, London, 1973, 273 pages.

Annual Reports in Organic Synthesis—1973, edited by R. Bryan Miller and Louis S. Hegedus, Academic Press, New York, San Francisco, London, 1974, 423 pages.

Annual Reports in Organic Synthesis—1974, edited by Louis S. Ilegedus and Stephen R. Wilson, Academic Press, New York, San Francisco, London, 1975, 397 pages.

Annual Reports in Organic Synthesis—1975, edited by R. Bryan Miller and L. G. Wade, Jr., Academic Press, New York, San Francisco, London, 1976, 555 pages.

Annual Reports in Organic Synthesis—1976, edited by R. Bryan Miller and L. G. Wade, Jr., Academic Press, New York, San Francisco, London, 1977, 449 pages.

Annual Reports in Organic Synthesis—1977, edited by R. Bryan Miller and L. G. Wade, Jr., Academic Press, New York, San Francisco, London, 1978, 457 pages.

Annual Reports in Organic Synthesis—1978, edited by L. G. Wade, Jr. and Martin J. O'Donnell, Academic Press, New York, San Francisco, London, 1979, 353 pages.

Annual Reports in Organic Synthesis—1979, edited by L. G. Wade, Jr. and Martin J. O'Donnell, Academic Press, New York, London, Toronto, Sydney, San Francisco, 1980, 461 pages.

Annual Reports in Organic Synthesis—1980, edited by L. G. Wade, Jr. and Martin J. O'Donnell, Academic Press, New York, London, Toronto, Sydney, San Francisco, 1981, 456 pages.

Annual Reports in Organic Synthesis—1981, edited by L. G. Wade, Jr. and Martin J. O'Donnell, Academic Press, New York, London, Paris, San Diego, San Francisco, Sao Paulo, Sydney, Tokyo, Toronto, 1982, 498 pages.

Annual Reports in Organic Synthesis—1982, edited by L. G. Wade, Jr. and Martin J. O'Donnell, Academic Press, New York, London, Paris, San Diego, San Francisco, Sao Paulo, Sydney, Tokyo, Toronto, 1983, 499 pages.

Annual Reports in Organic Synthesis—1983, edited by Martin J. O'Donnell and Louis Weiss, Academic Press, Orlando, San Diego, New York, London, Toronto, Montreal, Sydney, Tokyo, 1984, 493 pages.

Annual Reports in Organic Synthesis—1984, edited by Martin J. O'Donnell and Louis Weiss, Academic Press, Orlando, San Diego, New York, London, Toronto, Montreal, Sydney, Tokyo, 1985, 500 pages.

Annual Reports in Organic Synthesis—1985, edited by Martin J. O'Donnell and Eric F. V. Scriven, Academic Press, Orlando, San Diego, New York, London, Toronto, Montreal, Sydney, Tokyo, 1986, 513 pages.

Annual Reports on the Progress of Chemistry, Royal Chemical Society, Vol. 72—1975, Section B—Organic Chemistry, Vol. 79—1982, A. G. Davies and P. G. Garrau, 1983, 356 pages; Vol. 80—1983, A. G. Davies and P. G. Garrau, 1984, 403 pages; Vol. 81—1984, R. J. Garrau and J. M. Mellor, 1985, 400 pages; Vol. 82—1985, R. J. Garrau and J. M. Mellor, 1986, 424 pages.

Belker H., Berger W. et al., *Organicum, Practical Handbook of Organic Chemistry,* Pergamon Press Ltd, Oxford, New York, 1972, 747 pages.

Breslow Ronald, *Organic Reaction Mechanisms,* W. A. Benjamin, Inc., New York, Amsterdam, 1966, 232 pages.

Breslow Ronald, *Mécanismes des Réacrions Organiques,* French edition by J.-P. Guette, Y. Leroux and G. Martin, Ediscience, Paris, 1973, 294 pages.

Brown H. C., *Hydroboration,* W. A. Benjamin, Inc., New York, 1962, 290 pages.

Buehler Calvin A. and Pearson Donald E., *Survey of Organic Synthesis,* John Wiley & Sons, New York, London, Sydney, Toronto, Vol. 1, 1970, 1166 pages; Vol. 2, 1977, 1105 pages.

Carruthers W., *Some Modern Methods of Organic Synthesis,* Cambridge Chemistry Texts, Cambridge at the University Press, 1971, 399 pages.

Cragg Gordon M. L., *Organoboranes in Organic Synthesis,* Marcel Dekker, Inc., New York, 1973, 422 pages.

Danishefsky Sarah Etheredge, Danishefsky Samuel, *Progress in Total Synthesis,* Vol. 1, Appleton Century Crofts, Meredith Corporation, 1971, 265 pages.

Davies Stephen G., *Organotransition Metal Chemistry,* Applications to Organic Synthesis, Pergamon Press, Oxford, New York, Toronto, Sydney, Frankfurt, 1982, 411 pages.

De Barry Barnett E., *Mechanism of Organic Chemical Reactions,* Blackie & Son Ltd, London, Glasgow, 1956, 289 pages.

Dehmlow E. V., Dehmlow S. S., *Phase Transfer Catalysis,* Verlag Chemie, Weinheim, Deerfield Beach, Florida, Basel, 1980, 316 pages.

Deslongchamps P., Le Concept de Stratégie en Synthèse Organique, *Bull. Soc. Chim. France,* 1984, p. 349.

Fieser L. F., Fieser Mary, *Reagents for Organic Synthesis,* John Wiley & Sons Inc., NewYork, London, Sydney; Vol. 1, 1967, 1457 pages, Vol. 2, 1969, 538 pages, Vol. 3, 1972, 401 pages, Vol. 4, 1974, 660 pages, Vol. 5, 1975, 864 pages, Vol. 6, 1977, 765 pages, Vol. 7, 1979, 487 pages, Vol. 8, 1980, 602 pages, Vol. 9, 1981, 596 pages, Vol. 10, 1982, 528 pages, Vol. 11, 1984, 669 pages, Vol. 12, 1986, 643 pages.

General and Synthetic Methods, Specialist Periodical Reports, The Chemical Society, 8 volumes (1978–1986).

Gerrard W., *The Organic Chemistry of Boron,* Academic Press, London and New York, 1961, 308 pages.

Goto T., Hirata Y. and Stout G. H., *Problems in Advanced Organic Chemistry,* Holden Day Inc., San Francisco, London, Amsterdam.

Haines Alan H., *Methods for the Oxidation of Organic Compounds,* Alkanes, Alkenes, Alkynes and Arenes, Best Synthetic Methods, Academic Press, 1985, 388 pages.

Harrison Ian T. and Harrison Shuyen, *Compendium of Organic Synthetic Methods,* John Wiley & Sons, New York, Chichester, Brisbane, Toronto, Singapore, Vol. 1, 1971, 529 pages; Vol. 2, 1974, 437 pages; Vol. 3, 1977, 495 pages; Vol. 4, Leroy G. Wade, Jr., 1980, 497 pages; Vol. 5, Leroy G. Wade, Jr., 1984, 552 pages.

Henrici-Olive C., Olive S., *Coordination and Catalysis,* Monograph in Modern Chemistry, Verlag Chemie, Weinheim, New York, 311 pages.

House Herbert O., *Modern Synthetic Reactions,* Second Edition, W. A. Benjamin Inc., 1972, 856 pages.

International Union of Pure and Applied Chemistry (IUPAC), Organic Photochemistry, 2, International Symposium Enschede, 1967, Plenary Lectures, London, Butterworths, 1967, 200 pages.

Keller W. E., *Phase-Transfer Reactions,* Georg Thieme Verlag, Stuttgart, New York, Vol. 1, 1986, 415 pages, Vol. 2, 1987, 1067 pages.

Kirby A. J. and Warren S. G., *The Organic Chemistry of Phosphorus,* Elsevier Publishing Company, Amsterdam, London, New York, 1967, 404 pages.

Lefebvre G., Sajus L. and Teyssie Ph., *Catalyse par Complexes de Coordination,* Editions Technip, 1972, 147 pages.

Maitlis Peter M., *The Organic Chemistry of Palladium,* Academic Press, New York, London, Vol. 1, Metal Complexes, 1971, 319 pages; Vol. 2, Catalytic Reactions, 1971, 216 pages.

March J., *Advanced Organic Chemistry,* Reactions, Mechanisms and Structure, Third Edition, John Wiley & Sons, New York, Chichester, 1985, 1146 pages.

Mathieu J., Panico R., Weill-Raynal J., *Les Grandes Réactions de la Synthèse Organique,* Actualités Scientifiques et Industrielles, 1374, Hermann, Paris, 1975, 341 pages.

Mathieu J., Panico R., Weill-Raynal J., *L'Aménagement Fonctionel en Synthèse Organique,* Actualités Scientifiques et Industrielles, 1386, Hermann, Paris, 1977, 382 pages.

Mathieu J. and Allais A., *Cahiers de Synthèse Organique,* Méthodes et Tableaux d'Application, 12 volumes, Masson & Cie., Paris.

Monson Richard S. *Advanced Organic Synthesis,* Methods and Techniques, Academic Press, New York and London, 1971, 203 pages.

Organic Photochemistry; Edited by Orville L. Chapman, Marcel Dekker, Inc., New York, Vol. 1, 1967, 339 pages; Vol. 2, 1969, 230 pages. Edited by Orville L. Chapman, Marcel Dekker, Inc., New York, Vol. 3, 1973, 311 pages. Edited by Albert Padwa, Marcel Dekker, Inc., New York and Basel, Vol. 4, 343 pages; Vol. 5, 502 pages; Vol. 6, 438 pages; Vol 7, 498 pages; Vol. 8, 373 pages; Vol. 9, 353 pages.

Organic Reactions, Chief editor R. Adams, John Wiley & Sons, Inc., New York, London, Sydney, 34 volumes.

Organic Reactions in Liquid Ammonia, Herchel Smith, Gehrart Jander, Hans Spandau, C. C. Addison, Friedr. Vieweg & Sohn, Braunschweig Interscience Publishers, John Wiley & Sons, Inc., New York, London; 1963, 363 pages.

Organic Syntheses, Collective Volumes, John Wiley & Sons Inc., New York, London, Sydney (67 separate volumes).

Organic Syntheses via Metal Carbonyls, edited by Irving Wender and Piero Pino, John Wiley & Sons, New York, London, Sydney, Toronto, Vol. 2, 743 pages.

Organometallic Chemistry, edited by H. Zeiss, Reinhold Publishing Corp., New York, Chapman & Hall, Ltd., London, 411 pages.

Oxidation in Organic Chemistry, edited by Kenneth B. Wiberg, Organic Chemistry, A series of monographs, Part A, Academic Press, New York, London, 1965, 443 pages.

Oxidation in Organic Chemistry, edited by Walter S. Trahanovsky, Organic Chemistry, A series of monographs, Vol 5B, Academic Press, New York, London, 1973, 248 pages.

Oxidation in Organic Chemistry, edited by Walter S. Trahanovsky, Part D, Academic Press, New York, London, Paris, San Diego, San Francisco, Sao Paulo, Sydney, Tokyo, Toronto, 1982, 360 pages.

Oxidation of Organic Compounds, Proceedings of the International Oxidation Symposium, arranged by Stanford Research Institute, in San Francisco, Calif. Aug. 28–September 1, 1967; Frank R. Mayo, American Chemical Society, Washington, D.C. 1968; Vol. 1 (75): Liquid-Phase, Base-Catalyzed and Heteroatom Oxidations, Radical Initiation and Interactions, Inhibition, 364 pages; Vol. 2 (76): Gas-Phase Oxidations, Homogeneous and Heterogeneous Catalysis, Applied Oxidations and Synthetic Processes, 438 pages; Vol. 3 (77): Ozone Chemistry, Photo and Singlet Oxygen and Biochemical Oxidations, 310 pages.

Oxidation, Vol. 1, Techniques and Applications in Organic Synthesis, edited by Robert L. Augustine, Marcel Dekker, Inc., New York, 1969, 368 pages; Vol. 2, edited by Robert L. Augustine and David J. Trecker, 1971, 204 pages.

Pesez M. and Poirier P., *Méthodes et Réactions de l'Analyse Organique,* Déterminations générales et recherches fonctionnelles. Vol. 1: Méthodes de l'Analyse Générale; Vol. 2: Méthodes de caractérisation, 1953, 278 pages; Vol. 3: Réactions colorées et fluorescences, 1954, 298 pages. Masson & Cie, Paris.

Photochemistry Past Present and Future, Editor R. P. Wayne, Associate Editor J. D. Coyle, Elsevier Sequoia S. A., Lausanne, 1984, 97 pages.

Photoselective Chemistry, Part 2, edited by Joshua Jortner, Raphael D. Levine and Suuart A. Rice, An Interscience Publication, John Wiley and Sons, New York, Chichester, Brisbane, Toronto, 718 pages.

Progress in Organic Chemistry, J. W. Cook, editor, London, Butterworths Scientific Publications, Vol. 1, 1952, 287 pages, Vol. 2, 1953, 212 pages, Vol. 3, 1955, 273 pages, Vol. 4, 1958, 256 pages; J. W. Cook and W. Carruthers, editors, Vol. 5, 1961, 172 pages, Vol. 6, 1964, 256 pages, Vol. 7, 166 pages; W. Carruthers and J. K. Sutherland, editors, Vol. 8, 343 pages.

Ranganathan D. and Ranganathan S., *Challenging Problems in Organic Reaction Mechanisms,* Academic Press, New York and London, 1972, 160 pages.

Ranganathan S., *Fascinating Problems in Organic Reaction Mechanisms,* Holden Day Inc., San Francisco, London, Amsterdam, 2 volumes.

Razumovskii S. D., Zaikov G. E., *Ozone and Its Reactions, with Organic Compounds,* Studies in Organic Chemistry, 15, Elsevier, 1984, 403 pages.

Reduction with Complex Metal Hydrures, Norman G. Gaylord, Interscience Publishers Ltd., London, New York, 1956, 1046 pages.

Roberts J. D., Caserio M., *Problèmes de Chimie Organique,* French edition under the direction of J. Metzger, Ediscience, Paris, 1970, 517 pages.

Rylander Paul N., *Hydrogenation Methods,* Best Synthetic Methods, Academic Press, 1985, 193 pages.

Schudler Stanley R., Karo Wolf, *Organic Functional Group Preparations,* Organic Chemistry, A series of monographs, Vol. 12–III, Academic Press, New York, London, 1972, 496 pages.

Selections from Aldrichchimica Acta, 1968 through 1982, Aldrich, 1984, 354 pages.

Starks Charles M., Liotta Charles, *Phase Transfer Catalysis,* Academic Press, New York, San Francisco, London, 1978, 365 pages.

Stewart Ross, *Oxidation Mechanisms,* Applications to Organic Chemistry, W. A. Benjamin, Inc. New York, Amsterdam, 1964, 179 pages.

Structure and Mechanism in Organic Chemistry, Second Edition, C. K. Ingold, Cornell University Press, Ithaca and London, 1969, 1266 pages.

Transition Metal Hybrids, edited by Karl L. Muetterties, The Hydrogen Series, Marcel Dekker Inc., New York, Vol. 1, 1971, 343 pages; Vol. 2, Chemistry of Boron Hybrids.

Turney, T. A., Oxidation Mechanisms, Butterworths, London, 1965, 208 pages.

Warren Stuart, *Designing Organic Syntheses,* A Programmed Introduction to the Synthon Approach, John Wiley & Sons Inc., Chichester, New York, Brisbane, Toronto, 1978, 285 pages.

Warren Stuart, *Organic Synthesis: The Disconnection Approach,* 1982, John Wiley & Sons.

Weber W. P., Gokel G. W., Phase Transfer Catalysis in Organic Synthesis, Springer Verlag, Berlin, Heidelberg, New York, 1977, 280 pages.

Weiss Howard D., *Guide of Organic Reactions,* Burgess Publishing Company, 1970, 247 pages.

Wollweber Hartmund, *Dielsalder-Reaktion,* Georg Thieme Verlag, Stuttgart, 1972, 271 pages.

Articles Dealing with Computer-Assisted Synthesis

Corey E. J., *J. Pure Appl. Chem.,* 1967, 14, 19.

Corey E. J. and Wipke W. *T., Science,* 1969, 166, pp. 178–92.

Corey E. J., *J. Pure Appl. Chem.,* 1971, 2, pp. 45–68.

Corey E. J., *Quart. Rev. Chem. Soc.* 1971, 25, pp. 455–82.

Corey E. J., Long A. K. and Rubinstein S. D., *Science,* 1985, 228, pp. 408–18.

Hendrickson J. B., *Acc. Chem. Pres.* 1986, 19, pp. 274–81 (Several references concerning computer programs are cited in this article.).

Natural Substances

Amoore John E., *Molecular Basis of Odor,* Charles C. Thomas, Publishers, Springfield, 1970, 200 pages.

Analysis of Volatiles Methods Applications, Proceedings—International Workshop, Würzburg, Peter Schreier, editor, New York, 1984, XI, 469 pages.

Arctander Steffen, *Perfume and Flavor Materials of Natural Origin,* 1960, Elizabeth, N.J., 736 pages.

Bedoukian P. Z., *Perfumery and Flavoring Materials,* Annual Review Articles, 1945–1982, Allured Publishing Corporation, Wheaton, Ill., 1982, 437 pages.

Bedoukian P. Z., *Perfumery and Flavoring Materials,* 3rd Revised Edition, Allured Publishing Corporation, Wheaton, Ill., 1986, 465 pages.

Bericht Der Schimmel & Co, Aktingesellschaft Miltitz by Leipzig über "Atherische" Ole Riechstoffe, USW, Ausgabe 1928, 268 pages; 1940, 106 pages; Schimmel Report, 1945, 127 pages; 1948, 229 pages; 1944–47, 137 pages; 1947–48, 125 pages; Jahrgänge 1940–48, 93 pages; Ausgabe 1950, 214 pages; 1951, 240 pages; 1952–53, 236 pages; 1959, 264 pages; 1960, 258 pages; 1961, 249 pages; 1962, 373 pages; 1963–64, 372 pages; Miltitzer Berichte, Veb Chemiche Fabrik Miltitz, 1965–66, 350 pages; 1967–68, 299 pages; 1969, 191 pages; 1970, 163 pages; 1971, 165 pages; 1972, 199 pages; 1973, 206 pages; 1974, 217 pages, 1975, 235 pages; 1976, 199 pages; 1977, 288 pages; 1978, 270 pages; 1979, 188 pages; 1980, 365 pages; 1981, 491 pages; 1982, 333 pages; 1983, 270 pages; 1984, 293 pages; 1985, 346 pages; 1986, 306 pages.

Bulletin Semestriel de la Maison Schimmel & Cie (Fritzsche Frères) Miltitz (près Leipzig); London and New York, October 1909, 215 pages, April 1910, 188 pages.

Corbier B. and Teisseire P., *Contribution à la connaissance de l'Huile Essentielle de Sauge Sclarée—Présence de Germacrène* D. Recberches, n° 19, July 1974, P. 275.

Compilation of Odor and Taste Threshold Values Data, W. H. Stahl, editor, DS48, American Society for Testing and Materials, 1973, 250 pages.

Di Giacomo A., *Gli Oli Essenziali degli Agrumi,* La Editrice Universitaria Messina, 1966, 244 pages.

Durvelle J. P., *Nouveau Guide du Parfumeur,* J. Fritsch, editor, Paris, 1885, 449 pages.

Essential Oils, 1976–1977, B. M. Lawrence, editor. Allured Publishing Corporation, Wheaton, Ill., 165 pages.

Essential Oils, 1978, B. M. Lawrence, editor. Allured Publishing Corporation, Wheaton, Ill., 1979, 192 pages.

Essential Oils, 1979–1980, B. M. Lawrence, editor. Allured Publishing Corporation, Wheaton, Ill., 1981, 292 pages.

Essential Oils. A symposium sponsored by the Division of Agricultural and Food Chemistry at the 178th Meeting of the American Chemical Society. Washington, D.C., September 12, 1979. Edited by B. J. Mookherjee and C. J. Mussiman, Allured Publishing Corp. Wheaton, Ill., 1981, 272 pages.

Fenaroli G., *Le Sostanze Aromatiche Naturale,* V. Hoepli, editor, Milano, 1963, 1004 pages.

Flavor Research, Recent Advances, edited by R. Teranishi, R. A. Flath, H. Sugiwasa, Marcel Dekker Inc., New York and Basel, 1981, 381 pages.

Flavour '81 3rd Weurman Symposium Proceedings on the International Conference, Munich, April 28–30, 1981, Peter Schreier, editor, de Gruyter, Berlin, New York, 1981, XIV, 780 pages.

Formacek V., Kubeczka K.-H, *Essential Oils Analysis by Capillary Gas Chromatography & Carbon-13 NMR Spectra,* John Wiley & Sons, Chichester, New York, Brisbane, Toronto, Singapore, 1982, 373 pages.

Fragrance Chemistry: The Science of the Sense of Smell, edited by E. T. Theimer, Academic Press, 1982, 635 pages.

Fragrance and Flavor Substances, Proceedings of the Second International Haarmann & Reimer Symposium on Fragrance and Flavor Substances, September 24–25, 1979, New York, edited by Ronald H. R. Croteau, Washington State University Institute of Biological Chemistry and Biochemistry, Biophysics Program, Pullman, Washington, 1980, 200 pages.

Fragrances and Flavors Recent Developments, Chemical Technology Review, n° 156, edited by S. Torrey, Noyes Data Corporation, Park Ridge, N.J., 1980.

Gildemeister E. and Hoffmann Fr., Die Atherishen Ole, 4th edition revised and completed by W. Treibs and K. Bournot, 1956–1961, 7 books, 10 volumes, Akademie Verlag, Berlin.

Guenther E. S., *The Essential Oils,* Book 1, 1948, 427 pages; Book 2, 1949, 852 pages; Book 3, 1949, 777 pages; Book 4, 1950, 752 pages; Book 5, 1952, 507 pages; Book 6, 1952, 481 pages; D. Van Nostrand Company Inc., Princeton. N.J., Toronto, New York, London.

Gustation and Olfaction, edited by G. Ohlolf and Alan F. Thomas (Food Science and Technology, a series of monographs), Academic Press, London and New York, 1971, 275 pages.

Harborne J. B., Turner B. L., *Plant Chemosystematics,* Academic Press, London, San Diego, Austin, New York, Toronto, Boston, Sydney, Tokyo, 1984, 562 pages.

Ikan R., *Natural Products,* A Laboratory Guide, Academic Press Inc., New York, San Francisco, London, 1969, 301 pages.

International Union of Pure and Applied Chemistry (IUPAC), Chemistry of Natural Products, 5° International Symposium, London, 1968, Plenary Lectures, London, Butterworths, 1968, 547 pages.

James Ronald W., *Fragrance Technology Synthetic and Natural Perfumes,* Noyes Data Corporation, Park Ridge, New Jersey, London, 1975, 298 pages.

Jennings W. and Shibamoto T., *Qualitative Analysis of Flavor and Fragrance Volatiles by Glass Capillary Gas Chromatography,* Academic Press, 1980, 472 pages.

Klyne W. and Buckingham T., *Atlas of Stereochemistry. Absolute Configurations of Organic Molecules,* Chapman and Hall, London, 1974, 311 pages.

Kocovsky K., Turecek F. and Hajicek J., *Synthesis of Natural Products: Problems of Stereoselecrivity,* CRC Press Inc. Boca Raton, Florida, Vol. 1, 1986, 233 pages. Vol. 2, 1986, 284 pages.

Labaune L. M., *Parfumerie,* "Traité de Chimie Organique de V. Grignard," T. XXII, p. 1007ff, Masson, 1953.

Masada Y., *Analysis of Essential Oils by Gas Chromatography and Mass Spectrometry,* Wiley & Sons Inc., New York, London, Sydney, Toronto, 1976, 334 pages.

Mookherjee B. D. and Mussiman C. J., *Essential Oils,* A symposium sponsored by she Division of Agricultural and Food Chemistry at the 178th Meeting of the American Chemical Society, Washington, D.C., September 12, 1979, International Flavors and Fragrances Inc. Union Beach, N. J., 272 pages.

Natural Products Chemistry, K. Nakanishi, T. Goto, Shô-Itô, S. Natori, S. Nozoe, Academic Press Inc., New York, San Francisco, London, Kodansha, Ltd., Tokyo; Vol. 1, 1974, 562 pages; Vol. 2, 1975, 586 pages.

Naves Y.-R. and Masuyer G., *Natural Perfume Materials,* Gauthier-Villars, Paris, 1939.

Naves Y.-R., *Huiles Essentielles in traité de chimie organique de V. Grignard. T. XVI,* p. 467ff, Masson, Paris, 1949.

Naves Y.-R., *Technologie et Chimie des Parfums Naturels,* Masson, 1974, 326 pages.

Ohemotaxanomie der Pflanzen, R. Hegnauer Birhäser Verlag, Book 1, Tallophyten, Bryophyten, Pteridophyten und Gymnospermen, 1962, 517 pages; Book 2, Monocotyledoneae, 1963, 540 pages; Book 3, Dicotyledoneae I. Teil von Acanthaceae by Cyrillaceae, 1964, 743 pages; Book 4, Dicotyledoneae: Daphniphyllaceae Lythraceae, 1966, 551 pages; Book 5, Dicotyledoneae: Magnoliaceae Quiinaceae, 1969, 506 pages; Book 6, Dicotyledoneae: Rapflesiaceae Zygophyllaceae, 1973, 882 pages.

On Essential Oils, James Verghese, editor, Published Synthite, Synthese Industrial Chemicals Private Ltd. Kolenchery, India, 1986, 428 pages.

Perfume and Flavor Materials of Natural Origin, Steffen Arctander, 1960, chez l'auteur, Elizabeth, N. J., 736 pages.

Pesnelle P., Corbier B. and Teisseire P., Comparaison entre les essences de Géranium Bourbon et de Géranium Afrique au niveau de leur Constituants Sesquiterpeniques, *Recherches,* n° 18, July 197l, p. 45.

Pharmacognosy and Phytochemistry, 1st International Congress, Munich, 1970, edited by H. Wagner and L. Horhammer, Springer Verlag, Berlin, Heidelberg, New York, 197l, 386 pages.

Plattier M. and Teisseire P., Contribution à la Connaissance de l'Huile Essentielle de Cèdre de l'Atlas, *Recherches,* n° 19, July 1974, p. 131.

Progress in Essential Oil Research, 1984 (Proceedings of the International Symposium on Essential Oil, Holminden / Neuhaus, RFA, September 18–21, 1985) E.-J. Brunke, editor, de Gruyter, New York, 1986, 668 pages.

Progress in Flavor Research, 1984 (Development in Food Science 10) Proceedings of the 4th Weurman Flavor Research Symposium (Dourdan, France, 9–11 May 1984), edited by J. Adda, Elsevier, Amsterdam, Oxford, New York, Tokyo, 1985, 634 pages.

Progress in Phytochemistry, 1968, Vol. 1, 723 pages; 1970, Vol. 2, 512 pages; edited by L. Reinhold, Y. Liwschitz, Interscience Publishers, London, New York, Sydney; 1972, Vol. 3, 375 pages, edited by L. Reinhold, Y. Liwschitz, Interscience Publishers, London, New York, Sydney, Toronto; Vol. 5, 329 pages, Vol. 6, 294 pages, Vol. 7, 344 pages, edited by L. Reinhold, H. B. Harborne, T. Swain, Pergamon Press, Oxford, New York, Toronto, Sydney, Paris, Frankfurt.

Progress in the Chemistry of Organic Natural Products (Fortschritte der Chemie Organischer Naturstoffe), founded by L. Zechmeister, edited by W. Herz, H. Grisebach, G. W. Kirby, Springer Verlag, Wien, New York, 50 volumes published in 1986.

Rolet A., *Plantes à Parfums et Plantes Aromatiques,* Encyclopédie Agricole, Librairie J.-B. Baillière et Fils, 1918, 432 pages.

Semiochemistry Flavors and Pheromones, Proceedings of the American Chemical Society Symposium, August 1983, Washington D.C. T. E. Acree and D. M. Soderlund, editors, de Gruyter, Berlin, New York, 1985, XX, 289 pages.

Semmler F. W., *Die Atherischen Ole,* nach ihren Chemischen Bestand teilen; Eister Band Leipzig, Verlag Veit & Comp., 1906; 860 pages, von Allegemeiner teil Methan derivate; Zweiter Band Leipzig, Verlag von Veit & Comp., 1906, 612 pages, Hydriert-cyclische Verbindungen Kohlenwasserstoff,; Dritter Band Leipzig, Verlag von Veit & Comp., 1906, 824 pages, Sauerstoffhaltige Verbindungen der Hydriert Cyclische Reihe.

Srinivas Srivas R., *Atlas of Essential Oils,* New York, 1986, 1244 pages.

Substances Naturelles de Synthèse, Préparations et Méthodes de Laboratoire, J. Mathieu, A. Petit, P. Poirier and L. Velluz, Vol. 1, 141 pages; Vol. 2, 138 pages; Vol. 3, 156 pages; Vol. 4, 165 pages; Vol. 5, 206 pages; A. Allais, I. Mathieu, A. Petit, P. Poirier and L. Velluz, Vol. 6, 156 pages; Vol. 7, 157 pages; Vol. 8, 157 pages; Vol. 9, 186 pages; Vol. 10, 200 pages; Masson et Cie, Paris.

Teisseire P. and Galfre A., Contribution à la Connaissance de l'Huile Essentielle d'Ylang-Ylang 3e, *Recherches,* n° 19, July 1974, p. 269.

Teisseire P., Maupetit P. and Corbier B., Contribution à la Connaissance de l'Huile Essentielle de Patchouli, *Recherches,* n° 19, July 1974, p. 8.

Teisseire P., Corbier B. and Plattier M., Synthèses d'Undécatriènes, substances caractéristiques de l'Huile Essentielle de Galbanum, *Recherches,* n° 16, December 1967, p. 5.

Teisseire P., Contribution à la Connaissance de l'Huile Essentielle de Galbanum, *Recherches,* n° 14, December 1964, p. 89.

Teisseire P., Contribution à la Connaissance de l'Huile Essentielle de Sauge Sclarée, *Recherches,* n° 9, October 1959, p. 10.

Teisseire P. and Bernard P., L'Huile Essentielle de Sauge Sclarée, *Recherches,* n° 7, June 1957, p. 2.

Teisseire P., L'Huile Essentielle de Lavandun, *Recherches,* n° 4, August 1954, p. 51.

Teisseire P. and Bernard P., L'Huile Essentielle de Sauge Sclarée, *Recherches,* n° 5, June 1955, p. 32.

Teisseire P., Contribution à la Connaissance de l'Huile Essentielle de Palmarosa, *Recherches,* n° 3, October 1953, p. 33.

Teisseire P., Quelques Recherches Récentes dans l'Extraction des Essences par les Solvants Volatils, *Recherches,* n° 2, October 1952, p. 11.

The Total Synthesis of Natural Products, edited by John Ap Simon. Vol. 1, 1973, 603 pages; Vol. 2, 1973, 754 pages; Vol. 3, 1977, 566 pages; Vol. 4, 1981, 610 pages; Vol. 5, 1983, 550 pages; Vol. 6, 1984, 291 pages; Wiley Interscience, New York.

Vernin P., *Les Techniques Analytiques Modernes et leurs Applications en Parfumerie,* Book 1, Techniques Chromatographiques, Société P. Chauvet & Cie, éditeur, Seillans (Var) 1964, 152 pages; Book 2, Techniques Spectroscopiques U. V., I. R., R. M. N. and S. M., Société P. Chauvet & Cie, Seillans (Var) 1965, 220 pages.

Atti del 1° Convegno Internazionale di Studi e Ricerche Sulle Essenze, Reggio Calabria 27–29 March, 1956, Milano, 1957, 198 pages.

Rapport Du 3e Congrès International des Huiles Essentielles, 26–28 May 1964, Plovdiv, Bulgarie, Ogam, Verone (Italy), 317 pages.

International Congress on Essential Oils, Tbilisi 17–21 September 1968, Sommaire des compterendus, 232 pages.

Ve Congresso Internacional do Oleos Essenciais, Brasil 1971, Rio de Janeiro, 11–16 October 1971, Anais da Academia Brasiliera de Ciencias, Vol. 44 (Supplemento) 1972, edit. Mauro Taveira Magalhhaes e Salatiel Motta, Rio de Janeiro, 444 pages.

Aroma Encircles the World, Proceedings of VIIe International Congress of Essential Oils, 7–11 October, 1977, Kyoto, Japan, edited by Congress Committees of VIIe International Congress of Essential Oils, 1979, 552 pages.

Annales Techniques du VIIIe Congrès International des Huiles Essentielles, 12–17 October 1980, Cannes-Grasse (France), Grasse.

Fedarom, 1982, 714 pages, Imprimerie Bayeusaune, Bayeux.

IX International Congress of Essential Oils, Singapore, Malaisy, 13–17 March 1983, Book 1, Essential Oil Commercial Paper, 35 pages; Book 2, Essential Oil Technical Paper, 203 pages; Book 3, Essential Oil Technical Paper, 203 pages; Book 4, Essential Oil Technical Paper, 202 pages; Book 5, Essential Oil Technical Paper, 204 pages; Book 6, Essential Oil Technical Paper, 252 pages.

Terpenoid II Chemistry

Annual Reports "Terpenoids and Steroids" Published by Chemical Society London, 85 volumes.

Aspects of Terpenoid Chemistry and Biochemistry, Proceedings of the Phytochemical Society Symposium, April 1970, edited by T. W. Godwin, Academic Press, 1971, 441 pages.

Chemistry of Terpenes and Terpenoids (A Survey for Advanced Students and Research Workers) A. A. Newman, editor, Academic Press, London and New York, 1972, 449 pages.

Cyclopentanoid Terpene Derivatives, edited by W. I. Taylor and A. R. Battersby, Marcel Dekker, Inc., New York, 1969, 432 pages.

Encyclopaedia of the Terpenoids, John S. Glasby, 2 volumes, Vol. 1 A–H, 1982, 1322 pages; Vol. 2 I–Z, 1982, 2646 pages.

Handbook of Terpenoids, Sukh Dev, editor; Vol. 1, Monoterpenoids, Sukh Dev, Anubhar P. S., Narula, Jhillu Singh Yadar, CRC Press, 1982, 251 pages; Vol. 2, Monoterpenoids, Sukh Dev. A.P.S. Narula, J. S. Yadar, CRC Press Inc. Boca Raton, Florida, 1982, 515 pages.

Pinder A. R., *The Chemistry of the Terpenes,* Chapman and Hall Ltd., 1960, 223 pages.

Plattier M., Giudicelli M. and Teisseire P., Synthèse de la gammaméthyl-gammavinyl-gamma-butyrolactone, *Recherches,* n° 18, July 1971, p. 21.

Progress in Terpene Chemistry, Proceedings of the Conference on Terpene Chemistry, Grasse (France), 24–25 April 1986, edited by D. Joulain, Editions Frontières, Gif-sur-Yvette, 1986, 452 pages.

Shimizu B., *"Recherches sur la Synthèse s'Hétérocycles à caractère terpénique,"* Académie de Montpellier, Université des Sciences et Techniques du Languedoc, thesis submitted on 23 March 1973.

Simonsen J. L., *The Terpenes*—Vol. 1, The Simpler Acyclic and Monocyclic Terpenes and Their Derivatives (Second Edition revised by J. L. Simonsen and L. N. Owen), Cambridge, at the University Press, 1947, 479 pages; Vol. 2, The Dicyclic Terpenes and Their Derivatives, by Sir J. L. Simonsen (Second Edition revised by J. L. Simonsen and L. N. Owen), Cambridge, at the University Press, 1949, 619 pages; Vol. 3, The Sesquiterpenes, Diterpenes and Their Derivatives, by Sir J. Simonsen and D. H. R. Barton, Cambridge, at the University Press, 1952, 579 pages; Vol. 4, The Triterpenes and Their Derivatives Hydrocarbons, Alcohols, Hydroxy-aldehydes, Ketones and Hydroxy-ketones, by the Late Sir John Simonsen and W. C. J. Ross, Cambridge, at the University Press, 1957, 524 pages; Vol. 5, the Triterpenes and Their Derivatives Hydroxy Acids, Hydroxy Lactones, Hydroxyaldehydo Acids, Hydroxyketo Acids and the Stereochemistry of the Triterpenes, by the Late Sir John Simonsen and W. C. J. Ross, Cambridge, at the University Press, 1957, 662 pages.

Teisseire P., and Galfre A., Contribution à la Chimie de la Fenctione et des Fenchols, *Recherches,* n° 19, July 1974, p. 232.

Teisseire P., Shimizu B., Plattier M., Corbier B. and Rouiller P., Synthèses de Dérivés hétérocycliques azotes à squelette terpénique—Synthèse de la Fabianine, *Recherches,* n° 19, July 1974, p. 241.

Teisseire P., Galfre A., Plattier M., and Corbier B., Etudes Stéréochimiques dans la série du Pinane (2ème mémoire), *Recherches,* n° 16, December 1967, p. 59.

Teisseire P., Rouiller P. and Galfre A., Etudes Stéréochimiques dans la série du Pinane (3ème Mémoire), *Recherches,* n° 16, December 1967, p. 68.

Teisseire P., Galfre A., Plattier M., Rouiller P. and Corbier B., Etudes Stéré, ochimiques dans la série du Carane, *Recherches,* n° 16, December 1967, p. 119.

Teisseire P., Galfre A., Plattier M. and Corbier B., Action des Organo aluminiques sur les Pinocamptiones et l'Epoxyde d'alpha-Pinène, *Recherches,* n° 15, March 1966, p. 52.

Teisseire P., and Plattier M., Quelques Applications des Organo-aluminiques en Synthèse Organique, *Recherches,* n° 13, December 1963, p. 34

Terpene Chemistry, edited by James Verghese, Synthite Industrial Chemicals Private Ltd, Kolenchery, Kerala, Tata McGraw-Hill Publishing Comp. Ltd. New Delhi, 169 pages.

Terpenoids and Steroids; Vol. 1, Review of the Literature Published Between September 1969 and August 1970, senior reporter K. H. Overton and all, The Chemical Society, Burlington House, London, 1971, 557 pages; Vol. 2, September 1970–August 1971, 1972, 450 pages; Vol. 3, September 1971–August 1972, 1973, 527 pages; Vol. 4, September 1972–August 1973, 1974, 608 pages; Vol. 5, September 1973–August 1974, 1975, 390 pages; Vol. 6, September 1974–August 1975, 1976, 363 pages; Vol. 7, J. R. Hanson, September 1975–August 1976, 1977, 349 pages; Vol. 8, J. R. Hanson, September 1976–August 1977, 1978, 301 pages; Vol. 9, J. R. Hanson, September 1977–August 1978, 1979, 352 pages; Vol. 10, J. R. Hanson, September 1978–August 1979, 1981, 284 pages; Vol. 11, J. R. Hanson, September 1979–August 1980, 1982, 243 pages; Vol. 12, J. R. Hanson, September 1980–August 1981, 1983, 354 pages.

The Chemistry of Natural Products, Vol. 2, Mono- and Sesquiterpenoids, by P. de Mayo, Interscience Publishers Inc., New York, 1959, 320 pages.

The Chemistry of Natural Products, Vol. 3, the Higher Terpenoids, by P. de Mayo, Interscience Publishers Inc., New York, London, 1959, 239 pages.

The Total Synthesis of Natural Products, edited by John Ap Simon, Vol. Seven, John Wiley & Sons, New York, Chichester, Brisbane, Toronto, Singapore, 1968, 468 pages.

The Synthesis of Monoterpenes 1, Alan F. Thomas in The Total Synthesis of Natural Products, edited by John Ap Simon, Wiley Interscience, New York, 1973, Vol. 1, p. 1ff.

The Synthesis of Monoterpenes 2, Alan F. Thomas and Yvonne Bessière in The Total Synthesis of Natural Products, edited by John Ap Simon, Wiley Interscience, New York, 1981, Vol. 4, p. 451ff.

The Synthesis of Monoterpenes 3, 1980–1986, Alan F. Thomas and Yvonne Bessière in The Total Synthesis of Natural Products, edited by John Ap Simon, Wiley Interscience, New York, Vol. 7, p. 275ff.

The Total Synthesis of Natural Products, edited by John Ap Simon, Vol. Two, Wiley Interscience, New York, 1973, 754 pages.

Vezes M. and Dupont G., *Résines and Térébenthines,* Les Industries Dérivées, Encyclopédie de Chimie Industrielle, Publiée sous la direction de M. Matigon, J.-B. Baillière et Fils, Paris, 1924, 656 pages.

Von W. F. Erman, *An Encyclopedic Handbook,* Chemistry of the Monoterpenes, Part A-B, Marcel Dekker Inc., New York, Basel, 1985, 1709 pages.

Industrial Syntheses Starting from α- and β-pinene

Teisseire P., *La Chimie des Parfums,* Actualités Chimiques, first part, October 1977, pp. 7–19, second part, November 1977, pp. 9–18.

Teisseire P., *Les Grandes Voies de Synthèses des Terpénoides Aliphatiques in Progress in Terpene Chemistry,* edited by D. Joulain, Proceedings of the Conference on Terpene Chemistry, Grasse (France), 24–25 April 1986, Editions Frontières, pp. 1–39.

Total Syntheses of Aliphatic Terpenes of Industrial Importance

Kimel W., Sax N. W., *Amer. Pat.* 2661368, 1953.

Saucy G., Chopart Dit Jean L. H., Guex W., Ryser G. and Isler O., Ziir Kenntnis der Pseudo Jonon Synthese von Kimel & Sax, *Helv. Chim. Acta,* n° 19, 41, 160 (1958).

Saucy G., Marbet R., Lindlar H., and Isler O., Über eine neue Synthese von Citral und Verwandten Verbindungen, *Helv. Chim. Acta, VI,* n° 211–12 (1959) pp. 1945–59; *Chimia* 12, 326 (1958); *Angew. Chem.* 71, 81 (1959).

Saucy G. and Marbet R., Uber eine neuartige Synthese von beta-Ketoallen durch Reaktion von tertiaren Acetylencarbinolen mit Vinyläthern. Eine ergiebige Methode zur Darstellung des Pseudojonons und verwandter Verbindungen, *Helv. Chim. Acta,* Vol. 50, pp. 1158–67, 1967.

Teisseire P., Nouvelles Applications et Généralisation de la Réaction de Carroll, *Recherches,* n° 5, June 1955, p. 3.

Teisseire P., Bernard P. and Corbier B., Nouvelles Applications et Généralisation de la Réaction de Carroll, *Recherehes,* n° 6, June 1956, p. 30.

Teisseire P., Nouvelles Applications et Généralisation de la Réaction de Carroll, *Recherches,* n° 7, June 1957, p. 29.

Teisseire P., Sur Quelques Produits Secondaires de la Réaction de Carroll, *Recherches,* n° 10, December 1960, p. 18.

Teisseire P. and Corbier B., Etude par Chromatographie de Partition Vapeur-Liquide des Produits d'Hydrogénation des Pseudoionones, *Recherches,* n° 12, December 1962, p. 74.

Teisseire P. and Rinaldi M., Sur une Synthèse Univoque du (d-l)-Lavandulol, *Recherches,* n° 13, December 1963, p. 4.

Teisseire P. and Corbier B., Recherches sur "l'Oxyde de Rose" et sur des Substances Hétérocycliques voisines, *Recherches,* n° 13, December 1963, p. 78.

Teisseire P., Plattier M. and Corbier B., Synthèses Univoques de l'Isolavandulol, de l'Isolavandulal et du (d-l)-Lavandulal, *Recherches,* n° 14, December 1964, p. 44.

Teisseire P. and Corbier B., Synthèses des Tagétones et de la Dihydrotagétone, *Recherches,* n° 17, March 1969, p. 5.

Teisseire P., *Les Grandes Voies de Synthèses des Terpénoïdes Aliphatiques in Progress in Terpene Chemistry,* edited by D. Joulain, Proceedings of the Conference on Terpene Chemistry, Grasse (France), 24–25 April 1986, Editions Frontières, pp. 1–39.

Biogenesis of Natural Substances

1. General

A Ciba Foundation Symposium on the Biosynthesis of Terpenes and Sterols, E. E. W. Wolstenholme and Maeve O'Connor, editors, J. and A. Churchill Ltd, 1959, 311 pages.

Alberts Bruce, Bray Dennis, Lewis Julian, Raff Martin, Roberts Keith, Watson James D., *Biologie Moléculaire de la Cellule,* Traduction par Marianne Minkowski, Flammarion Médecine Sciences, 1983, 1146 pages + 1–35.

Banthorpe D. V., Bunton C. A., Cori O., Francis M. J. O., Correlation Between Loss of Prochiral Hydrogen and E, Z Geometry in Isoprenoid *Biosynthesis* 24, n° 2, pp. 251–52, 1985.

Biogenesis of Natural Compounds, edited by Peter Bernfeld, Pergamon Press, Oxford, London, New York, Paris, 1963, 930 pages.

Biosynthesis of Isoprenoid Compound, Vol. 1, edited by John W. Porter and Sandra L. Spurgeon, Wiley Interscience, New York, Chichester, Brisbane, Toronto, 1981, 558 pages.

Biosynthesis, A Review of the Literature Published During 1970 and 1971, T. A. Geissman, S. A. Brown, T. W. Goodwin, J. R. Harborne, E. Leete, H. H. Rees, The Chemical Society, Burlington House, London, 1972, 249 pages.

Javillier M., Polonovski M., Florkin M., Boulanger P., Lemoigne M., Roche J., Wurmser R., *Traité de Biochimie Générale;* Book I: Composition chimique des Organismes (First Fascicule) 1959, 706 pages, Book I: Composition chimique des Organismes (Second Fascicule) 1959, pp. 707–1475, Book II: Les agents de Synthèse et les Dégradations Biochimiques (First Fascicule) Vitamines, Olégoéléments, Hormones, 1962, 700 pages, Book II: Les agents de Synthèse et les Dégradations Biochimiques (Second Fascicule) Les Enzymes, 1963, 753 pages, Book III: Les Processeus Biochimiques de Synthèse et de Dégradation, 1967, 671 pages, Masson et Cie, Paris.

Julia M., *Biomimetic Prenylation Reactions* in "Organic Synthesis, Today and Tomorrow" (IUPAC), edited by B. M. Trost and C. R. Hutchinson, Pergamon Press, Oxford and Croteau R. New York, 1981, pp. 231–40.

Lehninger Albert L., *Biochimie, Bases Moléculaires de la Structure et des Fonctions Cellulaires,* Second Edition, French translation by P. Cartier and P. Kamoun, Flammarion Médecine Sciences, 1977, 1088 pages.

Poulter C. Dale, Argyle J. C., Mash E. A., Laskovics G. M., Wiggins P. L., and King C. R., *Inhibitors of Isoprenoid Biosynthesis,* Scientific Papers of the Institute of Organic and Physical Chemistry of Wroclaw Technical University, n° 22, 7, pp. 149–62.

Richards John H. and Hendrickson James B., *The Biosynthesis of Steroids, Terpenes and Acetogenins,* W. A. Benjamin, Inc., New York, Amsterdam, 1964, 416 pages.

Schreier P., *Chromatographic Studies of Biogenesis of Plant Volatiles,* On Alfred Hüthig Verlag, Heidelberg, Basel, New York, 1984, 171 pages.

Teisseire P., Biogénèse des Substances Naturelles, *Recherches,* n° 17, March 1969, p. 77.

Teisseire P., *Biogénèse des Substances Naturelles,* La France et ses Parfums, n° 65 (September–October) 1969, pp. 299–311.

Watson James D., *Biologie Moléculaire du Gène,* French Edition under the direction of François Gras, Ediscience, Paris, 1968, 492 pages.

Weiss Ulrich and Edwards J. Michael, *Biosynthesis of Aromatic Compounds,* Wiley Interscience, 1977.

2. Biosynthesis of Mono- and Sesquiterpenic Compounds

Akhila A., Banthorpe D. V. and Rowan M. G., Biosynthesis of Carvone in Mentha Spicata, *Phytochemistry,* 19, pp. 1433–37 (1980).

Ambid C., Moisseeff M. and Fallot J. Biogenesis of Monterpenes Bioconversion of Citral by a Cell Suspension Culture of Muscat Grapes, *Plant Cell Reports 1,* pp. 91–93 (1982).

Banthorpe D. V., Chrisov P. N., Pink C. R., and Watson D. G., Metabolism of Linaloyl, Neryl and Geranyl Pyrophosphates in Artemisia Annua, *Phytochemistry,* 22, n° 11, pp. 2465–68 (1983).

Banthorpe D. V., Long D. R. S. and Pink C. R., Biosynthesis of Geraniol and Related Monoterpenes in Pelargonium Graveolens, *Phytochemistry,* 22, n° 11, pp. 2459–63 (1983).

Banthorpe D. V., Ekundayo O. and Njar V. C. O., Biosynthesis of Chiral Alpha- and Beta-Pinenes in Pinus Species, *Phytochemistry,* 23, n° 2, pp. 291–94 (1984).

Croteau R., Felton M. and Ronald R. C., *Biosynthesis of Monoterpenes: Conversion of the Acyclic Precursors Geranyl Pyrophosphate and Neryl Pyrophosphate to the Rearranged Monoterpenes Fenchol and Fenchone by a Soluble Enzyme Preparation from Fennel (Foeniculum Vulgare),* Archives of Biochemistry and Biophysics, 200, n° 2, pp. 524–33 (1980).

Croteau R., *The Biosynthesis of Terpene Compounds,* Perfumer & Flavorist, pp. 35–59, (1980).

Croteau R., *The Biosynthesis of Terpene Compounds,* Soap, Perfume Cosmetics (SPC) August 1980, pp. 428–40 and 475.

Croteau R. and Martinkus Ch., Metabolism of Monoterpenes, Demonstration of (+) Neomenthyl-β-D-glucoside as a Major Metabolite of (–) Menthone in Peppermint (*Mentha Piperita*), *Plant Physiol.* 64, pp. 169–75 (1979).

Croteau R. and Felton N. M., Substrate Specificity of Monoterpenol Dehydrogenase from Foeniculum Vulgare and Tanacetum Vulgare, *Phytochemistry,* 19, pp. 1343–47 (1980).

White E. E., Biosynthetic Implications of Terpene Correlations in Pinus Contorta, *Phytochemistry,* 22, n° 6, pp. 1399–1405 (1983).

Won Erman F., *Biogenesis* of *Monoterpenoids* in "Chemistry of the Monoterpenes," An Encyclopedic Handbook, Part A., pp. 34–126 Marcel Dekker, New York, Basel, 1965.

3. Biosynthesis of Sterols, Phytosterols, and Triterpenes

Anding C., Brandt R. D., Ourisson G., Pryce R. J. and Rohmer M., *Some Aspects of the Biosynthesis,* Proc. Roy. Soc., London, B. 180, pp. 115–24 (1972).

Benveniste P., Hirth L. and Ourisson G., La Biosynthèse des Stérols dans les Tissus de

Tabac Cultivés in Vitro I, Isolement des Stérols et des Triterpènes, *Phytochemistry,* 5, pp. 31–44 (1966).

Cane David E., The Stereochemistry of Allylic Pyrophosphase Metabolism, *Tetrahedon* 36, pp. 1109–59 (1980).

Clayton R. B., Biosynthesis of Sterols, Steroids and Terpenoids, *Quart. Rev.* London, 19, 168ff. (1965).

Eschenmoser A., Ruzicka L., Jeger O., Arigoni D., Zur Kenntnis der Triterpene. Eine Stereochemische Interpretation der biogenetischen Isoprenregel bei den Triterpenen, *Helvetica Chimica Acta,* 38, 1890 (1955).

Goodwin Trevor W., Biosynthesis of Terpenoids, *Ann. Rev. Plant Physiol.* 30, pp. 369–404 (1979).

Goodwin Trevor W., *Biosynthesis of Plant Sterols and Other Triterpenoids in Biosynthesis of Isoprenoid Compounds, Vol. 1,* edited by John W. Porter & Sandra L. Spurgeon, John Wiley & Sons, Inc., 1981, p. 445ff.

Harding Kenn E., On the Stereochemistry of Biogenetic-like Olefin Cyclisations, *Bioorganic Chemistry,* 2, pp. 248–65 (1973).

Johnson Wm. B., Nonenzymic Biogenetic-like Olefin Cyclizations, *Acc. Chem. Res.* 1968, p. 1ff.

Phillips Gareth T. and Clifford Kenneth M., Stereochemistry of a Methylgroup Rearrangement during the Biosynthesis of Lanosterol, Eur. *J. Biochem.* 61, pp. 271–86 (1976).

Ress H. H., Goad L. J. and Goodwin T. W., Studies in Phytosterol Biosynthesis: Mechanism of Biosynthesis of Cycloartenol, *Biochem. J.* 107, pp. 417–25 (1968).

Sliwowski J., Mechanisms of Squalene Cyclization to Tetra- and Pentacyclic Triterpenes, *Post. Biochem.* 20, pp. 281–303 (1974).

Turowska Grazyna, Plant Sterols Biosynthesis, *Post Biochem,* 15, pp. 257–72 (1972).

Van Tamelen E. E. and Sharpless K. B., Positional Selectivity during Controlled Oxidation of Polyolefines, *Tetrahedron Letters* n° 28, pp. 2655–59 (1967).

Van Tamelen E. E., Bioorganic Chemistry: Total Synthesis of Tetra- and Pentacyclic Triterpenoids, *Accounts of Chemic Research* 8, pp. 152–58 (1975).

Van Tamelen E. E., The Role of Organic Synthesis in Bioorganic Chemistry, *Pure & Applied Chem.* 53, pp. 1259–70 (1981).

Van Tamelen E. E. and Nadeau R. G., Cyclization Studies with 14, 15-Oxidogeranyl-geranyl Acetate, *Bioorganic Chemistry,* 11, pp. 197–218 (1982).

Van Tamelen E. E., Storni A., Hessler E. J. and Schwatz M. A., Cyclization Studies with (+) 10, 11-Oxidofarnesyl Acetate, Methylfarnesate and Methyl (+)-oxidofarnesate, *Bioorganic Chemistry,* 11, pp. 133–70 (1982).

Some Aspects of the Chemistry of Sesquiterpenes and Diterpenes

Brun Hélène, *Etude de la Réactivité Photochimique de Cétones Sesquiterpéniques,* thesis submitted on 13 October 1979, Université Claude Bernard-Lyon I.

Brun Hélène, *Mécanisme de la Photoréacuvité de Cétones Sesquiterpéniques,* thesis submitted on 9 December 1983, Université Claude Bernard-Lyon I.

De Mayo P., *Mono and Sesquiterpenoid, Vol.* 2, Interscience Publishers Inc, New York, 1959, 320 pages.

Dumas David, *The Synthesis of Patchouli Alcohol,* thesis submitted on 5 July 1968, University of Pittsburgh.

Giraudi E., Plattier M. and Teisseire P., Synthèse des Cis- et trans-gamma-bisabolènes et du trans alpha-bisabolène, *Recherches,* n° 19, July 1974, p. 205.

Heathcock C. H., *The Total Synthesis of Sesquiterpenes* in The Total Synthesis of Natural Products, Vol. 2, John Ap Simon 1973.

Heathcock C. H. et al., *The Total Synthesis of Sesquiterpenes,* 1970–1979, in The Total Synthesis of Natural Products, Vol. 5, pp. 1–550, Wiley Intersciences, New York, 1983, 550 pages.

Helmlinger Daniel, *Réactions Transannulaires du Longifolène,* thesis submitted on 1 March 1969 (n° 506), Science Faculty of the University of Strasbourg.

Heymes A., Plattier M. and Teisseire P., Synthèses des Calaménènes et de l'alphacalacorène, *Recherches,* n° 19, July 1974, p. 214.

Lhomme J and Ourisson G., Le Longifolène, *Recherches,* n° 15, March 1966, p. 15.

Luu Bang, *Transpositions à étapes Multiples d'Epoxydes Sesquiterpéniques,* thesis submitted on 1 July 1972, University of Strasbourg.

Luu Bang and Guest I., Transpositions à Etapes Multiples d'Epoxydes Sesquiterpéniques, *Recherches,* n° 19, July 1974, p. 97.

Oberhansli W. E. and Sctionholzer P., *Recherches,* n° 19, July 1974, p. 62.

Ourisson G., Munavalli S., Ehret C., Sorm F., *Tables Internationales de Constantes Sélectionnées relatives aux Sesquiterpénoides,* Pergamon Press, Oxford, Paris, London, Edinburg, New York, Brunswick, 1966, 70 pages.

Pelerin G., *Vers de Nouvelles Voies d'accès au Norpatchoulénol,* thesis submitted on 11 March 1978, Faculty of Sciences and Technology of Saint-Jérome, Marseille.

Pelerin G., *Synthèse du Norpatchoulénol et d'Homologues,* thesis submitted on 11 Février 1980, Faculty of Sciences and Technology of Saint-Jérome, Marseille.

Pesnelle P., Les Epoxydes du Guaiol, *Recherches,* n° 15, March 1966, p. 33.

Pesnelle P. and Teisseire P., Isomérisations en série Sesquiterpénique. Transformation du Calarène et Aristolène, *Recherches,* n° 17, March 1969, p. 121.

Plattier M., Rouiller P. and Teisseire P., Structure du gamma-himachalène, *Recherches,* n° 19, July 1974, p. 145.

Plattier M., and Teisseire P., Synthèses, Structures et Configurations Absolues des Atlantones, *Recherches,* n° 19, July 1974, p. 153.

Ruegg R., Pfiffner A., Montavon M., Synthèse de Lancéols Lévogyres cis et trans, *Recherches,* n° 15, March 1966, p. 3.

Sorm Fr., and Dolejs L., *Guaianolides and Germacranolides* (Chimie des Substances Naturelles, collection edited by Edgar Lederer) Actualités Scientifiques et Industrielles 1309, Hermann, Paris, 1965, 153 pages.

Teisseire P., Plattier M., Wojnarovski W. and Ourisson G., Etudes Stéréochimiques dans la série du Cédrane, *Recherches,* n° 16, December 1967, p. 84.

Teisseire P. and Galfre A., Etudes Stéréochimiques dans la série du Ceédranee *Recherches,* n° 16, December 1967, p. 104.

Teisseire P., and Plattier M., Nouvelle Cétone Sesquiterpénique bicyclique isolée de l'Huile Essentielle de Cèdre de l'Atlas, *Recherches,* n° 19, July 1974, p. 167.

Teisseire P., Maupetit P., Corbier B. and Rouiller P., Norpatchoulénol, Etude Chimique, Structure et Configuration absolue, *Recherches,* n° 19, July 1974, p. 36.

Teisseire P., Pesnelle P., Corbier B., Plattier M. and Maupetit P., Synthèse du Norpatchoulénol, *Recherches,* n° 19, July 1974, p. 69.

Teisseire P., Hydroxylations du Patchoulol par Voie Biologique et Microbiologique, *Bull. Soc. Chim. France,* 1980, n° 1–2, pp. II-66–II-70.

Wolff G., Le Seychellène, *Recherches,* n° 19, July 1974, p. 85.

Yoshioka M., Mabry T. J., Sesquiterpene Lactones. Chemistry, NMR and Plant Distribution, University of Tokyo Press, 1973, 544 pages.

Substances with a Musk Odor

Abe S., Eto T. and Tsujito Y., Recent Advances in the Development of Macrocyclic Musk Compounds, *Cosmetics and Perfumery,* 88, 67, June 1973.

Bedoukian P. Z., *Perfumery and Flavoring Synthetics,* Third Revised Edition, Allured Publishing Corporation, Wheaton, Ill., 1986, pp. 301–48.

Le Musc (anonyme), *Nouvelles de naarden,* 1957, August, p. 2.

Mathur H. H., Bhattacharyya S. C., Macrocyclic Musk Compounds, *P.E.O.R.,* 57, 10, 629 (1966).

Stoll M., Les Produits à Odeur Musquée et l'Evolution de la Chimie Mégacyclique, *Chimia,* 2, 217 (1948).

Teisseire P., Sur une Synthèse de l'Acide Junipérique, *Recherches,* n° 11, December 1961.

Theimer E. T., Davies J. T., *Olfaction, Musk Odor and Molecular Properties,* Agricultural and Food Chemistry, 15, 6 (1967).

Van Thomel Frank, Über Riechstoffe des Holz-ambra-moschuskomplexes, *Chemiker Zeitung* 105, n° 9, 239 (1981).

Bibliography for the Appendices

The Wittig Reaction

Maercker Adalbert, *The Wittig Reaction,* Organic Reactions, Vol. 14, Chapter 3, pp. 270–490, John Wiley & Son Inc., New York, London, Sydney, 1965.

Functionalization of Unactivated Carbon Atoms

Mazur Yahuda, *Functionalization of Nonactivated Carbon Atoms,* IUPAC, Pure and Applied Chemistry, Vol. 41, p. 145ff., Butterworths, 1975.

Methods of Identifying Natural and Synthetic Compounds

Balabane M., Letolle R., Bayle J.-C. and Derbesy M., *$^{13}C/^{12}C$ et $^{2}H/^{1}H$ Déterminanon du Rapport $^{13}C/^{12}C$ et $^{2}H/^{1}H$ des différents Anétholes,* Parf. Cosm. Arômes, n° 49 (February–March), 27 (1983).

Bricout J., Koziet J., Derbesy M. and Beccat B., *Nouvelles possibilités de l'Analyse des Isotopes stables du Carbone dans le Contrôle de la Qualité des Vanilles,* Ann. Fals. Exp. Chimiques, 74, n° 803, 691 (1981).

Bricout J. and Koziet J., *Characterization of Synthetic Substances in Food Flavors by Isotopic Analysis, in Flavor of Foods and Beverages—Chemistry and Technology,* edited by G. Charalambous and G. Inglett, Academic Press, 1978.

Bricout J. and Koziet J., Détermination de l'Origine des Substances Arômatiques par Spectrographie de masse isotopique, *Ann. Fals., Exp. Chim.,* 69, n° 747, 845 (1976).

Bricout J., Fontes J.-C., and Merlivat L., Detection of Synthetic Vanillin and Vanilla Extracts by Isotopic Analysis, *Journ. of AOAC,* 57, n° 3, 713 (1974).

Derbesy M., Application de l'Analyse des Isotopes stables au Contrôle des Matières Aromatisantes, *Labo-Pharma-Problèmes et Techniques,* 32, n° 343, 467 (1984).

Bricout J. and Koziet J., Contrôle de la Qualité des Vanilles, Détermination des Abondances des Isotopes Stables du Carbone dans la Vanilline, *Labo-Pharma-Problèmes et Techniques,* n° 299, 490 (1980).

Hofman P. G. and Salb M., Isolation and Stable Ratio Analysis of Vanillin, *J. Agric. Food Chem.,* 27, n° 2, 352 (1979).

Krueger H. W. and Reesman R. H., *Carbon Isotope Analysis in Food Technology* in "Mass Spectrometry Reviews," pp. 205–36, John Wiley & Sons Inc. 1982.

Krueger D. A. and Krueger H. W., Carbon Isotopes in Vanillin and the Detection of Falsified "natural" Vanillin, *J. Agric. Food Chem.,* 31, 1265 (1983).

Krueger D. A. and Krueger H. W., Detection of Fraudulent Vanillin Labeled with ^{13}C in the Carbonyl Carbon, *J. Agric. Food Chem.,* 33, n° 3, 323 (1985).

Martin G. J. and Martin M. L., Le Fractionnement isotopique Spécifique de l'Hydrogène dans les Composés Naturels: Une Méthode de Contrôle de la Chaptalisation des Vins, *l'Actualité Chimique,* October 1982, p. 31.

Martin G. J. and Martin M. L., Deuterium Labelling at the Natural Abondance Level as Studied by High Field Quantitative ^2H NMR, *Tetrahedron* Letters, 22, n° 6, 3525 (1981).

Martin G. J., Martin M. L. and Mabon F., A New Method for the Identification of the Origin of Natural Products. Quantitative ^2H NMR at the Natural Abondance Level Applied to the Characterization of Anetholes, *J. Am. Chem. Soc.,* 104, 2658 (1982).

O'Leary M. H., Carbon Isotope Fractionation in Plants, *Phytochemistry,* 20, n° 4, 553 (1981).

Toulemonde B., Horman I., Egli H. and Derbesy M., Food-Related Application of High-Resolution NMR, Part II. Differentiation Between Natural and Synthetic Vanillin Samples Using ^2H-NMR, Helv. *Chim. Acta,* 66, 7, 2342 (1983).

Summary of Some Physiochemical Methods for Analyzing Natural Odorant Products

Bellanato J. and Hidalgo A., *Infrared Analysis of Essential Oils,* edited by Heyden & Son Ltd in Co-operation with Sadtler Research Laboratories Inc., 1971, 164 pages.

Teisseire P., Application de la Chromatographie en Phase Gazeuse à la Parfumerie, *Rechcrches,* n° 12, December 1962, p. 54.

Teisseire P., Influence des Méthodes Physico-chimiques Modernes sur l'Analyse des Huiles Essentielles, *Recherches,* n° 14, December 1964, p. 12.

Teisseire P., Quelques Aspects de la Chromatographie en Phase Gazeuse dans l'Industrie des Matières Odorantes Utilisées en Parfumerie, *Recherches,* n° 17, March 1969, p. 37.

INDEX

Acenaphthene derivatives, 353
Acetaldehyde, 392–393
Acetals, 389–392
Acetic acid derivatives, 144
Acetic units, 144–145
Acetoacetyl-CoA, 167–171
Acetogenic substances, 144
Acetone, 108, 109
Acetylene, 108, 109
Acid-catalyzed reactions, 85–92, 102, 103
Acidic media, 29–30
Acyloin, 297–305
Adenosine, 150–151, 155–156
Adenosine diphosphate, 151, 152
Adenosine monophosphate, 151
Adenosine triphosphate, 151–152
Alcohols of lilac, 31–32
Aleuritic acid, 301, 303
Alkyl sodium compounds, 387
Alkylcarbonium ions, 79
Alkylvinyl ethers, 393–394
Allylic alcohol, 395, 397–399
Allylic amines, 137, 139
Aluminum hydrides, 386–387
Ambral, 339
Ambrette, 288–289
Ambrette absolute, 289
Ambrette butter, 289
Ambrettolide, 289
Amine oxides, 54–55
Amino acids, 146
Amyl salicyclate, 15
Androstenol, 288
Anethole, 421–423
Angelica, 288
Annulenes, 328
Anti-Sytzeff olefin, 49
Apoenzyme, 149
Apollan-11-ol, 223, 224, 408
Arbuzov-Razumov reaction, 377
Aristone, 381
Artabsine, 231
Artemisia alcohol, 32, 33–36
Autoxidation, 99

Baeyer strain, 196
Baeyer-Villiger reaction, 9, 324
Baker, muscone synthesis of, 321–323
Barton reaction, 405–407
Benzene, Kekulian formula for, 7
Benzenic nitro musks, 331–336
Benzenic non-nitro musks, 339–342
Bergamotenes, 140, 409
Betaine, formation of, 367–373
in Wittig reaction, 364–365
Beure d'ambrette, 289
Bicyclic derivatives, 188
Biomolecules, origin of, 145–147
Birch, 144–145
Birch reaction, 183–190
Blomquist, macrocyclic ketone synthesis of, 295–297
Borneol, 9, 82, 186–187
Boron, 386–387
Bouveault and Blanc reaction, 299
Brannock homologation method, 314
Brannock reaction, 395
Breslow, 414–416
Bromohydrin, 220–222
Butyndiol, 115–117

C_4 cycle, 419
Cahn, Ingold, Prelong rule, 360, 361
Calvin, 147
Calvin cycle, 419
Camphane, 64, 65
Camphene, 186–187, 193, 195

Camphor, 6, 9, 76–77
  rearrangements in, 82
  synthesis of, 88–92
Camphor of patchouli, 232
Carane, 64, 65
  derivatives of, 70–76
Carbocyclic ketones, 312–314
Carbon, tetravalence of, 8
Carbon atoms, unactivated, functionalization of, 401–416
Carbon compounds, musk odor of, 330
Carbon isotopes, 418–420
Carbonium ions, as reaction intermediates, 77–78
  non-classic, 79–80
  vs. alkylcarbonium ions, 79
Carotenoids, 20, 21
Carroll reaction, 118–119
  generalized, 127–128
Carveol, 42, 43
Carvone, 39–42, 102
Caryolanol, 226, 228
Caryophyllene, 219–223, 268, 269, 270–271
α-Caryophyllenic alcohol, 408–409
Cations, 79–81
Caucalol diacetate, 223–226
$^{13}C/^{12}C$ ratio, 418, 429
8,14-Cedranolide, 407–408
Celestolide, 343, 345
Cembrene, 279–286
Chamazulene, 230–231
Characterization, 429
Chemical separation, 426
Chirality, external, 359
Chromatography, 10–11, 21
  adsorption, 427
  gas, 427, 428
trans-Chrysanthemic acid, 60–63
Chugaev reaction, 53
Cinnamic acid, 12–13
Citral, as substrate, 108
  consumption of, 131–132
  radioactive identification of, 417
  radiocarbon analysis of, 418
  synthesis of, 131–134
Citronellal, 108
  derivatives of, 96–99
  synthesis of, 138, 142
Citronellol, derivatives of, 96–99
  synthesis of, 137, 138
Civet, 287, 288
Civettol, 289
Civettone, 301, 303
Claisen reaction, inverse, 118
Coenzyme(s), 149, 160
Coenzyme A, 152–153, 155–163
  biosynthesis of, NADP in, 158–163
Copabornane, 240–245
Copaborneol, 187
  synthesis of, 240–245, 257–260
Copacamphane, 240–245
Copaisoborneol, 240–245
Cope reaction, 54–55
Cope rearrangement, 207–210
Corey, E. J., caryophyllene synthesis of, 268–271
  α-humulene synthesis of, 264–268, 269
  longifolene synthesis of, 245–248
  santalene synthesis of, 248–249, 251–253
  α-santalol synthesis of, 254–256
Corticosteroids, 414–416
Coupled reactions, 169–171
Crassulacean acid mechanism, 419
Cryophyllene, 226–229
Curcumene, 183–190
Cyathoclina lyrata, 32
Cyclization, in acidic media, 29–30
  intramolecular, 327–329
Cyclocopacamphene, 240–245
Cyclodecanone, 307–308
  in macrocyclic ketone synthesis, 312–319
Cyclogermacrene, 200–201
Cyclohexadecanolide, 325–327
Cyclopentadecane, 325–327
Cyclopentadecanolide, 289–290
Cyclopentadecanone, 313–314
Cyclopentane, 196–197
Cyclopentanyl monoterpene derivatives, 56–59
Cyclopropyl ketones, 377, 379
Cyperene, 239, 240
Cytochromes, 414

Danishevsky, patchoulol synthesis of, 269, 271
Davanone, 171, 174
Dehydolinalyl acetate, 125
Dehydrolinalool, 123–134
  in citral synthesis, 132–134
  in linalool preparation, 123–127
  in substrate syntheses, 108, 109
Dehyrohalogenation, thermal, 55–56
Delepine reaction, 342
Dendrolasine, 171, 175, 176
Deuterium (D)/hydrogen (H) ratio, 421, 429
Deuterium nuclear magnetic resonance spectroscopy, 423
α,ω-Diacetylenic esters, 327–329
Dialkylaluminum hydrides, 386–387

Dialkylborane hydrides, 386
Diastereoisomer, 358
Diazomethane, 312–313
Dibal, 71, 74
Diels-Alder reaction, 25–27, 37, 38, 277
Diethylgeranylamine, 137
Dihydrocoumarin, 355, 356
Dihydromyrcene, 96–97
Dihydrotagetone, 395, 397
Dimerization reaction, 25, 26
1,3-Diphosphoglycerate, 161–163
Distillation, 3–4, 426–427
Diterpene(s), 22
  macrocyclic, 191–286
  14-member ringed, 279–286
  synthesis of, 395, 396
Dumas, patchoulol synthesis of, 269, 271

Elemenone, 206–207
Elemol, 205–206, 260–264
Elephantine, 213
Elephantopine, 213
Elimination reactions, 47–49
Enamine(s), 315
  in sesquicarene syntheses, 181–182
  isomerization of, 138, 142
  mesomeric forms of, 385
Enantiotopic molecules, 357–358
Enfleurage, 4
Enol ethers, 385–399
  acetal reaction with, 389–392
  A-X compound reactions with, 386
  cationic polymerization of, 385
  H-X compound reactions with, 385–386
  in carbon-carbon bond, 389–393
  mesomeric forms of, 385
  rearrangements of, 393–395
Enzymatic reactions, 414–415
Enzymatic reductions, 360
Enzyme(s), classification of, 148–149
Enzyme-catalyzed reactions, 163–171
Enzymology, 148–171
Epoxidation reactions, 70–76
Epoxycaryophyllene, 228, 229
Epoxyisoyophyllene, 228, 229
Epoxysulfones, 317–319
Eremophilane, 214, 215
Erman, santalol synthesis by, 254, 255
Eschenmoser, macrocyclic ketone synthesis of, 307–312
Essence of cataire, 58
Essential oils, 108
  annual production of, 107, 108
  cinnamic acid in, 12
  history of, 3–7
  norterpenoids in, 11
  phenylpropanoids in, 12
  terpenoids in, 11, 12
Esterification, 426
Estragole, 421–423
α,β-Ethylenic aldehydes, 389
Ethynylation, catalytic, 115–117
  in substrate syntheses, 109–118
  semi-catalytic, 112–115
  stoichiometric, 110–112
Eucalyptus dives, 46, 47
Exaltolide, 328–329
Exaltone, 307–309, 313–314
Extraction, 426
  hydrocarbon, 5
  with volatile solvents, 11

β-Farnesene, 171, 174
Farnesol, 21, 176
Fenchanes, 64, 65
Fenchenes, 89, 91
Fenchone, 90–92
Floral extraction, 4–5
para-Formaldehyde, 375–376
Frantz method, 116–117
Free radicals, 401
Furanoelemadienes, 210

Gaiane, 198, 229–240
Gaianolide, 230
Galaxolide, 354
Gas chromatography, 428
Geijerene, 206
Geraniol, 21, 137
  allylic migration in, 27–28
  as substrate, 108
  cyclization of, 29
Geranium oils, 108
Geranylacetone, 125
Geranylgeraniol, 21
Germacrane(s), 198–216
  classification of, 199
  1,5-cyclodecadienic system of, 200
  isolation of, 199, 200
Germacranolides, 199–200
Germacrene(s), Cope rearrangement in, 207–210
  derivatives of, 205–207
  ionic mechanisms in, 203
  photochemical mechanisms in, 203
  trans-annular reactions in, 212
Germacrene A, 201

Germacrene B, epoxides of, 206–207
in sesquiterpene biosynthesis, 204–205
isolation of, 201, 202
Germacrene C, 201, 202
Germacrene D, 202–203
Glaser reaction, 327
Glycolysis, 161
Goldfarb, macrocyclic ketone synthesis by, 305–307

Haller and Bauer reaction, 90–92
Heliotropine, 13–14
Hemachalene, 193, 194
Heterocyclization reactions, 412–413
Heteroproteins, 149
Heterosides, 56
$^2H/^1H$ ratio, 420
Hoch, homologation method of, 317, 318
Hoffmann olefin, 48
Hoffmann-elimination reactions, 48
Hoffmann-La Rôche, 119–121, 134
Hoffmann-La Rôche reaction, 128–131
Hofmann-Löffler-Freytag reaction, 401–405
H_2O-KOH system, 110–112
Holoproteins, 149
Homologation, in macrocyclic ketone synthesis, 312–319
Hormones, vs. pheromones, 60
Horner reaction, 376, 377
Humuladienone, 217
Humulane, 216–223
α-Humulane, 217–219
as caryophyllene precursor, 219–223
conformations of, 219, 220
α-Humulene, 264–268, 269
Humulenone, 217
Humulol, 217
Hünig, homologation method of, 317, 318
Hunsdiecker, macrocyclic ketone synthesis of, 295, 296
Hydrindacene derivatives, 349–350
Hydroboration reactions, 70–76
Hydrocarbons, dienic, 37
Hydrogen, 420–423
Hydrogenolysis reaction, 299
Hydroperoxidation reactions, 99–107
Hydroperoxide, 99, 329–331
Hydroxyacids, depolymerization of, 324–325
Hydroxycitronellal, 15, 98
Hydroxypentadecanoic acid, 290

Indanic nitro musks, 336–338
Indanic non-nitro musks, 342–345
Infrared spectroscopy, 428–429
Insecticides, 61–64
Intramolecular cyclization, 327–329
Ionic reactions, 52
Ionones, 15, 29
Iridoid glucosides, 56–57
Iridoids, 56–59
Iris, essential oil of, 104
Irones, 104–106
Isobutylene, 121–122
Isocamphanes, 64, 65
Isocaraone, 70–76
Isocaryophyllene, 228
Isocaucalol, 223, 225
Isochromanic non-nitro musks, 353–356
Isolavandulal, 377, 378
Isolavandulol, 377, 378
Isoprene, 122, 143, 343
Isoprene structure rule, 143–144
Isoprene unit, 144
Isoprenoid coupling, 19–20
Isoprine biogenesis rule, 20
Isopulegol, 42, 44

Japanese hops, 217
Julia, Marc, 125
α-santalene synthesis of, 256–260

Kaurene, 395, 396
Ketones, 380
Kröhnke reaction, 342

Lead tetracetact, 411
Lignin, 12
Limonene, 37, 38, 41, 43
Linalool, 137
acid-catalyzed cyclization of, 30
as substrate, 108
optically active, 106–107
preparation of, 123–127
Linalyl acetate, 31–32
Lithium acetylide, 114–115
Loganine, 57
Longiborneal, 193, 194
Longicyclene, 193, 194
Longifolene, 193, 194, 195
camphene and, 193, 195
synthesis of, 245–248
Longipinene, 193, 194
Lycopene, 20

Macrocyclic alkynones, 319–321
Macrocyclic diesters, 330

Macrocyclic ketones, oxidation of, 323–324
  synthesis of, 290–323
    Blomquist method of, 295–297
    cyclododecanone in, 312–319
    Eschenmoser method of, 307–312
    Goldfarb/Taits method of, 305–307
    Hunsdiecker method of, 295, 296
    Prelong/Stoll method of, 297–305
    ring homologation in, 312–319
    Ruzicka method of, 290–293, 294
    Tanabe method of, 307–312
    Utimoto method of, 319–323
    Ziegler method of, 293–295
Macrocyclic lactones, preparation of, 324–325
  synthesis of, 323–331
    hydroperoxide fragmentation in, 329–331
    intramolecular cyclization in, 327–329
    Story method of, 325–327
Mass spectrometry, 428
Mass spectroscopy, 418–423
Matheus, 147
Meerwein, 77
Mentha arvensis, 42
trans-para-Menthane, 49, 50
Menthol(s), 42, 44
  as substrate, 108
  configurations of, 45
  derivatives of, 49–51, 55–56
  elimination reactions in, 47–49
  industrial production of, 47
  synthesis of, 46, 138, 142
  thermal eliminations in, 53
Metacresol, 335, 336
Metal acetylides, 113–115
Metamenthane derivatives, 55–56
Metaxylene, 335
Methyl chrysanthemate, 380
Methyl octine carbonate, 15
Methylbutenol, 118–121
Methylheptenone, 69
  in substrate syntheses, 108, 109
  synthesis of, methods for, 118–123
Mevalonic acid, 166–167
Michaelis-Arbuzov reaction, 376
Mirrington, patchoulol synthesis of, 269, 272
Monosaponification, 357–358
Monoterpene(s), 19
  acyclic, 24–36
    Diels-Alder reactions in, 25–27
    irregular linkage in, 32
    photochemical reactions in, 27–29
  bicyclic, 64–83
    types of, 64–65
  cis β-ocimene form of, 24, 25
  classification of, 24
  cyclobutyl, 59–60
  derivatives of, 60–64
  isoprene unit linking in, 22–24
  monocyclic, 37–64
  β-myrcene form of, 24, 25
  structure of, 10, 22
  trans β-ocimene form of, 24, 25
Mori, homologation method of, 316–317
Moskene, 336, 342
Musc en graine, 287
Musc en poche, 287
Muscone, 289
  configuration of, 322–323
  synthesis of, 315–317
    Baker method of, 321–323
Muscopyridine, 289, 309–312
Musizine, 144
Musk, aromatic, 331–356
    nitro, 331–339
  benzenic nitro, 331–336
  benzenic non-nitro, 339–342
  history of, 14–15
  indanic nitro, 336–338
  indanic non-nitro, 342–345
  isochromanic non-nitro, 353–356
  odor/molecular structure of, 332–335
  odor/nuclear variation of, 346–349
  odor/structural substitutes of, 344–345
  teralinic non-nitro, 345–349
  tetralinic nitro, 338–339
  tricyclic non-nitro, 349–353
Musk 89, 354
Musk ambrette, 332, 335, 336
Musk B, 331
Musk Baur, 331
Musk ketone, 332, 335
Musk odor, 287–356
  chemistry of, 12
  odorant gland production of, 287
  substances with, 289–290
Musk tibetine, 332, 335
Musk Toluene, 331
Musk xylene, 332, 335
Muskrat, odorant glands of, 287, 288
Myrcene, 25
  as substrate, 94–96
  in terpenoid synthesis, 137–140
  pyrolysis of, 93–94

β-Myrcene, 24, 25
  dimerization reactions in, 25, 26
  in oxygenated terpenes, 27–29

NAD, 160–161
  in coenzyme A biosynthesis, 158–163
  reduction of, 163–171
NADP, 160–161
  in coenzyme A biosynthesis, 158–163
  reduction of, 163–171
Näf, patchoulol synthesis of, 274–275
Nametkin rearrangement, 82
Natural odorants, characterization of, 429
  chemical separation of, 426
  gas chromatography of, 428
  infrared spectroscopy of, 428–429
  mass spectrometry of, 428
  physical separation of, 426–428
  physico-chemical analysis of, 425–429
Natural substances, acetic acid derivatives in, 143–144
  biogenesis of, 143–190
  isoprene structure of, 143–144
Neomenthols, derivatives of, 49, 51–56
Nepetalactone, 58–59
Nerol, 108, 137
Nerolidol, 126–127
Ngaione, 171, 174
Nicotine, 403–404
Nitric acid, 14–15
Nitrogen, 417
Nitrosochlorides, 38–39
Norcyperene, 239, 240
Norpatchoulenol, 237–240
  synthesis of, 275–279
Nozaki, homologation method of, 316–317

Ocimene, 24–26
allo-Ocimene *6*, 26–27
Odorant value, 425–426
Ohloff, patchoulol synthesis of, 274–275
Olefins, in Wittig reaction, 366–369
  nonstabilized phosphoranes in, 370–373
  stablized phosphoranes in, 369–370
$^{18}O/^{16}O$ ratio, 420
Oppolzer, norpatchoulenol synthesis of, 277–279
Organic compounds, identification of, 417–423
Organo-magnesium compounds, 387–388
Organo-mercury compounds, 388–389
Organo-metallic catalysts, 137, 141, 142
Organo-metallic compounds, 387–389
Orsellinic acid, 144–145
Orthomenthane derivatives, 55–56
Oxalactones, 330–331
Oxaphosphetane, 364–365
Oxidation, distant, 414–415
Oxiranes, 379, 380
Oxygen, 420–423

Pantetheine phosphate, 155–158
Pantothenic acid, 153–155
Paracelsus, Quinta essentia theory of, 3
Paracyclophanes, 304–305
Parham, cyclododecanone homologation method of, 314
Patchoulene, 233–235
α-Patchoulene, 237
Patchouli alcohol, 232
Patchouli oil, 231–232
Patchoulol, 232
  structure of, 232–233, 236
  synthesis of, 233–239, 269–275
Patchoulyl acetate, 237, 238
Pavan, 57–58
Pavlovskaya, 147
Pennyroyal oil, 46
Peracids, 323–324
Perfume history, 1–17
Perfumery, substrates of, 108
Perkin method, 13
Phantolid, 342, 343
Phenol, musk odor of, 355–356
Phenylpropenoids, 420–421
Pheromones, 59–60
Phosphines, 365
Phosphoranes, in betine formation, 365–366
  in betine reversibility, 368–369
  in olefin formation, 369–370
  Wittig reaction and, 374
Photochemical reactions, in acyclic monoterpenes, 27–29
  in carvone, 39–40, 42
Photocyclocitral, 56–59
Photo-oxygenation, 41, 43
Photosynthesis, 419
Physical separation, 426–428
Piers, sesquiterpene syntheses by, 240–245
Pinacol-rearrangement, 10
Pinane(s), 8, 64, 65, 76
  derivatives of, 76–83
    reaction intermediate in, 77
  hydroperoxidation of, 106–107
trans-Pinane oxide, 102
α-Pinene, 8–9, 96–106
β-Pinene, 96–99

Pinene(s), acid-catalyzed reactions in, 85–92
bornyl esters from, 89
hydroperoxidation reactions in, 99–107
pyrolysis reactions of, 92–99
Piperitone, 46, 47
Piperonal, 13–14
Piria reaction, 290–293, 294
Pitzer strain, 196
Plants, CO_2 fixation of, 419
Plinols, 30
Pliny the Elder, 2
Polyacetic rule of Birch, 145
Polyesters, 324–325
Pommades, 4–5
Ponnamperuma, 147
Potassium hydroxide, 110, 112–113
Prins reaction, 103
Prochirality, 357–362
molecular structure in, 357–358
nomenclature of, 359–360
trigonal systems in, face designation in, 361–362
Propargyl alcohol, 115–117
Propionaldehyde, 392–393
Pseudogaianolides, 231, 232
Pseudoionone(s), synthesis of, 127–132
Pulegone, 46, 47
Pullman, 147
Pyrethric acid, 61–64
Pyrethrosine, 199
Pyridine nucleotide dehydrogenases, 158–159
Pyrolysis, 92–99, 102–106

Quinta essentia theory, 3

Radioactivity, 417
Reaction intermediate, 77–79
Reimer-Tiemann reaction, 14
Rhone-Poulence, 132–134
Rule of Markownikoff, 66
Ruzicka, L., 12, 21, 143–144
macyrocyclic ketone synthesis by, 290–293, 294

Sabinene, 66–67
Santalane, 248–260
Santalene, 10
Santalol, 15, 254, 255, 257–260
Santolinyl structures, 32–33
Saucy-Marbet I reaction, 132
Saucy-Marbet II reaction, 128–131
Saytzeff olefin, 47, 48
Schmalzl, patchoulol synthesis of, 269, 272
Sclary sage oils, 108
Semi-hydrogenation reaction, 117–118
Sesquicarene, 176–181
Sesquiterpene(s), caryophyllene, 219–229
chemistry of, 191–286
copabornane, 240–245
cyclizations in, 191–193
gaiane, 229–240
germacrane, 198–216
himachalene, 193, 194
humulane, 216–223
interconversions in, 191–193
iosprenic, 215
isoprene coupling in, 20
longiborneal, 193, 194
longicyclene, 193, 194
longifolene, 245–248
longipinene, 193, 194
medium-sized rings of, 191–193, 195–198
santalane, 248–260
steric interactions in, 191–192
structure of, 10, 22
synthesis of, 240–279
trans-annular reactions in, 192
types of, 193
valence angles in, 196
Sesquiterpenoids, 171–175
6-2 Shift, 83
Sinesal, 176, 397–399
Sodium acetylide, 114–115
Sommelet reaction, 342
Spectrometry, mass, 428
Spectroscopy, DNMR, 423
infrared, 428–429
Squalene, 20, 21
Stereochemical doctrines, 8
Stereoisomerism, 357
Stevens reaction, 34
Story, cyclohexadecanolide synthesis of, 325–327
Strain theory, 8
Substrates, 108
Sulfoxides, 378–380
Sulfur, 377–381, 417
Synthesis, 16–17

Tagetones, 395, 397
Taits, macrocyclic ketone synthesis by, 305–307
Tanabe, macrocyclic ketone synthesis of, 307–312
Telomerization, 134–137
Teralinic non-nitro musks, 345–349

Terpene(s), 6–7
biosynthesis of, 21
cyclic, 8
rearrangements in, 82–83
structure of, 10
Terpene und Camphor, 7
Terpenoids, 19–83
acetic acid in, 165
aliphatic, 107–142, 134–137
biosynthesis of, 171–190, 420–421
in essential oils, 11, 12
isoprenoid coupling of, 19–20
nomenclature in, 21–36
structure of, 21–24
synthesis of, 134–142
acetylenic, 108, 109
by telomerization, 134–137
industrial, 16
myrcene in, 137
Terpine(s), acid-catalyzed reactions in, 85–88
dehydration of, 86
electrocyclic opening of, 37, 38
Tetracyclic derivatives, 188, 190
Tetrahydronaphthindanone, derivatives of, 350, 352–353
Tetralinic hydrocarbons, 338–339
Thermal eliminations, 52–54
Thermal syn-eliminations, 54–56
Thermolysis, 211
Thiophene, 305–307
Thorpe reaction, 293–295
Thuyane, 64, 65
derivatives of, 66–69
Thuyol, 67, 68
Thuyone, 8, 68
Tibal, in caranone reduction, 70–71, 73
Torreyal, 171, 175
Tricyclic derivatives, 188, 189
Tricyclic non-nitro musks, 349–353
acenaphthene derivatives in, 353
hydrindacene derivatives in, 349–350, 351
tetrahydronaphthindanone derivatives in, 350, 352–353
Tricycloekasantalal, 376
Trigonal systems, face designation in, 361–362
Triterpenes, 20
Turpentine oil, 70
annual production of, 107
in derivative manufacture, 76
in terpine manufacture, 86
Indian, 193

Utimoto, macrocyclic ketone synthesis of, 319–323

Valencane, 214, 215
Van Emster, 77
Vanilla, history of, 14
Vanillin, natural vs. synthetic, 419, 420
Verbenol, acid-catalyzed reactions of, 102, 103
hydroperoxidation of, 100–102
pyrolysis of, 102–106
transformation of, 102–106
Verley-Meerwein-Ponndorf reaction, 47
Versalide, 346
Vetispirane, 214, 215, 216
Vitamin B_3, 153–155
Von Baeyer, Adolf, 7–9, 77, 327

Wagner-Meerwein rearrangement, 10, 20, 80–83
generalization of, 78
in camphor synthesis, 89–92
in patchoulol, 236
Wallach, Otto, 6–7, 21, 76
Whitmore, 78, 87
Wieland-Meschler ketone, 247
Wittig reaction, 34, 363–381, 374
application of, 373–377
as elimination reaction, 363–364
olefins formed in, 366–373
reactions related to, 376
sulfur in, 377–381
triphenylphosphine oxide in, 374–375
with carbonyl compound, 364–366
Wittig reagent, 364
Wittig-Schlosser reaction, 373–374
Wolff-Kischner reaction, 103

Yamada, patchoulol synthesis of, 273
Yamamoto, homologation method of, 316–317
Yomogi alcohol, 32–33

Ziegler, macrocyclic ketone synthesis of, 293–295
α-Zingiberene, 181–182